Neurobiological Effects of Sex Steroid Hormones contains chapters by some of the most productive and innovative researchers in the area of neurobiological regulation of reproductive behaviors. The common theme of this volume is the systems approach used by many of its contributors to study gonadal steroid (sex steroid) interactions with neural circuits. The book integrates current results from molecular biological, morphological, and physiological methods. There are three main parts: the first defines the circuitry involved in reproductive behaviors, the second describes the transduction of sex steroid information into neural signals, and the third deals with the cellular and molecular events underlying sex steroid activation of the central nervous system involved with reproductive behavior. This book will be invaluable for neuroscientists interested in reproductive neuroendocrinology.

NEUROBIOLOGICAL EFFECTS OF SEX STEROID HORMONES

NEUROBIOLOGICAL EFFECTS OF SEX STEROID HORMONES

Edited by

PAUL E. MICEVYCH
University of California, Los Angeles

RONALD P. HAMMER, JR.
Tufts University

CAMBRIDGE
UNIVERSITY PRESS

Published by the Press Syndicate of the University of Cambridge
The Pitt Building, Trumpington Street, Cambridge, CB2 1RP
40 West 20th Street, New York, NY 10011-4211, USA
10 Stamford Road, Oakleigh, Melbourne 3166, Australia

First published 1995

Printed in the United States of America

Library of Congress Cataloging-in-Publication Data

Neurobiological effects of sex steroid hormones / edited by Paul E.
Micevych, Ronald P. Hammer, Jr.

p. cm.

Includes index.

ISBN 0-521-45430-1 (hc)

1. Hormones, Sex – Physiological effect.
2. Psychoneuroendocrinology. I. Micevych, Paul E. II. Hammer,
Ronald P.
[DNLM: 1. Sex Hormones – physiology. 2. Steroids – physiology.
3. Central Nervous System – physiology. 4. Reproduction – physiology.
5. Sex Behavior – physiology. WK 900 N4939 1995]
QP572.S4N48 1995
612.6 – dc20
DNLM/DLC
for Library of Congress   94-26902
                                                                CIP

A catalog record for this book is available from the British Library.

ISBN 0-521-45430-1 Hardback

Contents

Contributors

Rula Abbud   Department of Physiology, University of Pittsburgh

Thomas R. Akesson   Department of Veterinary Comparative Anatomy, Physiology, and Pharmacology, Washington State University

Arthur P. Arnold   Department of Physiological Science, University of California, Los Angeles

Barbara Attardi   Department of Physiology, University of Pittsburgh

Etienne-Emile Baulieu   Unité de Recherches sur les Communications Hormones, INSERM

Kathie Berghorn   Department of Physiology, University of Pittsburgh

Jeffrey D. Blaustein   Department of Psychology, University of Massachusetts, Amherst

Sun Cheung   Department of Anatomy and Reproductive Biology, University of Hawaii

Geert J. De Vries   Department of Psychology, University of Massachusetts, Amherst

Gary Dohanich   Department of Psychology and Neuroscience, Tulane University

Loretta M. Flanagan   Department of Animal Biology, University of Pennsylvania

Mona Freidin   Laboratory of Neurobiology and Behavior, Rockefeller University

Ronald P. Hammer, Jr.   Departments of Psychiatry, Pharmacology, and Experimental Therapeutics, and Anatomy and Cellular Biology, Tufts University

Glenn I. Hatton   Department of Neuroscience, University of California, Riverside

Gloria E. Hoffman   Department of Physiology, University of Pittsburgh

Elaine M. Hull   Department of Psychology, State University of New York at Buffalo

Wei-Wei Le Department of Physiology, University of Pittsburgh

Christine A. Lisciotto Institute of Animal Behavior, Rutgers, State University of New Jersey

Karl F. Malik Laboratory of Cell Biology, National Institute of Mental Health

Bruce S. McEwen Laboratory of Neuroendocrinology, Rockefeller University

John M. Meredith Department of Neuroscience and Behavior, University of Massachusetts, Amherst

Paul E. Micevych Department of Neurobiology, University of California, Los Angeles

Joan I. Morrell Institute of Animal Behavior, Rutgers, State University of New Jersey

Sarah W. Newman Department of Anatomy and Cell Biology, University of Michigan

Donald Pfaff Laboratory of Neurobiology and Behavior, Rockefeller University

Paul Popper Department of Surgery, University of California, Los Angeles, School of Medicine

Catherine A. Priest Department of Anatomy and Cell Biology, University of California, Los Angeles

Paul Robel Unité de Recherches sur les Communications Hormones, INSERM

Barney A. Schlinger Department of Physiological Science, University of California, Los Angeles

Richard B. Simerly Division of Neuroscience, Oregon Regional Primate Research Center, and University of Oregon

M. Susan Smith Oregon Regional Primate Research Center

Marc J. Tetel Department of Neuroscience and Behavior, University of Massachusetts, Amherst

C. Dominique Toran-Allerand Department of Anatomy and Cell Biology, Columbia University

Christine K. Wagner Institute of Animal Behavior, Rutgers, State University of New Jersey

Ruth I. Wood Department of Anatomy and Cell Biology, University of Michigan

Pauline Yahr Department of Psychobiology, University of California, Irvine

Dedication

The editors dedicate this volume to Professor Emeritus Charles H. Sawyer and Professor Roger A. Gorski, in the Department of Anatomy and Cell Biology of the University of California, Los Angeles, School of Medicine. These individuals, each in his own way, with his own style and character, have charted and influenced new directions in the field of reproductive neuroendocrinology. We believe it fitting that these two scientists, who are such good friends, should be honored together here.

Charles "Tom" Sawyer is the embodiment of the phrase "a gentleman and a scholar." Tom has led us toward our goal by his example of enthusiasm, scientific endeavor, and genuine humility. Tom came to UCLA in 1951 from Duke University, where, with John W. Everett, Joseph E. Meites, and W. Henry Hollingshead, he published a series of papers on the central control of hormone secretion and reproductive function. While at UCLA, Tom established one of the leading neuroendocrine laboratories. Many of the fundamental principles of neural control of the anterior pituitary were elucidated in his laboratory. He and a series of distinguished visiting scientists, postdoctoral fellows, and students published more than 350 papers on the electrophysiology of neuroendocrine neurons, steroidal stimulation of the limbic system and hypothalamus, and the neurobiology of the reticular activating system. His laboratory was the first to report the phenomenon of paradoxical sleep. His seminal studies on the role of neurotransmitters in the regulation of luteinizing hormone release are among the most important on this topic. His seminal discoveries continue to inspire current researchers. During a career spanning more than 40 years, Tom's research has established and defined the field of neuroendocrinology. He has asked and answered some of the most basic questions regarding the interface of neuroendocrine interactions. He also trained many of the leaders in the field of neuroendocrinology, a generation of scientists who have advanced his spirit of research throughout the world.

It has been a great honor for both of us to work and teach side by side with Tom. He has been supportive of us as faculty colleagues, and we are pleased to thank him on behalf of many others.

Roger Gorski is a true pioneer and explorer who has endeavored throughout his career to characterize fundamental physiological and structural sex differences in the mammalian brain. As he is fond of saying, "I've spent a lifetime continuing to work on my dissertation." His thoughts on and efforts toward elucidating the influence of sex steroid hormones on the brain have directly influenced much of the research that appears in this book. Perhaps Roger's greatest contribution was the discovery and characterization of the sexual dimorphic nucleus of the medial preoptic area (SDN-POA) in the rat. The SDN-POA captured the imagination of neuroscientists and provided an elegant model for hormone actions due, in no small measure, to Roger's abilities to present this important discovery in a meaningful way; his approach soon enticed morphologists, neurochemists, developmental biologists, and behavioral scientists to investigate the SDN-POA. Sex differences were discovered in myriad species, and the sexually dimorphic distribution of neurotransmitters/modulators and receptors was documented. This flurry of activity was led by the efforts of visiting scientists, postdoctoral fellows, and students in the Gorski laboratory. With his colleague Laura Allen, Roger has described several significant morphological sex differences in the human brain, and he continues to pursue the issues of steroid hormone–induced differentiation of the SDN-POA.

Roger has been both a mentor and a guide to us. In fact, we are converts to the field of reproductive neuroendocrinology and like so many of our colleagues have been inspired along this path by Roger's work and words. We would like to express our appreciation for his vision and guidance.

Preface

Many books are subject to the fundamental questions "Why this topic?" and "Why now?" Scientific texts are perhaps most susceptible because they often present similar topics. As a partial answer to these questions, we paraphrase P. B. Medawar in his *Advise to a Young Scientist*: We have tried to prepare the kind of book that we ourselves would like to read and have as a reference.

In recent years, the field of reproductive neuroendocrinology has experienced a renaissance brought about by the application of cellular and molecular biological techniques. We have made significant progress in understanding the mechanisms that underlie central nervous system control of reproductive behavior. This progress has been well documented at various meetings and in individual papers. We felt it was necessary, therefore, to offer a collection of essays by some of those who have contributed to this renaissance. We hasten to add that the chapters in this volume do not necessarily reflect all of the vital issues of behavioral neuroendocrinology. Rather, they represent brief reviews by and current data from a number of productive scientists in this field.

Because of a limitation of space, several important topics are not discussed or are only briefly presented in this volume. These include the spinal nucleus of the bulbocavernosus system, cell membrane steroid receptors, interactions of steroids with γ-aminobutyric acid receptors, the songbird neural circuitry, as well as the insect and amphibian models of reproduction and metamorphosis. Each of these models has proved to be extremely useful for studying the effects of sex steroid hormones on the nervous system. We hope that future volumes will be devoted to them. This book was originally meant to focus on brain circuits and factors involved in the regulation of sexual behavior in mammalian species and thus these other systems were not included.

The contributors were asked to present the most interesting and novel results of their work and not to engage in a comprehensive review of the literature. The volume attempts to integrate the results gleaned from various molecular biological approaches with those derived from more traditional techniques that have been used to investigate the neurobiology of reproductive behavior. The text emphasizes a "systems" approach to the study of sex steroid actions on the brain. The focus on neural circuitry is an important feature, because it highlights the importance of examining cellular or molecular events within the appropriate system. The book is divided into three parts: the first is an introduction to the sex steroid–sensitive circuitry of the mammalian brain; the second deals with the interaction of sex steroid hormones with specific neurochemically defined circuits in the hypothalamus and limbic system; the third describes the cellular events involved in the synthesis, metabolism, and processing of steroid hormone input to the brain.

Part I introduces the concept of a steroid-responsive system and defines the circuitry of the interconnected hypothalamic and limbic cell groups. The chapters in this part discuss the morphological and neurochemical sex differences that may underlie differences in the reproductive behaviors of males and females. They deal with studies that defined the circuits responsible for the integration of vomeronasal and sex steroid inputs to determine reproductive behaviors, as well as research that has begun to elucidate the behavioral function of specific cell groups and their connections within these circuits. The authors examine the relationship between the expression of steroid receptor mRNA, binding of ligands, and behavior as a means of integrating molecular and cellular events with the response of an animal to alterations of sex steroid hormones.

In Part II, the focus is on the mechanisms underlying the regulation of male and female reproductive behaviors by specific neuroactive molecules. Consideration of neuropeptides has been extremely productive in explaining the steroid-induced regulation of transcription, behavior, and receptor mechanisms. A common theme has emerged from these studies: Neuropeptides modulate behavior in a steroid-dependent, site-specific manner. Sex steroids regulate not only the levels of mRNA and subsequently peptide levels in presynaptic neurons but also the cognate receptors in their target structures. Studies of neuropeptide circuits have revealed that sex steroids regulate expression in very specific sets of neurons that are apparently involved in the control of reproductive behaviors from the regulation of lordosis to maternal behavior, suggesting a finely tuned system of regulation of brain function. Although this section is oriented toward steroid–neuropeptide interactions, additional chapters are devoted to cholinergic and dopaminergic regulation.

Part III highlights a series of fundamental cellular events, from the synthesis of sex steroid hormones to actions of steroids on neurotrophins. These chapters describe studies that have revealed that the brain can synthesize *de novo* steroids (termed *neurosteroids*), which in some species are an important source of circulating estrogen (via aromatization of testosterone to estrogen). The mechanisms by which sex steroids activate cells is discussed in chapters that deal with the molecular mechanism of direct steroid interaction with specific response elements, such as the steroid induction of Fos protein or neurotrophins and their receptors, and the regulation of direct cell–cell interactions and communication. These studies demonstrate the rich diversity of steroid interactions with mechanisms involved in cellular metabolism and point out the importance of using a wide variety of techniques and systems to understand the action(s) of sex steroids in the brain.

This volume, then, provides a rapid tour through the brain circuitry, neurochemistry, and cellular mechanisms that are influenced and induced by sex steroid hormones, and that thereby initiate a complex cascade of events that is ultimately expressed as reproductive behavior. Some of the discussions overlap in their presentation of the circuitry or mechanisms proposed. We apologize for this repetition, which would be apparent only to those who read the book from start to finish. Since most readers of this edited volume will probably read chapters individually, we feel that this is not a concern. In fact, it is fascinating that such a variety of experimental approaches, observations, and results engender such common conclusions. This suggests that the study of sex steroid regulation of the brain has reached a new plateau of understanding.

Acknowledgments

We are all too aware that this book did not edit itself; we had an extraordinary amount of assistance. Among those who deserve special recognition are Dr. Catherine Priest for her helpful comments, Ms. Krista Vink, who did a wonderful job of producing the index list, and Mss. Gerda Goette and Kathleen Kaya, who shepherded the manuscripts through various revisions. At Cambridge University Press, we are deeply grateful for the understanding and concern displayed by Dr. Robin Smith throughout the process. He even called to determine whether the Los Angeles earthquake of 1994 had damaged any of the contributions. Finally, we would like to thank all of the contributors for their forbearance. Those whose chapters were subjected to editorial revision were unfailingly cheerful and helpful. Interacting with the contributors has been one of the most positive aspects of our experience.

On a more somber note, while we were putting the last chapters together, we received word that Thomas Akesson had died in Pullman, Washington. Tom had been a postdoctoral fellow and colleague at UCLA before taking a position as an assistant professor at Washington State University. Many of us knew Tom and appreciated his detailed morphological analyses of steroid-concentrating cells and circuits. His work focused on determining the neurochemistry of steroid receptive cells in the limbic system and hypothalamus. His contribution to this book is the last manuscript he wrote. We will miss him.

Part I

Sex steroid–responsive circuits regulating male and female reproductive behaviors

1

Hormonal influence on neurons of the mating behavior pathway in male hamsters

RUTH I. WOOD AND SARAH W. NEWMAN

Introduction

Steroid hormones produced in the gonads are a prerequisite for mating behavior in the male Syrian hamster, as in most male mammals. While a substantial body of early research was directed toward defining the hormonal requirements (both quantity and identity) for this behavior, more recent studies have focused on hormone metabolism, receptor distribution, and mechanisms of steroid action in the central nervous system (CNS). This chapter explores the transduction of hormonal signals by steroid receptor–containing neurons to facilitate sexual behavior in the male hamster.

Hormonal requirements for copulation

Copulation in the male Syrian hamster

The sequence of copulation in the male Syrian hamster has been reviewed previously (Siegel 1985; Sachs and Meisel 1988). However, a brief description of the behaviors expressed during mating would be helpful for understanding the critical role that hormones play in maintaining this activity. Figure 1.1 illustrates sexual behavior over a 10-minute period in a sexually experienced male. Initial contact with a receptive female is characterized by investigation of the female's head and flank, followed by extensive sniffing and licking of the anogenital region. Through this activity, the male receives chemosensory stimulation via the vomeronasal organ and olfactory mucosa. The ability to perceive chemosensory signals and the integrity of the neural pathways that transmit chemosensory stimuli are essential for copulation, because disruption of chemosensory cues immediately and permanently abolishes mating (reviewed by Sachs and Meisel 1988). In a sexually experienced male, anogenital investigation of the female is followed shortly by a series of mounts and intromissions, interspersed with brief (1- to

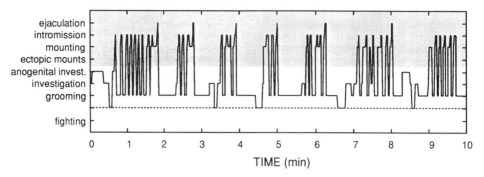

Figure 1.1. Mating behavior in a sexually experienced adult male hamster. Behavior was recorded every second for 10 minutes immediately after the introduction of a receptive female. Baseline indicates no interaction with the female; shaded area highlights copulatory behavior, including mounts, intromissions, and ejaculations.

2-second) grooming of the penis and perineum. This culminates after three to six intromissions in the first ejaculation. Ejaculation is distinguished by the reflexive thrusting action of the pelvis during the final intromission. Afterward, in the post-ejaculatory interval, there is a longer bout of self-grooming and anogenital investigation, and the cycle is repeated. When exposed to a single receptive female, the male may ejaculate 10 or more times during a 20- to 30-minute period until sexual satiety.

Hormonal control of male sexual behavior

In general, the endocrine control of sexual behavior in the hamster is similar to that in males of other mammalian species. The following is a brief summary of the normal steroid environment in the male hamster and the hormonal conditions necessary for sexual behavior, which will serve as a basis for discussing the neural mechanisms that transduce steroid signals from the gonads. Although the importance of gonadal steroids in the expression of sexual behavior has long been recognized, much remains to be learned regarding the specific pathways and transmitters in the brain, and the cellular mechanisms by which steroids stimulate copulation in the male.

Control of steroid production. The production of steroids in the interstitial (Leydig) cells of the testis is stimulated by the pulsatile release of luteinizing hormone (LH) from the anterior pituitary. Testosterone is the principal steroid product of the testis, but measurable quantities of estrogen and dihydrotestosterone are also present in the circulation (Handa et al. 1986). Based on data from other animal models, it appears that LH pulses are driven by pulses of gonadotropin-releasing hormone (GnRH), which traverse the hypothalamohypophyseal portal vasculature upon release from GnRH neuron terminals in the median eminence (Jackson et al. 1991). In addition to their stimulation of mating behavior, gonadal steroid hormones suppress the further release of LH in a classic negative-feedback loop (reviewed by Bartke 1985). Thus, the frequency of LH pulses, and presumably of GnRH pulses, increases shortly after the elimination of endogenous steroids by castration. Replacement of exogenous steroids at physiological levels can reverse this effect.

This relationship between LH and steroid production is not constant throughout the male's lifetime. The responsiveness of the hypothalamopituitary axis to steroid feedback can be modified by signals from the internal and external environment. For example, sensitivity to steroid action is altered by day length in hamsters (Tamarkin et al. 1976; Turek 1977) and by nutritional status in a variety of species (Pirke and Spyra 1981). Hamsters normally reproduce during the long days of spring and summer, when food is plentiful. During the short days of winter, steroid feedback sensitivity on gonadotropin secretion increases, thereby suppressing further steroid production (reviewed by Bartke 1985). This leads to testicular regression and azoospermia. At the same time, the ability of steroids to stimulate mating behavior decreases (Miernicki et al. 1990). Thus, although circulating concentrations of endogenous steroids are low in the male exposed to short days, the effects of photoperiod on gonadotropin secretion and sexual behavior are not equivalent to castration. This illustrates the complex relationship between peripheral steroid levels and their effects in the CNS. However, the neural basis for changes in steroid responsiveness is not known for any species.

Hormonal requirements for mating. In a normal male hamster, chronic steroid exposure during adulthood permits the expression of sexual behavior, and this behavior declines slowly when endogenous steroids are reduced (as in the male exposed to inhibitory short-day photoperiod) or eliminated (by orchidectomy; Morin and Zucker 1978). Figure 1.2 illustrates sexual behavior in a representative male at 2, 6, and 8 weeks after castration. Note that anogenital investigation and mounting persist after castration for a longer time than intromissive or ejaculatory behavior. In a sexually naive male, mating behavior declines even more rapidly after castration (Lisk and Heimann 1980). Furthermore, orchidectomy before puberty virtually eliminates mating in the adult (Miller et al. 1977). These observations highlight the interaction of steroids and sexual experience in the male. Copulatory behavior in a castrated male can be maintained after orchidectomy by immediate replacement of exogenous steroids, or even reinstated in a long-term castrate that has ceased to copulate (Sachs and Meisel 1988). However, restoration of sexual behavior following long-term castration or exposure to short photoperiod is not immediate, requiring high levels of steroids in the circulation for several weeks.

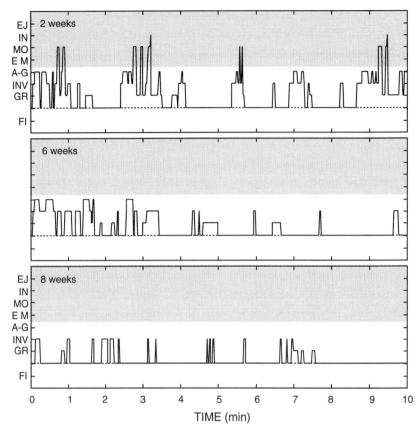

Figure 1.2. Mating behavior in a sexually experienced adult male hamster at 2, 6, and 8 weeks after castration. At each time point, behavior was recorded every second for 10 minutes immediately after the introduction of a receptive female. Baseline indicates no interaction with the female; shaded area highlights copulatory behavior, including ectopic mounts (E M), mounts (MO), intromissions (IN), and ejaculations (EJ). Other behaviors recorded during each test included grooming (GR), investigation of the female (INV), anogenital investigation (A-G), and fighting (FI).

Androgen versus estrogen effects on male copulatory behavior. One question that has been a subject of continued interest and considerable debate is whether the hormonal control of sexual behavior in the male hamster and other species is determined by endogenous androgens or estrogens. Although testosterone is the dominant steroid secreted by the testes, it can be converted peripherally or in the CNS to dihydrotestosterone (DHT) through the actions of 5α-reductase, or to estradiol via the aromatase enzyme. These potent testosterone metabolites bind to neural androgen and estrogen receptors, respectively. In studies of castrated males given exogenous steroid treatment, the ability of estradiol or DHT alone to maintain sexual behavior is clearly dose-dependent (Whalen and DeBold 1974; DeBold and Clemens 1978). However, estimating physiological doses of these hormones is difficult because of the potential for local conversion from testosterone in the brain. Thus, measurement of serum concentrations of DHT and estradiol will tend to underestimate their availability to neural tissues. Nonetheless, a variety of experimental studies make it clear that either DHT or estradiol alone can maintain sexual behavior in castrated males (Whalen and DeBold 1974; Payne and Bennett 1976; DeBold and Clemens 1978; Powers et al. 1985), with the caveat that estradiol-treated males do not intromit or ejaculate because the male's peripheral structures require androgenic stimulation. Because both androgens and estrogens are behaviorally effective, it seems likely that, in the normal adult male brain, testosterone, DHT, and estradiol each contribute to steroid facilitation of copulation. In support of this concept, the combination of DHT and estradiol is more effective in maintaining sexual behavior in a castrated male than either hormone alone (DeBold and Clemens 1978). The implication of these observations using the peripheral hormone replacement paradigm to maintain sexual behavior is that both androgen and estrogen receptor–containing neurons in the brain can mediate hormonal effects on copulation.

Interactions of hormonal and chemosensory cues. Additional evidence that steroids play a long-term role in copulation derives from the observation that mating behavior is not dependent on changing concentrations of circulating testosterone. Male hamsters mate normally when endogenous steroids are replaced at constant physiological levels by steroid-filled Silastic capsules. Although there is presumably some minimal level of steroid concentration sufficient for copulation, the evidence suggests that this "threshold" may be unique for each steroid-dependent element of sexual behavior (chemoinvestigation, grooming, mounts, intromissions, etc.). In particular, testosterone replacement at moderate levels in castrated males can induce copulation but does not restore chemoinvestigatory behavior (Powers et al. 1985).

Although changes in serum steroid levels are not essential for sexual behavior, the concentrations of endogenous steroids in circulation are far from static. Testosterone is normally released episodically in the intact male in response to pulsatile stimulation by LH (Ellis and Desjardins 1982). Furthermore, mating or exposure to female hamster vaginal secretion acutely elevates testosterone concentrations in circulation (Macrides et al. 1974), a response presumably mediated by neural pathways from the olfactory bulbs to GnRH neurons in the preoptic area and hypothalamus. The neural circuitry underlying this pheromonally mediated neuroendocrine response is not known in the hamster, although the response is dependent on fibers in the lateral olfactory tract (LOT; Macrides et al. 1976), and does not require an intact, GnRH-containing terminal nerve (Wirsig-Weichmann 1993). Based on work in other rodents, it can be assumed that transection of the LOT severs critical efferents from the main and/or accessory olfactory bulbs,

rather than afferents to the bulbs from limbic areas, since removal of the vomeronasal organ in mouse (Wysocki et al. 1983) and destruction of the olfactory mucosa in rat (Larsson et al. 1982) prevent testosterone increases after exposure of
the male to female stimuli.

Not only do chemosensory cues stimulate circulating steroid concentrations,
but the hormonal status of the male also determines his interest in chemoinvestigation of the female. This suggests the existence of reciprocal interactions
between chemosensory and hormonal signals to control sexual behavior. Castration decreases investigation of female hamster vaginal secretion, but attraction to
chemosensory stimuli can be reinstated with high levels of exogenous steroids
(Powers et al. 1985). These interactions of chemosensory and hormonal cues
suggest integration of central neural pathways mediating these two classes of
stimuli.

Steroid-responsive networks in the brain

Chemosensory pathways

Projections of the vomeronasal organ and olfactory mucosa. In most vertebrate
species, neural pathways that process chemosensory information from the nasal
cavities arise from two discrete populations of receptor neurons in the olfactory
mucosa and vomeronasal organ. The function of olfactory and vomeronasal receptors is regulated by autonomic innervation of nasal vasculature (Meredith and
O'Connell 1979), and the sensations arising from these receptors are supple-
mented by the activity of the trigeminal afferents (Silver 1987) and, perhaps, by
nervus terminalis fibers (Wirsig and Leonard 1986). Although these other neural
systems innervating the nasal cavities can influence olfactory and vomeronasal
function, experimental evidence suggests that the olfactory, and particularly the
vomeronasal, pathways through the amygdala to the hypothalamus play a central
role in the initiation of neuroendocrine responses and social behaviors (Halpern
1987; Wysocki and Meredith 1987).

From their cell bodies in the mucosae, axons of the olfactory and vomeronasal
receptor neurons project through the ethmoidal cribiform plate into the cranial
cavity, where they terminate separately in the glomerular layers of the main and
accessory olfactory bulbs, respectively (Barber and Raisman 1974). Here, they
synapse on dendrites of mitral and tufted cells, the projection neurons of the
bulbs, which form the origin of the major efferent fiber system in the LOT. Olfactory fibers in the LOT and vomeronasal fibers in the accessory olfactory tract
travel in separate but adjacent fiber bundles to largely segregated terminal fields
on the ventral surface of the brain (Scalia and Winans 1975). Figure 1.3 illustrates
not only this segregation, but also the disparity in size of the olfactory and

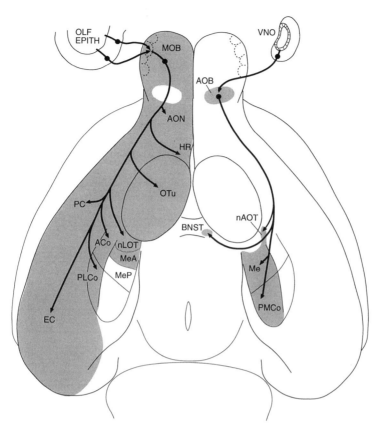

Figure 1.3. Diagram of the ventral surface of the hamster brain. Shaded areas indicate brain regions that receive efferent projections of the olfactory system, via the main olfactory bulb (*left*), and vomeronasal system, via the accessory olfactory bulb (*right*). Abbreviations: ACo, anterior cortical nucleus of the amygdala; AOB, accessory olfactory bulb; AON, anterior olfactory nucleus; BNST, bed nucleus of the stria terminalis; EC, entorhinal cortex; HR, hippocampal rudiment; Me, medial nucleus of the amygdala; MeA, medial nucleus of the amygdala, anterior division; MeP, medial nucleus of the amygdala, posterior division; MOB, main olfactory bulb; nAOT, nucleus of the accessory olfactory tract; nLOT, nucleus of the lateral olfactory tract; OLF EPITH, olfactory epithelium; OTu, olfactory tubercle; PC, piriform cortex; PLCo, posterolateral cortical nucleus of the amygdala; PMCo, posteromedial cortical nucleus of the amygdala; VNO, vomeronasal organ.

vomeronasal terminal fields. As seen in this schematic view of the rodent brain, the main olfactory bulbs project widely to the allocortex on the ventral surface. They provide input to several different systems, including the piriform cortex, which gives rise to pathways responsible for olfactory discrimination (Sapolsky and Eichenbaum 1980); the olfactory tubercle of the ventral striatum; the lateral entorhinal cortex, which is the gateway to the ventral hippocampal formation; and the corticomedial amygdala. In contrast, the vomeronasal fibers of the accessory olfactory tract project only to the corticomedial amygdala and through the stria terminalis to a small cluster of cells in the bed nucleus of the stria terminalis (BNST) (Davis et al. 1978). It is through their connections to the amygdala that the olfactory and vomeronasal systems gain access to the medial preoptic area (MPOA) and hypothalamus (Kevetter and Winans 1981a; Maragos et al. 1989), brain regions that collectively control neuroendocrine function and social behavior.

Recent studies of the connections of the corticomedial amygdala provide evidence for discrete anatomical and functional circuits arising within this region. An important framework for describing and distinguishing these circuits is the concept of the "extended amygdala" developed by Alheid and Heimer (1988) to describe the organization of a ring of interconnected structures in the limbic system. The relationship of these structures was outlined first by Johnston in 1923 and more recently described in detail by de Olmos et al. (1985; for review see Heimer and Alheid 1991). According to this concept, the medial and central nuclei of the amygdala form part of a continuous ring of gray matter within the brain (Fig. 1.4). On the opposite side of this ring from the amygdala is the BNST, and connecting the bed nucleus with the amygdala are two strings of cells, one of which follows the stria terminalis and the other the ventral amygdalofugal pathway through the sublenticular substantia innominata. Taken together, these structures comprise the extended amygdala.

De Olmos and colleagues (1985) have described two rings within this system. One includes the central nucleus of the amygdala and the lateral part of the BNST. The other includes the medial and posteromedial cortical nuclei of the amygdala together with the medial and intermediate parts of the BNST. The latter ring encompasses the "vomeronasal amygdala" of Kevetter and Winans (1981a), but according to de Olmos et al. (1985) the structures of the "olfactory amygdala" (Kevetter and Winans 1981b) are not considered part of the extended amygdala. However, recent studies suggest that within the extended amygdala of the rodent there are two circuits associated with the medial amygdaloid nucleus (Me), in addition to the circuit of the central nucleus. This revised model of the extended amygdala is based on investigations delineating the separate projections of the anterior and posterior parts of the Me in the hamster (Gomez and Newman 1992) and studies of connections of the corticomedial amygdaloid nuclei surrounding the Me in the rat (Canteras et al. 1992a,b). The two medial circuits, one associated with the anterior Me (MeA), the other with the posterior Me (MeP), are defined by the discrete neural connections that link their subunits and by the differences in density of their direct neural inputs from the main and accessory olfactory bulbs. Furthermore, this new model includes both the olfactory and the vomeronasal amygdalae of Kevetter and Winans (1981a,b) in the extended amygdala. Finally, in the hamster and rat, neuronal populations that produce steroid receptors and a variety of specific neurotransmitters are preferentially distributed within one or the other of these two medial circuits.

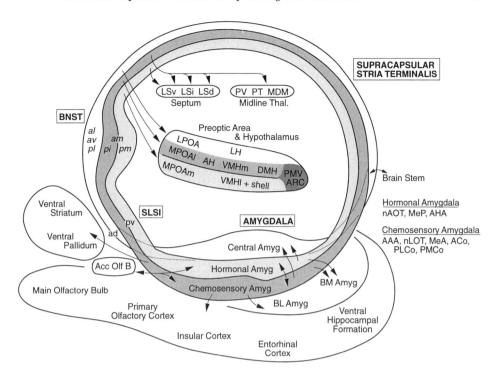

Figure 1.4. Neural circuitry through the extended amygdala depicting the various component subnuclei and their principal connections. Light shading indicates the circuitry of the hormonal amygdala, including nAOT, MeP, and AHA. These areas are connected with BNSTam and BNSTpm through the supracapsular stria terminalis and SLSI. This circuit has projections to MPOAm, VMHl and shell, PMV, and ARC. Medium shading depicts the chemosensory amygdala, which encompasses the anterior amygdaloid area, nLOT, MeA, ACo, PLCo, and PMCo. These regions, together with BNSTam and BNSTpi, project into MPOAl, AH, VMHm, and DMH. Common projections of these hormonal and chemosensory circuits, PMV and ARC, are indicated by dark shading. The circuitry of the central nucleus of the amygdala and BNSTpl is unshaded. Arrows depict connections of these subcircuits with other brain areas and interconnections between subcircuits. Abbreviations: ACo, anterior cortical nucleus of the amygdala; AH, anterior hypothalamus; ARC, arcuate nucleus of the hypothalamus; BNST, bed nucleus of the stria terminalis; DMH, dorsomedial nucleus of the hypothalamus; LH, lateral hypothalamus; LPOA, lateral preoptic area; LSd, LSi, LSv, lateral septum, dorsal, intermediate, and ventral; MDM, mediodorsal nucleus of the thalamus, medial division; MeA, medial nucleus of the amygdala, anterior division; MeP, medial nucleus of the amygdala, posterior division; MPOAl, m, medial preoptic area, lateral part and medial part; nAOT, nucleus of the accessory olfactory tract; nLOT, nucleus of the lateral olfactory tract; PLCo, posterolateral cortical nucleus of the amygdala; PMCo, posteromedial cortical nucleus of the amygdala; PMV, ventral premammillary nucleus; PT, paratenial nucleus of the thalamus; PV, paraventricular nucleus of the thalamus; SLSI, ad, pv, sublenticular substantia innominata, anterdorsal and posteroventral area; VMHl, m, ventromedial nucleus of the hypothalamus, lateral and medial.

Figure 1.4 illustrates not only the key nuclei within the extended amygdala that belong to these two medial circuits, but also their numerous connections with other chemosensory and limbic areas. The complementary projections of these two circuits into the MPOA and hypothalamus are illustrated schematically in

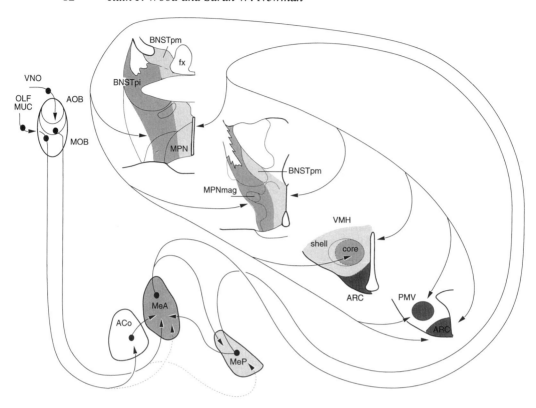

Figure 1.5. Connections of the hormonal and chemosensory subcircuitry of the extended amygdala with nuclei of the preoptic area and hypothalamus. Brain regions that encompass the hormonal subcircuit are depicted by the lightly shaded areas. Neural circuitry through which chemosensory signals are transmitted is shown in medium shading. Brain regions that receive projections from both subcircuits (i.e., PMV and ARC) are denoted by dark shading. Abbreviations: ACo, anterior cortical nucleus of the amygdala; AOB, accessory olfactory bulb; ARC, arcuate nucleus of the hypothalamus; BNSTpi, pm, bed nucleus of the stria terminalis, posterointermediate and posteromedial division; fx, fornix; MeA, medial nucleus of the amygdala; MeP, posterior nucleus of the amygdala; MOB, main olfactory bulb; MPN, medial preoptic nucleus; MPNmag, manocellular medial preoptic nucleus; OLF MUC, olfactory mucosa; PMV, ventral premammillary nucleus; VMH, ventromedial nucleus of the hypothalamus; VNO, vomeronasal organ.

Figure 1.4 and in more anatomical detail in Figure 1.5. The first of these adjacent but largely separate circuits originates from the division of the extended amygdala that includes the MeA, the posterolateral (PLCo) and posteromedial (PMCo) cortical nuclei, and the posterointermediate division of the BNST (BNSTpi in hamster, Gomez and Newman 1992; posterointermediate part of the medial division in rat, de Olmos et al. 1985; transverse, interfascicular, and part of the principal division in rat, Ju and Swanson 1989; posterior intermediate division in rat, Moga et al. 1989). This is the "chemosensory amygdala" of Figure 1.4. It is characterized by extensive olfactory and vomeronasal input and by widespread connections to other chemosensory and limbic areas of the forebrain. The MeA receives abundant vomeronasal afferents from the accessory olfactory bulb and modest olfactory bulb afferents. The PLCo receives olfactory information from the main olfactory bulb and projects to the MeA and the PMCo, which, like the MeA, receives substantial vomeronasal input from the accessory olfactory bulb. Through their efferent projections, these structures of the extended amygdala reach not only widespread olfactory target areas on the ventral telencephalon, but also the lateral part of the MPOA and the medial part of the ventromedial nucleus of the hypothalamus (VMH), projections that are complementary to those of the second circuit, as shown in Figure 1.5.

The second circuit within the medial extended amygdala, the network of the MeP, is much more restricted. In addition to the MeP itself, this subcircuit includes the amygdalohippocampal area and the posteromedial BNST (BNSTpm, Gomez and Newman 1992; posteromedial part of the medial division, de Olmos et al. 1985; principal division, Ju and Swanson 1989; posterior medial division, Moga et al. 1989). Compared with the substantial olfactory and vomeronasal system connections through the MeA, the subcircuitry through the MeP has limited reciprocal connections with the accessory olfactory bulbs and no direct input from the main olfactory bulbs. Furthermore, the projections of this second medial circuit are restricted to the medial subdivisions of the MPOA (primarily the anteroventral periventricular and medial preoptic nuclei, MPN) and the lateral subdivision of VMH (the ventrolateral portion of this nucleus in the rat). Although these two networks arising in the MeA and MeP do share overlapping projections to the ventral premammillary nucleus (PMV), ventral lateral septum (LSv), and lateral areas of the arcuate nucleus of the hypothalamus (ARC; Fig. 1.5), it is not known whether they connect with the same populations of cells in these areas. Thus, the circuit through the MeP and BNSTpm appears to exist in parallel with the "chemosensory circuit" through the MeA and BNSTpi, and the separate connections of these two networks suggest that they may have distinctly different functions.

Distribution of steroid receptors along chemosensory pathways. One of the key distinguishing features of the subcircuitry including the MeP and BNSTpm is the striking abundance of steroid receptor–containing neurons. The location of neurons having receptors for androgens and estrogens has been delineated using a variety of techniques, including *in situ* hybridization (Simerly et al. 1990), steroid autoradiography (Krieger et al. 1976; Doherty and Sheridan 1981; Wood et al. 1992), and immunocytochemistry (Wood and Newman 1993a), and the results of these different methods produce a consistent picture of steroid receptor distribution in the brain. It has long been recognized that receptors for androgens and estrogens are heavily concentrated and widely distributed in the limbic system. Furthermore, as noted by Cottingham and Pfaff (1986), the hormone-sensitive nuclei belong to a limited, tightly integrated circuit of nuclei that project primarily to one another.

However, a detailed examination of the distribution of steroid receptor–containing neurons within the amygdala, BNST, MPOA, and hypothalamus suggests that receptors for gonadal steroid hormones are not evenly scattered throughout these nuclei. Instead, the majority of hormone-responsive cells are concentrated within the MeP and its projection areas, the "hormonal amygdala" of Figure 1.4 (Wood et al. 1992; Wood and Newman 1993a). An example of this pattern is illustrated in Figure 1.6, which depicts the distribution of androgen receptor–immunoreactive neurons in the extended amygdala, MPOA, and hypothalamus of the male hamster brain. In this example, androgen-sensitive neurons are preferentially concentrated in the MeP, and in its principal targets, the BNSTpm and medial MPOA, especially the MPN. Fewer steroid receptors are present in the subcircuit composed of the MeA, BNSTpi, and lateral MPOA, which relays the majority of direct chemosensory inputs. Caudally, in the VMH, the preferential localization of steroid receptors in the shell and lateral subdivision of this nucleus remains faithful to the projections of the MeP. These observations of the distribution and connections of steroid receptor–containing neurons in the hamster limbic system elaborate the concept proposed by Cottingham and Pfaff (1986) of a hormone-responsive network in the brain. They reveal that steroid receptors are concentrated preferentially within discrete interconnected subnuclei of the larger hormonal circuitry, thereby constituting a hormone-responsive subcircuit of exquisite detail.

Although Figure 1.6 presents only the distribution of androgen receptors within the circuitry of the extended amygdala, androgens and estrogens are both capable of stimulating copulation in the male hamster and appear to act in concert to control sexual behavior. Thus, it is necessary to consider that both androgen and estrogen receptor–containing neurons might transduce hormonal information to control male mating behavior. The distributions of these two classes of steroid receptors show considerable overlap, suggesting that individual neurons in certain brain regions may have the potential to respond to both signals. To address this question, we recently co-localized androgen and estrogen receptors using laser scanning confocal microscopy (Fig. 1.7; Wood and Newman 1993c). Outside the chemosensory pathways essential for copulation, the majority of steroid receptor–containing neurons were labeled for either androgen (in LSv and PMv) or estrogen receptors (in the amygdalohippocampal area). However, double-labeled neurons are found in the MeP and BNSTpm, with numerous cells in the MPN, consistent with the observation that both androgens and estrogens facilitate male sexual behavior.

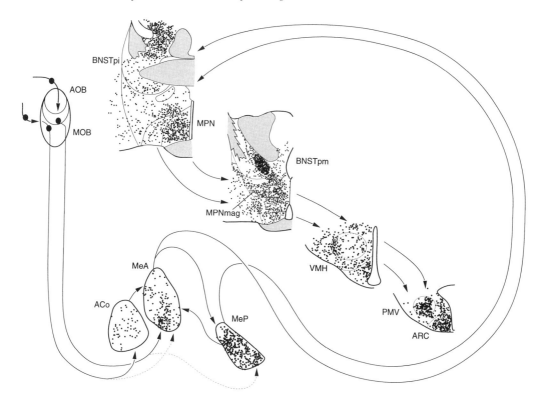

Figure 1.6. Distribution of androgen receptor–immunoreactive neurons in the medial amygdala, BNST, MPOA, and hypothalamus of an intact adult male Syrian hamster. Dots indicate individual androgen receptor–immunoreactive nuclei. Coronal sections from brains fixed in 4% *para*-formaldehyde were stained with androgen receptor antibody (Prins et al. 1991) and visualized using NiCl-enhanced diaminobenzidine, as described in Wood and Newman (1993a). For abbreviations, see caption to Figure 1.5.

Separate hormonal and chemosensory networks?

Although in the foregoing discussion we emphasized the distinguishing features of the chemosensory and hormonal networks, it is important to acknowledge that these circuits are not isolated from one another. First, there are connections linking adjacent elements of the two circuits, such as reciprocal connections between the MeA and MeP (Gomez and Newman 1992). Second, these two circuits project to the central and basomedial amygdaloid nuclei, as well as to selected areas of the hypothalamus already discussed (PMv and ARC; Gomez and Newman 1992). Thus, the potential exists for integration of chemosensory and hormonal signals necessary for mating behavior through the interconnections or common targets of these networks. Current evidence, however, favors the potential importance of the interconnections rather than the common targets of these areas. At present there is no evidence that lesions of the central or basomedial amygdaloid

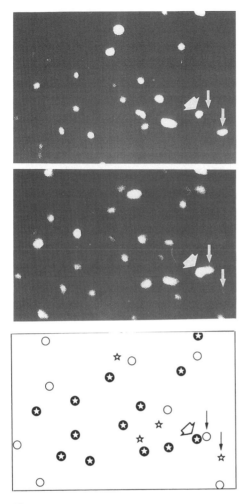

Figure 1.7. Co-localization of androgen and estrogen receptor immunoreactivity in the MPOA of the male Syrian hamster brain. *Top*: Androgen receptor–immunoreactive neurons were stained with androgen receptor antibody (Prins et al. 1991), labeled with Cy5, and visualized with a 647-nm krypton–argon laser. *Middle*: Estrogen receptor–containing neurons were identified with estrogen receptor antibody, labeled with Cy3, and viewed with a 568-nm krypton–argon laser. *Bottom*: Schematic diagram of single- and double-labeled steroid receptor–containing neurons. Circles represent neurons singly labeled for androgen receptors; estrogen receptor–immunoreactive neurons are identified by stars; circled stars indicate neurons that contain receptors for both androgens and estrogens.

nuclei or the PMv or ARC disrupt mating behavior. It is also important to emphasize that although the hormonal circuit contains a greater number of androgen and estrogen receptors, the neurons of the chemosensory circuit are not devoid of steroid receptors (Wood et al. 1992; Wood and Newman 1993a).

Neural coordination of male mating behavior

The concept of dual neural circuitry – that is, chemosensory and hormonal networks – provides a framework in which to evaluate the behavioral role of specific groups of hormone-sensitive neurons. Although systemic replacement of steroids in castrated males has been used to determine the hormonal requirements for copulation, this paradigm cannot reveal which brain regions are involved in the behavior. To investigate this question we have used specific brain lesions, intracerebral steroid implants, and immediate early gene expression to unravel the neural pathways through which steroids control copulation.

Effects of specific brain lesions. The results of lesioning selective areas along these pathways lend support to the concept of distinct neural circuits mediating chemosensory and hormonal cues. Lesions of the olfactory and vomeronasal pathways to the MeA, or of its projections through the BNST to the caudal MPOA, immediately and permanently abolish male sexual behavior in the hamster. Bilateral lesions that eliminate this behavior include combined deafferentation of the olfactory and vomeronasal input to the bulbs (Winans and Powers 1977), removal of the olfactory bulbs (Murphy and Schneider 1970), section of the LOT (Devor 1973), damage to the MeA (Lehman and Winans 1982; Lehman et al. 1983), lesions including the BNSTpi or centered in the caudal, lateral MPOA (Eskes 1984; Powers et al. 1987). In contrast, lesions confined to the MeP, the BNSTpm, or the medial part of the MPOA might alter mating behavior, but fail to eliminate it (Lehman et al. 1983; Powers et al. 1987). Thus, in keeping with behavioral observations from chemosensory deafferentation or castration, lesions that disrupt essential neural transmission through the central "chemosensory circuitry" have more abrupt and severe effects than lesions that damage brain areas mediating the long-term hormonal facilitation of copulation.

Effects of intracerebral testosterone implants. One interpretation of the behavioral effects of lesions in the MeP–BNSTpm network is that the structures of the hormone circuit are not actively involved in copulation; rather, they might participate in steroid-dependent functions other than sexual behavior. However, a very different conclusion emerges from studies using specific steroid replacement within this circuitry. Steroid microimplants can be placed into discrete brain regions of castrated males under stereotaxic guidance. In this manner, the effects of androgens and estrogens on specific brain nuclei can be studied in the absence of systemic hormonal effects. In male hamsters 12 weeks after castration, brain cannulae providing testosterone stimulate mounting when positioned adjacent to sexually relevant hormone-responsive neurons (Fig. 1.8). Unfortunately, it is not possible to elicit intromissions or ejaculations in these males because their periph-

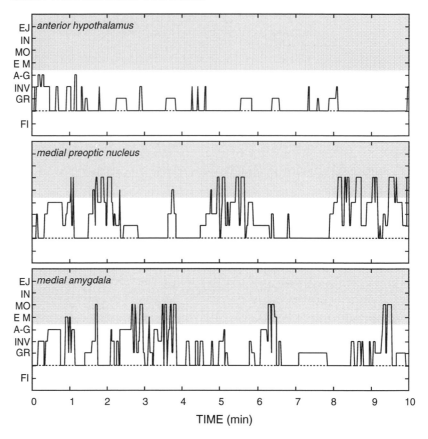

Figure 1.8. Mating behavior in three representative orchidectomized male Syrian hamsters after intracerebral testosterone implantation. For each male, behavior was recorded every second for 10 minutes immediately after the introduction of a receptive female. Baseline indicates no interaction with the female; shaded area highlights copulatory behavior, including ectopic mounts (E M), mounts (MO), intromissions (IN), and ejaculations (EJ). Other behaviors recorded during each test included grooming (GR), investigation of the female (INV), anogenital investigation (A-G), and fighting (FI). *Top*: Implants placed in brain regions adjacent to the mating behavior circuitry (as in the anterior hypothalamus) fail to stimulate copulatory behavior. However, implants that stimulate steroid receptor–containing neurons in the BNST and MPOA (*middle*) or the MeA (*bottom*) induce mounting in long-term castrates.

eral tissues (i.e., penis, accessory sex glands) lack hormonal stimulation. The selectivity of this approach is verified by the failure of implants to induce sexual behavior when placed in nearby brain areas that are outside the hormone network (e.g., anterior hypothalamus; Fig. 1.8). This method has provided abundant evidence from a number of species that steroid-responsive neurons in the BNST and MPOA transduce hormonal information to control male copulation (reviewed by Sachs and Meisel 1988; Lisk and Bezier 1980). In addition, a few reports suggest that hormonal stimulation of the Me, and possibly LSv, also induces sexual behavior (Baum et al. 1982). In our study comparing steroid stimulation of the

BNST and MPOA or Me, implants of crystalline testosterone in either of these areas facilitated male sexual behavior. (It is difficult to stimulate BNST or MPOA selectively because of their close proximity.) To estimate which specific brain regions are influenced by the hormone implant, we utilized the sensitivity of androgen receptor immunoreactivity to androgens. Androgen receptor immunoreactivity is faint or absent in brain regions distant from the steroid-filled cannula, similar to brains from castrated males (Wood and Newman 1993b). However, in the brain areas surrounding the cannula tip that have received androgenic stimulation, androgen receptor immunoreactivity is robust and confined to the cell nucleus, as observed in the intact male. This technique provides a sensitive morphological measure of androgen action and has revealed that testosterone implants can influence androgen receptor–containing neurons up to 1 mm from the tip of the cannula. Testosterone implants stereotaxically placed in the BNST and MPOA facilitated sexual behavior via steroid receptor–containing neurons in the MPN and BNSTpm, immediately caudal to the anterior commissure (Fig. 1.9). In the Me, effective implants activated androgen receptors in the dorsal MeP, adjacent to the optic tract (Fig. 1.9).

Androgen versus estrogen action in the brain. Although testosterone is an excellent hormonal stimulus to identify the behaviorally relevant sites of hormonal stimulation, it cannot reveal the specific hormonal signal for copulation, because testosterone is capable of conversion to either androgen or estrogen. The results of limited studies comparing intracerebral implants of estradiol or DHT suggest that androgens and estrogens have unique sites of action to facilitate mating. In general, estradiol is active in the same sites as testosterone. Estradiol implants in the Me can stimulate mating in the male rat (Rasia-Filho et al. 1991), just as we found in the hamster with testosterone. Lisk and Greenwald (1983) restored mating behavior in castrated male hamsters with intracerebral implants of estradiol, but not with DHT, in the region of the BNST and MPOA. However, Baum et al. (1982) facilitated copulation in male rats with implants of DHT placed in the Me and LSv, but not in the MPOA. If hormone actions are similar in these two rodent species, these observations suggest that DHT may be behaviorally ineffective in the MPOA but may act preferentially in other areas of the mating behavior pathway.

Several conclusions can be drawn from studies using intracerebral hormone implants to activate sexual behavior. First, it is reasonable to expect that, in the intact male, steroid hormones normally act at multiple points along the mating behavior pathway, although the exact hormonal signal (i.e., androgen or estrogen) at each site may differ. In addition, the observation that lesions in the MeP or BNSTpm do not prevent the initiation of copulation does not necessarily imply that these brain regions are unimportant for mating behavior.

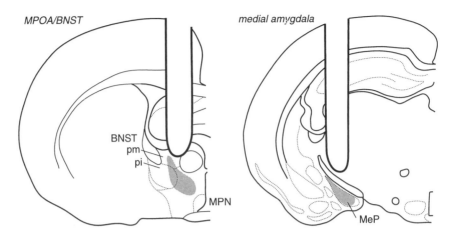

Figure 1.9. Drawings of coronal sections through the hamster brain at the level of BNST and MPOA (*left*) and MeP (*right*) to illustrate the location of intracerebral testosterone implants that stimulate mating behavior. Heavy lines indicate the path of the 22-gauge stainless steel cannula, which was filled with crystalline testosterone. Shaded area indicates androgen receptor neurons activated by the implant (see text for details). For abbreviations see caption to Figure 1.5.

Neuronal activation during mating. The foregoing brain lesion and intracerebral steroid replacement studies provide evidence that chemosensory and hormonal cues essential for mating are transmitted through subcircuits of the extended amygdala. However, these findings leave open the question of whether these two networks are functionally integrated. Although anatomical connections between them provide the opportunity for interactions, recent evidence obtained by using the expression of immediate early genes to identify activated neurons has provided the first indication of functional integration. Fos, the protein product of the immediate early gene c-*fos*, can be detected immunocytochemically in the cell nucleus of specific groups of neurons following mating behavior and is presumed to reflect increased activity of those neurons during mating. Reports of the activation of central pathways in the male hamster brain by female hamster vaginal secretions (Fiber et al. 1993) and by mating behavior itself (Kollack and Newman 1992) have demonstrated that areas of the extended amygdala in both the "hormonal" and "chemosensory" networks were significantly activated. Surprisingly, in these studies several structures of the hormonal circuit, such as the MeP, were more strongly activated than those of the chemosensory circuit.

The striking overlap of Fos immunostaining with the distribution of androgen receptor–containing neurons (Figs. 1.6 and 1.10) suggests that androgen receptor–containing neurons in the Me, BNST, and MPOA may be directly activated by mating. This possibility was investigated by means of fluorescent double labeling of these two antigens (Fig. 1.11; Wood and Newman 1993a). The co-

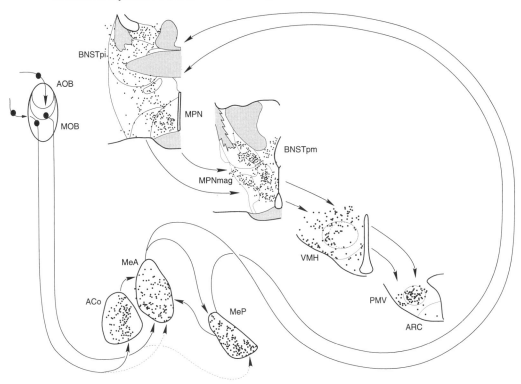

Figure 1.10. Distribution of Fos-immunoreactive neurons in the medial amygdala, BNST, MPOA, and hypothalamus of an intact adult male Syrian hamster. Dots indicate individual Fos-immunoreactive nuclei. Coronal sections from brains fixed in 4% *para*-formaldehyde were stained with Fos antibody and visualized using NiCl-enhanced diaminobenzidine, as described in Wood and Newman (1993a). For abbreviations see caption to Figure 1.5.

localization of Fos with androgen receptors was selective. Androgen receptors were present in 20–40% of Fos-immunoreactive neurons in the corticomedial amygdala, whereas double labeling of Fos-immunostained neurons increased to 70% in the magnocellular MPN. These results provided the first direct evidence that androgen receptor–containing neurons are activated by mating behavior – that is, that they are part of the neural circuitry underlying copulation. Furthermore, they show that mating selectively and differentially activates androgen receptor–positive cells in different parts of the mating behavior pathway.

It is tempting to assume that testosterone released during mating (Macrides et al. 1974) is responsible for Fos activation of androgen receptor–containing neurons. However, studies in rats have shown that the general pattern of Fos immunoreactivity after mating is not altered if steroids are held at a constant physiological level in castrated males by means of a Silastic implant (Baum and Wersinger 1993). In fact, the presence of circulating gonadal steroids is not required for

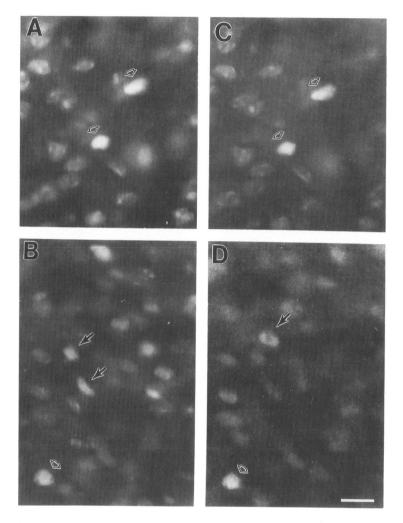

Figure 1.11. Photomicrographs showing double labeling for RITC fluorescence of androgen receptor immunoreactivity (A, B) and FITC fluorescence of Fos immunoreactivity (C, D) in the BNST from a representative male hamster. Arrows indicate neurons immunoreactive to both Fos and androgen receptors; arrowheads identify neurons labeled for only a single antigen. Scale bar, 20 μm. Reprinted with permission from Wood and Newman (1993a).

mating-induced Fos expression in short-term castrates. Thus, it appears that other stimuli, including chemosensory activation (Fiber et al. 1993), are responsible for c-*fos* induction in androgen receptor–containing neurons during mating.

Separate pathways, different mechanisms?

These methods provide different perspectives on the functions of the parallel chemosensory and hormonal subcircuits through the extended amygdala. However, to envision a coherent model for chemosensory and hormonal integration, it must be considered that these two stimuli, both of which are essential for mating behavior, are transduced by entirely different neuronal mechanisms. Chemosensory cues activate neural impulses along a discrete chain of neurons. The serial arrangement of this network is revealed by the lesion studies cited earlier. Any break in the chain prevents the transmission of neural signals generated by chemosensory stimulation, thereby resulting in immediate cessation of the behavior. By contrast, hormonal signals are mediated by a distributed neural network. The observation that lesions in the hormone-responsive circuitry do not eliminate the behavior suggests that a lesion of any single area within this hormone network is not sufficient to eliminate sexual behavior because of the redundancy (stability) built into this highly interconnected circuitry (Cottingham and Pfaff 1986). This is supported by the observation that steroid replacement at any one of a variety of sites in this network in castrated male hamsters (Lisk and Bezier 1980) and other rodents (Baum et al. 1982; Rasia-Filho et al. 1991) is sufficient to restore normal mating behavior. The multiple sites of steroid action, together with the long-term nature of steroid effects, suggest that hormonal stimulation of mating behavior is achieved not through direct neuronal excitation, but rather through trophic effects on steroid-responsive neurons.

Cellular mechanisms of steroid action

To understand how steroids control mating behavior in the male hamster, it is necessary to consider the effects of steroids on individual neurons. Steroids have diverse actions on receptor-positive neurons in different brain regions. These include alterations in neuron size or morphology, effects on the production of neurotransmitters or neurotransmitter receptors, as well as the regulation of steroid receptors themselves. This final section reviews key elements of steroid action on neurons in the mating behavior pathway in the Syrian hamster. In general, long-term steroid exposure appears to serve a "maintenance" function for steroid receptor–containing neurons, replenishing neurotransmitter stores and maintaining neuronal processes and connections. However, the specific effects of steroids on any individual neuron are related to the unique function and connections of that cell. Most steroid actions are presumably mediated through the binding of the activated hormone–receptor complex to DNA, thereby activating specific genes. Unfortunately, relatively little is known about specific steroid effects on DNA in the brain to control behavior, although current research is focused on understanding this general mechanism.

Effects of steroids on steroid receptors

Intracellular partitioning of steroid receptors. In steroid receptor–containing neurons, the presence of ligand influences both the location and number of steroid receptors within the cell. Recent immunocytochemical studies have demonstrated that castration causes repartitioning of both androgen and estrogen receptors from an exclusively nuclear localization in the presence of ligand to a distribution across both nuclear and cytoplasmic domains in the absence of steroids (Fig. 1.12). This finding relates to the ongoing controversy regarding the intracellular location of receptors for steroid hormones. Originally, based on cell fractionation studies, a "two-step" model of steroid receptor action was proposed wherein the free receptors were present in the cell cytoplasm and were translocated into the cell nucleus upon steroid binding (Gorski et al. 1968). This view was revised in 1984 on the basis of studies by King and Greene (1984) and Welshons et al. (1984) demonstrating that the majority of both bound and unbound estrogen receptors were present within the cell nucleus. According to this more recent model, the binding of the receptor to its ligand induces a tight association of the hormone–receptor complex with nuclear chromatin, whereas unbound receptors are more readily displaced from the cell nucleus (Gorski et al. 1986). Thus, the abundant steroid receptors measured in preparations of cell cytosol have been considered an artifact of cell fractionation.

In the past few years, the question of cytoplasmic steroid receptors has been revisited, and a new model of receptor localization is emerging. The presence of unbound receptor within the cell cytoplasm has now been demonstrated immunohistochemically for all major classes of steroid hormones (estrogen and progesterone: Blaustein et al. 1992; glucocorticoid: Ahima and Harlan 1991; androgen: Wood and Newman 1993b). Such observations are nonetheless compatible with the predominantly nuclear localization proposed by King and Greene (1984) and Welshons et al. (1984) because the cytoplasmic steroid receptor immunoreactivity is faint compared with that in the cell nucleus. Thus, the revised model of steroid receptor localization, incorporating a limited population of unbound receptor in the cell cytoplasm, represents a compromise between the earlier "two-step" hypothesis (Gorski et al. 1968) and the more recent findings of King and Greene (1984) and Welshons et al. (1984).

Autoregulation of steroid receptors in the brain. Although repartitioning of steroid receptors between the cell nucleus and cytoplasm in response to ligand is common to all major classes of steroid receptors, androgens and estrogens appear to have opposite effects on the number of their respective receptors in the brain. Estrogen autoregulation of the estrogen receptor has been studied extensively in females, particularly in the VMH. The results of these studies indicate that estro-

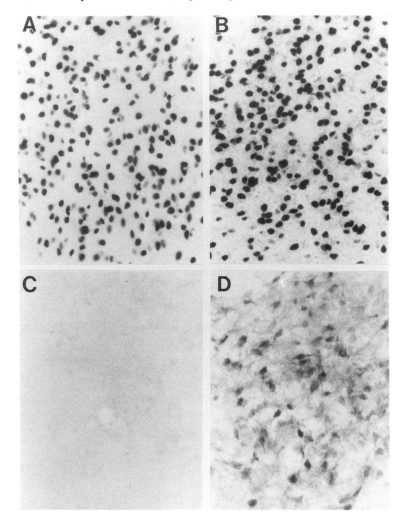

Figure 1.12. Photomicrographs of androgen receptor immunoreactivity in the ventral pre-mammillary nucleus (PMv) from intact (A, B) and orchidectomized male hamsters (C, D). Brain sections were exposed to diaminobenzidine for 6 (A, C) or 60 minutes (B, D). Scale bar, 50 μm. Reprinted with permission from Wood and Newman (1993b).

gen receptors are downregulated by estrogen in a typical negative-feedback rela-tionship, as demonstrated by receptor binding (Cidlowski and Muldoon 1978; McGinnis et al. 1981) and immunocytochemistry (Blaustein and Turcotte 1989; Blaustein 1993). Furthermore, studies using *in situ* hybridization and Northern blotting have shown that this is accompanied by a reduction in estrogen receptor mRNA (Lauber et al. 1991; Simerly and Young 1991). Taken together, these data indicate that estrogen inhibits the transcription of estrogen receptor message, thereby reducing the number of receptors available for ligand binding.

Much less is known about autoregulation of androgen receptors in the brain.

However, from data available thus far, it seems that androgens may have effects opposite to those of estrogens, in that they appear to increase the number of their receptors in the brain. This conclusion is based on reports of decreased nuclear androgen receptor occupancy in rat brain after orchidectomy (Krey and McGinnis 1990) and reductions in total cellular androgen receptor binding in hamster brain upon exposure to short photoperiod (Prins et al. 1990), as well as on results of immunocytochemistry studies in our laboratory (Wood and Newman 1993b). In addition to observing cytoplasmic immunoreactivity in castrated males, we found that androgen receptor staining intensity was greatly reduced (Fig. 1.12). Although the data available thus far are consistent with upregulation of androgen receptors by androgen, this conclusion must await confirmation via more quantitative methods.

Recently, we have investigated the effects of endocrine manipulations on the androgen receptor using changes in the intensity and intracellular location and of androgen receptor immunoreactivity as an index of receptor–ligand binding. Reduced immunoreactivity and cytoplasmic localization of receptors in males castrated for 2 weeks was restored to intact levels of nuclear staining after treatment of orchidectomized males for 8 hours with physiological levels of exogenous testosterone (Wood and Newman 1993b). Likewise, long-term exposure to testosterone or DHT maintained nuclear androgen receptor staining (Fig. 1.13), whereas exposure to low physiological concentrations of estrogen was not effective in this regard. In estradiol-treated castrates, androgen receptor immunostaining was faint and present in the cell cytoplasm as well as the nucleus (Wood and Newman 1993b). Using a more physiological model, we investigated the effects of short-day photoperiod on androgen receptor staining (Fig. 1.14). In intact males exposed to short-day photoperiod for 10 weeks, staining intensity was reduced to levels approximating those in castrated males. However, despite the reduction in the amount of staining, no labeling was observed in the cell cytoplasm, thereby supporting the concept that the effects of inhibitory photoperiod are not equivalent to castration. In addition, testosterone supplementation of short-day males restored immunoreactivity to levels equivalent to intact males, suggesting that the effects of short photoperiod on receptor immunoreactivity were due primarily to low endogenous testosterone levels, and not to the effects of photoperiod per se. The foregoing observations confirm and extend previous observations from biochemical studies on the regulation of neural androgen receptors. Furthermore, they suggest a cellular mechanism to account for the reduction in behavioral responsiveness to androgen after castration.

Steroid effects on neuronal morphology and transmitters

According to the standard model for hormone action in the CNS, binding of the hormone–receptor complex in the neuronal cell nucleus alters protein production

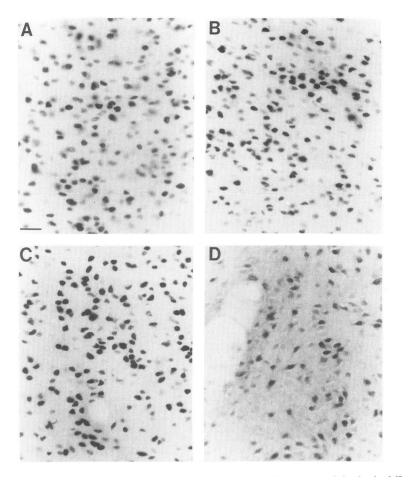

Figure 1.13. Photomicrographs of androgen receptor immunoreactivity in the MPN of a representative intact male hamster (A) and of orchidectomized males given physiological levels of exogenous gonadal steroids (B–D). Males received Silastic capsules of testosterone (B), DHT (C), or estradiol (D). Scale bar, 25 μm. Reprinted with permission from Wood and Newman (1993b).

within steroid-sensitive neurons. Numerous studies have investigated the hypothesis that proteins regulated by this mechanism include both intrinsic structural proteins and proteins associated with neurotransmitter release (e.g., neuropeptides and enzymes essential for neurotransmitter synthesis; see Part II, this volume). Considerable indirect support for this hypothesis has come both from studies of hormone-dependent sex differences during CNS development and from evidence for plastic changes after hormonal manipulations in the adult (Arnold and Breedlove 1985; Arnold 1990).

In an effort to delineate the role of sex steroids in facilitating and maintaining adult male sexual behavior, we and others have focused our attention on hormonal regulation of neuronal morphology and neurotransmitter production in the adult

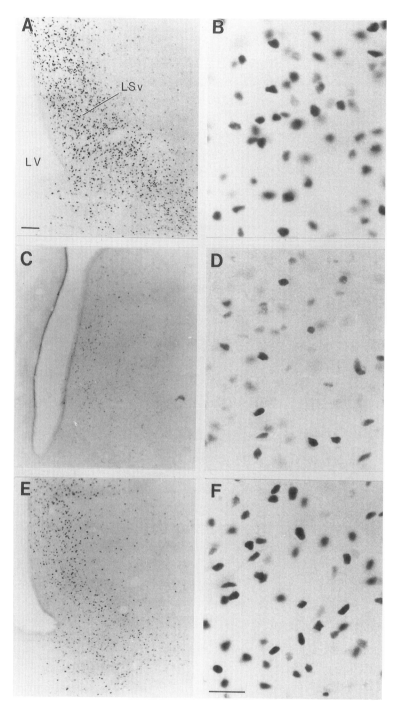

Figure 1.14. Photomicrographs of androgen receptor immunoreactivity in the lateral septum of a representative male hamster under long-day photoperiod at low (A) and high (B) magnification and of males exposed to short day lengths for 10 weeks (C–F). Males exposed to inhibitory photoperiod remained untreated (C, D) or received supplemental testosterone via Silastic implant (E, F). Scale bars, 100 μm (A, C, E) and 25 μm (B, D, F). Reprinted with permission from Wood and Newman (1993b).

brain. Therefore, in this discussion we will consider only limited examples of evidence for developmentally based sexual dimorphism in the extended amygdala and its projections to the lateral septum, preoptic area, and hypothalamus.

Hormone influence on dendritic morphology. Gonadal steroids have been shown to effect a variety of transient and permanent changes in the architecture of neurons in the hormonal circuit. In the Syrian hamster brain, early studies by Greenough et al. (1977) using quantitative analysis of Golgi-impregnated neurons provided evidence for sexual dimorphism in the preoptic area. Gonadectomized males and females treated similarly with exogenous hormones nevertheless showed differences in orientation of the dendritic fields in the MPOA. This long-term effect on neural morphology is presumably due to gonadal steroid exposure during development. In contrast to this stable dimorphism, more recent studies in females have demonstrated rapid changes in total dendritic length (Meisel and Luttrell 1990) in the VMH. These changes, documented over the 4-day hamster estrous cycle, represent transient remodeling of neural structure in response to acute changes in steroid concentrations.

Gonadal steroids have also been shown to influence neuronal morphology in the male Syrian hamster. We analyzed Golgi-impregnated neurons in the Me in males 12 weeks after castration, at a time when mating behavior was reduced to only sporadic mounting. Structural changes were observed in neurons of the MeP, but not the MeA. In the MeP, the mean somal area, the mean highest dendritic branching level, and the percentage of neurons with tertiary branches were significantly decreased in castrates compared with intact control animals (Fig. 1.15; Gomez and Newman 1991). Furthermore, replacement therapy with exogenous testosterone or estradiol, but not DHT, prevented these changes from occurring in orchidectomized males. It is interesting that the limited and very specific neuronal changes observed in the MeP after castration in hamsters were similar to those reported in the spinal nucleus of the bulbocavernosus in rats (Sasaki and Arnold 1991), another androgen-sensitive area of the CNS.

One puzzling finding of our study was that DHT treatment in castrates maintained sexual behavior in these males without discernible restorative effects on neuronal structure in the MeP. We recognize that the specific sites of action of DHT along the hormonal circuit have not yet been definitively established. One interpretation of these findings is that the neuronal characteristics altered by castration are inconsequential for the role of MeP neurons in mating behavior, or that the specific MeP neurons analyzed are unnecessary for the behavior to occur. It is also possible that the dendritic changes we observed in the MeP were not characteristic of changes elsewhere in the hormonal circuit (i.e., that DHT replacement affected neurons in the BNST and/or the MPOA differently than those in the MeP), and that the redundancy in this circuitry compensated for the deleterious

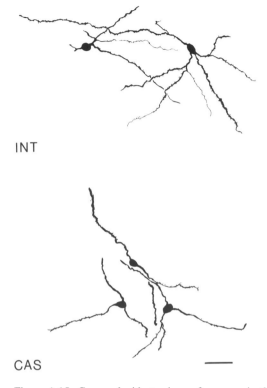

INT

CAS   ———

Figure 1.15. Camera lucida tracings of neurons in the posterior division of the medial nucleus of the amygdala from an intact (INT) male and a male that had been castrated (CAS) 12 weeks earlier. Scale bar, 50 μm.

effects on the neurons of the MeP. Our understanding of steroid effects on neural structure to control sex behavior is currently hampered by the scarcity of data. The recent advent of new technologies for sophisticated reconstruction of neuronal morphology, combined with new information about the functional roles of individual cell groups, should facilitate this aim.

Hormonal regulation of neurotransmitter production. Evidence for gonadal hormone influence on the production of neurotransmitters and neuromodulators is now abundant in the literature (see the chapters in Part II of this volume). This influence has been documented in adult males and females of numerous species. However, unlike the consistency across different mammalian species in the anatomy of neural circuits that form the substrate for mating, the neurotransmitters and neuromodulators involved in this pathway appear to be highly variable and often species-specific. This is especially true of neuropeptides, but is also recognized increasingly in what are traditionally considered the "classical" neurotransmitters such as the catecholaminergic systems. An example of this species variability is represented by our observations that dynorphin peptides are abundant in the MeP, BNSTpm, and MPN of the male hamster brain (Fig. 1.16), but are essentially absent in these same nuclei in the male rat (Neal and Newman 1989). In contrast, these three areas are densely populated by neurons that produce cholecystokinin (CCK) in the male rat (Micevych et al. 1988a, b), and production of CCK in these cells is dependent on circulating testosterone (Simerly and Swanson 1987; Micevych et al. 1988b), whereas CCK has not been localized in any of these cell groups in the male hamster (unpublished observations).

One neuropeptide that is found in abundance in this same circuitry in both rat and hamster is substance P (SP; Fig. 1.16). However, in this case the two species appear to differ in the regulation of peptide production. In the rat brain, SP is co-localized with CCK in neurons of the MeP, BNSTpm, and MPN, yet castration does not affect the preprotachykinin mRNA content in these nuclei (Simerly 1990). In contrast, in the hamster, long-term castration inhibits the production of SP, particularly in the BNSTpm and MPOA (Kream et al. 1987; Swann and Newman 1992), and SP immunoreactivity is restored by exogenous testosterone (Swann and Newman 1992). Although 50–75% of these SP neurons also contain leumorphin (Neal et al. 1989), we do not yet know if testosterone has a differential effect on the production of these two peptides.

As already noted, species differences have also been observed in the so-called classical neurotransmitter systems. Tyrosine hydroxylase (TH), the rate-limiting enzyme in catecholamine production, is found in numerous neuronal areas of the forebrain outside the classically defined catecholamine systems (Hokfelt et al. 1984). While these nontraditional cell groups have been found in a variety of species (Asmus and Newman 1993b), the Syrian hamster is the only species in which this enzyme has been localized in cells of the medial nucleus of the amygdala (Davis and Macrides 1983; Vincent 1988), where two distinct populations of TH cells in the MeA and MeP can be labeled both with antibodies against TH (Asmus et al. 1992) and with *in situ* hybridization for TH mRNA (Asmus and Newman 1993b). While neither of these populations also contains dopamine β-hydroxylase or phenylethanolamine-*N*-methyltransferase (the first of which synthesizes norepinephrine, the second epinephrine), neurons in the MeP, but not those in the MeA, can be immunolabeled with antibodies against dopamine (Asmus et al. 1992). Furthermore, 75% of the TH neurons in the MeP also contain androgen receptors, whereas only 33% of those in the MeA show androgen receptor immunoreactivity (Asmus and Newman 1993a).

Consistent with the hypothesis that distinctive "chemosensory" and "hormonal" circuits exist within the extended amygdala of the hamster, additional species-specific TH-immunoreactive neurons that have been localized within the BNSTpm and the lateral part of the MPOA (Fig. 1.16) have characteristics in common with the TH neurons of the MeP and MeA, respectively. Eighty percent of the neurons in the BNSTpm, like those of the MeP, produce androgen recep-

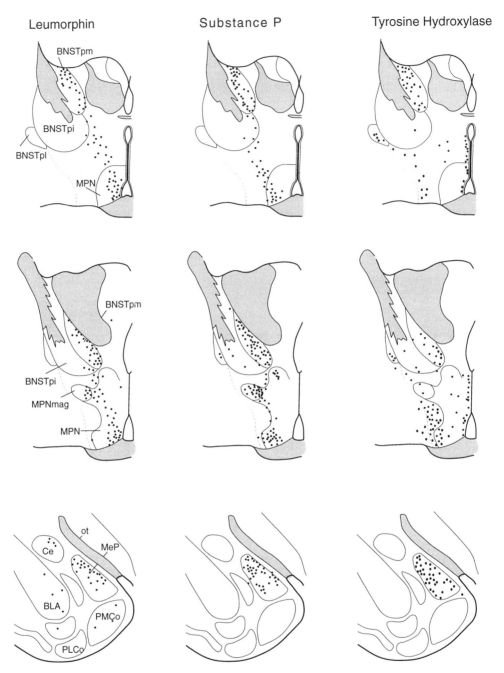

Figure 1.16. The distribution of leumorphin-, substance P–, and tyrosine hydroxylase–immunoreactive neurons in the BNST terminalis, preoptic area, and caudal medial amygdala of the male Syrian hamster brain. Abbreviations: BLA, basolateral nucleus of the amygdala; BNSTpl, bed nucleus of the stria terminalis, posterolateral division; Ce, central nucleus of the amygdala; ot, optic tract. Also see captions to Figures 1.4 and 1.5.

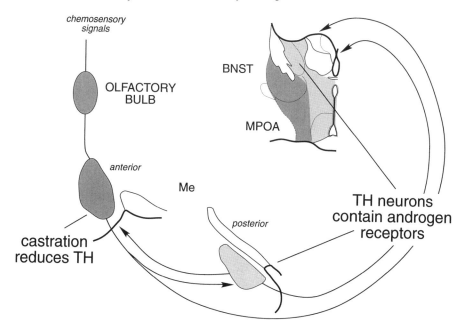

Figure 1.17. Hormonal regulation of tyrosine hydroxylase (TH) neurons along the mating behavior pathways in the male Syrian hamster brain. In the hormonal subcircuit (light shading), numerous TH neurons have androgen receptors in MeAp and BNSTpm, suggesting that these cells have the potential to respond to steroid signals from the gonads. However, castration reduces the number of TH neurons only in MeAa (dark shading).

tors and dopamine, whereas those of the lateral MPOA, in the same chemosensory circuit with the MeA, show less than 5% co-localization with androgen receptors and no dopamine immunostaining (Asmus and Newman 1993a).

These observations led to the hypothesis that TH production in the MeP and BNSTpm would be directly affected by changes in circulating testosterone levels. Although this hypothesis may be correct, castration had no effect on immunohistochemically detectable levels of TH in these cells at 2, 4, 6, or 12 weeks after orchidectomy. Thus, the role of the androgen receptors in the metabolism of these cells remains to be elucidated. In contrast, in the MeA, where only one-third of the cells contain androgen receptors, the number of TH neurons was reduced by 60% 6 weeks after castration and returned to intact values after 12 weeks (Fig. 1.17). We interpret these transient changes, and the observation that castration altered TH production in neurons apparently lacking androgen receptors, to suggest indirect effects of steroids on these TH cells.

The neurons in the MeA receive direct and indirect olfactory and vomeronasal inputs from the main and accessory olfactory bulbs, as well as inputs from structures of the hormonal circuit such as the MeP. Furthermore, lesions in this part of the MeA mimic olfactory bulbectomy; they drastically reduce chemoinvestigatory behavior and completely eliminate copulation in the male Syrian hamster. Taken together, the data reviewed here suggest that this part of the extended amygdala and its projections to the MPOA are essential for processing chemosensory signals that initiate sexual responses in these animals. However, it is very clear that trophic influences of hormones, which are equally essential for long-term maintenance of the behavior, are mediated through adjacent circuits of the extended amygdala in the MeP and its projection targets. The mechanisms by which these sustaining effects of hormones are achieved remain to be resolved.

Acknowledgments

We thank Ms. R. Kaye Brabec, Ms. Lorita Dudus, and Ms. Carrie Cartwright of the Reproductive Sciences Program Morphology Core Facility (P30 HD18258) for assistance with tissue processing. This work was supported by USPHS Research and Training Awards NS20629 and HD07514.

References

Ahima, R. S., and Harlan, R. E. (1991). Differential corticosteroid regulation of Type II glucocorticoid receptor-like immunoreactivity in the rat central nervous system: Topography and implications. *Endocrinology* 129: 226–236.

Alheid, G. F., and Heimer, L. (1988). The functional-anatomical organization of the basal forebrain: New perspectives of special relevance for neuropsychiatric disorders. *Neuroscience* 27: 1–39.

Arnold, A. P. (1990). Hormonally induced synaptic reorganization in the adult brain. In *Hormones, Brain and Behavior in Vertebrates*, ed. J. Balthazart, pp. 82–91. Basel: Karger.

Arnold, A. P., and Breedlove, S. M. (1985). Organizational and activational effects of sex steroids on brain and behavior: A reanalysis. *Horm. Behav.* 19: 469–498.

Asmus, S. E., Kincaid, A. E., and Newman, S. W. (1992). A species-specific population of tyrosine hydroxylase-immunoreactive neurons in the medial amygdaloid nucleus of the Syrian hamster. *Brain Res.* 575: 199–207.

Asmus, S. E., and Newman, S. W. (1993a). Tyrosine hydroxylase neurons in the male hamster chemosensory pathway contain androgen receptors and are influenced by gonadal hormones. *J. Comp. Neurol.* 331: 445–457.

Asmus, S. E., and Newman, S. W. (1993b). Tyrosine hydroxylase mRNA-containing neurons in the medial amygdaloid nucleus and the reticular nucleus of the thalamus in the Syrian hamster. *Mol. Brain Res.* 20: 267–273.

Barber, P. C., and Raisman, G. (1974). An autoradiographic investigation of the projection of the vomeronasal organ to the accessory olfactory bulb in the mouse. *Brain Res.* 81: 21–30.

Bartke, A. (1985). Male hamster reproductive endocrinology. In *The Hamster: Reproduction and Behavior*, ed. H. I. Siegel, pp. 73–98. New York: Plenum Press.

Baum, M. J., Tobet, S. A., Starr, M. S., and Bradshaw, W. G. (1982). Implantation of dihydrotestosterone propionate into the lateral septum or medial amygdala facilitates copulation in castrated male rats given estradiol systemically. *Horm. Behav.* 16: 208–223.

Baum, M. J., and Wersinger, S. R. (1993). Equivalent levels of mating-induced neural c-fos immunoreactivity in castrated male rats given androgen, estrogen, or no steroid replacement. *Biol. Reprod.* 48: 1341–1347.

Blaustein, J. D. (1993). Estrogen receptor immunoreactivity in rat brain: Rapid effects of estradiol injection. *Endocrinology* 132: 1218–1224.

Blaustein, J. D., Lehman, M. N., Turcotte, J. C., and Greene, G. (1992). Estrogen receptors in dendrites and axon terminals in the guinea pig hypothalamus. *Endocrinology* 131: 281–290.

Blaustein, J. D., and Turcotte, J. C. (1989). Estrogen receptor-immunostaining of neuronal cytoplasmic processes as well as cell nuclei in guinea pig brain. *Brain Res.* 495: 75–82.

Canteras, N. S., Simerly, R. B., and Swanson, L. W. (1992a). Connections of the posterior nucleus of the amygdala. *J. Comp. Neurol.* 324: 143–179.

Canteras, N. S., Simerly, R. B., and Swanson, L. W. (1992b). Projections of the ventral premammillary nucleus. *J. Comp. Neurol.* 324: 195–212.

Cidlowski, J. A., and Muldoon, T. G. (1978). The dynamics of intracellular estrogen receptor as influenced by 17β-estradiol. *Biol. Reprod.* 18: 234–246.

Cottingham, S. L., and Pfaff, D. (1986). Interconnectedness of steroid hormone-binding neurons: Existence and implications. In *Current Topics in Neuroendocrinology*, Vol. 7: *Morphology of Hypothalamus and Its Connections*, ed. D. Ganten and D. Pfaff, pp. 223–249. Berlin: Springer.

Davis, B. J., and Macrides, F. M. (1983). Tyrosine hydroxylase immunoreactive neurons and fibers in the olfactory system of the hamster. *J. Comp. Neurol.* 214: 427–440.

Davis, B. J., Macrides, F., Young, W. M., Schneider, S. P., and Rosene, D. L. (1978). Efferents and centrifugal afferents of the main and accessory olfactory bulbs in the hamster. *Brain Res. Bull.* 3: 59–72.

DeBold, J. F., and Clemens, L. G. (1978). Aromatization and the induction of male sexual behavior in male, female, and androgenized female hamsters. *Horm. Behav.* 11: 401–413.

de Olmos, J. S., Alheid, G. F., and Beltramino, C. A. (1985). The amygdala. In *The Rat Nervous System*, Vol. 1: *Forebrain and Midbrain*, ed. G. Paxinos, pp. 223–334. London: Academic Press.

Devor, M. (1973). Components of mating dissociated by lateral olfactory tract transection in male hamsters. *Brain Res.* 64: 437–441.

Doherty, P. C., and Sheridan, P. J. (1981). Uptake and retention of androgen in neurons of the brain of the golden hamster. *Brain Res.* 219: 327–334.

Ellis, G. B., and Desjardins, C. J. (1982). Male rats secrete luteinizing hormone and testosterone episodically. *Endocrinology* 110: 1618–1627.

Eskes, G. A. (1984). Neural control of the daily rhythm of sexual behavior in the male golden hamster. *Brain Res.* 293: 127–141.

Fiber, J. M., Adames, P., and Swann, J. M. (1993). Pheromones induce c-fos in limbic areas regulating male hamster mating behavior. *NeuroReport* 4: 871–874.

Gomez, D. M., and Newman, S. W. (1991). Medial nucleus of the amygdala in the adult Syrian hamster: A quantitative Golgi analysis of gonadal hormonal regulation of neuronal morphology. *Anat. Rec.* 231: 498–509.

Gomez, D. M., and Newman, S. W. (1992). Differential projections of the anterior and posterior regions of the medial amygdaloid nucleus in the Syrian hamster. *J. Comp. Neurol.* 317: 195–218.

Gorski, J., Toft, D., Shyamala, G., Smith, D., and Notides, A. (1968). Hormone receptors: Studies on the interaction of estrogens with the uterus. *Rec. Prog. Horm. Res.* 24: 45–80.

Gorski, J., Welshons, W. V., Sakai, D., Hansen, J., Walent, J., Kasis, J., Shull, J., Stack, G., and Campen, C. (1986). Evolution of a model of estrogen action. *Rec. Prog. Horm. Res.* 42: 297–329.

Greenough, W., Carter, C. S., Steerman, C., and DeVoogd, T. (1977). Sex differences in dendritic patterns in hamster preoptic area. *Brain Res.* 126: 63–72.

Halpern, M. (1987). The organization and function of the vomeronasal system. *Annu. Rev. Neurobiol.* 10: 325–362.

Handa, R. J., Reid, D. L., and Resko, J. A. (1986). Androgen receptors in brain and pituitary of female rats: Cyclic changes and comparisons with the male. *Biol. Reprod.* 34: 293–303.

Heimer, L., and Alheid, G. (1991). Piecing together the puzzle of basal forebrain anatomy. In *The Basal Forebrain: Anatomy to Function*, ed. T. C. Napier, P. W. Kalivas, and I. Hanin, pp. 1–42. New York: Plenum Press.

Hokfelt, T., Martensson, R., Bjorklund, A., Kleinau, S., and Goldstein, M. (1984). Distributional maps of tyrosine-hydroxylase-immunoreactive neurons in the rat brain. In *Handbook of Chemical Neuronatomy*, Vol. 2: *Classical Transmitters in the CNS*, Part I, ed. A. Bjorklund and T. Hokfelt, pp. 277–379. Amsterdam: Elsevier.

Jackson, G. L., Kuehl, D., and Rhim, T. J. (1991). Testosterone inhibits gonadotropin-releasing hormone pulse frequency in the male sheep. *Biol. Reprod.* 45: 188–194.

Johnston, J. B. (1923). Further contributions to the study of the evolution of the forebrain. *J. Comp. Neurol.* 35: 337–481.

Ju, G., and Swanson, L. W. (1989). Studies on the cellular architecture of the bed nucleus of the stria terminalis in the rat: I. Cytoarchitecture. *J. Comp. Neurol.* 280: 587–602.

Kevetter, G. A., and Winans, S. S. (1981a). Connections of the corticomedial amygdala in the golden hamster, I: Efferents of the "vomeronasal amygdala." *J. Comp. Neurol.* 197: 81–98.

Kevetter, G. A., and Winans, S. S. (1981b). Connections of the corticomedial amygdala in the golden hamster, II: Efferents of the "olfactory amygdala." *J. Comp. Neurol.* 197: 99–112.

King, W. J., and Greene, G. L. (1984). Monoclonal antibodies localize oestrogen receptor in the nuclei of target cells. *Nature* 307: 745–747.

Kollack, S. S., and Newman, S. W. (1992). Mating behavior induces selective expression of Fos protein within the chemosensory pathways of the male Syrian hamster brain. *Neurosci. Lett.* 143: 223–228.

Kream, R. M., Clancy, A. N., Kumar, M. S. A., Schoenfeld, T. A., and Macrides, F. (1987). Substance P and luteinizing hormone releasing hormone levels in the brain of the male golden hamster are both altered by castration and testosterone replacement. *Neuroendocrinology* 46: 297–305.

Krey, L. C., and McGinnis, M. Y. (1990). Time courses of the appearance/disappearance of nuclear androgen + receptor complexes in the brain and adenohypophysis following testosterone administration/withdrawal to castrated male rats: Relationships with gonadotropin secretion. *J. Steroid Biochem.* 35: 403–408.

Krieger, M. S., Morrell, J. I., and Pfaff, D. W. (1976). Autoradiographic localization of estradiol-concentrating cells in the female hamster brain. *Neuroendocrinology* 22: 193–205.

Larsson, K., Dessi-Fulgheri, F., and Lupo, C. (1982). Social stimuli modify the neuroendocrine activity through olfactory cues in the male rat. In *Olfaction and Endocrine Regulation*, ed. W. Breipohl, pp. 51–62. London: IRI Press.

Lauber, A. H., Mobbs, C. V., Muramatsu, M., and Pfaff, D. W. (1991). Estrogen receptor messenger RNA expression in rat hypothalamus as a function of genetic sex and estrogen dose. *Endocrinology* 129: 3180–3186.

Lehman, M. N., Powers, J. B., and Winans, S. S. (1983). Stria terminalis lesions alter the temporal pattern of copulatory behavior in the male golden hamster. *Behav. Brain Res.* 8: 109–128.

Lehman, M. N., and Winans, S. S. (1982). Vomeronasal and olfactory pathways to the amygdala controlling male hamster sexual behavior: Autoradiographic and behavioral analyses. *Brain Res.* 240: 27–41.

Lisk, R. D., and Bezier, J. L. (1980). Intrahypothalamic hormone implantation and activation of sexual behavior in the male hamster. *Neuroendocrinology* 30: 220–227.

Lisk, R. D., and Greenwald, D. P. (1983). Central plus peripheral stimulation by androgen is necessary for complete restoration of copulatory behavior in the male hamster. *Neuroendocrinology* 36: 211–217.

Lisk, R. D., and Heimann, J. (1980). The effects of sexual experience and frequency of testing on retention of copulatory behavior of the male golden hamster. *Psychonom. Sci.* 29: 288–290.

Macrides, F., Bartke, A., Fernandez, F., and D'Angelo, W. (1974). Effects of exposure to vaginal odor and receptive females on plasma testosterone in the male hamster. *Neuroendocrinology* 15: 355–364.

Macrides, F., Firl, A. C. Jr., Schneider, S. P., Bartke, A., and Stein, D. C. (1976). Effect of one-stage or serial transection of the lateral olfactory tracts on behavior and plasma testosterone levels in male hamsters. *Brain Res.* 109: 97–109.

Maragos, W. F., Newman, S. W., Lehman, M. N., and Powers, J. B. (1989). Neurons of origin and fiber trajectory of amygdalofugal projections to the medial preoptic area in Syrian hamsters. *J. Comp. Neurol.* 280: 59–71.

McGinnis, M. J., Krey, L. C., MacLusky, N. J., and McEwen, B. S. (1981). Characterization of steroid receptor levels in intact and ovariectomized rats: An examination of the quantitative, temporal and endocrine factors which influence the neuroendocrine efficacy of an estradiol stimulus. *Neuroendocrinology* 33: 158–161.

Meisel, R. L., and Luttrell, V. R. (1990). Estradiol increases the dendritic length of ventromedial hypothalamic neurons in female Syrian hamsters. *Brain Res. Bull.* 25: 165–168.

Meredith, M., and O'Connell, R. J. (1979). Efferent control of stimulus access to the hamster vomeronasal organ. *J. Neurophysiol.* 286: 301–316.

Micevych, P. E., Akesson, T., and Elde, R. (1988a). Distribution of cholcystokinin-immunoreactive cell bodies in the male and female rat, II: Bed nucleus of the stria terminalis and amygdala. *J. Comp. Neurol.* 269: 381–391.

Micevych, P. E., Matt, D. W., and Go, V. L. W. (1988b). Concentrations of cholecystokinin substance P and bombesin in discrete regions of male and female rat brain: Sex differences and estrogen effects. *Exp. Neurol.* 100: 416–425.

Miernicki, M., Pospichal, M. W., and Powers, J. B. (1990). Short photoperiods affect male hamster sociosexual behaviors in the presence and absence of testosterone. *Physiol. Behav.* 47: 95–106.

Miller, L. L., Whitsett, J. M., Vandenbergh, J. G., and Colby, D. R. (1977). Physical and behavioral aspects of sexual maturation in male golden hamsters. *J. Comp. Physiol. Psychol.* 91: 245–259.

Moga, M. M., Saper, C. B., and Gray, T. S. (1989). Bed nucleus of the stria terminalis: Cytoarchitecture, immunohistochemistry, and projection to the parabrachial nucleus in the rat. *J. Comp. Neurol.* 283: 315–332.

Morin, L., and Zucker, I. (1978). Photoperiodic regulation of copulatory behavior in the male hamster. *J. Endocrinol.* 77: 249–258.

Murphy, M., and Schneider, G. E. (1970). Olfactory bulb removal eliminates mating behavior in the male golden hamster. *Science* 167: 302–303.

Neal, C. R. Jr., and Newman, S. W. (1989). Prodynorphin peptide distribution in the forebrain of the Syrian hamster and rat: A comparative study with antisera against dynorphin A, dynorphin B and the C-terminus of the prodynorphin precursor molecule. *J. Comp. Neurol.* 288: 353–386.

Neal, C. R. Jr., Swann, J. M., and Newman, S. W. (1989). The colocalization of substance P and prodynorphin immunoreactivity in neurons of the medial preoptic area, bed nucleus of the stria terminalis and medial nucleus of the amygdala of the Syrian hamster. *Brain Res.* 496: 1–13.

Payne, A. P., and Bennett, N. K. (1976). Effects of androgens on sexual behavior and somatic variable in the male golden hamster. *J. Reprod. Fert.* 47: 239–244.

Pirke, K. M., and Spyra, B. (1981). Influence of starvation on testosterone-luteinizing hormone feedback in the rat. *Acta Endocrinol.* 96: 413–421.

Powers, J. B., Bergondy, M. L., and Matochik, J. A. (1985). Male hamster sociosexual behaviors: Effects of testosterone and its metabolites. *Physiol. Behav.* 35: 607–616.

Powers, J. B., Newman, S. W., and Bergondy, M. L. (1987). MPOA and BNST lesions in male Syrian hamsters: Differential effects on copulatory and chemoinvestigatory behaviors. *Behav. Brain Res.* 23: 181–195.

Prins, G. S., Bartke, A., and Steger, R. W. (1990). Influence of photoinhibition, photostimulation and prolactin on pituitary and hypothalamic nuclear androgen receptors in the male hamster. *Neuroendocrinology* 52: 511–516.

Prins, G. S., Birch, L., and Greene, G. L. (1991). Androgen receptor localization in different cell types of the adult rat prostate. *Endocrinology* 129: 3187–3199.

Rasia-Filho, A. A., Peres, T. M. S., Cubilla-Gutierrez, F. H., and Lucion, A. B. (1991). Effect of estradiol implanted in the corticomedial amygdala on the sexual behavior of castrated male rats. *Brazil. J. Med. Biol. Res.* 24: 1041–1049.

Sachs, B. D., and Meisel, R. I. (1988). The physiology of male sexual behavior. In *The Physiology of Reproduction*, ed. E. Knobil and J. Neill, pp. 1393–1485. New York: Raven Press.

Sapolsky, R. M., and Eichenbaum, H. (1980). Thalamocortical mechanisms in odor-guided behavior, II: Effects of lesions of the mediodorsal thalamic nucleus and frontal cortex on odor preferences and sexual behavior in the hamster. *Brain Behav. Evol.* 17: 276–290.

Sasaki, M., and Arnold, A. P. (1991). Androgenic regulation of dendritic trees of motoneurons in the spinal nucleus of the bulbocavernosus: Reconstruction after intracellular iontophoresis of horseradish peroxidase. *J. Comp. Neurol.* 308: 11–27.

Scalia, F., and Winans, S. S. (1975). The differential projections of the olfactory bulb and accessory olfactory bulb in mammals. *J. Comp. Neurol.* 161: 31–56.

Siegel, H. I. (1985). Male sexual behavior. In *The Hamster: Reproduction and Behavior*, ed. H. I. Siegel, pp. 191–206. New York: Plenum Press.

Silver, W. L. (1987). The common chemical sense. In *Neurobiology of Taste and Smell*, ed. T. E. Finger and W. L. Silver, pp. 65–87. New York: Wiley.

Simerly, R. B. (1990). Hormonal control of neuropeptide gene expression in sexually dimorphic olfactory pathways. *Trends Neurosci.* 13: 104–110.

Simerly, R. B., Chang, C., Muramatsu, M., and Swanson, L. W. (1990). Distribution of androgen and estrogen receptor mRNA-containing cells in the rat brain: An in situ hybridization study. *J. Comp. Neurol.* 294: 76–95.

Simerly, R. B., and Swanson, L. W. (1987). Castration reversibly alters levels of cholecystokinin immunoreactivity within cells of three interconnected sexually dimorphic forebrain nuclei in the rat. *Proc. Natl. Acad. Sci. (USA)* 84: 2087–2091.

Simerly, R. B., and Young, B. J. (1991). Regulation of estrogen receptor messenger ribonucleic acid in rat hypothalamus by sex steroid hormones. *Mol. Endocrinol.* 5: 424–432.

Swann, J. M., and Newman, S. W. (1992). Testosterone regulates substance P in neurons of the medial nucleus of the amygdala, the bed nucleus of the stria terminalis and the medial preoptic area of the male golden hamster. *Brain Res.* 590: 18–28.

Tamarkin, L., Hutchison, J. S., and Goldman, B. D. (1976). Regulation of serum gonadotropins by photoperiod and testicular hormone in the Syrian hamster. *Endocrinology* 99: 1528–1533.

Turek, F. W. (1977). The interaction of photoperiod and testosterone in regulation of serum gonadotropin levels in castrated male hamsters. *Endocrinology* 101: 1210–1215.

Vincent, S. R. (1988). Distributions of tyrosine hydroxylase-, dopamine-β-hydroxylase-, and phenylethanolamine-*N*-methyltransferase-immunoreactive neurons in the brain of the hamster (*Mesocricetus auratus*). *J. Comp. Neurol.* 268: 584–599.

Welshons, W. V., Lieberman, M. E., and Gorski, J. (1984). Nuclear localization of unoccupied oestrogen receptors. *Nature* 307: 747–749.

Whalen, R. E., and DeBold, J. F. (1974). Comparative effectiveness of testosterone, androstenedione and dihydrotestosterone in maintaining mating behavior in the castrated male hamster. *Endocrinology* 95: 1674–1679.

Winans, S. S., and Powers, J. B. (1977). Olfactory and vomeronasal deafferentation of male hamsters: Histological and behavioral analyses. *Brain Res.* 126: 325–344.

Wirsig, C. R., and Leonard, C. M. (1986). The terminal nerve projects centrally in the hamster. *Neuroscience* 19: 709–717.

Wirsig-Weichmann, C. R. (1993). Nervus terminalis lesions, I: No effect on pheromonally induced testosterone surges in the male hamster. *Physiol. Behav.* 53: 251–255.

Wood, R. I., Brabec, R. K., Swann, J. M., and Newman, S. W. (1992). Androgen and estrogen receptor-containing neurons in chemosensory pathways of the male Syrian hamster brain. *Brain Res.* 596: 89–98.

Wood, R. I., and Newman, S. W. (1993a). Mating activates androgen receptor-containing neurons in the chemosensory pathways of the male Syrian hamster brain. *Brain Res.* 614: 65–77.

Wood, R. I., and Newman, S. W. (1993b). Intracellular partitioning of androgen receptor immunoreactivity in the brain of the male Syrian hamster: Effects of castration and steroid replacement. *J. Neurobiol.* 24: 925–938.

Wood, R. I., and Newman, S. W. (1993c). Androgen and estrogen receptors coexist within neurons of the limbic system in the Syrian hamster. *Soc. Neurosci. Abstr.* 19: 820.

Wysocki, C. J., Katz, Y., and Bernhard, P. (1983). Male vomeronasal organ mediates female-induced testosterone surges in mice. *Biol. Reprod.* 28: 917–922.

Wysocki, C. J., and Meredith, M. (1987). The vomeronasal system. In *Neurobiology of Taste and Smell*, ed. T. E. Finger and W. L. Silver, pp. 125–150. New York: Wiley.

2

Neural circuitry for the hormonal control of male sexual behavior

PAULINE YAHR

Role of the medial preoptic area in the hormonal control of male sexual behavior

Mating behaviors of males provide excellent examples of the neurobiological effects of sex steroid hormones. In mammals (Larsson 1979; Michael and Bonsall 1979; Sachs and Meisel 1988), birds (Balthazart 1983; Silver et al. 1979), and other vertebrates (Crews 1979; Crews and Silver 1985; Kelley and Pfaff 1978), testosterone (T) promotes the sexual activity of adult males. Gonadally intact males are more likely to mount receptive females, and to copulate to ejaculation, than are castrated males, unless the castrates are given T. Although these stimulatory effects of T result largely from its action on the brain (Kelley and Pfaff 1978; Larsson 1979; Sachs and Meisel 1988), T also acts on the penis and spinal neurons to affect copulatory performance (Breedlove 1984; Hart 1978; Hart and Leedy 1985; Sachs 1983).

Within the brain, one of the most important areas mediating the effects of T on male sex behavior is the medial preoptic area (MPOA) or the MPOA–anterior hypothalamus (AH) continuum. The MPOA–AH contains many cells that accumulate T or its metabolites, estradiol (E) and dihydrotestosterone (DHT) (Kelley and Pfaff 1978; Luine and McEwen 1985; Michael and Bonsall 1990). However, the conclusion that the MPOA–AH mediates many of the effects of T on male sexual behavior is based on the behavioral changes produced by manipulations of this area (Sachs and Meisel, 1988). Implanting T or E into the MPOA–AH restores mounting in castrated males, and lesioning the MPOA–AH eliminates sexual behavior in males that are exposed to circulating T.

Attempts to identify specific MPOA cell groups that are essential for male sexual behavior

In addition to its role in male sexual behavior, the MPOA–AH affects other functions sensitive to gonadal steroids. These include social displays (Kelley and Pfaff

1978; Yahr 1983), female sexual behavior (Pfaff 1980), maternal behavior (Numan 1988), and secretion of prolactin (Pan and Gala 1985a,b) and gonadotropin (Feder 1981). Thus, MPOA–AH cell groups that respond biochemically or histochemically to T may, or may not, be related to male sexual behavior. Similarly, knowing that a cell group is sexually dimorphic, that it differentiates under the control of T, or that its development is altered by factors that disrupt sexual differentiation, such as prenatal stress (Ward and Ward 1985), does not ensure that the cell group affects male sexual behavior. These features are common to many functions that respond to gonadal steroids in adulthood and thus cannot implicate one function over another. Sometimes, however, one can capitalize on differences in the dose– or time–response relationships underlying these processes to determine which functions are not likely to be controlled by a particular cell group (e.g., see Gorski et al. 1978).

Until recently, few attempts were made to identify specific cell groups in the MPOA–AH that could account for its effects on male sexual behavior. Such an endeavor may have been discouraged by an early report (Heimer and Larsson 1966) that the cessation of sexual activity in male rats given large electrolytic lesions of the MPOA–AH could not be replicated in any of 42 males given smaller lesions. However, a smaller portion of the MPOA–AH was implicated in the control of male sexual behavior by van de Poll and van Dis (1979). They showed that male rats stopped mating after receiving lesions that included only the caudal MPOA and the most rostral part of the AH. Similarly, male hamsters stopped copulating after receiving electrolytic lesions that consistently included a small area in the caudal MPOA (Powers et al. 1987). The common area of damage overlapped a group of large cells that extended from the lateral edge of the medial preoptic nucleus (MPN).

By lesioning specific cell groups, we and others have tried to identify more directly the MPOA cells necessary for male sexual behavior in quail (Balthazart and Surlemont 1990a; Balthazart et al. 1992) and gerbils (Yahr and Gregory 1993). The cell group in quail that appears to be responsible for the effects of the MPOA on male sexual behavior is the medial preoptic nucleus, designated the POM (Viglietti-Panzica et al. 1986). Bilateral electrolytic lesions of the POM greatly reduce mount attempts in sexually experienced, T-treated male quail, even when only 10–50% of the nucleus is destroyed (Balthazart and Surlemont 1990a; Balthazart et al. 1992). Lesions that are near the POM but do not overlap it, or that overlap less than 10% of it, do not affect mating. Sexual behavior can also be suppressed in sexually experienced, T-treated male quail by bilateral implantation into the POM of androstatrienedione (ATD), which blocks the conversion of T to E; tamoxifen, which blocks the action of E; or flutamide, which blocks the action of DHT (Balthazart and Surlemont 1990b). When ATD implants were placed outside the POM, T induction of sexual behavior was not curtailed. Conversely,

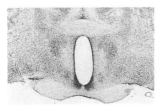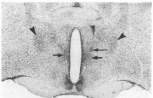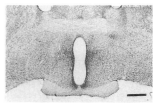

Figure 2.1. Photomicrographs of thionin-stained coronal sections through the SDA of three male gerbils. *Left*: A male given bilateral cell-body lesions of the mSDA that spared most of the lSDA. *Center*: A sham-operated male. The arrowheads, the larger arrows, and the small arrow point to the lSDA, mSDA, and SDApc, respectively. The SDApc was present bilaterally in this male but was not present on the left side of this section. The track produced by the infusion cannula is also visible on the right side of this section, dorsolateral to the SDApc. *Right*: A male given bilateral cell-body lesions of the lSDA that spared most of the mSDA. Scale bar, 0.5 mm.

sexual activity can be restored in castrated male quail by unilateral implants of T or bilateral implants of diethystilbestrol, a synthetic form of E, that are in or very close to the POM (Balthazart and Surlemont 1990a,b; Balthazart et al. 1992). Testosterone implants that are more than 0.2 mm from POM are not effective.

In gerbils, two MPOA cell groups have been implicated in the control of male sexual behavior, both of which are components of the sexually dimorphic area (SDA) in the caudal MPOA (Commins and Yahr 1984a,b). One of them, the medial SDA (mSDA), is an oblong cell group that parallels the third ventricle and is just lateral to the periventricular nucleus and above the suprachiasmatic nuclei (Fig. 2.1, center). Based on similarities in location, histochemical characterization, steroid receptor distribution, and neural connections (Commins and Yahr 1984a,b, 1985; De Vries et al. 1988; Finn et al. 1993), the mSDA appears to be homologous to the medial MPN (MPNm) in rats (Simerly and Swanson 1986, 1988; Simerly et al. 1984; Simerly et al. 1990). The other MPOA cell group involved in the production of male sexual behavior in gerbils is the lateral SDA (lSDA), an ovoid cell group dorsolateral to the mSDA and near the anterior commissure. Based on location and cell size, it might be homologous to the magnocellular MPN of hamsters (Powers et al. 1987).

Both the mSDA and the lSDA are essential for copulation in male gerbils. As summarized in Figure 2.2A, when either cell group is lesioned using *N*-methyl DL-aspartate (NMA), sexually experienced male gerbils stopped copulating despite exposure to exogenous T (Yahr and Gregory 1993). The majority of males given vehicle infusions mounted and ejaculated during the same period of time. Thus, these NMA lesions, illustrated in Figure 2.1, produced deficits in male sexual behavior that were as severe as those produced by much larger radiofrequency lesions of the MPOA–AH (Yahr et al. 1982) or those produced in rats by much larger cell-body lesions of the MPOA–AH (Hansen et al. 1982).

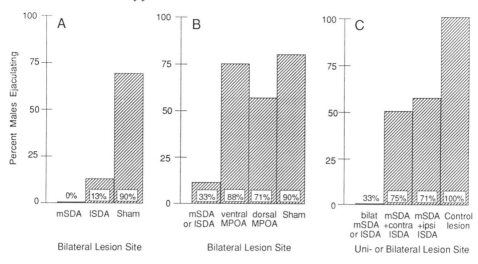

Figure 2.2. The data shown were obtained 2–3 weeks after lesion surgery; number at the bottom of each bar represents the percentage of males that mounted during this period. (A) Effects of bilateral cell-body lesions of the mSDA or lSDA on the copulatory behavior of sexually experienced, T-treated male gerbils. (B) Effects of bilateral lesions of the SDA (mSDA or lSDA), or of the dorsal or ventral MPOA at the level of the SDA, on the copulatory behavior of sexually experienced, T-treated male gerbils. (C) Effects of bilateral lesions of the mSDA or lSDA, of unilateral mSDA lesions made ipsilaterally (ipsi) or contralaterally (contra) to a lesion of the lSDA, and of caudal MPOA lesions (control lesions) on the copulatory behavior of sexually experienced, T-treated male gerbils.

Because mSDA and lSDA lesions produced comparable effects, the concern arose that similar lesions anywhere in the caudal MPOA might also eliminate mating behavior. A follow-up study indicated that the deficits observed were specific to the SDA (Yahr and Gregory 1993). Testosterone-treated male gerbils were given NMA lesions in the mSDA or lSDA, above the mSDA and medial to the lSDA (dorsal MPOA), or below the lSDA and lateral to the mSDA (ventral MPOA). Although the dose of NMA used was lower than in the earlier experiment, the lesions of the SDA still produced severe deficits in sexual behavior, as illustrated in Figure 2.2B. Males given dorsal or ventral MPOA lesions displayed fewer copulatory acts per minute than sham-operated control animals, but these groups did not differ otherwise. As in the earlier study, the deficits seen among the SDA-lesioned males were due primarily to a failure to initiate sexual behavior (i.e., a failure to display mounts with thrusting).

Because the mSDA and lSDA are connected reciprocally (De Vries et al. 1988; Finn et al. 1993), the similarities in their effects on male sexual behavior raised the possibility that they might affect mating via a common pathway. A second possibility was that mSDA and lSDA lesions exerted similar effects because they destroyed similar amounts of the SDA as a whole. A third possibility was that the mSDA and lSDA affect mating independently (i.e., via separate neural pathways), since each is connected to many areas of the brain. Because the connections between the mSDA and lSDA are almost entirely uncrossed, it was hypothesized that these possibilities could be distinguished by lesioning the pathway(s) between the mSDA and lSDA bilaterally but asymmetrically (i.e., at different levels on the two sides of the brain). If the mSDA affects mating via its projections to the lSDA, or vice versa, then making an mSDA lesion on one side of the brain and an lSDA lesion on the other should mimic the effects of bilateral lesions of either cell group, but making the same two lesions ipsilaterally should not. In contrast, if the second hypothesis is correct, then a unilateral lesion of the mSDA, plus a unilateral lesion of the lSDA, should disrupt mating as effectively as bilateral lesions of either area, whether the two lesions are ipsi- or contralateral to each other, since about half of the SDA would be damaged in each case. However, if the mSDA and lSDA affect mating independently, then neither contralateral nor ipsilateral lesions of the mSDA and lSDA should mimic the effects of bilateral lesions of either.

As illustrated in Figure 2.2C, the resulting data favor the third hypothesis (Yahr and Gregory 1993). Sexually experienced, T-treated male gerbils either were given one infusion of NMA aimed at the left or right mSDA and another aimed at the contralateral lSDA or were given both infusions ipsilaterally. In addition, during histological examination, males were identified in which either the mSDA or lSDA had been lesioned bilaterally, or in which one or both lesions missed the intended target. Bilateral lesions of either the mSDA or lSDA virtually eliminated mating behavior; only one of three males with such lesions mounted a female 2–3 weeks after surgery. In contrast, most males with unilateral lesions of the two cell groups remained sexually active, whether the two lesions were on the same or opposite sides of the brain. Males in which at least half of the mSDA and lSDA were damaged took longer to ejaculate than males with less or no damage to the SDA, but since this deficit was comparable in ipsi- and contralaterally lesioned males, it apparently reflects total damage to the SDA rather than damage to connections between the mSDA and lSDA. The observation that two MPOA cell groups are independently necessary for male sexual behavior suggests that the MPOA plays an even more complex role in the control of this behavior than was previously known.

Interestingly, a third component of the SDA, the SDA pars compacta (SDApc), is not necessary for male sexual behavior. The SDApc is a small, dense cell group that lies in, or on the medial edge of, the caudal part of the dorsal mSDA (Fig. 2.1; Commins and Yahr 1984a). It is present in both sexes at birth but disappears in most females by 2 weeks of age, unless the females are exposed to T neonatally (Ulibarri and Yahr 1988, 1993; Yahr 1988). Thus, in adult gerbils, it is the most dimorphic part of the SDA. Based on location, histochemical characterization, hormone receptor distribution, and neural connections (Commins and Yahr 1984a,b, 1985; De Vries et al. 1988; Finn et al. 1993), the SDApc appears to be homologous to the central MPN (MPNc) of rats (Simerly and Swanson 1986, 1988; Simerly et al. 1984; Simerly et al. 1990), indicating that it would also be homologous to the caudal part of the sexually dimorphic nucleus (SDN) of the rat MPOA (Bloch and Gorski 1988).

Like the rat SDN (Arendash and Gorski 1983), the gerbil SDApc does not appear to play a role in the control of male sexual behavior in sexually experienced males given exogenous T. In the first experiment in which we made NMA lesions of the mSDA or lSDA, most lesioned males had complete lesions of the SDApc as well. Thus, those data were consistent with the possibility that the SDApc was involved. However, in the follow-up study of site specificity, the SDApc was also destroyed in some of the males with lesions in the dorsal MPOA. Some of those males in which no SDApc could be identified on either side of the brain continued to ejaculate. This agrees well with the effects of lesions in rats that destroy most or all of the SDN, and hence most or all of the MPNc. In sexually naive male rats, bilateral, electrolytic lesions of the SDN retard the development of male sexual behavior (De Jonge et al. 1989), while in female rats given T, they decrease mounting (Turkenburg et al. 1988). Another highly dimorphic nucleus in the MPOA–AH, the male nucleus of ferrets (Tobet et al. 1986), also has little effect on sexual behavior in sexually experienced, T-treated males (Cherry and Baum 1990).

Activation of the SDA and areas connected to it during sexual activity

More recently, we have utilized immunocytochemical staining for the protein product of the c-*fos* oncogene to identify cell groups that are activated during mating (Heeb and Yahr 1993). Since the mSDA and lSDA are each connected to more than 70 areas of the brain (De Vries et al. 1988; Finn et al. 1993), identifying areas that are activated during mating might help to determine which SDA efferents and afferents modulate the effects of the SDA on male sexual behavior. Areas in which the number of c-*fos*-positive cell nuclei is greater in sexually experienced males that copulate to ejaculation an hour before perfusion than in sexually experienced males that remain in their home cage include the mSDA, the lSDA, the posterodoral nucleus of the preoptic area (PdPN), the ventral part of the lateral septal nucleus (LSv), the caudal part of the medial bed nucleus of the stria terminalis (caudal BSTm), the posterodorsal part of the medial nucleus of the amygdala (MeA), the amygdalohippocampal area (AHi), the ventral premammillary nucleus (PMV), the retrorubral field of the midbrain tegmentum (RRF), and the subparafasicular nucleus of the thalamus. The latter area appears to correspond to the portion of the central tegmental field in male rats that shows heighted c-*fos* expression during mating (Baum and Everitt 1992).

In some of these areas, heightened neural activity, as assessed by the number of

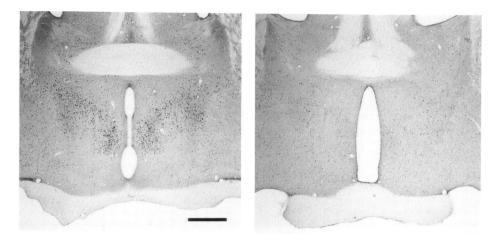

Figure 2.3. Photomicrographs of coronal sections through the SDA of two sexually experienced male gerbils showing the distribution of cell nuclei that stain immunocytochemically for the protein product of the c-*fos* oncogene. *Left*: A male that copulated to ejaculation an hour before perfusion. *Right*: A male that was with a sexually receptive female while the male on the left was copulating, but that did not mount. Scale bar, 0.3 mm.

cell nuclei that stain positively for c-*fos*, appears to vary with specific aspects of sexual behavior or with exposure to sex-related stimuli. For example, in the LSv, more c-*fos*-positive cells were seen in males that were placed in the arena in which they had previously encountered receptive females than in males that were merely handled. The number of labeled cells in the LSv did not increase further if the male copulated or not (M. M. Heeb and P. Yahr, unpublished results). In other areas, such as the mSDA, PdPN, and MeA, the number of labeled cells did not differ between males that mounted or intromitted and those that merely investigated the female, although it was higher among males that ejaculated (Fig. 2.3). The labeling seen in the MeA after ejaculation resembled that seen in rats (Baum and Everitt 1992) and hamsters (Kollack and Newman 1992), in which distinct clusters of labeled cells appeared on the lateral edge of the cell group. In still other areas, such as the PMV or RRF, the number of labeled cells was not significantly greater among males that copulated than among those that were merely handled.

Importance of the projection from the SDA to the RRF for male sexual behavior

The RRF is one of the few areas of the brain that receives a projection from the SDA but does not reciprocate (De Vries et al. 1988; Finn et al. 1993). The SDA efferents to the RRF arise primarily from the lSDA (Finn et al. 1993). At the level shown (Fig. 2.4A), nearly all of the labeled cells that could be retrogradely labeled from the RRF are within the lSDA. In more rostral sections, however,

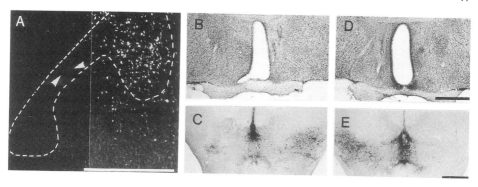

Figure 2.4. (A) Fluorescence photomicrograph showing the right SDA of a male gerbil that had been given an injection of Fluoro-Gold in the right RRF. The dashed lines show the outline of the SDA as estimated from an adjacent section that was stained with thionin. The arrowheads point to the SDApc. As shown, most of the cells that were retrogradely labeled by Fluoro-Gold were in the lSDA. Reproduced from Finn et al. (1993) with permission. (B, D) Bright-field photomicrographs of thionin-stained coronal sections through the SDA of two male gerbils that were given unilateral NMDA lesions of the SDA on the right (B) or left (D) sides of the brain. (C, E) Bright-field photomicrographs of coronal sections through the RRF of the males shown in B and D, respectively. These sections were stained immunocytochemically for tyrosine hydroxylase to visualize the A8 cells. The male in C was given an infusion of NMA in the left RRF; the male in E was given an infusion of 6-OHDA in the right RRF. Thus, in both cases, the RRF infusion was contralateral to the SDA infusion. As shown, NMA and 6-OHDA produced comparable damage to the A8 cells, but they produced differing effects on male copulatory behavior (see Fig. 2.5B). Scale bars, 0.5 mm.

labeled cells were also seen between the mSDA and lSDA and in the mSDA itself. The projections from the SDA to the RRF, which is the site of the A8 dopamine–containing cell bodies, were examined further because the RRF overlaps the dorsolateral tegmentum, which has connections with the MPOA that are essential for mounting in rats. When these connections are destroyed bilaterally by one unilateral electrolytic lesion in the MPOA and another in the contralateral, dorsolateral tegmentum, sexually experienced male rats stop mounting; however, when the connections are destroyed unilaterally, sexual activity persists (Brackett and Edwards 1984).

In gerbils, more specific lesions of the SDA and RRF produce the same deficits. Unilateral lesions made with *N*-methyl D-aspartate (NMDA) of the source of the SDA–RRF pathway eliminated mating in most males when combined with contralateral, but not ipsilateral, lesions of the RRF (Finn and Yahr 1994; see Fig. 2.5A). The deficits associated with lesions of the SDA–RRF pathway, like the deficits associated with lesions of the SDA, are seen only when the lesions are bilateral. As before, the deficits in sexual behavior occur despite the fact that the males were sexually experienced and exposed to exogenous T, and the deficits were due primarily to a disruption of mounting behavior (Finn and Yahr 1994).

Having shown that the projection from the SDA to the RRF was essential for male sexual behavior, we then asked whether the behavioral effects mediated by

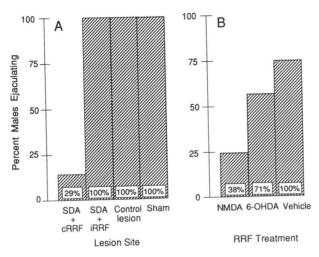

Figure 2.5. (A) Effects of unilateral cell-body lesions in the SDA, made ipsi- (i) or contra-laterally (c) to cell-body lesions in the RRF, on the copulatory behavior of sexually expe-rienced, T-treated male gerbils. Control lesions missed one or both of their targets. The data shown are for tests done 1–3 weeks after surgery. (B) Effects of unilateral cell-body lesions of the SDA, made contralaterally to NMDA or 6–OHDA lesions in the RRF, on the copulatory behavior of sexually experienced, T-treated male gerbils.

this pathway involved A8 dopamine cells, other RRF cells, or both. To assess this, we compared the effects of lesions in the RRF made with NMDA or 6-hy-droxydopamine (6-OHDA), a neurotoxin that is selective for cells that contain catecholamines. To limit its neurotoxic effect on noradrenergic neurons, we pre-treated the males that were given 6-OHDA with desmethylimipramine (Breese and Traylor 1971).

All males were given unilateral NMDA lesions at the source of the SDA–RRF pathway. In the RRF contralateral to the SDA lesion, they were given infusions of NMA, 6-OHDA, or one of the toxin vehicles (Finn and Yahr 1994; see Fig. 2.5B). When the A8 cells were lesioned, mating behavior did not decline as severely as it did when other RRF cells were damaged as well. Most males in which the SDA and A8 cell groups were destroyed asymmetrically continued to mount and to intromit, although they were less likely to intromit and to ejaculate than were vehicle-treated controls. In contrast, most males in which the SDA and all RRF cell types were destroyed asymmetrically stopped mounting receptive females.

γ-Aminobutyric acid as a transmitter in SDA cells that project to the RRF

Since the mSDA and lSDA are connected to so many areas of the brain, these cell groups, like the MPOA as a whole, may subserve a number of hormone-sensitive functions. Our discovery that the SDA cells that project to the RRF are essential for sexual behavior in male gerbils means that retrograde labeling from the RRF can help us to identify the behaviorally relevant cells. Recently, we used this approach to determine whether SDA cells that can be stained immunocyto-chemically for γ-aminobutyric acid (GABA) or its synthetic enzyme, glutamic acid decarboxylase (GAD), are good candidates for influencing male sexual be-havior (N. Hoffman and P. Yahr, unpublished data).

Cells throughout the SDA stained positively for GAD/GABA, whereas SDA cells that project to the RRF (labeled with Fluoro-Gold) were again found primar-ily in the ipsilateral lSDA. Interestingly, nearly all of the lSDA cells that were positive for Fluoro-Gold were also positive for GAD/GABA (Fig. 2.6). While lSDA cells that were positive for GAD/GABA but not for Fluoro-Gold were also apparent, very few lSDA cells were positive for Fluoro-Gold but not for GAD/GABA. Thus, it is likely that the SDA–RRF projection that is essential for male sexual behavior utilizes GABA as its neurotransmitter.

Forebrain connections of the SDA and male sex behavior

Another approach we have used to study the circuitry for male sexual behavior has been to identify the sources of SDA afferents that are severed by knife cuts that disrupt mating (Yahr and Jacobsen 1994). To determine whether the mSDA and/or lSDA could affect mating via laterally projecting axons, as suggested for the MPOA of rats (Paxinos 1974; Paxinos and Bindra 1972, 1973; Scouten et al. 1980; Szechtman et al. 1978), we examined the behavioral effects of bilateral, parasagittal knife cuts that were made just lateral to either the mSDA or lSDA. To determine whether these cuts actually severed SDA efferents, and to determine which SDA afferents were severed by the behaviorally effective cuts, we also made cuts using a knife coated with horseradish peroxidase (HRP), a retrograde tracer that readily enters cut axons (Mesulam 1982; Scouten et al. 1982).

Medial knife cuts (those just lateral to the mSDA) virtually eliminated copula-tion (Fig. 2.7A). In contrast, the behavior of males with lateral cuts (those just lateral to the lSDA) did not differ significantly from that of sham-operated con-trols. In addition, medial cuts with an HRP-coated knife labeled more cells in the mSDA than did lateral cuts. HRP from medial knife cuts also labeled more cells than HRP from lateral cuts in the LSv, the posterodorsal MeA, the AHi, the encapsulated portion of the bed nucleus of the stria terminalis (BSTe), which lies within the caudal BSTm, and the PMV (Fig. 2.8). The behavioral and tracing data suggest that medial cuts disrupt mating by severing laterally projecting efferents of the mSDA, by combined effects on the efferents of the mSDA and lSDA, and/or by reducing the afferent input to the SDA from the LSv, posterodorsal MeA, AHi, PMV, and/or BSTe.

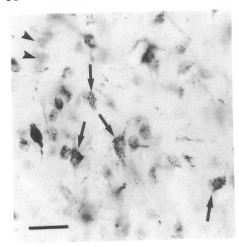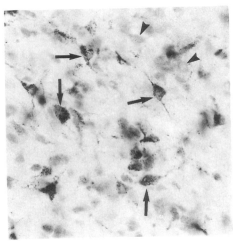

Figure 2.6. Bright-field photomicrographs of sections through the lSDA of two male gerbils that were given Fluoro-Gold in the ipsilateral RRF. The sections were processed immunocytochemically to visualize cells that contain GAD immunoreactivity and to visualize cells that were labeled retrogradely with Fluoro-Gold. The arrows point to some of the cells labeled by both techniques. The arrowheads point to some of the cells that were immunoreactive for GAD but did not contain Fluoro-Gold. Scale bar, 0.03 mm.

To test more directly whether the SDA affects mating via its connections with the LSv, the caudal BSTm, or the posterodorsal MeA and adjacent AHi (MeA–AHi), we again utilized asymmetric cell-body lesions (Sayag et al. 1994). Sexually experienced, T-treated male gerbils were given unilateral infusions of NMA aimed at either the SDA or the caudal BSTm. Those in which the first infusion was aimed at the SDA were given a second infusion aimed at the contra- or ipsilateral LSv, caudal BSTm, or MeA–AHi. Those in which the first infusion was aimed at the caudal BSTm were given a second infusion aimed at the contra- or ipsilateral MeA–AHi. Additional groups were formed during histological examination after the actual locations of the lesions were identified. Lesions that missed their telencephalic target often damaged other cell groups of interest. As a result, a total of 17 distinct lesion groups were examined..

Lesions that bilaterally disconnected the SDA from the caudal BSTm (i.e., contralateral lesions of the SDA and caudal BSTm) disrupted mating as effectively as bilateral lesions of the SDA (compare Fig. 2.7A with Fig. 2.2). Most males with these bilateral pathway lesions stopped mounting females by 2–3 weeks after surgery. Ipsilateral lesions that disconnected the SDA and caudal BSTm unilaterally did not decrease sexual activity; these males displayed as much copulatory behavior as males in which neither lesion, or only the SDA lesion, hit its target.

Some of the telencephalic lesions damaged more than one cell group connected

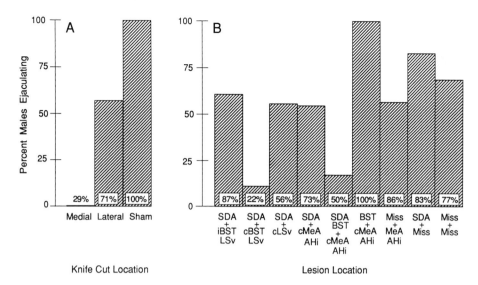

Figure 2.7. (A) Effects of bilateral, parasagittal knife cuts made just lateral to the mSDA (medial) or lSDA (lateral) on the copulatory behavior of sexually experienced, T-treated male gerbils. (B) Effects of unilateral cell-body lesions in the SDA and/or caudal BSTm when combined with ipsi- (i) or contralateral (c) lesions of the LSv, caudal BSTm and LSv, or MeA–AHi, on the copulatory behavior of sexually experienced, T-treated male gerbils. Also shown are the effects of lesions that missed one or both of their intended targets.

to the SDA. In particular, most of the males with caudal BSTm lesions also had ipsilateral LSv lesions. However, two observations make it unlikely that the LSv lesions accounted for the behavioral deficits observed. First, males that had contralateral SDA and LSv lesions that left the caudal BSTm intact displayed more sexual activity than males that had contralateral lesions of the SDA and caudal BSTm (Fig. 2.7B). Second, males that had contralateral SDA and caudal BSTm lesions, but that did not have lesions in the LSv, performed as poorly as males that had lesions of both the LSv and caudal BSTm contralateral to a lesion in the SDA. Thus, while connections between the SDA and LSv could influence male sexual behavior, they appear to be less important than connections between the SDA and caudal BSTm. Whether the BSTe is more important than other parts of the caudal BSTm in regard to mating behavior could not be determined from this study, since nearly all of the caudal BSTm lesions involved lesions of the BSTe.

It is not clear whether the connections between the SDA and caudal BSTm that are essential for male sexual behavior involve SDA afferents, efferents, or both. However, caudal BSTm cells that are critical for mating via their connections with the SDA do not merely relay information from the MeA–AHi. This possibility had been suggested (Sachs and Meisel 1988) because of the similarities in the effects of bilateral, electrolytic lesions in the BST and amygdala of rats. In

52   *Pauline Yahr*

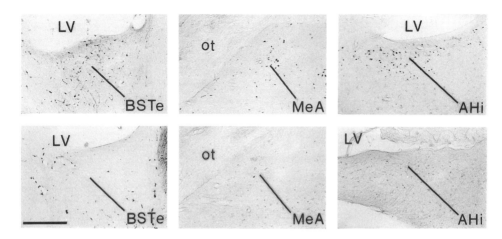

Figure 2.8. Bright-field photomicrographs showing the differential retrograde labeling seen in the BSTe, MeA, and AHi after parasagittal knife cuts were made just lateral to the mSDA (top row) or just lateral to the lSDA (bottom row) using a knife coated with HRP. The behavioral effects of such cuts are shown in Figure 2.7A. LV denotes lateral ventricle; ot, optic tract. Scale bar, 0.3 mm.

rats, lesions of the corticomedial amygdala that leave most of the MeA intact retard ejaculation (Giantonio et al. 1970; Harris and Sachs 1975), as do BST lesions that involve primarily the rostral BST (Emery and Sachs 1976; Giantonio et al. 1970). Since the corticomedial amygdala projects to the BST (Krettek and Price 1978), which projects to the caudal MPOA (Simerly and Swanson 1986; Swanson and Cowan 1979), BST and corticomedial amygdala lesions could produce similar effects by disrupting a common pathway. In gerbils (G. De Vries, P. Finn, and P. Yahr, unpublished observation), as in rats (Canteras et al. 1992; Krettek and Price 1978; Ottersen 1980; Swanson and Cowan 1979), the MeA–AHi projects to the caudal BSTm, particularly the BSTe. Yet if bilaterally disconnecting the caudal BSTm from the SDA affects sexual behavior by disrupting the transmission of information from the MeA–AHi, one would predict that bilaterally disconnecting the MeA–AHi from either the SDA or the caudal BSTm should mimic the behavioral effects of bilaterally disconnecting the caudal BSTm and the SDA from each other. However, as shown in Figure 2.7B, it does not. The behavior of males with contralateral lesions of the MeA–AHi and caudal BSTm did not differ from that of control groups (e.g., males in which the MeA–AHi was lesioned but in which the other lesion missed its target or males in which both lesions missed their targets). Males with contralateral lesions of the MeA–AHi and SDA (with or without lesions of the caudal BSTm) ejaculated less often than did males with only SDA or MeA–AHi lesions, but these groups did not differ in mounting behavior or intromission. In regard to connections between the

SDA and MeA–AHi that might affect ejaculation, it is unclear whether SDA afferents, efferents, or both, are involved. However, if this pathway is polysynaptic, it does not involve a relay in the caudal BSTm, since bilaterally disconnecting the MeA–AHi from the caudal BSTm did not mimic the effects of bilaterally disconnecting the MeA–AHi from the SDA. Clearly, there is still much to be learned about the neural circuitry for the hormonal control of male sexual behavior.

References

Arendash, G. W., and Gorski, R. A. (1983). Effects of discrete lesions of the sexually dimorphic nucleus of the preoptic area or other medial preoptic regions on the sexual behavior of male rats. *Brain Res. Bull.* 10: 147–154.

Balthazart, J. (1983). Hormonal correlates of behavior. In *Avian Biology*, Vol. 7, ed. D. S. Farner, J. R. King, and K. C. Parkes, pp. 221–365. New York: Academic Press.

Balthazart, J., and Surlemont, C. (1990a). Copulatory behavior is controlled by the sexually dimorphic nucleus of the quail POA. *Brain Res.* 25: 7–14.

Balthazart, J., and Surlemont, C. (1990b). Androgen and estrogen action in the preoptic area and activation of copulatory behavior in quail. *Physiol. Behav.* 48: 599–609.

Balthazart, J., Surlemont, C., and Harada, N. (1992). Aromatase as a cellular marker of testosterone action in the preoptic area. *Physiol. Behav.* 51: 395–401.

Baum, M. J., and Everitt, B. J. (1992). Increased expression of c-*fos* in the medial preoptic area after mating in male rats: Role of afferent inputs from the medial amygdala and midbrain central tegmental field. *Neuroscience* 50: 627–646.

Bloch, G. J., and Gorski, R. A. (1988). Cytoarchitectonic analysis of the SDN–POA of the intact and gonadectomized rat. *J. Comp. Neurol.* 275: 604–612.

Brackett, N. L., and Edwards, D. A. (1984). Medial preoptic connections with the midbrain tegmentum are essential for male sexual behavior. *Physiol. Behav.* 32: 79–84.

Breedlove, S. M. (1984). Steroid influences on the development and function of a neuromotor system. *Prog. Brain Res.* 61: 147–170.

Breese, G. R., and Traylor, T. D. (1971). Depletion of brain noradrenaline and dopamine by 6-hydroxydopamine. *Br. J. Pharmacol.* 42: 88–99.

Canteras, S. S., Simerly, R. B., and Swanson, L. W. (1992). Connections of the posterior nucleus of the amygdala. *J. Comp. Neurol.* 324: 143–179.

Cherry, J. A., and Baum, M. L. (1990). Effects of lesions of a sexually dimorphic nucleus in the preoptic/anterior hypothalamic area on the expression of androgen- and estrogen-dependent sexual behaviors in male ferrets. *Brain Res.* 522: 191–203.

Commins, D., and Yahr, P. (1984a). Adult testosterone levels influence the morphology of a sexually dimorphic area in the Mongolian gerbil brain. *J. Comp. Neurol.* 224: 132–140.

Commins, D., and Yahr, P. (1984b). Acetylcholinesterase activity in the sexually dimorphic area of the gerbil brain: Sex differences and influences of adult gonadal steroids. *J. Comp. Neurol.* 224: 123–131.

Commins, D., and Yahr, P. (1985). Autoradiographic localization of estrogen and androgen receptors in the sexually dimorphic area and other brain regions of the gerbil brain. *J. Comp. Neurol.* 231: 473–489.

Crews, D. (1979). Endocrine control of reptilian reproductive behavior. In *Endocrine Control of Sexual Behavior*, ed. C. Beyer, pp. 167–222. New York: Raven Press.

Crews, D., and Silver, R. (1985). Reproductive physiology and behavior interactions in nonmammalian vertebrates. In *Handbook of Behavioral Neurobiology*, Vol. 7: *Reproduction*, ed. N. Adler, D. Pfaff, and R. W. Goy, pp. 101–182. New York: Plenum Press.

De Jonge, F. H., Louwerse, A. L., Ooms, M. P., Evers, P., Endert, E., and van de Poll, N. E. (1989). Lesions of the SDN–POA inhibit sexual behavior of male Wistar rats. *Brain Res. Bull.* 23: 483–492.

De Vries, G. J., Gonzales, C. L., and Yahr, P. (1988). Afferent connections of the sexually dimorphic area of the hypothalamus of male and female gerbils. *J. Comp. Neurol.* 271: 91–105.

Emery, D. E., and Sachs, B. D. (1976). Copulatory behavior in male rats with lesions in the bed nucleus of the stria terminalis. *Physiol. Behav.* 17: 803–806.

Feder, H. H. (1981). Experimental analysis of hormone actions on the hypothalamus, anterior pituitary, and ovary. In *Neuroendocrinology of Reproduction*, ed. N. T. Adler, pp. 243–278. New York: Plenum Press.

Finn, P. D., De Vries, G. J., and Yahr, P. (1993). Efferent projections of the sexually dimorphic area of the gerbil hypothalamus: Anterograde identification and retrograde verification in males and females. *J. Comp. Neurol.* 338: 491–520.

Finn, P. D., and Yahr, P. (1994). Projection of the sexually dimorphic area of the gerbil hypothalamus to the retrorubral field is essential for male sex behavior: Role of A8 and other cells. *Behav. Neurosci.* 108: 362–378.

Giantonio, G. W., Lund, N. L., and Gerall, A. A. (1970). Effect of diencephalic and rhinencephalic lesions on the male rat's sexual behavior. *J. Comp. Physiol. Psychol.* 73: 38–46.

Gorski, R. A., Gordon, J. H., Shryne, J. E., and Southam, A. M. (1978). Evidence for a morphological sex difference within the medial preoptic area of the rat brain. *Brain Res.* 148: 333–346.

Hansen, S., Köhler, C., Goldstein, M., and Steinbusch, H. V. M. (1982). Effects of ibotenic acid-induced neuronal degeneration in the medial preoptic area and the lateral hypothalamic area on sexual behavior in the male rat. *Brain Res.* 239: 213–232.

Harris, V. S., and Sachs, B. D. (1975). Copulatory behavior in male rats following amygdaloid lesions. *Brain Res.* 86: 514–518.

Hart, B. L. (1978). Hormones, spinal reflexes and sexual behavior. In *Biological Determinants of Sexual Behaviour*, ed. J. B. Hutchison, pp. 319–347. New York: Wiley.

Hart, B. L., and Leedy, M. G. (1985). Neurological bases of male sexual behavior. In *Handbook of Behavioral Neurobiology*, Vol. 7: *Reproduction*, ed. N. Adler, D. Pfaff, and R. W. Goy, pp. 373–422. New York: Plenum Press.

Heeb, M. M., and Yahr, P. (1993). c-*fos* in the sexually dimorphic area (SDA) of the gerbil hypothalamus, and related areas, during sexual activity. *Neurosci. Abstr.* 19: 1019.

Heimer, L., and Larsson, K. (1966). Impairment of mating behavior in male rats following lesions in the preoptic–anterior hypothalamic continuum. *Brain Res.* 3: 248–263.

Kelley, D. B., and Pfaff, D. W. (1978). Generalizations from comparative studies on neuroanatomical and endocrine mechanisms of sexual behaviour. In *Biological Determinants of Sexual Behaviour*, ed. J. B. Hutchison, pp. 225–254. New York: Wiley.

Kollack, S. S., and Newman, S. W. (1992). Mating behavior induces selective expression of Fos protein within the chemosensory pathways of the male Syrian hamster brain. *Neurosci. Lett.* 143: 223–228.

Krettek, J. E., and Price, J. L. (1978). Amygdaloid projections to subcortical structures within the basal forebrain and brainstem in the rat and cat. *J. Comp. Neurol.* 178: 225–254.

Larsson, K. (1979). Features of the neuroendocrine regulation of masculine sexual behav-

Endocrine Control of Sexual Behavior, ed. C. Beyer, pp. 77–163. New York: Raven Press.

Luine, V. N., and McEwen, B. S. (1985). Steroid hormone receptors in brain and pituitary. In *Handbook of Behavioral Neurobiology*, Vol. 7: *Reproduction*, ed. N. Adler, D. Pfaff, and R. W. Goy, pp. 665–721. New York: Plenum Press.

Mesulam, M.-M. (1982). Principles of horseradish peroxidase neurohistochemistry and their applications for tracing neural pathways: Axonal transport, enzyme histochemistry and light microscopic analysis. In *Tracing Neuronal Connections with Horseradish Peroxidase*, ed. M.-M. Mesulam, pp. 1–151. New York: Wiley.

Michael, R. P., and Bonsall, R. W. (1979). Hormones and the sexual behavior of rhesus monkeys. In *Endocrine Control of Sexual Behavior*, ed. C. Beyer, pp. 279–302. New York: Raven Press.

Michael, R. P., and Bonsall, R. W. (1990). Androgens, the brain and behavior in male primates. In *Hormones, Brain and Behavior in Vertebrates*, ed. J. Balthazart, pp. 15–26. Basel: Karger.

Numan, M. (1988). Maternal behavior. In *The Physiology of Reproduction*, ed. E. Knobil, J. D. Neill, L. L. Ewing, G. S. Greenwald, C. L. Market, and D. W. Pfaff, pp. 1569–1645. New York: Raven Press.

Ottersen, O. P. (1980). Afferent connections to the amygdaloid complex of the rat and cat, II: Afferents from the hypothalamus and the basal telencephalon. *J. Comp. Neurol.* 194: 267–289.

Pan, J. T., and Gala, R. R. (1985a). Central nervous system regions involved in the estrogen-induced afternoon prolactin surge, I: Lesion studies. *Endocrinology* 117: 382–387.

Pan, J. T., and Gala, R. R. (1985b). Central nervous system regions involved in the estrogen-induced afternoon prolactin surge, II: Implantation studies. *Endocrinology* 117: 388–395.

Paxinos, G. (1974). The hypothalamus: Neural systems involved in feeding, irritability, aggression, and copulation in male rats. *J. Comp. Physiol. Psychol.* 87: 110–119.

Paxinos, G., and Bindra, D. (1972). Hypothalamic knife cuts: Effects on eating, drinking, irritability, aggression, and copulation in the male rat. *J. Comp. Physiol. Psychol.* 79: 219–229.

Paxinos, G., and Bindra, D. (1973). Hypothalamic and midbrain neural pathways involved in eating, drinking, irritability, aggression, and copulation in rats. *J. Comp. Physiol. Psychol.* 82: 1–14.

Powers, J. B., Newman, S. W., and Bergondy, M. L. (1987). MPOA and BNST lesions in male Syrian hamsters: Differential effects on copulatory and chemoinvestigatory behaviors. *Behav. Brain Res.* 23: 181–195.

Sachs, B. D. (1983). Potency and fertility: Hormonal and mechanical causes and effects of penile actions in rats. In *Hormones and Behaviour in Higher Vertebrates*, ed. J. Balthazart, E. Pröve, and R. Gilles, pp. 86–110. Berlin: Springer.

Sachs, B. D., and Meisel, R. L. (1988). The physiology of male sexual behavior. In *The Physiology of Reproduction*, ed. E. Knobil, J. D. Neill, L. L. Ewing, G. S. Greenwald, C. L. Market, and D. W. Pfaff, pp. 1393–1485. New-York: Raven Press.

Sayag, N., Hoffman, N. W., and Yahr, P. (1994). Telencephalic connections of the sexually dimorphic area of the gerbil hypothalamus that influence male sex behavior. *Behav. Neurosci.* 108: 743–757.

Scouten, C. W., Burrell, L., Palmer, T. and Cegavske, C. F. (1980). Lateral projections of the medial preoptic area are necessary for androgenic influence on urine marking and copulation in rats. *Physiol. Behav.* 25: 237–243.

Scouten, C. W., Harley, C. W., and Malsbury, C. W. (1982). Labeling knife cuts: A new method for revealing the functional anatomy of the CNS demonstrated on the noradrenergic dorsal bundle. *Brain Res. Bull.* 8: 229–232.

Silver, R., O'Connell, M., and Saad, R. (1979). Effect of androgens on the behavior of birds. In *Endocrine Control of Sexual Behavior*, ed. C. Beyer, pp. 223–278. New York: Raven Press.

Simerly, R. B., Chang, C., Muramatsu, M., and Swanson, L. W. (1990). Distribution of androgen and estrogen receptor mRNA-containing cells in the rat brain: An *in situ* hybridization study. *J. Comp. Neurol.* 294: 76–95.

Simerly, R. B., and Swanson, L. W. (1986). The organization of the neural inputs to the medial preoptic nucleus of the rat. *J. Comp. Neurol.* 246: 312–342.

Simerly, R. B., and Swanson, L. W. (1988). Projections of the medial preoptic nucleus: A *Phaseolus vulgaris* leukoagglutinin anterograde tract-tracing study in the rat. *J. Comp. Neurol.* 270: 209–242.

Simerly, R. B., Swanson, L. W., and Gorski, R. A. (1984). Demonstration of a sexual dimorphism in the distribution of serotonin-immunoreactive fibers in the medial preoptic nucleus of the rat. *J. Comp. Neurol.* 225: 151–166.

Swanson, L. W., and Cowan, W. M. (1979). The connections of the septal region in the rat. *J. Comp. Neurol.* 186: 621–656.

Szechtman, H., Caggiula, A. R., and Wulkan, D. (1978). Preoptic knife cuts and sexual behavior in male rats. *Brain Res.* 150: 569–591.

Tobet, S. A., Zahniser, D. J., and Baum, M. J. (1986). Sexual dimorphism in the preoptic/anterior hypothalamic area of ferrets: Effects of adult exposure to sex steroids. *Brain Res.* 364: 249–257.

Turkenburg, J. L., Swaab, D. F., Endert, E., Louwerse, A. L., and van de Poll, N. E. (1988). Effects of lesions of the sexually dimorphic nucleus on sexual behavior of testosterone-treated female Wistar rats. *Brain Res. Bull.* 21: 215–224.

Ulibarri, C. M., and Yahr, P. (1988). Role of neonatal androgens in sexual differentiation of brain structure, scent marking, and gonadotropin secretion in gerbils. *Behav. Neurol. Biol.* 49: 27–44.

Ulibarri, C. M., and Yahr, P. (1993). Ontogeny of the sexually dimorphic area of the gerbil hypothalamus. *Dev. Brain Res.* 74: 14–24.

van de Poll, N. E., and van Dis, H. (1979). The effect of medial preoptic–anterior hypothalamic lesions on bisexual behavior of the male rat. *Brain Res. Bull.* 4: 505–511.

Viglietti-Panzica, C., Panzica, G. C., Fiori, M. G., Calcagni, M., Anselmetti, G. C., and Balthazart, J. (1986). A sexually dimorphic nucleus in the quail preoptic area. *Neurosci. Lett.* 64: 129–134.

Ward, I. L., and Ward, O. B. (1985). Sexual differentiation: Effects of prenatal manipulations in rats. In *Handbook of Behavioral Neurobiology*, Vol. 7: *Reproduction*, ed. N. Adler, D. Pfaff, and R. W. Goy, pp. 77–98. New York: Plenum Press.

Yahr, P. (1983). Hormonal influences on territorial marking behavior. In *Hormones and Aggressive Behavior*, ed. B. B. Svare, pp. 145–175. New York: Plenum Press.

Yahr, P. (1988). *Pars compacta* of the sexually dimorphic area of the gerbil hypothalamus: Postnatal ages at which development responds to testosterone. *Behav. Neurol. Biol.* 49: 118–124.

Yahr, P., Commins, D., Jackson, J. C., and Newman, A. (1982). Independent control of sexual and scent marking behaviors of male gerbils by cells in or near the medial preoptic area. *Horm. Behav.* 16: 304–322.

Yahr, P., and Gregory, J. E. (1993). The medial and lateral cell groups of the sexually dimorphic area of the gerbil hypothalamus are essential for male sex behavior and act via separate pathways. *Brain Res.* 631: 287–296.

Yahr, P., and Jacobsen, C. H. (1994). Hypothalamic knife cuts that disrupt mating in male gerbils sever efferents and forebrain afferents of the sexually dimorphic area. *Behav. Neurosci.* 108: 735–742.

3

Estrogen receptor mRNA: neuroanatomical distribution and regulation in three behaviorally relevant physiological models

JOAN I. MORRELL, CHRISTINE K. WAGNER,
KARL F. MALIK, AND CHRISTINE A. LISCIOTTO

Introduction

Gonadal steroid hormone receptors are ligand-activated transcription factors that are part of a complex superfamily of such factors (Evans 1988). These receptors alter the transcription of genes containing specific promoter or enhancer sequences. In this way, estrogen, acting through the estrogen receptor (ER), alters the transcription of genes in the cells where the receptors are found.

Neurons that contain gonadal steroid hormone receptors are a key to the mechanisms through which these hormones govern the behavioral and neuroendocrine processes underlying reproduction (Morrell et al. 1975; Morrell and Pfaff 1983). A complex and only partly understood cascade of events occurs subsequent to the genomic regulation initiated by these ligand-activated transcription factors (Yamamoto 1985). The presence of gonadal steroid hormone receptors in neurons has been documented by means of steroid hormone autoradiography, biochemical assays for binding, and immunocytochemistry (Blaustein and Olster 1989; DonCarlos et al. 1991; Giordano et al. 1991; Morrell et al. 1992). Now the tools of molecular biology provide a means of investigating the mRNA from which gonadal steroid hormone receptors are translated.

The differential sensitivity of brain regions to steroid hormones is based on the combination of the regional location of neurons containing these receptors, the number of neurons containing the receptors per brain region, and the number of receptors per neuron. The degree of sensitivity to steroid hormones is not a static property of the brain as there is increasing evidence that the endocrine and behavioral status of the adult mammal can govern the sensitivity of brain regions to steroid hormones by regulating either the number of neurons expressing the receptors or the amount of receptor per neuron (Hnatczuk et al. 1994; Koch and Ehret 1989; Pearson et al. 1993; Simerly and Young 1991).

Of course, the amount of ligand-occupied receptor (hence functional receptor) is directly dependent on the amount of ligand circulating. In addition, the amount of ligand-occupied receptor is dependent on the amount of receptor made available through the neuron's protein-synthetic pathways. The amount of receptor protein available is regulated either by modification of translational processes or through genomic processes, whereby the amount of mRNA available for transcription of additional receptor protein is regulated (Giordano et al. 1991; Hnatczuk et al. 1994; Lisciotto and Morrell 1993; Migliaccio et al. 1989; Orti et al. 1992). Theoretically, the steroid hormone receptor expressed within neurons in different brain regions might be regulated at any one of the steps in the protein-synthetic pathway, thereby providing a considerable number of regulatory options through which to alter the brain's sensitivity to steroid hormones.

We have used *in situ* hybridization histochemistry to examine relative differences in the steady-state level of ER mRNA. We have chosen this method over other molecular methods because it is possible to gather data with a high degree of spatial resolution. As a consequence, small cell groups can be selectively examined, even within brain regions such as the preoptic area that have many specialized subdivisions with multiple specific functions. This is important since, within any one cell group, intermingled neurons are heterogeneous with respect to connectivity, chemical content, and ultimate function. An additional important advantage of *in situ* hybridization histochemistry over other methods for measuring mRNA is that individual animals can be analyzed and additional measures (e.g., circulating hormone levels) or behavior can be correlated with the mRNA levels.

Utility of in situ *hybridization histochemistry*

We are using *in situ* hybridization histochemistry to determine the relative steady-state levels of ER mRNA across experimental groups. A significant number of experiments have established the conditions that must be fulfilled in order to use *in situ* hybridization histochemistry as a quantitative tool to compare the relative amounts of mRNA across experimental conditions (Angerer et al. 1987; Lawrence and Singer 1985; McCabe et al. 1986). These conditions have been met in our *in situ* hybridization histochemistry experiments. Under our empirically determined, optimal conditions, signal detection is at saturation neither with respect to the capacity of the emulsion to detect radioactivity nor with regard to the capacity of our image analysis system to count grains. We have also accomplished the conventional control experiments to establish the specificity of the signal (Lisciotto and Morrell 1993).

We have, therefore, achieved reasonable conditions for analysis of the relative quantity of neuronal ER mRNA in the brain across experimental conditions. This is quite analogous to the conventional use of Northern or slot–blot analysis to compare the relative densities of signals across lanes in a gel to determine the relative change in steady-state levels of mRNA. These methods do not enable one to determine the absolute level or concentration of endogenous mRNA being measured.

Although we do not know the minimum number of molecules of mRNA that must be present before signal is detected with *in situ* hybridization, we have detected the same number of neurons with our *in situ* hybridization histochemistry for ER mRNA that we detected with estradiol autoradiography and with ER immunocytochemistry. This is strong proof that our hybridization histochemistry conditions are sufficiently sensitive to detect the complete subset of neurons containing ER.

In spite of the limitations on absolute quantification that are inherent in *in situ* hybridization histochemistry, many publications attest to its usefulness for detecting relative differences in the amount of mRNA in view of its sensitivity and high spatial resolution (Chowden-Breed et al. 1989; Malik et al. 1991; Sherman et al. 1986; Shughrue et al. 1992; Swanson and Simmons 1989; Toranzo et al. 1989; Wise et al. 1990; Wray et al. 1989; Zoeller et al. 1988).

Overall experimental goals

This chapter presents three experiments in which we explored the effect of the animal's endocrine status on the relative steady-state ER mRNA content of neurons in the forebrain of the rat and guinea pig using *in situ* hybridization histochemistry. Each model was selected to answer specific questions as described in the individual sections. Nevertheless, the data from these models can be compared with one another.

Experiments on male rats employ a model that we have come to consider the "engineer's" model because it utilizes an experimentally induced endocrine condition not found in nature – namely, gonadectomy. We have taken care to replace the normally circulating testosterone at a physiological level in the testosterone-treated gonadectomized males for comparison with the nontreated gonadectomized males. This model was created to test the limits of the physiological system. We also used a model of physiological, behaviorally relevant hormone replacement whereby estrogen and progesterone, sufficient to stimulate sexual behavior, were administered to gonadectomized female guinea pigs. The third model examined ER mRNA across pregnancy in the female rat. During this condition many different hormones act on the brain, and any alterations in ER mRNA cannot simply be accounted for as regulation of the steroid hormone receptor by its autologous ligand.

For all three experiments, a cRNA probe complementary to the ER mRNA, encoding primarily the steroid binding domain of the ER, was subcloned and

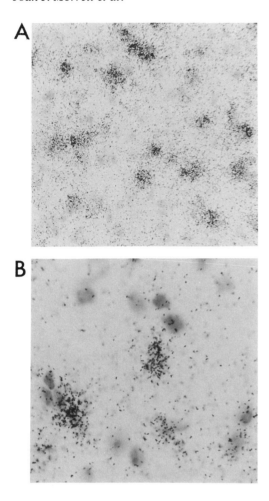

Figure 3.1. (A) Low- and (B) high- (different fields) magnification bright-field photomicrographs of representative neurons containing ER mRNA in the PePOA of the adult female guinea pig. Probe concentration was 1.5×10^7 cpm/ml; autoradiograms were exposed for 4 weeks; sections were counterstained with methylene blue, allowing the detection of cell nuclei. Reduced silver grains (small black grains) are seen in the cytoplasm around the cell nuclei (light gray in this black and white print).

transcribed (Lisciotto and Morrell 1993). The rat ER cDNA (Koike et al. 1987) was generously supplied by Dr. M. Muramatsu (University of Tokyo). The procedure for *in situ* hybridization histochemistry used in these experiments is a slight modification of Simerly et al. (1990) and Simmons et al. (1989).

The majority of the cells in the brain that contain gonadal steroid hormone receptors have the morphology of neurons (Fig. 3.1) and have extensive axonal projections to other brain regions (Morrell et al. 1992; Silverman et al. 1991). Therefore, throughout this chapter we will refer to cells in the brain that contain ER mRNA as neurons.

Regulation of ER mRNA by steroid hormones in males

In the rat, male sexual behavior wanes following castration, and the time course for testosterone-induced reinstatement of male sexual behavior depends on the time without steroid hormone. One month after gonadectomy, reinstatement of male reproductive behavior requires 5–10 days of testosterone circulating at physiological levels (McGinnis and Dreifuss 1989; McGinnis et al. 1989; Sachs and Meisel 1988). These slow changes are in sharp contrast to the rapid loss of female sexual behavior after ovariectomy, and its rapid return, usually in 24–48 hours, with estrogen and progesterone replacement (Pfaff 1980). The mechanism for the initial lack of sensitivity to testosterone following long-term testosterone deprivation is not known. We hypothesized that alterations in behavioral sensitivity to steroid hormone are related to the altered amounts of steroid hormone receptor in the central nervous system (CNS).

Experimental goals

We examined the expression of steroid receptor mRNA following short- and long-term gonadectomy and then following testosterone replacement. Specifically, we examined ER mRNA because in male rats the conversion of testosterone to estrogen via aromatization in the medial preoptic nucleus (MPN) and the bed nucleus of the stria terminalis (BNST) is a critical step in the stimulation of male reproductive behavior (Lisciotto and Morrell 1990; Pfaff and Keiner 1973; Roselli et al. 1985; Sachs and Meisel 1988; Sar and Stumpf 1973, 1977).

In situ hybridization histochemistry was used in two studies to examine the steady-state levels of ER mRNA in the MPN and BNST of male rat brains after (1) depletion of gonadal steroid hormones for 3 days and (2) depletion of gonadal steroid hormones for 6 weeks followed by testosterone replacement in the physiological range for 3, 10, or 21 days (Figs. 3.2 and 3.3). The methodological details have been previously described (Damassa et al. 1976; Lisciotto and Morrell 1993).

Regulation

Testosterone depletion for 3 days did not significantly alter the number of neurons that contained ER mRNA in either the MPN or the BNST (Fig. 3.2a), although there was a nonsignificant trend for gonadectomized males to have more labeled neurons than intact males. Gonadectomy, however, did result in a dramatic and statistically significant increase in the relative amount of ER mRNA per neuron, as measured by the number of grains per cell, within both the MPN and BNST

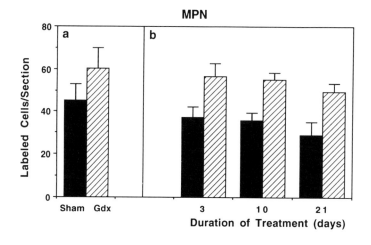

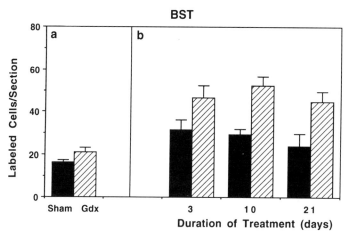

Figure 3.2. Effect of gonadectomy (Gdx) and testosterone replacement on the mean number of ER mRNA–containing neurons in the medial preoptic nucleus (MPN, top graph) and bed nucleus of the stria terminalis (BST, bottom graph). The number of ER mRNA–containing neurons was determined within a sampling field of 0.06 mm^2 within each region using an automated image analysis system (Bioquant, R&M Biometrics Inc., Nashville, Tennessee). Solid bars represent gonadally intact (a) or testosterone-replaced (b) male rats. Striped bars represent short-term Gdx (a) males or long-term Gdx males with cholesterol (control) implants (b). Panel a: Short-term Gdx study. Comparison of the mean ($\pm$ SEM) number of ER mRNA–containing neurons in the MPN and BNST of gonadally intact rats and rats that were gonadectomized 3 days before sacrifice ($n=6$ for each treatment group). The effect was not statistically significant (student's t-test). Panel b: Long-term Gdx study. Comparison of the mean ($\pm$ SEM) number of ER mRNA–containing neurons in the MPN and BNST of rats that were gonadectomized for 6 weeks, then received testosterone or cholesterol for 3, 10, or 21 days ($n=6$ for each treatment group). Two-way ANOVA; Duncan's multiple range test as a post hoc did not reveal any effect.

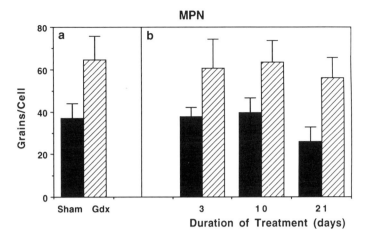

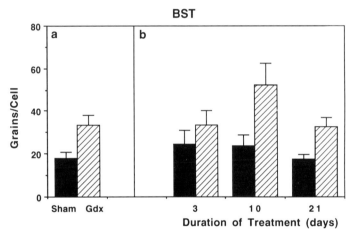

Figure 3.3. Effect of gonadectomy (Gdx) and testosterone replacement on the relative amount of ER mRNA per neuron in the medial preoptic nucleus (MPN, top graph) and bed nucleus of the stria terminalis (BST, bottom graph). The relative amount of ER mRNA per neuron is determined by a calculation of the area (μm^2) covered by reduced silver grains for each cellular profile (grains per cell). The mean grains per cell within a sampling field represents the relative amount of ER mRNA per neuron. Grain counting was carried with the aid of the automated image analysis system (Bioquant). Panel a: Short-term Gdx significantly increased the mean ($\pm$ SEM) amount of ER mRNA per cell (as measured by grains per cell) in both MPN and BNST (student's t-test). Panel b: Following long-term Gdx, testosterone treatment significantly reduced the relative amount of ER mRNA in the MPN and BNST compared with the control (cholesterol) treatment. There was no effect of the duration of testosterone treatment on the number of grains per cell (two-way ANOVA). Bar codes are as in Figure 3.2.

(Fig. 3.3a). In each of these two brain regions the mean number of grains per cell in the gonadectomized males was nearly twofold that of the sham castrated males.

In male rats that had been gonadectomized for 6 weeks, testosterone replacement rapidly reduced the number of ER mRNA–containing neurons in both the

MPN and the BNST (Fig. 3.2b). Compared with cholesterol-treated males, testosterone-treated males had significantly fewer ER mRNA–containing neurons. There was no effect of the duration of hormone treatment on the number of ER mRNA–containing neurons.

The pattern of downregulation of ER mRNA was also observed in the measure of grains per cell (Fig. 3.3b). There was a significant effect of hormone treatment on the relative amount of ER mRNA per neuron, such that testosterone-treated male rats had significantly fewer grains per cell in neurons of the MPN and BNST compared with cholesterol-treated rats. In addition, there was no effect of the duration of hormone treatment on the number of grains per labeled neuron in either the MPN or BNST.

These studies have shown that gonadectomy increases the steady-state levels of ER mRNA within the MPN and BNST of male rats, and that testosterone replacement decreases the ER mRNA content of the MPN and BNST. Specifically, gonadectomized male rats have almost twice as much ER mRNA in the MPN and BNST than intact male rats. The increase in ER mRNA occurs rapidly, and the elevated levels of ER mRNA appear to be sustained as long as testosterone is absent. Male rats in the short-term study (gonadectomized for 3 days) had roughly the same number of labeled neurons and the same number of grains per neuron as cholesterol-treated male rats in the long-term study that had been gonadectomized for 6 weeks.

Regulation within 3 days

Even in male rats that have been without circulating gonadal steroid hormones for 6 weeks, testosterone replacement rapidly decreases the steady-state levels of ER mRNA. Within 3 days of testosterone replacement, ER mRNA levels and number of ER mRNA–containing neurons return to the lower level observed in intact males. Once the ER mRNA level is downregulated with the return of testosterone, the steady-state level remains fairly constant as long as testosterone is present.

These data do not support the hypothesis that delayed and gradual restoration of male reproductive behavior following long-term gonadectomy and testosterone replacement is related to slow changes in the expression of ER. Following long-term gonadectomy, behavioral sensitivity to testosterone or dihydrotestosterone (DHT) can take as long as 10 days to return, while restoration of steady-state levels of ER mRNA occurs rapidly, within 3 days or less. Moreover, the direction of the change in ER mRNA content after gonadectomy (short or long term) would not predict behavioral insensitivity. Following gonadectomy, ER mRNA levels are upregulated, while male rats are apparently behaviorally insensitive to testosterone and its estrogenic metabolite.

Regulation of receptors by ligand as a generalized phenomenon

The upregulation of receptor by the absence of its ligand is a generalized biological phenomenon, extending from neurotransmitters and their receptors to insulin and its well-characterized receptor (Goldfine 1987). The upregulation of receptor is usually associated with increased sensitivity to ligand at the cellular level. However, where behavioral sensitivity is concerned, the issue is more complicated. A number of other factors may contribute to the slow recovery of male reproductive behavior. For example, unmetabolized testosterone, as well as DHT, in combination with estrogen also play a role in the expression of male reproductive behavior (Baum and Vreeburg 1973; McGinnis and Dreifuss 1989). There is evidence that androgen receptor mRNA levels may increase transiently following gonadectomy (Quarmby et al. 1990), but then decrease with prolonged gonadal hormone deprivation (McGivern et al. 1992). Thus, there may be a temporal correlation between behavioral sensitivity to testosterone and the expression of androgen receptor mRNA.

Much of the functional activity of testosterone requires its metabolic conversion to estrogen and/or DHT. The regulation of ER mRNA by testosterone, described here, may be a form of homologous regulation; that is, locally formed estrogen (see Chapter 13 by Schlinger and Arnold, this volume) may regulate ER mRNA. In ovariectomized female rats, exogenous estrogen replacement decreases ER mRNA in the ventromedial and arcuate nuclei of the hypothalamus (Lauber et al. 1990; Simerly and Young 1991). The MPN and BNST, examined in the studies described here, have an abundance of aromatase activity (Roselli and Resko 1987). It is possible, therefore, that the regulation of ER mRNA by testosterone (metabolized to estrogen) in these brain regions is a form of homologous regulation. Alternatively, locally formed DHT or unmetabolized testosterone may heterologously regulate ER mRNA. Although little is known about heterologous regulation of ER mRNA, there is evidence that DHT downregulates ER mRNA in ZR-75-1 human breast cancer cells (Poulin et al. 1989).

Regulation of ER mRNA in the brain of female rats during pregnancy

During pregnancy in the female rat there are many hormonal and behavioral changes that are unique to this reproductive state. The dynamic nature of the brain during pregnancy can provide information on the neural control of these naturally occurring changes.

One of the primary behavioral changes observed during pregnancy and the postpartum period in the rat is the onset of maternal behavior. Maternal behavior comprises a group of specific behaviors that occur spontaneously only in the pregnant female and begin only hours before parturition. In the rat they consist of nest building, nursing, pup retrieval, and maternal aggression toward an intruder (reviewed by Numan 1988). Estrogen is known to play a role in the induction of maternal behavior (Siegel and Rosenblatt 1975), and it has been well documented that the MPN is one site where estrogen acts to induce such behavior (Adieh et al. 1987; Numan et al. 1977). Lesion studies have shown that the rostral aspect of the MPN plays a greater role in the display of maternal behavior than the caudal portion (Gray and Brooks 1984).

It has been hypothesized that, within specific brain regions, the sensitivity of neurons to circulating estrogen may increase during pregnancy through fluctuations in ER content. In fact, studies using steroid hormone binding assays have determined that binding of radiolabeled estradiol to the nuclear fraction of a homogenate of the MPN was increased during pregnancy (Giordano et al. 1991).

Experimental goal

The objective of the following study was to determine whether the unique endocrine state of pregnancy might alter the sensitivity of the MPN to estrogen via changes in ER mRNA levels. *In situ* hybridization histochemistry was used to measure relative steady-state levels of ER mRNA within the rostral and caudal aspects of the MPN in five groups of rats, those in diestrus 1, day 8, 16, or 22 of pregnancy, and postpartum day 1.

Change in ER mRNA level during pregnancy

The results of this experiment demonstrate that relative steady-state levels of ER mRNA are altered over the course of pregnancy within the MPN. The relative amount of mRNA per neuron within the MPN was significantly elevated between the nonpregnant state and day 8 of pregnancy. Within the rostral MPN, this was a transitory rise in ER mRNA per neuron that was reduced on day 16 and remained lower through the rest of pregnancy and the postpartum period.

The mean number of grains per neuron within the rostral MPN and the caudal MPN is shown in Figure 3.4. In both subregions of the MPN, the mean number of grains per neuron was significantly higher in females on day 8 of pregnancy than in nonpregnant, diestrous controls ($p \leq .05$) (Fig. 3.4A, C). This level was significantly reduced between day 8 and day 16 of pregnancy in the rostral MPN ($p \leq .05$), while in the caudal MPN the number of grains per neuron was not significantly reduced by day 16.

Comparisons between the rostral and caudal MPN reveal some interesting differences. In nonpregnant, diestrous females there is a significantly greater amount

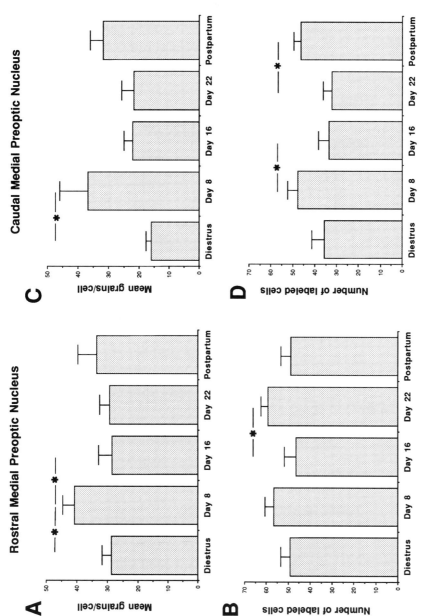

Figure 3.4. The mean number of grains per cell ($\pm$SEM) (A and C) and the mean number of ER mRNA–containing cells ($\pm$SEM) (B and D) within the rostral or caudal MPN in virgin females in diestrus 1, females on day 8, day 16, or day 22 of pregnancy, or females on the day after parturition, postpartum day 1. (Asterisk signifies $p \leq .05$; student's t-test. Since the number of t-tests performed was equal to $\alpha - 1$, where α represents the number of groups per experiment, these comparisons can be made without correcting for family-wise error by modifying the probability level.) The autoradiograms were exposed for 2 weeks.

of ER mRNA per neuron within neurons of the rostral MPN than in the caudal MPN ($p \leq .001$). However, this difference in the number of grains per neuron between rostral and caudal is absent at all of the time points examined during pregnancy and postpartum (compare Figs. 3.4A and C).

Changes in the number of neurons expressing ER mRNA during pregnancy

The number of neurons expressing ER mRNA within the MPN was also altered during pregnancy in both the rostral and caudal MPN, but these regions differed as to when the changes occurred (Fig. 3.4). In the rostral MPN, the number of labeled neurons was significantly increased between days 16 and 22 of pregnancy ($p \leq .05$), a time when the number of grains per neuron remained completely unaltered. In the caudal MPN, the number of labeled neurons was significantly reduced between days 8 and 16 of pregnancy ($p \leq .05$), but was significantly increased between day 22 of pregnancy and postpartum day 1 ($p \leq .01$).

Changes in the number of labeled neurons may represent a biological recruitment of neurons to begin expressing ER mRNA or may reflect an increase in the number of neurons reaching the level of technical detectability. Moreover, Simerly et al. (1990) determined that the protocol used in this experiment saturates available ER mRNA in the rat brain. In addition, if the change in neuron number was simply an increase in the mean number of grains per neuron to the point where more neurons reached detectability, one would predict a corresponding increase in the number of grains for all neurons. However, this is clearly not the case in the rostral MPN. Therefore, it is our working hypothesis that there is an increase in the number of neurons expressing ER mRNA.

Correlations with other measures

Estrogen levels remain relatively low throughout pregnancy until day 16, when they begin to rise, with peak levels occurring on day 22 (Shaikh 1971). Changes in the amount of ER mRNA per neuron and increases in the number of neurons expressing ER mRNA within the MPN earlier in pregnancy may be a mechanism by which this region becomes more sensitive to the rising levels of estrogen at the end of pregnancy, the time when maternal behavior is induced by estrogen (Siegel and Rosenblatt 1975).

Although steroid binding assays demonstrated dramatic changes in the levels of nuclear fraction estrogen binding within the MPN (Giordano et al. 1991), they could not distinguish subregions of the MPN that appear to be more or less important for maternal behavior. Using *in situ* hybridization histochemistry, we were able to distinguish and analyze on a neuron-by-neuron basis the rostral and caudal aspects of the MPN.

Differences in the time course of changes in ER mRNA during pregnancy within the neurons of the rostral and caudal aspects of the MPN suggest that these two regions have different estrogen-regulated functions during pregnancy. Earlier studies demonstrated that lesions further in the rostral portion of the MPN are more effective at disrupting maternal behavior than lesions placed more caudally in the MPN, implying that the rostral MPN is the critical region for the induction of maternal behavior.

Regional specificity

There also appears to be a high degree of regional specificity with regard to changes in ER mRNA levels during pregnancy. While such levels were altered in both the rostral and caudal MPN, they remained completely unaltered in the arcuate nucleus (data not shown) throughout all the time points measured. This is interesting in light of findings that the levels of ER measured by steroid binding assays increased during pregnancy in both the MPN and the arcuate. This suggests that the mechanisms regulating functional ER levels during pregnancy may be specific to the brain regions examined.

Functional consequences

The present results are consistent with the hypothesis that pregnancy may alter the sensitivity of the MPN to estrogen, thereby facilitating the estrogen-induced onset of maternal behavior. While a causal experiment remains to be performed, the timing of various regulatory events during pregnancy correlates well with predictions based on this hypothesis.

Changes in the amount of ER mRNA per neuron within both the rostral and caudal MPN occur on day 8 of pregnancy and then remain lower throughout the remaining time points measured. Figure 3.5 shows the temporal relationship between changes in ER mRNA levels in the MPN demonstrated by the present experiment and changes in nuclear ER binding determined by Giordano et al. (1991). In addition, the relationships of both these measures to circulating estrogen levels (Shaikh 1971) and the onset of maternal behavior are shown.

Assuming that increases in mRNA precede changes in receptor protein, which, in turn, result in increased sensitivity of the neurons to estrogen, Figure 3.5 demonstrates that the time course for these changes is consistent with the hypothesis that the MPN becomes more sensitive to estrogen around the time that circulating estrogen levels are increasing dramatically. An increase in sensitivity may occur as a result of a rise in the levels of functional receptor protein caused by an

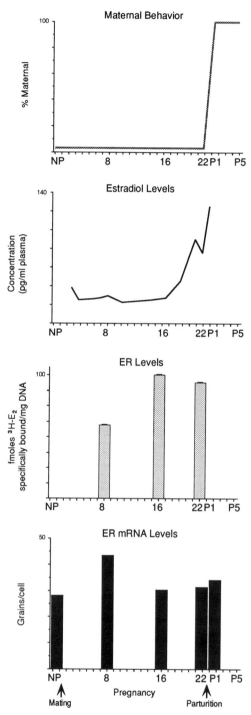

Figure 3.5. Schematic graphs showing the temporal relationship between ER mRNA levels in the MPN, ER levels in the MPN as measured by *in vitro* steroid binding assays (from Giordano et al. 1991), blood estradiol levels (from Shaikh 1971), and the onset of

increase in the amount of ER mRNA available for translation. This increased sensitivity to estrogen in a brain region critical for the induction of maternal behavior may in turn facilitate the occurrence of the behavior.

ER mRNA in the forebrain of the female guinea pig after gonadal steroid hormone treatment

Our quantitative studies on ER mRNA content focused on specific forebrain regions of the female guinea pig that contain ER-immunoreactive neurons (Blaustein and Turcotte 1989; DonCarlos et al. 1991; Malik, Feder, and Morrell 1993a; Warembourg et al. 1989). The vast majority of the neurons that contain ER immunoreactivity were found in the preoptic area, hypothalamus, and specific limbic structures such as the BNST and the corticomedial portion of the amygdala. These neuronal cell groups also contain the highest binding levels of estradiol by biochemical and steroid hormone autoradiographic methods (Walker and Feder 1977b; Warembourg 1977).

In addition to containing ERs, the regions selected for examination, the MPN, the periventricular preoptic nucleus (PePOA), the ventral lateral nucleus (VL), and the infundibular nucleus (Inf) of the medial basal hypothalamus, are important in the regulation of the behavioral and neuroendocrine events of reproduction in the guinea pig (Blaustein and Feder 1979, 1980; Blaustein and Olster 1989; Malik et al. 1993b). In the guinea pig, the nomenclature for the neuronal cell groups in the medial basal hypothalamus is at slight variance with that used for the rat (DonCarlos et al. 1989, 1991; Silverman et al. 1991). The VL in the guinea pig is equivalent in neuroanatomical position, neuroendocrine markers, and function to the ventromedial nucleus of the rat. The Inf in the guinea pig is referred to as the arcuate nucleus in the rat.

One of the best-known effects of estrogen on the brain is the induction of progestin receptors in specific, limited brain regions. The brain regions we focused on also have substantial progestin-binding capacity and a large number of neurons that contain immunoreactive progestin receptor (Blaustein and Feder 1979; DonCarlos et al. 1989; Warembourg 1978; Warembourg et al. 1986; also see Chapter 14 by Blaustein et al., this volume). Neurons in these same regions respond to estrogen treatment with a substantial increase in progestin receptor protein content (Thornton et al. 1986). Since Warembourg et al. (1989) have

Caption for Figure 3.5 (*cont.*) maternal behavior during the prepregnancy (NP, not pregnant), pregnancy, and postpartum states (P1, P5) of a female rat. The rise in ER mRNA levels occurs on day 8 of pregnancy before the rise in ER binding levels on day 16. The rise in ER levels occurs just at the time blood estradiol levels are rising and reaching their peak on day 22 of pregnancy. The presumably estrogen-induced onset of maternal behavior occurs abruptly and completely on day 22 of pregnancy just before parturition.

demonstrated the co-localization of ER and progesterone receptor (PR) immunoreactivity in many neurons in these same brain regions, it is very likely that estrogen acting on its specific receptor alters by genomic mechanisms in the same neurons the expression of PR. Thus, altered expression of PR in these brain regions is an example of one specific function of ER in neurons of these brain regions. Presumably a cascade of cellular events supports major alterations in the behavioral or neuroendocrine status of the animal, which in turn supports reproduction.

Experimental goals

Whereas considerable work has been done on the estrogen binding and ER-immunoreactive content of these forebrain regions in the guinea pig, to date no work has been published on their ER mRNA content. Our goal was to determine the neuroanatomical distribution of the neurons containing ER mRNA in the guinea pig and to begin examining whether the number of neurons with ER mRNA and the relative amount of ER mRNA per neuron were altered when circulating estradiol and progestin levels were altered. We chose to examine the effect of estrogen followed by progestin, a steroid hormone regimen that is well established as necessary and sufficient to stimulate female sexual behavior in the ovariectomized female guinea pig (Blaustein and Olster 1989; DonCarlos et al. 1989; Malik et al. 1993b). We have also examined the relative abundance of PR and ER immunoreactivity under these hormonal conditions (DonCarlos et al. 1989). Our primary goal was to understand the possible regulation of ER mRNA under behaviorally relevant conditions. We also examined the effects of shorter-term alterations in serum levels of estradiol to investigate the time course and nature of ER mRNA regulation by its cognate ligand.

Neuroanatomical distribution

The vast majority of neurons in the guinea pig forebrain that contained ER mRNA were located in the MPN, PePOA (Fig. 3.1), VL, and Inf of the medial basal hypothalamus, BNST, cortical and medial nuclei of the amygdala, and lateral septum. In addition, a small number of ER mRNA–containing neurons were found in specific preoptic, hypothalamic, and limbic cell groups, including the supraoptic and paraventricular nuclei, lateral preoptic area, and lateral hypothalamus. Only brain regions of the guinea pig previously found to contain neurons that bind estradiol (Keefer 1981; Lieberburg et al. 1980; Walker and Feder 1977a,b; Warembourg 1977) and to contain ER-immunoreactive neurons (Blaustein and Turcotte 1989; Cintra et al. 1986; DonCarlos et al. 1991; Silverman et al. 1991; Warembourg et al. 1989) contained neurons that expressed ER mRNA. Estrogen receptor mRNA–expressing neurons were rare in the thalamus, neocortex, and basal ganglia. Overall, in the guinea pig, both the number of neurons that expressed ER mRNA and the neuroanatomical distribution of these neurons are virtually identical to the number and distribution of neurons that bind radiolabeled estradiol and contain ER immunoreactivity.

Density of ER mRNA–containing neurons across regions

We analyzed the number of ER mRNA–containing neurons in four areas: the MPN, PePOA, VL, and Inf of the medial basal hypothalamus (Fig. 3.6B).

The number of ER mRNA–containing neurons varied across the brain regions examined; the rank order from higher to lower numbers is PePOA > Inf > VL > MPN. These data are represented in the histogram in Figure 3.6B. The PePOA tended to have a higher number (per constant grid area) of ER mRNA–expressing neurons than the three other brain regions, and the difference reached statistical significance in one experiment. The Inf consistently tended to have a higher density of ER mRNA–expressing neurons than both the MPN and the VL (Fig. 3.6B); however, this trend was not statistically significant.

The number of grains per labeled neuron was also found to vary across these brain regions, indicating that the relative cellular abundance of ER mRNA varies across these populations (Fig. 3.6A). When the mean number of grains per neuron was calculated for each region, they could be rank-ordered from higher to lower as VL > Inf > PePOA > MPN. The only statistically significant difference was the larger mean number of grains per labeled neuron in the VL than in the MPN.

Overall, our ranking of brain regions with respect to the density of ER mRNA correlates well with the density of ER-immunoreactive neurons (DonCarlos et al. 1991; Malik et al. 1993a). The PePOA had the most dense collection of neurons containing both ER mRNA and ER immunoreactivity. Since there were differences across brain regions in the neuronal content of both the immunoreactive receptor protein and the mRNA from which the protein is translated, a reasonable extension of this is the hypothesis that both the levels of protein and the mRNA for ER are regulated in the different brain regions. This implies that different means of regulation and different amounts of regulation might take place at the protein and the mRNA level.

Effect of estrogen and progestin treatments

The number of neurons containing ER mRNA found in ovariectomized females was not altered by hormonal treatment with estrasiol benzoate or by estradiol benzoate followed by progestin (regime that facilitates sexual receptivity). There

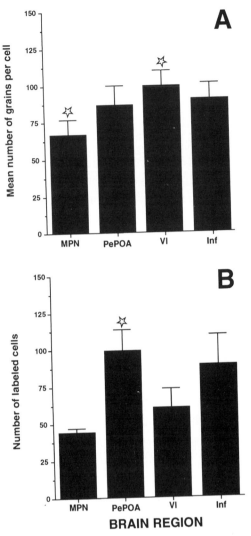

Figure 3.6. A. Histogram of the mean ($\pm$ SEM) number of grains per ER mRNA–containing neuron in four specific nuclei of the female guinea pig brain. B. Histogram of the mean number ($\pm$ SEM) of neurons containing ER mRNA per 0.0625 mm^2 in the same specific nuclei. Comparisons were made with ANOVA, followed by Duncan's multiple range test, significantly different at $p \leq .05$.

was a reproducible trend suggesting a decrease in the number of neurons that express ER mRNA in the animals treated with estrogen followed by progestin, the very group that would have been sexually receptive (Table 3.1). The mean number of grains per labeled neuron was also only minimally altered by these hormonal treatments. Moreover, frequency distribution analysis of the number of grains per neuron in each area did not yield any significant differences across the hormone conditions (data not shown).

Table 3.1. *Number of ER mRNA–containing neurons and relative abundance of ER mRNA per neuron in the brain of the female guinea pig after estradiol benzoate and progestin treatment*

Brain region	Treatment	Mean grains per labeled cell (SEM)	Mean number of labeled cells (SEM)
MPN	Oil	76.12 (18.93)	47.75 (14.22)
	EB	69.85 (16.41)	61.63 (6.28)
	EB + P	71.12 (34.97)	26.89 (17.77)
PePOA	Oil	112.85 (30.48)	121.25 (30.01)
	EB	109.99 (19.90)	128.38 (22.24)
	EB + P	88.77 (44.74)	67.50 (38.83)
VI	Oil	75.00 (13.97)	47.62 (8.98)
	EB	87.15 (27.07)	45.88 (11.56)
	EB + P	117.22 (6.51)	24.75 (14.29)
Inf	Oil	82.76 (15.64)	75.75 (20.93)
	EB	79.83 (22.44)	72.13 (21.78)
	EB + P	99.95 (8.06)	53.38 (30.82)

Note: There were three treatment groups (four animals per group): oil as vehicle control; estradiol benzoate (EB) alone (10 μg per animal), animal killed 44 hours later; EB plus progestin (P), 10 μg of EB, then 40 hours later 500 μg of P, animal killed 4 hours later. This hormone regime facilitates sexual receptivity in the female guinea pig (DonCarlos et al. 1989; Malik et al. 1993b). All hormone-treated females had plasma estradiol levels of approximately 15 + 2 pg/ml, measured by radioimmunoassay (Hazleton Washington, Inc., Baltimore). All other analytic details as in captions to Figures 3.2 and 3.3.

Analysis of all four brain regions showed that the number of neurons expressing ER mRNA and the number of grains per labeled neuron were also not significantly altered earlier during the course of estradiol treatment. Table 3.2 shows that the number of neurons with ER mRNA and the relative abundance of the mRNA per neuron was not altered 2, 4, or 24 hours after estradiol administration.

On the basis of our previous work on ER mRNA regulation in the rat forebrain and that of others (Lauber et al. 1990; Lisciotto and Morrell 1993; Simerly and Young 1991), we expected that estradiol would downregulate the amount of ER mRNA in the forebrain of ovariectomized adult guinea pigs. It is possible that we detected no clear downregulation of the ER mRNA because the time course and nature of the estradiol- and progestin-induced alterations are, in principle, different in the guinea pig than in the rat. These two species are, in fact, quite different in their ovarian cycles and steroid hormone–dependent behaviors. The guinea pig has a much longer estrous cycle than the rat because it forms a functional corpora lutea (Feder 1981).

There is an alternative explanation for the difference between our predicted and actual findings. While the subtle increases in circulating estradiol produced by our

Table 3.2. *Number of ER mRNA–containing neurons and relative abundance of ER mRNA per neuron in the brain of the female guinea pig after estradiol treatment*

Brain region	Treatment (hours after injection)	Mean grains per labeled cell (SEM)	Mean number of labeled cells (SEM)
MPN	V (2)	88.73 (25.07)	44.25 (2.50)
	E$_2$ (2)	74.33 (19.71)	52.00 (19.69)
	E$_2$ (4)	57.55 (4.73)	43.12 (15.76)
	E$_2$ (24)	64.88 (16.36)	42.62 (11.07)
PePOA	V (2)	104.82 (21.64)	99.25 (14.56)
	E$_2$ (2)	105.87 (16.60)	86.50 (33.61)
	E$_2$ (4)	96.11 (33.35)	81.00 (28.13)
	E$_2$ (24)	90.60 (31.01)	81.88 (15.66)
VL	V (2)	126.61 (25.81)	59.75 (13.36)
	E$_2$ (2)	115.47 (10.97)	44.25 (16.80)
	E$_2$ (4)	89.21 (9.53)	45.62 (7.01)
	E$_2$ (24)	93.72 (11.26)	47.38 (12.04)
Inf	V (2)	101.85 (23.23)	88.63 (20.19)
	E$_2$ (2)	123.33 (12.76)	81.00 (28.27)
	E$_2$ (4)	70.02 (15.17)	82.12 (15.44)
	E$_2$ (24)	94.75 (25.05)	88.88 (20.31)

Note: Females were injected with 10 μg estradiol (E$_2$) and killed 2 [vehicle-treated (V) animals were also killed at this time], 4, or 24 hours after injection. This hormone regime was not designed to induce sexual receptivity; the purpose was rather to examine the time course of the effect of the cognate ligand. Estradiol-17β has a shorter time course of action than the esterified form, estradiol benzoate (Walker and Feder 1977a,b; Wilcox et al. 1984). All hormone-treated females had plasma E$_2$ levels of approximately 15 ± 2 pg/ml. All other analytic details as in captions to Figures 3.2 and 3.3

exogenous injections of steroid were sufficient to produce sexual behavior, they may be insufficient to alter ER mRNA levels. This logically implies that sexual behavior in the guinea pig is not dependent on alterations in ER at the level of mRNA within the 44-hour period it takes for estrogen and progestin to induce the cascade of neuronal events that support this behavior.

In our experiments and those from other laboratories using rats, there is a downregulation of ER mRNA after an increase in circulating estradiol or testosterone. In each of these experiments, exogenous treatments resulted in more substantial replacement of circulating steroid hormones, because a significantly higher dose of exogenous hormone or a longer time course of hormone treatment was used (Lisciotto and Morrell 1991; Simerly and Young 1991). It is possible that our increases in circulating estrogen (and progestin) levels in the guinea pig were too subtle to alter ER mRNA in the brain.

Theoretical summary

It may be useful to look back at these three sets of experiments and consider what we have learned. First, we must recall that each of the three methods by which we can examine steroid hormone receptors reveals information about a different aspect of the cell biology of the neuron's steroid hormone receptor complement. The method with the longest history of use enables us to examine the amount or specificity of binding of estradiol to its own receptor. Unfortunately, the methods most commonly used for binding assays do not provide information about total binding capacity. While it can be useful to consider together data from both nuclear and cytosolic binding assays (Gorski et al. 1993), these do not sum to total binding, and many endocrine states await analysis by a method with which one can precisely measure total capacity of binding (Walters et al. 1993). Newer methods enable us to examine the amount of receptor protein and whether that protein may be phosphorylated at key amino acids. Using the methods reported in this chapter, we determined a measure of the transcriptional processing of the nucleotide sequence from which ER protein is derived.

Determination of the subset of estrogen-sensitive neurons

In the context of the gonadal steroid hormone–induced alterations of neuronal events that are important for the neuroendocrine and behavioral events of reproduction, we take the position that the neurons containing specific nuclear receptors are the critical starting point for hormonally mediated events. We would argue that using a third method to measure the neuroanatomical location and number of neurons that contain these receptors supports the idea that we are indeed examining the complete subset of ER-containing neurons that are most likely to be significantly effected by estrogen. Thus, collectively the data show that the neurons that bind estradiol contain immunoreactive ER protein and contain mRNA from which the ER sequences are found in the same neuroanatomical location and in essentially the same number. While polymerase chain reaction (PCR) methods might theoretically allow the detection of a few copies of receptor per neuron in other brain regions, the weight of the current data argues that the subset of estrogen-sensitive neurons of first-order biological importance are already demonstrated.

These arguments are specific to the intracellular steroid hormone receptors and do not extend to potential membrane receptors for gonadal steroid hormones that have periodically been suggested by others (Tischkau and Ramirez 1993). When considering the possible structure of such membrane receptors, one would certainly want to consider the models of other membrane receptors (neurotransmitter receptors) that are complex membrane-spanning proteins with intra- and extra-cellular domains of particular functional importance. If such membrane receptors for steroid hormones exist and are biologically functional, it seems very likely that they are entirely different from the intracellular receptors on which we have focused our attention. Furthermore, since the membrane receptor is likely to be an entirely different protein, this membrane receptor could be present in a subset of neurons that may or may not overlap with the subset that contains the intracellular gonadal steroid hormone receptor.

Regulation of estrogen receptors in neurons

We have for some time theorized that one way of regulating the brain's sensitivity to gonadal steroid hormones is to alter either the number of neurons that express the gonadal steroid hormone receptors or the amount of receptor per neuron. In fact, it appears to us that Mother Nature has already adopted this general strategy by sharply limiting the neuroanatomical location and number of neurons that contain these receptors. We propose that a titering of the number of neurons that contain ER and/or the number of receptors per neuron might be used to achieve neuroanatomically precise, function-specific steroid hormone regulation of neuronal events in the face of a single circulating concentration of ligand–estradiol. Thus, it would be possible for brain regions to have different functional or temporal courses of estrogen sensitivity. This might be critical to the sequencing of reproductive events that are controlled by the action of steroid hormones on the brain.

From the cell biological perspective, regulation of the number of neurons expressing ER and the number of receptors per neuron could theoretically be achieved by regulating transcriptional, post-transcriptional, translational, and/or post-translational processing of the ER. It is also theoretically possible that ER-mediated events are altered by other transcription factors acting at any of these levels of cellular events, but we will not discuss this here. We have shown in these experiments that circulating gonadal hormones *can* alter the amount of ER mRNA expressed in particular brain regions. While it remains to be determined whether these steady-state mRNA levels are altered by changes in transcription rate, mRNA stability, or rate of degradation, our data indicate that regulation of the amount of ER does take place at this level.

The most dramatic effect on ER mRNA is caused by the elimination of circulating gonadal hormones by castration. This substantial increase in the amount of ER mRNA could be a means by which neurons increase their sensitivity to estradiol in order to maximize their possible response to any ligand that becomes available. This might be a situation in which a low threshold of response to the ligand is initially necessary. Once ligand is available at physiological levels, the neuron's sensitivity need not be maximal. In fact, such great sensitivity (low threshold) might interfere with the subtle summation of additional regulatory signals – for example, alterations that might be signaled by afferent neuronal connections. While the nature of these afferent signals is not well understood, it has been suggested that alteration of the phosphorylation state of the receptor and hence its availability for ligand binding (Migliaccio et al. 1989), or even its binding to the genome in an unoccupied state, is possible (Power et al. 1991). We propose that, under physiological conditions, the amount of functional steroid hormone receptor might be determined by circulating hormones that include but are not limited to gonadal steroid hormones, as well as the afferent input from other neurons. The threshold number of functional receptors needed for a particular steroid hormone receptor–mediated event might be reached by a summation of receptors expressed as a result of circulating hormones and receptors influenced by afferent neuronal signals. We believe that this is the case represented in our pregnancy study; that is, changes in both circulating hormone levels and the neuronal activity of the CNS could be summing to regulate the amount of ER mRNA found in different brain regions during the course of pregnancy. Comparing our data on ER mRNA with total estradiol binding throughout pregnancy is not currently possible. However, data from nuclear binding assays of estradiol suggest that ER-mRNA is not the only cellular level at which alterations are made in the amount of receptor during pregnancy. That is translational, post-translational events, and even receptor phosphorylation might contribute to the regulation of the functional receptor pool across pregnancy.

In this chapter we have focused on a step in the production of functional ER that is a necessary precursor to the total ER protein and the level of receptor available for biologically active binding of the ligand. While the relationship of total receptor binding capacity to its protein and to the mRNA that is its precursor has not been well investigated thus far, our data indicate that the level of mRNA for the ER is regulated by the hormonal status.

Acknowledgments

The work presented in this chapter was supported by USPHS Awards HD 22983 to J. I. Morrell; HD 04467 to H. H. Feder and J. I. Morrell; National Research Service Awards to C. A. Lisciotto and C. K. Wagner with J. I. Morrell as sponsor; the Johnson and Johnson Discovery Award to J. I. Morrell; and grants from the Charles and Joanna Busch Foundation to J. I. Morrell.

References

Adieh, H. B., Mayer, A. D., and Rosenblatt, J. S. (1987). Effect of brain antiestrogen implants on maternal behavior and on postpartum estrus in pregnant rats. *Neuroendocrinology* 46: 522–531.

Angerer, L. M., Stoler, M. H., and Angerer, R. C. (1987). In situ hybridization with RNA probes: An annotated recipe. In *In Situ Hybridization: Applications to Neurobiology*, ed. K. L. Valentino, J. H. Everwine, and J. D. Barchas, pp. 42–70. New York: Oxford University Press.

Baum, M. J., and Vreeburg, J. T. M. (1973). Copulation in castrated male rats following combined treatment with estradiol and dihydrotestosterone. *Science* 182: 283–285.

Blaustein, J. D., and Feder, H. H. (1979). Cytoplasmic progestin-receptors in guinea pig brain: Characteristics and relationship to the induction of sexual behavior. *Brain Res.* 169: 481–497.

Blaustein, J. D., and Feder, H. H. (1980). Nuclear progestin receptors in guinea pig brain measured by an in vitro exchange assay after hormonal treatments that affect lordosis. *Endocrinology* 106: 1061–1069.

Blaustein, J. D., and Olster, D. H. (1989). Gonadal steroid hormone receptors and social behaviors. In *Advances in Comparative and Environmental Physiology*, Vol. 3, ed. J. Balthazart, pp. 31–104. Berlin: Springer.

Blaustein, J. D., and Turcotte, J. C. (1989). Estradiol-induced progestin receptor immunoreactivity is found only in estrogen receptor-immunoreactive cells in the guinea pig brain. *Neuroendocrinology* 49: 454–461.

Chowden-Breed, J. A., Steiner, R. A., and Clifton, D. K. (1989). Sexual dimorphism and testosterone dependent regulation of somatostatin gene expression in the periventricular nucleus of the rat brain. *Endocrinology* 125: 357–362.

Cintra, A., Fuxe, K., Harfstrand, A., Agnati, L. F., Miller, L. S., Greene, G. L., and Gustafsson, J-A. (1986). On the cellular localization and distribution of estrogen receptors in the rat tel- and diencephalon using monoclonal antibodies to human estrogen receptor. *Neurochem. Int.* 8: 587–595.

Damassa, D. A., Kobashigawa, D., Smith, E. R., and Davidson, J. M. (1976). Negative feedback control of LH by testosterone: A quantitative study in male rats. *Endocrinology* 99: 736–742.

DonCarlos, L. L., Greene, G. L., and Morrell, J. I. (1989). Estrogen plus progesterone increases progestin receptor immunoreactivity in the brain of ovariectomized female guinea pigs. *Neuroendocrinology* 50: 613–623.

DonCarlos, L. L., Monroy, E., and Morrell, J. I. (1991). Distribution of estrogen receptor-immunoreactive cells in the forebrain of the female guinea pig. *J. Comp. Neurol.* 305: 591–612.

Evans, R. M. (1988). The steroid and thyroid hormone receptor superfamily. *Science* 240: 889–895.

Feder, H. H. (1981). Estrous cyclicity in mammals. In *Neuroendocrinology of Reproduction*, ed. N. T. Adler, pp. 279–329. New York: Plenum Press.

Giordano, A. L., Siegel, H. I., and Rosenblatt, J. S. (1991). Nuclear estrogen receptor binding in microdissected brain regions of female rats during pregnancy: Implications for maternal and sexual behavior. *Physiol. Behav.* 50: 1263–1267.

Goldfine, I. D. (1987). The insulin receptor: Molecular biology and transmembrane signaling. *Endocrine Rev.* 8: 235–255.

Gorski, J., Furlow, J. D., Murdoch, F. E., Fritsch, M., Kaneko, K., Ying, C., and Maylayer, J. R. (1993). Perturbations in the model of estrogen receptor regulation of gene expression. *Biol. Reprod.* 48: 8–14.

Gray, P. and Brooks, P. J. (1984). Effect of lesion location within the medial preoptic–anterior hypothalamic continuum on maternal and male sexual behaviors in female rats. *Behav. Neurosci.* 98: 703–711.

Hnatczuk, O. C., Lisciotto, C. A., DonCarlos, L. L., Carter, S., and Morrell, J. I. (1994). Estrogen receptor immunoreactivity in specific brain areas of the prairie vole

(*Microtus ochrogaster*) is altered by sexual receptivity and genetic sex. *J. Neuroendocrinol.* 6: 89–100.

Keefer, D. A. (1981). Nuclear retention characteristics of [³H]estradiol by cells in four estrogen target regions of the rat brain. *Brain Res.* 229: 224–229.

Koch, M., and Ehret, G. (1989). Immunocytochemical localization and quantification of estrogen binding cells in the male and female (virgin, pregnant, lactating) mouse brain. *Brain Res.* 489: 101–112.

Koike, S., Masaharu, S., and Muramatsu, M. (1987). Molecular cloning and characterization of rat estrogen receptor cDNA. *Nucl. Acids Res.* 15: 2499–2513.

Lauber, A. H., Romano, G. J., Mobbs, C. V., and Pfaff, D. W. (1990). Estradiol regulation of estrogen receptor messenger ribonucleic acid in rat mediobasal hypothalamus: An in situ hybridization study. *J. Neuroendocrinol.* 2: 605–611.

Lawrence, J. B., and Singer, R. H. (1985). Quantitative analysis of in situ hybridization methods for detection of actin expression. *Nucl. Acids Res.* 13: 1777–1799.

Lieberberg, I., MacLusky, N., and McEwen, B. S. (1980). Cytoplasmic and nuclear estradiol-17β binding in male and female rat brain: Regional distribution, temporal aspects, and metabolism. *Brain Res.* 193: 487–503.

Lisciotto, C. A., and Morrell, J. I. (1990). Androgen-concentrating neurons in the forebrain project to the midbrain in rats. *Brain Res.* 516: 107–112.

Lisciotto, C. A., and Morrell, J. I. (1991). In situ hybridization for estrogen receptor mRNA in male rat brain following long-term testosterone deprivation and replacement. *Soc. Neurosci. Abstr.* 17: 1411.

Lisciotto, C. A. and Morrell, J. I. (1993). Circulating gonadal steroid hormones regulate estrogen receptor mRNA in the male rat forebrain. *Mol. Brain Res.* 20: 79–90.

Malik, K. F., Feder, H. H., and Morrell, J. I. (1993a). Estrogen receptor immunostaining in the preoptic area and medial basal hypothalamus of estradiol benzoate- and prazosin-treated female guinea pigs. *J. Neuroendocrinol.* 5: 297–306.

Malik, K. F., Morrell, J. I., and Feder, H. H. (1993b). Effects of clonidine and phentolamine infused into the medial preoptic area and medial basal hypothalamus of the guinea pig. *Neuroendocrinology* 57: 177–188.

Malik, K. F., Silverman, A. J., and Morrell, J. I. (1991). Gonadotropin-releasing hormone mRNA in the rat: Distribution and neuronal content over the estrous cycle and after castration of males. *Anat. Rec.* 231: 457–466.

McCabe, J. T., Morrell, J. I., Ivell, R., Schmale, H., Richter, D., Pfaff, D. W. (1986). In situ hybridization technique to localize rRNA and mRNA in mammalian neurons. *J. Histochem. Cytochem.* 34: 45–50.

McGinnis, M. Y., and Dreifuss, R. M. (1989). Evidence for a role of testosterone–androgen receptor interactions in mediating masculine sexual behavior in male rats. *Endocrinology* 124: 618–624.

McGinnis, M. Y., Mirth, M. C., Zebrowski, A. F., and Dreifuss, R. M. (1989). Critical exposure time for androgen activation of male sexual behavior in rats. *Physiol. Behav.* 46: 159–165.

McGivern, R. F., Bollnow, M. R., O'Keefe, J. A., and Handa, R. J. (1992). Increases in estrogen receptor mRNA in the medial preoptic area of the male rat following long-term castration. *Soc. Neurosci. Abstr.* 18: 893.

Migliaccio, A., Domenico, M. D., Green, S., DeFalco, A., Kajtaniak, E. L., Blasi, F., Chambon, P., and Auriccio, F. (1989). Phosphorylation on tyrosine of in vitro synthesized human estrogen receptor activates its hormone binding. *Mol. Endocrinol.* 3: 1061–1069.

Morrell, J. I., Corodimas, K. P., DonCarlos, L. L., and Lisciotto, C. A. (1992). Axonal projections of gonadal steroid receptor-containing neurons. In *Neuroprotocols: A Companion to Methods in Neuroscience*, Vol. 1: *Feedback Mechanisms*, ed. S. P. Kalra, pp. 4–15. New York: Academic Press.

Morrell, J. I., Kelley, D. B., and Pfaff, D. W. (1975). Sex steroid binding in the brains of vertebrates: Studies with light-microscopic autoradiography. In *Brain–Endocrine Interactions*, Vol. 2: *The Ventricular System*, ed. K. M. Knigge, D. E. Scott, H. Kobayashi, and S. Ishii, pp. 230–256. Basel: Karger.

Morrell, J. I., and Pfaff, D. W. (1983). Immunocytochemistry of hormone receiving cells in the central nervous system. In *Methods in Enzymology*, Vol. 103: *Hormone Action: Neuroendocrine Peptides*, ed. Michael Conn; pp. 639–662. New York: Academic Press.

Numan, M. (1988). Maternal behavior. In *The Physiology of Reproduction*, ed. E. Knobil, and J. Neill, pp. 1569–1645. New York: Raven Press.

Numan, M., Rosenblatt, J. S. and Komisaruk, B. R. (1977). Medial preoptic area and onset of maternal behavior in the rat. *J. Comp. Physiol. Psychol.* 91: 146–164.

Orti, E., Bodwell, J. E., and Munk, A. (1992). Phosphorylation of steroid hormone receptors. *Endocrinol. Rev.* 13: 105–128.

Pearson, P. L., Ross, L. R., and Jacobson, C. D. (1993). Differential down-regulation of estrogen receptor-like immunoreactivity by estradiol in the female Brazilian opossum brain. *Brain Res.* 617: 171–175.

Pfaff, D. W. (1980). *Estrogens and Brain Function*. New York: Springer.

Pfaff, D. W., and Keiner, M. (1973). Atlas of estradiol-concentrating cells in the central nervous system of the female rat. *J. Comp. Neurol.* 151: 121–158.

Poulin, R., Simard, J., Labrie, C., Petitclerc, L., Dumont, M., Lagace, L., and Labrie, F. (1989). Down-regulation of estrogen receptors by androgens in the ZR-75-1 human breast cancer cell line. *Endocrinology* 125: 392–399.

Power, R. F., Mani, S. K., Codina, J., Conneely, O. M., and O'Malley, B. (1991). Dopaminergic and ligand-independent activation of steroid hormone receptors. *Science* 254: 1636–1639.

Quarmby, V. E., Yarbough, W. G., Lubahn, D. B., French, F. S., and Wilson, E. M. (1990). Autologous down-regulation of androgen receptor messenger ribonucleic acid. *Mol. Endocrinol.* 4: 22–28.

Roselli, C. E., Horton, L. E., and Resko, J. A. (1985). Distribution and regulation of aromatase activity in the rat hypothalamus and limbic system. *Endocrinology* 117: 2471–2477.

Roselli, C. E., and Resko, J. A. (1987). The distribution and regulation of aromatase activity in the central nervous system. *Steroids* 50: 495–508.

Sachs, B. D., and Meisel, R. L. (1988). The physiology of male sexual behavior. In *The Physiology of Reproduction*, ed. E. Knobil, and J. Neill, pp. 1393–1485. New York: Raven Press.

Sar, M., and Stumpf, W. E. (1973). Autoradiographic localization of radioactivity in the rat brain after injection of 1,2-[^{3}H] testosterone. *Endocrinology* 92: 251–256.

Sar, M., and Stumpf, W. E. (1977). Distribution of androgen target cells in rat forebrain and pituitary after [^{3}H]dihydrotestosterone administration. *J. Steroid Biochem.* 8: 1131–1135.

Shaikh, A. A. (1971). Estrone and estradiol levels in the ovarian venous blood from rats during the estrous cycle and pregnancy. *Biol. Reprod.* 5: 297–307.

Sherman, T. G., McKelvey, J. F., and Watson, S. J. (1986). Vasopressin mRNA regulation in individual hypothalamic nuclei: A Northern and in situ analysis. *J. Neurosci.* 6: 1685–1694.

Shughrue, P. J., Bushnell, C. D., and Dorsa, D. M. (1992). Estrogen receptor messenger ribonucleic acid in female rat brain during the estrous cycle: A comparison with ovariectomized females and intact males. *Endocrinology* 131: 381–388.

Siegel, H. I., and Rosenblatt, J. S. (1975). Hormonal basis of hysterectomy-induced maternal behavior during pregnancy in the rat. *Horm. Behav.* 6: 211–222.

Silverman, A. J., DonCarlos, L. L., and Morrell, J. I. (1991). Ultrastructural characteristics of estrogen receptor-containing neurons of the ventrolateral nucleus of the guinea-pig hypothalamus. *J. Neuroendocrinol.* 3: 623–634.

Simerly, R. B., Chang, C., Muramatsu, M., and Swanson, L. W. (1990). Distribution of androgen and estrogen receptor mRNA-containing cells in the rat brain: An *in situ* hybridization study. *J. Comp. Neurol.* 294: 76–95.

Simerly, R. B., and Young, B. J. (1991). Regulation of estrogen receptor messenger ribonucleic acid in rat hypothalamus by sex steroid hormones. *Mol. Endocrinol.* 5: 424–432.

Simmons, D. M., Arriza, J. L., and Swanson L. W. (1989). A complete protocol for in situ hybridization of messenger RNAs in brain and other tissue with radiolabeled single-stranded RNA probes. *J. Histotechnol.* 12: 169–181.

Swanson, L. W., and Simmons D. M. (1989). Differential steroid hormone and neural influences on peptide mRNA levels in CRH cells of the paraventricular nucleus: A hybridization histochemical study in the rat. *J. Comp. Neurol.* 285: 413–435.

Thornton, J. E., Nock, B., McEwen, B. S., and Feder, H. H. (1986). Estrogen induction of progestin receptors in microdissected hypothalamic and limbic nuclei of female guinea pigs. *Neuroendocrinology* 43: 182–188.

Tischkau, S. A., and Ramirez, V. D. (1993). A specific membrane binding protein for progesterone in rat brain: Sex differences and induction by estrogen. *Proc. Natl. Acad. Sci. (USA)* 90: 1285–1289.

Toranzo, D., Dupont, E., Simard, J., Labrie, C., Couet, J., Labrie, F., and Pelletier, G. (1989). *Mol. Endocrinol.* 3: 1748–1756.

Walker, W. A., and Feder, H. H. (1977a). Inhibitory and facilitatory effects of various antiestrogens on the induction of female sexual behavior by estradiol benzoate in guinea pig brain. *Brain Res.* 134: 455–466.

Walker, W. A., and Feder, H. H. (1977b). Anti-estrogen effects on estrogen accumulation in brain cell nuclei: Neurochemical correlates of estrogen action on female sexual behavior in guinea pigs. *Brain Res.* 134: 467–478.

Walters, M. J., Brown, T. J., Hochberg, R. B., and MacLusky, N. J. (1993). In vitro autoradiographic visualization of estrogen receptors in the central nervous system of the rat: Rapid, selective measurement of occupied receptors using an iodinated estrogen ligand. *J. Histochem. Cytochem.* 51: 103–132.

Warembourg, M. (1977). Radioautographic localization of estrogen-containing cells in the brain and pituitary of the guinea pig. *Brain Res.* 123: 357–367.

Warembourg, M. (1978). Radioautographic study of the brain and pituitary after [^{3}H] progesterone injection into estrogen-primed ovariectomized guinea pigs. *Neurosci. Lett.* 7: 1–5.

Warembourg, M., Jolivet, A., and Milgrom, E. (1989). Immunohistochemical evidence of the presence of estrogen receptors and progestin receptors in the same neurons of the guinea pig hypothalamus and preoptic area. *Brain Res.* 480: 1–15.

Warembourg, M., Logeat, F., and Milgrom, E. (1986). Immunocytochemical localization of progesterone receptor in the guinea pig central nervous system. *Brain Res.* 384: 121–131.

Wilcox, J. N., Barclay, S. R., and Feder, H. H. (1984). Administration of estradiol-17β in pulses to female guinea pigs: Self-priming effects of estrogen on brain tissues mediating lordosis. *Physiol. Behav.* 32: 483–488.

Wise, P. M., Scarbrough, K., Weiland, N. G., and Larson, G. H. (1990). Diurnal pattern of proopiomelanocortin gene expression in the arcuate nucleus of proestrus, ovariectomized, and steroid-treated rats: A possible role in cyclic luteinizing hormone secretion. *Mol. Endocrinol.* 4: 886–892.

Wray, S., Zoeller, R. T., and Gainer, H. (1989). Differential effects of estrogen on luteinizing hormone-releasing hormone gene expression in slice explant cultures prepared from specific rat forebrain regions. *Mol. Endocrinol.* 3: 1197–1206.

Yamamoto, K. (1985). Steroid receptor regulated transcription of specific genes and gene networks. *Annu. Rev. Genet.* 19: 209–252.

Zoeller, R. T., Seeburg, P. H., and Young, W. S. (1988). In situ hybridization histochemistry for messenger ribonucleic acid (mRNA) encoding gonadotropin-releasing hormone (GnRH) mRNA in female rat brain. *Endocrinology* 122: 2570–2577.

4

Hormonal regulation of limbic and hypothalamic pathways

RICHARD B. SIMERLY

The expression of sexually differentiated patterns of behavior is a characteristic of many vertebrate species and often correlates with sex differences in the relative abundance of neurons in brain regions thought to control such behaviors. In general, two fundamental processes determine the number of neurons that survive into adulthood. First, changes in the number of neuroblasts formed in the ventricular zone can occur in response to mechanisms that are intrinsic to a particular population of cells or that are controlled by extrinsic factors such as cell–cell interactions, neuronal growth factors, and circulating hormones. Second, similar extrinsic cellular and hormonal factors can determine the number of cells of a particular lineage that reach their permanent destination, establish appropriate connections, and achieve a regionally specific functional phenotype. Sex steroid hormones can affect both processes but appear to exert their most pronounced influences on neuronal development during a restricted perinatal critical period (Arnold and Gorski 1984; Breedlove 1986; Goy and McEwen 1980; Harris and Levine 1965; Rhees et al. 1990). Thus, treatment of female neonates with sex steroids during the first few postnatal days alters the number of neurons residing in certain nuclei, as well as the morphology, synaptology, and neurotransmitter expression of individual neurons (Arai et al. 1986; Arnold and Jordan 1988; De Vries 1990; Gorski 1985; Raisman and Field 1973; Simerly 1989, 1991). Although abundant evidence indicates that neonatal exposure to sex steroids induces changes in cell number by promoting the survival of steroid-sensitive cells (Arnold and Jordan 1988; Breedlove 1986; Sengelaub et al. 1989), such sexual dimorphism does not appear to be produced by this mechanism alone. For example, the anteroventral periventricular nucleus of the preoptic region (AVPv) is approximately twice as large in female rats as in male rats, and treatment of neonatal females with testosterone decreases the number of neurons in this nucleus, suggesting that perinatal steroid hormones might also promote the death of neurons during the neonatal period (Bleier et al. 1982; Simerly 1989, 1991; Simerly et al. 1985b).

An important function of limbic and hypothalamic regions is to integrate complex signals at the level of the individual cells that ultimately contribute to the elaboration of coordinated actions in response to specific sensory cues from the environment (Simerly 1995). Therefore, a clear understanding of the cellular and molecular basis for signal transduction underlying this process is of fundamental importance for clarifying the ways in which environmental influences affect the wide variety of behavioral and physiological mechanisms mediated by the brain. Although changes in synaptic density and morphology have been documented in sexually dimorphic regions (Arai et al. 1986; Matsumoto 1992; McEwen et al. 1991; Naftolin et al. 1992), neuroanatomical evidence indicates that the basic neuronal circuits in the adult brain are essentially stable and that many of the key elements of neuronal plasticity, the ability of the brain to change over time, are related to molecular events that influence synaptic efficacy, including the transmitter status of individual cells (Black et al. 1987). In adult animals, both hormonal and synaptically mediated changes in neurotransmitter levels have been identified in a variety of functional neural systems, with the most marked changes taking place in neurons that contribute to pathways involved in reproduction (Armstrong and Montminy 1993; De Vries 1990; Harlan 1988; Micevych and Ulibarri 1992; Romano et al. 1987; Simerly 1991).

Gonadal steroid hormones appear to influence neural function and development through cellular mechanisms that are similar to those characterized in peripheral tissues. The sex hormones estrogen and progesterone, like all steroid and thyroid hormones, mediate their actions on the brain primarily by binding to their cognate receptors, which then function as ligand-activated nuclear transcription factors that alter the expression of specific sets of hormone-responsive genes (Evans 1988). Thus, the initial event in the acquisition of hormone sensitivity appears to be the expression of the appropriate complement of hormone receptors by specific subsets of neurons. Accordingly, the developmental fate of neurons that express these receptors early in life may be profoundly influenced by changes in levels of circulating hormones. In the same way, the expression of sex steroid receptors by neurons in specific neural pathways of mature animals may endow these circuits with the capacity to alter neurotransmission in response to acute endocrine changes. However, not all hormonal effects on neural pathways are affected genomically. Rapid influences of gonadal steroids are unlikely to involve acute changes in gene expression (Harrison et al. 1989; Kelly 1982; Tischkau and Ramirez 1993), but the cellular mechanisms underlying these intriguing effects remain a mystery. In this chapter, the organization of several sexually dimorphic pathways in the forebrain will be summarized and the regulation of hormone receptor gene expression in these pathways will be discussed in the context of possible cellular mechanisms underlying hormonal modulation of the activity of neural circuits controlling reproductive function. Although the following account

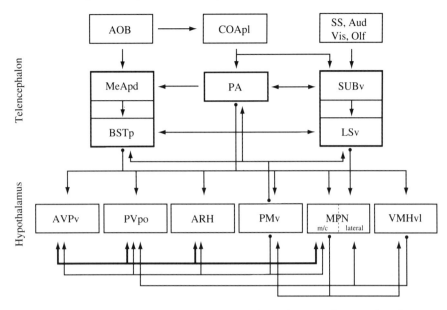

Figure 4.1. A schematic diagram summarizing the major neural connections between sexually dimorphic nuclei in the forebrain of the rat. The pathways are arranged to stress the organization of sensory pathways to the hypothalamus and the strong, largely bidirectional connections between periventricular nuclei (heavy line between the AVPv, PVpo, ARH) and the medial part of the MPN. The PMv provides the major feedback projection from sexually dimorphic nuclei in the hypothalamus to telencephalic regions that relay sensory information from the periphery. Abbreviations: AOB, accessory olfactory bulb; ARH, arcuate nucleus (n.) of the hypothalamus; Aud, auditory information; AVPv, anteroventral periventricular n.; BSTp, principal n. of the bed nuclei of the stria terminalis; COApl, cortical n. of the amygdala (posterior part, lateral zone); LSv, lateral septal n. (ventral part); MeApd, medial n. of the amygdala (posterodorsal part); MPN, medial preoptic n. (m/c, medial and central parts); PA, posterior n. of the amygdala (also known as the amygdalohippocampal zone); PMv, ventral premammillary n.; PVpo, preoptic periventricular n.; SUBv, ventral subiculum; VMHvl, ventromedial hypothalamic n. (ventrolateral part).

relies almost exclusively on findings obtained in rats, similar model systems have been described for other vertebrate species, which offer unique experimental opportunities (Arnold and Gorski 1984; DeVoogd 1991; Panzica et al. 1987; Tobet and Fox 1992; also see Chapter 1 by Wood and Newman, Chapter 2 by Yahr, and Chapter 3 by Morrell et al., this volume).

Sexually dimorphic forebrain circuitry

Numerous regions in the rodent forebrain undergo sexual differentiation and ultimately comprise a sexually dimorphic circuit (Fig. 4.1; Arai 1981; Canteras et al. 1992a; Cottingham and Pfaff 1986; Simerly and Swanson 1986, 1988). Key participants in this sexually dimorphic circuit represent nodal points in the neural pathways underlying reproductive function. For example, the central part of the medial preoptic nucleus is several times larger in male than in female rats (Bloch and Gorski 1988a; Dodson et al. 1988; Simerly et al. 1984). Lesions that include this discrete brain region result in diminished male sexual behavior (Hansen et al. 1982; see also Chapter 2 by Yahr, this volume). Similarly, the ventromedial nucleus of the hypothalamus is sexually dimorphic and plays a critical role in mediating female copulatory behavior (Pfaff and Schwartz-Giblin 1988). In addition, the AVPv is larger in female than in male rodents and represents a nodal point in the circuitry underlying the neural control of gonadotropin secretion (as discussed later). The identification of anatomical sexual dimorphisms in several forebrain regions of the rat suggests that sexual differentiation in these regions is induced by gonadal steroids and may relate to neural mechanisms underlying sex-specific reproductive functions, such as copulatory behavior and gonadotropin secretion (Arnold and Breedlove 1985; Barraclough 1979; Beatty 1979; Breedlove 1992; Gerall et al. 1992; Gorski 1985; Goy and McEwen 1980; McEwen 1983). However, attempts to ascribe a particular function to each sexually dimorphic cell group have met with only limited success (see Chapter 2 by Yahr, this volume). One reason that these cell groups have proved to be refractory to conventional functional analysis might lie in the complex organization of their neural circuitry. The anatomical relationships between sexually dimorphic nuclei in the rodent forebrain suggest that they comprise an integrated neural system of cell groups that develop under the influence of sex steroid hormones and also are sensitive to the effects of circulating gonadal steroids in adulthood. Although the precise developmental processes underlying the sexual differentiation of limbic and hypothalamic pathways remain unknown, an improved understanding of the importance of these pathways in the expression of sex-specific patterns of behavior is emerging.

The medial preoptic nucleus

The medial preoptic nucleus (MPN) lies at the center of the sexually dimorphic forebrain circuit and has been implicated directly in the neural control of male copulatory behavior (Hansen et al. 1982). The MPN should be viewed as a sexually dimorphic nuclear complex composed of three major parts: a cell-sparse lateral part, a cell-dense medial part (MPNm), and a compact, very cell-dense central part (MPNc) that is embedded within the MPNm (Simerly et al. 1984). Each of these subnuclei has a unique chemoarchitecture (Simerly et al. 1986) and displays a markedly different pattern of projections to regions affecting the neuroendocrine, autonomic, and somatomotor responses that are important components of sexually dimorphic functions (Simerly and Swanson 1988). The MPNc is the most dimorphic component of the MPN and sends its densest projections to other sexually dimorphic forebrain nuclei (Simerly and Swanson 1988) such as the lateral septal nucleus and the ventrolateral part of the ventromedial nucleus of the hypothalamus, both of which contain neurochemical sexual dimorphisms and are thought to influence female copulatory behavior (De Vries et al. 1985, 1986; Pfaff and Schwartz-Giblin, 1988). The MPNc also sends dense projections to the principal nucleus of the bed nuclei of the stria terminalis (BNSTp; Ju and Swanson, 1989), posterodorsal part of the medial nucleus of the amygdala (MeApd), and ventral premammillary nucleus, which have a greater number of neurotransmitter-specific neurons in male rats than in female rats and appear to be involved in the regulation of sexually dimorphic behaviors and gonadotropin secretion (Oro et al. 1988; Akesson 1993; Segovia and Guillamón 1993; see also Chapter 9 by Akesson and Micevych, this volume). The MPNm is also larger in male than in female rats, and it sends its strongest projections to regions of the hypothalamus involved in the regulation of hormone secretion from the anterior pituitary such as the AVPv, the preoptic periventricular nucleus, the arcuate nucleus, and the paraventricular nucleus (Simerly and Swanson 1988; Swanson 1986). Thus, in addition to being a nodal point in the neural pathways mediating male copulatory behavior, the MPN may influence other sexually dimorphic functions, including the control of gonadotropin release and female copulatory behavior, by virtue of its projections to sexually dimorphic nuclei that are key components of neural pathways involved in the regulation of these sex-specific functions (Fig. 4.2).

The major neural inputs to the MPN originate in other sexually dimorphic cell groups. Injection of retrograde tracer into the MPN labels numerous neurons in the ventral part of the lateral septum, the arcuate nucleus and ventrolateral part of the ventromedial nucleus of the hypothalamus, the AVPv, BNSTp, MeApd, and ventral premammillary nucleus. The MPN also receives a strong input from the posterior nucleus of the amygdala (amygdalohippocampal zone), which projects to each part of the sexually dimorphic circuit (Canteras et al. 1992a) and therefore may be sexually dimorphic itself. Each of the inputs to the MPN is distributed topographically within its three subdivisions, with projections from the AVPv, arcuate, and ventral premammillary nuclei terminating primarily in the MPNm, while the ventral part of the lateral septal nucleus and the ventrolateral part of the ventromedial hypothalamic nucleus provide inputs primarily to the lateral part of MPN (MPNl) (Canteras et al. 1992b, in press; Simerly and Swanson 1986). In addition, the BNSTp and MeApd send their densest projections to the MPNc and MPNm (Simerly et al. 1989; Simerly and Swanson 1986). This separation of inputs effectively divides afferent sensory influences between the medial (including the MPNc) and lateral components of the MPN (Fig. 4.1); thus, olfactory influences are relayed from the accessory olfactory bulb (AOB) by the MeApd and BNSTp to the MPNm including MPNc (see Chapter 1 by Wood and Newman, this volume). Other sensory influences (auditory, visual, somatosensory,

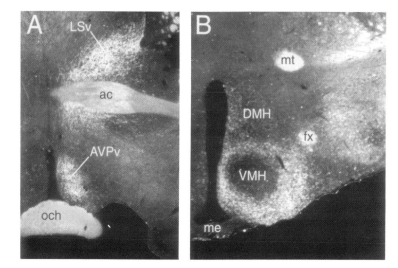

Figure 4.2. Dark-field photomicrographs illustrating the appearance and density of *Phaseolus vulgaris* leukoagglutinin-labeled projections from the MPN. Ascending projections (A) provide strong inputs to the anteroventral periventricular nucleus (AVPv) and the ventral part of the lateral septal nucleus (LSv); descending projections (B) form dense terminal fields in the arcuate nucleus and ventrolateral part of the ventromedial nucleus (VMH). Abbreviations: ac, anterior commisure; fx, fornix; me, median eminence; mt, mamillothalamic tract; och, optic chiasm (×40).

and primary olfactory) may reach the MPN via multisynaptic pathways to the ventral subiculum and the ventral part of the lateral septal nucleus, both of which project to the MPNl (Canteras and Swanson 1992; Simerly and Swanson 1986).

Hormonal modulation of gene expression in sexually dimorphic pathways

The impact of steroid hormones on these sexually dimorphic pathways is demonstrated by cellular changes that occur in two well-characterized sexually dimorphic forebrain circuits: the vomeronasal pathway, which relays olfactory information to the hypothalamus, and the AVPv, which is a critical part of the neural circuitry underlying control of gonadotropin secretion.

The vomeronasal pathway

Olfactory cues exert profound effects on several goal-oriented behaviors, such as aggressive behavior, maternal behavior, and reproduction, and experimental evidence suggests that pheromonal information affecting these functions originates in the vomeronasal organ (Johns 1986; Vandenbergh 1988). The vomeronasal organ provides primary afferent fibers to the accessory olfactory bulb, which relays this information to the MeApd and BNSTp, both of which provide strong inputs to the MPN (reviewed by Simerly 1990). Olfactory information conveyed by this pathway has been implicated in a variety of essential functions, including estrous cyclicity, regulation of male and female copulatory behavior, onset of puberty, and control of maternal behavior (reviewed by Halpern 1987; Segovia and Guillamón 1993; Wysocki 1979, 1987; Vandenbergh 1988). For example, damage to the vomeronasal system can block changes in ovulation or gonadotropin secretion that occur normally following exposure of female rats to male conspecifics and their associated odors (Johns 1986; Schachter et al. 1984). Similarly, lesions involving the AOB reduce copulatory behavior in male rats (Sachs and Meisel 1988) and increase lordosis responses in females (Pfaff and Schwartz-Giblin 1988).

Each cell group along this vomeronasal pathway is larger in male than in female rats and differentiates under the influence of circulating androgens during the perinatal period (del Abril et al. 1987; Segovia and Guillamón 1993). The MeApd and BNSTp are among those forebrain regions that contain the highest density of cells expressing estrogen and androgen receptors in adults (Pfaff and Keiner 1973; Sar and Stumpf 1975; Simerly et al. 1990), indicating that the activity of cells along this pathway could be sensitive to the regulatory effects of gonadal steroids in adulthood, as well as early in development. One possible mechanism underlying hormonal regulation of olfactory influences on normal brain function is the control of neuropeptide gene expression in neurons along the vomeronasal pathway. The neuropeptide cholecystokinin (CCK) has been implicated in the regulation of reproductive behavior and the control of gonadotropin secretion (Babcock et al. 1988; Bloch et al. 1989; Hashimoto and Kimura 1986; Kimura et al. 1987). An elegant series of studies by Li et al. (1992a, b) in mice suggests that CCK in the vomeronasal pathway mediates hormone-dependent olfactory influences on reproductive function. These authors observed that estrogen infusion into the region of the MeApd, or infusion of CCK into the medial preoptic area, increased activation of dopaminergic neurons in the arcuate nucleus following stimulation of the AOB. In addition, infusion of a CCK-B receptor antagonist into the medial preoptic area decreased the activity of arcuate dopamine neurons and blocked olfactory influences on pregnancy.

The MeApd contains a sexually dimorphic population of CCK-immunoreactive cells that differentiate under the influence of perinatal sex steroids, and intracellular levels of CCK appear to be regulated by physiological changes in circulating gonadal hormones (Micevych et al. 1988; Oro et al. 1988; Simerly and Swanson 1987a). Although a substantial percentage of CCK-immunoreactive cells co-express substance P immunoreactivity (Simerly et al. 1989), comparable changes in levels of substance P were not detected in adult animals, suggesting that circulating sex steroids might regulate differentially the expression of these co-localized

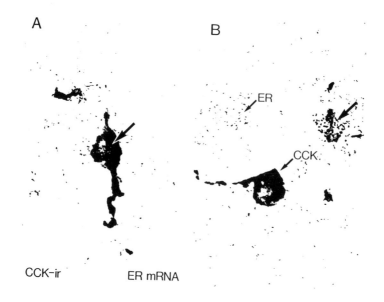

Figure 4.3. A combined immunohistochemical–*in situ* hybridization method was used to co-localize CCK immunoreactivity and ER mRNA in MeApd. Singly labeled CCK-immunoreactive (containing cobalt-intensified diaminobenzedine reaction product) and ER mRNA–containing (with above-background densities of overlying silver grains) cells are identified. Doubly labeled cells are indicated by large arrows ($\times$ 500).

neuropeptides in the MeApd. Combined immunohistochemical–*in situ* hybridization results indicate that CCK-immunoreactive neurons in the MeApd express ER mRNA (Fig. 4.3; see also Popper et al., Chapter 7, in this volume).

Because mRNA expression closely reflects neuropeptide gene regulation, *in situ* hybridization was used to examine the influence of estrogen on the levels of preproCCK (pCCK) and preprotachykinin (PPT) mRNA, which encode the preprohormones for CCK and substance P. The results of these studies (Simerly et al. 1989) were remarkably consistent with the earlier immunohistochemical data and suggest that estrogen regulates intracellular levels of CCK by regulating its biosynthesis. Comparable changes were not observed for PPT mRNA, which further suggests that circulating hormones differentially regulate pCCK and PPT gene expression, thereby altering the relative amounts of CCK and substance P co-expressed within individual neurons of the MeApd that project to the hypothalamus. If it is assumed that the synaptic release of co-expressed neuropeptides is proportional to cellular content, then a cellular model proposed by Swanson (1983) predicts that the functional impact of this release depends on the complement of receptors expressed on postsynaptic neurons, and will result in the preferential activation of neurons that express high levels of CCK receptors, relative to those that express primarily substance P receptors. Thus, the gene-specific regulation of neuropeptide expression in presynaptic neurons, and the concomitant regulation of postsynaptic receptors, could allow circulating gonadal hormones to influence the flow of sensory information through the vomeronasal pathway to regions of the hypothalamus that mediate sexually dimorphic functions such as gonadotropin secretion (Simerly 1990).

Neuroendocrine control of gonadotropin secretion

Like the vomeronasal pathway, the neural circuits that control the secretion of gonadotropin-releasing hormone (GnRH), which in turn controls the release of luteinizing hormone (LH) and ovulation, appear to be sensitive to the regulatory influences of gonadal steroid hormones during development and in adulthood. Neonatal steroids determine whether the adult pattern of gonadotropin secretion will be cyclic, producing the periodic surges of LH that lead to ovulation in female rats, or relatively constant, as occurs in male animals (Barraclough 1979; Gerall and Givon 1992). Moreover, the preovulatory surge of gonadotropin secretion typical of adult females is dependent on cyclic changes in the levels of estrogen and progesterone in the blood (Fink 1988). Only a subpopulation of GnRH neurons is sexually dimorphic, but these cells do not express estrogen receptors (Fig. 4.4; Shivers et al. 1983; Wray and Hoffman 1986). Although the hormonal regulation of GnRH gene expression remains controversial (Park et al. 1990; Roberts et al. 1989), ovarian steroids clearly regulate the release of GnRH and agents that activate cellular second messengers produce marked changes in levels of GnRH mRNA (Wray et al. 1993). Therefore, it has been proposed that the hormonal regulation of GnRH gene expression and secretion is mediated by hormone-sensitive inputs to these neurons from other regions of the forebrain, such as the arcuate nucleus or AVPv (reviewed by Fink 1988; Terasawa et al. 1980; Weiner et al. 1988).

The anteroventral periventricular nucleus

The AVPv is located just caudal to the vascular organ of the lamina terminalis and is unusual among sexually dimorphic nuclei in that it is larger in females than in males (Bleier et al. 1982; Bloch and Gorski 1988b). It appears to be a nodal point in the neural pathways regulating gonadotropin secretion since it receives massive projections from the MeApd and BNSTp (Simerly et al. 1989), and electrolytic lesions of the AVPv block spontaneous ovulation, induce persistent vaginal estrus, and abolish the preovulatory surge of LH and prolactin in female rats (Ronnekleiv and Kelly 1986; Wiegand and Terasawa 1982). The AVPv contains numerous cells that express estrogen receptors (Pfaff and Keiner 1973; Simerly et al.

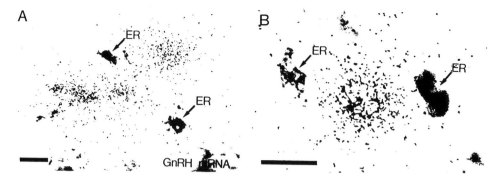

Figure 4.4. High-power photomicrographs showing the appearance and independence of labeling over GnRH mRNA–containing neurons and ER-immunoreactive cells in the rostral part of the preoptic region. GnRH mRNA–containing neurons were localized with an ^{35}S-labeled cRNA probe derived from a 331-bp Bam HI-Hind III restriction fragment of the rat hypothalamic GnRH cDNA isolated by Dr. J. Adelman (Oregon Health Sciences University). ER-Immunoreactive nuclei were stained using antisera H222 obtained from Abbott Laboratories. Scale bars, 25 mm.

1990), which is consistent with the findings that implants of estradiol (Akema et al. 1984) or antiestrogens (Peterson and Barraclough 1989) into the region of the AVPv cause dramatic changes in the level of circulating LH, and that the firing rate of neurons in the AVPv can be altered by systemic injections of estrogen (Kubo et al. 1975). In addition, the AVPv contains numerous cells that accumulate progesterone (Warembourg 1978; Warembourg et al. 1986) and express progesterone receptor mRNA (Simerly 1993). The AVPv does not contain GnRH perikarya (Simerly and Swanson 1987b; Terasawa and Davis 1983), but it appears to provide direct projections to regions of the rostral forebrain where many GnRH-containing cells are located (Simerly and Swanson 1988). The AVPv also sends projections to the arcuate nucleus, which contains dopaminergic and β-endorphin-containing neurons that are thought to participate in regulating the release of GnRH into the portal vasculature (Weiner et al. 1988). The results of retrograde transport studies indicate that this pathway is sexually dimorphic, with an approximately eightfold greater number of AVPv neurons projecting to the arcuate nucleus in female rats relative to males (Reinoso and Simerly 1991).

The AVPv contains sexually dimorphic populations of tyrosine hydroxylase (TH)-, enkephalin- (Simerly et al. 1985a, b; 1988), and dynorphin-immunoreactive neurons (Gu and Simerly, in press). Both dopamine and opioid peptides have been implicated in the neural control of gonadotropin secretion (reviewed by Gerall and Givon 1992; Weiner et al. 1988). These sexual dimorphisms are also apparent at the mRNA level, suggesting that perinatal gonadal steroids influence levels of TH, dynorphin, or enkephalin in the adult AVPv by determining the number of cells that are capable of expressing sufficient quantities of TH, prodynorphin (PDYN), or preproenkephalin (PPE) mRNA, rather than by generating sex-specific alterations of post-translational mechanisms. The AVPv contains three to four times as many TH and PDYN mRNA–containing cells in female rats than in males. These sexual dimorphisms appear to be dependent on perinatal levels of sex steroids, since orchidectomy of newborn males increases, and treatment of newborn females with testosterone decreases, the number of TH and PDYN mRNA–containing cells within the AVPv of adult animals (Simerly 1989, 1991). In contrast, the AVPv contains more enkephalinergic neurons in males than in females, and treatment of female neonates with testosterone on the day of birth results in numbers of enkephalinergic neurons in the AVPv that are typical of normal male rats (Simerly 1991; Simerly et al. 1988). Thus, postnatal exposure to androgen completely sex-reverses the development of at least three biochemically distinct populations of neurons in the AVPv and appears to exert both positive and negative effects on their development. Although an increase in cell number, such as observed for PPE mRNA–containing neurons, is generally thought to be due to enhanced survival of hormone-sensitive cells, the testosterone-induced reduction in the number of TH and PDYN neurons in the AVPv suggests that exposure to sex steroids during the perinatal period also can exert the opposite effect. The observation of enhanced neuronal loss in the AVPv during the first week of life in androgen-treated female rats (Murakami and Arai 1989) is consistent with this interpretation. Moreover, the demonstration of hormone-reversible neurochemical sexual dimorphisms at the mRNA level indicates that fundamental alterations of pretranslational mechanisms might underlie the sexual differentiation of neurotransmitter phenotype in the AVPv. However, it is not clear whether the hormone-induced changes in the number of TH, PPE, and PDYN mRNA–containing neurons are due to differential effects of sex steroids on cell survival or whether they are the result of permanent alterations in the expression of these genes.

In adult animals, circulating sex steroids appear to downregulate TH expression in the AVPv of male and female rats, indicating that testosterone and/or estrogen can exert a sustained influence on the biosynthetic activity of this sexually dimorphic population of dopaminergic cells (Simerly 1989). In contrast, estrogen induces expression of PDYN mRNA in the AVPv, but only in female rats, suggesting that these hormone-sensitive neurons might not be present in males (Simerly 1991). Recently, we have determined that levels of TH mRNA are correlated inversely with the level of plasma estradiol during the estrous cycle and that progesterone treatment has no effect on the level of TH expression in estrogen-primed females. The level of PDYN mRNA also appears to fluctuate during the estrous cycle, with a slight reduction on the afternoon of proestrus that is followed by a sharp increase on the day of estrus (A. Carr and R. Simerly, unpublished observations). Although it is possible that the high level of progesterone present during this period might contribute to the observed increase in PDYN mRNA, we did not observe that treatment of estrogen-primed (2 days) female rats with progesterone for 3 or 27 hours significantly altered PDYN expression. Thus, the increased level of PDYN mRNA observed on the day of estrus in normally cycling animals might represent a delayed response to the high level of estradiol present during proestrus, or it could reflect the regulatory influence of other cellular factors that have yet to be identified. Together, these data suggest that the hormonal regulation of TH and PDYN gene expression is both cell and tissue specific and, therefore, may be due to indirect regulation by nuclear trans-acting factors that are expressed differentially in the hypothalamus. The delayed time course of PDYN induction that is observed across the estrous cycle and the absence of consensus estrogen regulatory enhancer–like elements in the proximal 3 kb of the 5′ flanking sequence of PDYN (J. Douglass, personal communication) are also consistent with regulation by such an indirect mechanism. Alternatively, the regulation of TH and PDYN mRNA by estrogen might be mediated trans-synaptically by one of several inputs to the AVPv from other hormone-sensitive regions of the forebrain (Armstrong and Montminy 1993). Nevertheless, the induction of PDYN gene expression in the AVPv by estradiol in females, but not males, is consistent with the notion that PDYN mRNA–containing neurons in the AVPv represent a sexually differentiated component of the neural system that mediates hormonal feedback on the release of GnRH from the hypothalamus. Similarly, the inhibition of TH gene expression by estradiol during proestrus could represent the inhibition of a tonic suppressive effect of dopamine on the secretion of LH. Such postulated regulatory influences may be conveyed directly to GnRH neurons or may be mediated by neurons in the arcuate nucleus of the hypothalamus that contain β-endorphin or dopamine, which are thought to inhibit the release of GnRH and prolactin (see Neill 1980; Swanson 1986; Weiner et al. 1988). In this regard, it is notable that numerous dynorphin-immunoreactive nerve terminals appear to make synaptic contacts with dopaminergic neurons in the arcuate nucleus (Fitzsimmons et al. 1992) and that dopaminergic terminals from the region of the AVPv provide direct inputs to GnRH neurons (Hovath et al. 1993).

In contrast to the effects of sex steroids on TH and PDYN gene expression, sex steroids do not appear to exert a regulatory influence on the expression of PPE mRNA in the AVPv of either sex. While one might speculate that this sexually dimorphic population of enkephalinergic neurons represents a male-specific tonic inhibitory influence on the neural mechanisms responsible for inducing the preovulatory surge of gonadotropin secretion seen in females, this possibility remains to be studied. It is interesting that both estradiol and progesterone appear to exert regulatory influences on PPE gene expression in the ventromedial nucleus of the hypothalamus (Priest et al., in press), but only in female rats (Romano et al. 1988, 1989b; see also Hammer et al. 1993). Moreover, this regulation correlates

with changes in ENK immunoreactivity in a discrete part of the preoptic periventricular nucleus (Watson et al. 1986). These results suggest, therefore, that the
effects of sex steroids on PPE mRNA–containing neurons are regionally specific,
both during the neonatal period and in adulthood. Several other populations of
neurotransmitter-specific neurons have been identified in the AVPv and many of
these might also be sexually dimorphic (Simerly and Swanson 1987b), as indicated by the recent demonstration of sexually dimorphic populations of neurotensin and calcitonin gene-related peptide-containing neurons in the AVPv that show
distinct patterns of hormonal regulation (Alexander et al. 1989, 1991; Herbison,
1992; Herbison and Dye 1993).

Distribution of neurons that express steroid hormone receptors

In contrast to the distribution of neurons that express adrenal steroid receptors,
neurons that express gonadal steroid receptors are restricted largely to discrete
brain regions, with the highest densities located in the hypothalamus or in parts of
the limbic system that project to the hypothalamus (reviewed by Simerly 1993).
Moreover, there is considerable overlap between the distributions of neurons that
express androgen (AR), estrogen (ER), and progesterone (PR) receptors. The
MPN is a major target of gonadal hormone action, as indicated by the high densities of neurons that concentrate gonadal steroids (Pfaff and Keiner 1973; Sar and
Stumpf 1975) and neurons that express ER, AR, and PR mRNA or immunoreactivity (Blaustein et al. 1988; Clancy et al. 1992; Fuxe et al. 1987; Simerly 1993;
Simerly et al. 1990; Warembourg et al. 1986). The neurons that express these
receptors are distributed differentially within the MPN (Micevych 1991; Simerly
1993; Simerly et al. 1990), suggesting that each subdivision may be affected
differently by circulating hormones. Moreover, the sexually dimorphic regions
that supply strong afferent innervation to the MPN are among the regions of the
forebrain that contain the highest density of hormone-sensitive neurons (Pfaff and
Keiner 1973; Sar and Stumpf 1975; Simerly et al. 1990). Thus, the AVPv, arcuate nucleus, ventrolateral part of the ventromedial hypothalamic nucleus (VMH),
ventral premammillary nucleus, ventral part of the lateral septal nucleus, BNSTp,
MeApd, and posterior nucleus of the amygdala are among those regions that
contain the highest density of hormone-sensitive cells as well as the highest density of cells that project to the MPN. Therefore, the activity of cells within the
MPN can be modulated directly, by hormonal feedback acting through ER and
AR expressed within its cells, and indirectly, by steroid-sensitive information
relayed trans-synaptically to it through afferent pathways.

However, gonadal steroid receptors are not expressed uniformly throughout the
sexually dimorphic circuit. The highest densities of neurons that express ER (Fig.
4.5) and PR (Fig. 4.6) mRNA are found in regions involved in the regulation of

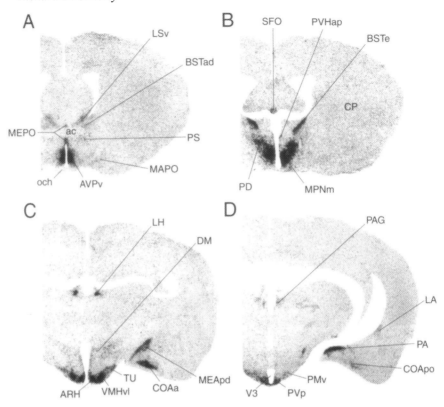

Figure 4.5. Autoradiographic images (8-day exposure) showing the distribution of ER mRNA expression in the rat brain. Abbreviations: ac, anterior commissure; BSTad, anterodorsal nucleus (n.) of bed nuclei of the stria terminalis; BSTe, encapsulated (principal) n. of bed nuclei of the stria terminalis; COa, cortical n. of the amygdala (anterior part); COApo, cortical n. of the amygdala (posterior part); CP, caudoputamen; DM, dorsomedial hypothalamic n.; LA, lateral n. of the amygdala; LH, lateral habenula; MAPO, magnocellular preoptic n.; MEPO, median preoptic n.; PAG, periaqueductal gray; PD, posterodorsal preoptic n.; PS, parastrial n.; PVHap, paraventricular n. of the hypothalamus (anterior parvicellular part); PVp, posterior periventricular n.; och, optic chiasm; SFO, subfornical organ; TU, tuberal n.; V3, third ventricle; see also legend to Figure 4.1.

gonadotropin secretion from the anterior pituitary, such as the AVPv, MPNm, and arcuate nucleus (Simerly 1993; Simerly et al. 1990). Neurons that express AR mRNA (Fig. 4.7) appear to predominate in the central part of the MPN and ventral premammillary nucleus, providing feedback sites for androgens to act on information received from, and projecting to, the medial and posterior nuclei of the amygdala (Canteras et al. 1992a,b; Micevych 1991; Simerly and Swanson 1986). In addition AR, ER, and PR are expressed differentially across the subdivisions of the VMH (Romano et al. 1989a; Simerly 1993; Simerly et al. 1990). Most of the neurons that express ER and PR are localized to the ventrolateral part of the VMH, while cells that express the AR are concentrated in its dorsomedial

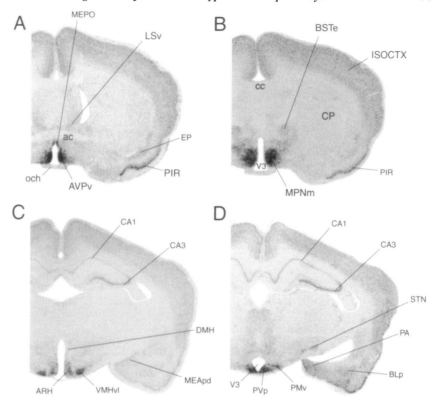

Figure 4.6. Autoradiographic images (8-day exposure) showing the distribution of PR mRNA expression in the forebrain of the rat. Abbreviations: ac, anterior commissure; AVPv, anteroventral periventricular nucleus (n.); ARH, arcuate n. of the hypothalamus; CA1, CA3, fields of Ammon's horn; cc, corpus callosum; DMH, dorsomedial hypothalamic n.; EP, endopiriform n.; ISOCTX, isocortex; LSv, lateral septal n. (lateral part); MEApd, medial n. of the amygdala (posterodorsal part); MEPO, median preoptic n.; MPNm, medial preoptic n. (medial part); och, optic chiasm; PIR, piriform cortex; STN, subthalamic n.; VMHvl, ventromedial n. of the hypothalamus (ventrolateral part); V3, third ventricle.

part. Since the ventrolateral part of the nucleus is thought to participate in mediating copulatory responses in female rats, the localization of estrogen- and progesterone-sensitive cells to this cell group is consistent with a predominant role for the ER and PR in this function (Etgen 1984). Similarly, the preferential expression of the AR by neurons in the dorsomedial part of the VMH suggests that projections from this nucleus to regions involved in the initiation of feeding and other behaviors might be modulated by androgen (reviewed by Canteras et al. 1994). In the telencephalon, the greatest density of neurons that express sex steroid receptors are found in those parts of the amygdala and septum that project to the medial and periventricular zones of the hypothalamus. Although both types of receptor appear to be expressed abundantly in these regions, in general AR-con-

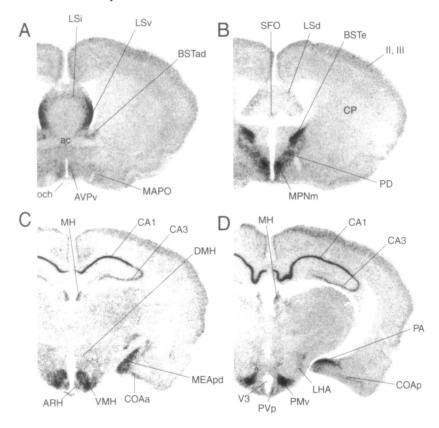

Figure 4.7. Autoradiographic images (8-day exposure) showing the distribution of AR mRNA expression in the rat brain. Abbreviations: ac, anterior commissure; BSTad, ante-rodorsal nucleus (n.) of bed nuclei of the stria terminalis; BSTe, encapsulated (principal) n. of bed nuclei of the stria terminalis; COa, cortical n. of the amygdala (anterior part); COAp, cortical n. of the amygdala (posterior part); CP, caudoputamen; DMH, dor-somedial hypothalamic n.; LHA, lateral hypothalamic area; LSd, lateral septal n. (dorsal part); LSi, lateral septal n. (intermediate part); MH, medial habenula; MAPO, magno-cellular preoptic n.; PD, posterodorsal preoptic n.; PVp, posterior periventricular n.; och, optic chiasm; SFO, subfornical organ; V3, third ventricle; II, III, layers of isocortex; see also legend to Figure 4.1.

taining neurons seem to predominate. Thus, the MeApd, BNSTp, and ventral part of the lateral septal nucleus all appear to contain more neurons that express AR mRNA than ER mRNA (Simerly et al. 1990). PR mRNA is also expressed by neurons in the MeA (Romano et al. 1989a; Simerly 1993), but its functional significance remains unexplored.

Sex steroid hormone receptors are expressed at low levels early in develop-ment, but the levels increase during the perinatal period (Lieberburg et al. 1980; MacLusky and Brown 1989; Meaney et al. 1985; Toyooka et al. 1989; Vito and Fox 1982), and sex differences in receptor expression have been reported in sev-eral forebrain nuclei of adult rats. Thus, the levels of AR binding and AR mRNA expression appear to be greater in the MeApd, BNSTp, preoptic periventricular nucleus, and VMH in male than in female rats (Roselli 1991; Simerly et al. 1990). In contrast, the levels of ER and PR binding and mRNA expression in the ventromedial hypothalamic nucleus, preoptic periventricular nucleus, and MPN appear to be higher in females than in males (Brown et al. 1987, 1988; Lauber et al. 1991a,b; Rainbow et al. 1982; Shughrue et al. 1992; Simerly et al. 1990). This may be because the levels of circulating steroids differ between male and female rats and sex differences in hormone receptor expression in intact animals may be due to differential regulation of gene expression.

Sex differences in the number of neurons that express sex hormone receptors and differences in receptor levels within individual cells in sexually differentiated pathways provide an additional level of cellular regulation that may contribute to the expression of sex-specific behaviors. In addition, sexually dimorphic nuclei in the rodent forebrain typically display high levels of activity for the enzyme aromatase, which is responsible for the conversion of testosterone to estradiol (see Chapter 13 by Schlinger and Arnold, this volume). High levels of aromatase activity in the MeApd, BNSTp, MPN, AVPv, and VMH indicate that ER might play a particularly important role in the activation of cellular function in these regions. Moreover, levels of aromatase activity are significantly greater in males and are increased by androgen in the BNSTp, periventricular preoptic area, MPN, and VMH of adult animals (Roselli and Resko 1993). Thus, sexually dimorphic patterns of hormone receptor expression might be augmented by regionally specific, steroid-sensitive sex differences in steroid metabolism.

Regulation of hormone receptor gene expression

The regulation of sex steroid receptor expression by circulating sex hormones influences the overall sensitivity of sexually dimorphic pathways to neural activation. In both male and female rats, circulating gonadal steroid hormones appear to upregulate the level of AR mRNA expression in the MeApd and downregulate cellular levels of ER mRNA expression in the MeApd and AVPv (Simerly 1993). Therefore, hormonal control of biosynthesis of the receptors that regulate pCCK mRNA expression in the vomeronasal pathway, or TH and PDYN mRNA expression in the AVPv, provides a possible feedback mechanism through which circulating hormones autoregulate neuropeptide gene expression. However, these regulatory influences need not be constant throughout the life of the animal. For example, we have observed that although gonadectomy increases AR mRNA expression and decreases ER mRNA expression in the MeAp acutely, the levels of both were significantly reduced 28 days after gonadectomy (Simerly 1993). Thus, differential regulatory effects of sex steroids on the expression of their receptors can occur over time, and might be particularly important during key developmen-

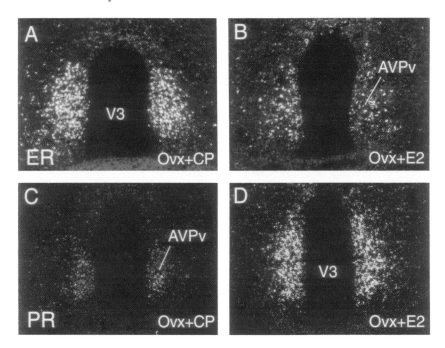

Figure 4.8. Dark-field photomicrographs illustrating suppression of ER mRNA (A and B) and induction of PR mRNA (C and D) in the AVPv of an adult female rat as seen with *in situ* hybridization. Ovx, ovariectemized; CP, control pellet; E2, estradiol.

tal periods such as puberty, neurogenesis, and the critical period for sexual differentiation. Similarly, estrogen upregulates the expression of PR mRNA in the VMH, MPN, and AVPv (Fig. 4.8; Park-Sarge et al. 1992; Romano et al. 1989a), which could represent cellular events underlying estrogen-dependent effects of progesterone on female copulatory behavior and gonadotropin secretion, respectively.

Although the overall distributions of steroid hormone receptors and the patterns of autologous regulation are similar in male and female animals, sex-related differences in the magnitude and temporal organization of regulatory responses have been demonstrated. The induction of AR mRNA expression in the MeApd is more pronounced in male rats and is maintained for a longer period than in female rats (Simerly 1993). The downregulation of ER mRNA expression in the ventrolateral part of the VMH displays a different time course in females (Simerly and Young 1991), and the suppression of ER mRNA expression and induction of PR mRNA expression in the arcuate nucleus are detectable only in female rats (Lauber et al. 1991b; Simerly and Young 1991).

The regulation of steroid hormone receptor gene expression may be regionally specific, having the effect of enhancing the impact of changes in circulating hormones in some regions, while dampening them in others. In contrast to the upregulation of AR mRNA expression observed in the MeApd in response to testosterone (Simerly 1993), similar treatments appear to have the opposite effect in the MPN (Burgess and Handa 1993). Moreover, in cycling female animals, differences in the pattern of hormonal regulation of receptor mRNA have been observed in components of the sexually dimorphic circuit. For instance, levels of ER mRNA expression are suppressed in the MPN on the day of proestrus, yet appear to be elevated in the VMH during the same period (Shughrue et al. 1992). At present, a clear understanding of the cellular mechanisms underlying such regionally specific regulation is lacking, but a combination of trans-acting factors, which may exert synergistic or antagonistic effects on gene expression by acting on multiple enhancer elements in the promoter region of hormone-sensitive genes, is probably involved. Alternatively, such cell-specific patterns of regulation might be due to complex protein–protein interactions between factors, such as those demonstrated for corticosteroid receptors and the proto-oncogenes c-*fos* and c-*jun* (Diamond et al. 1990; Pearce and Yamamoto 1993). In addition, the "A" form of the PR appears to act as a repressor of the "B" form independent of DNA binding (Tung et al. 1993; Vegeto et al. 1993), suggesting that similar interactions might occur between receptor subtypes. Furthermore, the expression of secondary transcription factors may display a complex temporal organization (e.g., high levels during the perinatal period, low levels during puberty), thereby controlling the regulatory influences of circulating hormones on the expression of their cognate receptors during discrete critical periods. In addition to genomic regulatory mechanisms, levels of steroid hormones might be responsive to other signals. For example, steroid hormone receptor binding can be altered by neural influences that, presumably, are mediated by synaptic transmission (Black et al. 1986); removal of the olfactory bulbs increases estrogen binding in the MeApd (McGinnis et al. 1985), and estrogen binding in the hypothalamus is altered by deafferentation (Carrillo and Sheridan, 1980; McGinnis et al. 1982). In addition, similar trans-synaptic mechanisms might underlie the regulatory effects of pharmacological manipulations of catecholaminergic pathways on levels of estrogen binding in the hypothalamus (reviewed by Blaustein and Olster 1989). These observations are consistent with recent *in vitro* data which suggest that expression of steroid hormone receptors undergoes ligand-independent regulation (Power et al. 1991). Clearly, a great deal remains to be elucidated before we can fully appreciate the molecular mechanisms underlying hormonal regulation of neural function and development. However, at present it is clear that gonadal steroid hormones control the development of neural circuits in the mammalian forebrain and not only regulate the activity of neurons that express steroid hormone receptors, but also influence the expression of these receptors, thereby fine-tuning the responsiveness of hormone-sensitive pathways.

Conclusions

Limbic and hypothalamic regions play essential roles in ensuring the survival of mammalian species by coordinating the expression of sexual and maternal behaviors. In order to accomplish this important function, information from endocrine, visceral, and somatomotor systems must be integrated with sensory cues from the environment to produce appropriate adaptive responses (see Swanson 1987). The intimate connections between the hypothalamus and the pituitary gland provide an effective means of coordinating stimulus-specific behavior with appropriate endocrine responses. Moreover, the accepted role of limbic structures in memory processes suggests that these regions are involved in associating reflex actions with learned behaviors. Since these responses are generally different in male and female animals, it is not surprising that the neural substrates mediating these behaviors are sexually differentiated. However, the widespread occurrence of sex differences in the mammalian forebrain is surprising. Although additional sex differences will probably continue to be identified, a sexually dimorphic neural system of interconnected nuclei has been defined in the rodent forebrain that is involved in coordinating appropriate reproductive behaviors with the internal endocrine state of the animal in response to sex-specific sensory cues. Each component of this sexually dimorphic circuit appears to develop under the influence of sex steroid hormones secreted during the perinatal period, and each region contains a large number of neurons that express receptors for these same hormones in adulthood.

Because of the high degree of interconnection between elements of this sexually dimorphic circuit, it is difficult to assign particular functions to any single component. Similarly, it is difficult to interpret clearly the functional impact of a particular neural sex difference, since the final effect of these complex pathways depends on hodological relationships with other sexually dimorphic structures, interrelated cellular signaling pathways, and fluctuating regulatory influences caused by changes in levels of circulating hormones. Nevertheless, certain patterns have emerged that may prove useful in clarifying the functional organization of neural circuits controlling the expression of sexually dimorphic functions. Sexual dimorphisms in the brain tend to be relative and quantitative rather than absolute; they do not depend on the unique presence or absence of a morphological or biochemical feature in one sex. Moreover, the basic connections between the sexually dimorphic regions appear to be similar, but are quantitatively different in males and females. It is reasonable to assume that neural pathways derived from structures that are larger in males will have a greater impact on regions receiving afferent connections from such a pathway in males than in females. For example, the greater number of neurons in the MeApd and BNSTp in males should have a more pronounced impact on the activities of neurons in the MPN and ventromedial hypothalamic nuclei in male rats, consistent with the proposed stimulatory role of the MeApd and BNSTp on male copulatory behavior and their inhibitory influence on female copulatory behavior (Micevych and Ulibarri 1992; Pfaff and Schwartz-Giblin 1988; Sachs and Meisel 1988). Similarly, the greater number of neurons in the AVPv of female rats would lead one to predict that the regulatory impact of these cells on the activity of neurosecretory cells would be greater in females. However, given the high degree of convergence and bidirectionality that characterize connections between sexually dimorphic nuclei, and the accompanying sex differences in the ratio of presynaptic to postsynaptic neurons in sexually dimorphic pathways, these predictions must be verified experimentally.

Sensory information appears to reach sexually dimorphic nuclei in the hypothalamus by different routes, although connections between these pathways suggest a certain amount of convergence. Thus, olfactory stimuli are relayed from the AOB to the hypothalamus by the MeApd and BNSTp, with major terminal fields found in the AVPv, MPNm, MPNc, ventral premammillary nucleus, and ventrolateral part of the VMH. As reviewed earlier, these projections indicate that olfactory stimuli will have a profound effect on neural pathways mediating reproductive behavior and gonadotropin secretion. Other sensory information, including primary olfactory, auditory, visual, and somatosensory stimuli, reach the hypothalamus by way of the ventral subiculum and its projections to the ventral part of the lateral septal nucleus. However, in contrast to the more medially directed olfactory pathways, projections from the lateral septum primarily affect the lateral parts of the hypothalamus that are involved in modulating behavioral state, a suggestion consistent with the observed impact of these sensory modalities on reproductive behavior (Larsson 1979; Swanson 1987). Therefore, both physiological and anatomical findings indicate that olfaction is the major sensory modality influencing the sexually dimorphic circuit. Although there is a convergence of olfactory information transmitted by the vomeronasal system with other sensory information at the level of the MPN, its topographical arrangement indicates that these sensory influences are largely conveyed to different parts of the forebrain.

The sexually dimorphic nuclei that receive olfactory information, either directly from the MeApd and BNSTp or from the MPN, contain the highest density of neurons that express gonadal steroid receptors. Thus, it is likely that the impact of olfactory stimuli on the activity of these regions is modulated by changes in circulating levels of steroid hormones. Moreover, the hormonal amplification of sensory information reaching nodal points in the neural circuits mediating reproductive function may complement, or diminish, the direct action of the same hormones on the activity of sexually dimorphic pathways. The functional result of this neurohumeral integration is an increased probability that appropriate preprogrammed patterns of reproductive function will be generated in response to specific sensory cues. Thus, the influence of perinatal steroid hormones on the architecture of sexually dimorphic sensory pathways may have lasting effects on the display of reproductive functions, and by altering the impact of specific stimuli and sensory activation during development the expression of sexually dimorphic functions appears to be modulated dynamically by the same hormones in adulthood.

Despite considerable progress in characterizing sex differences in the mammalian brain, we currently understand very little about the cellular mechanisms underlying their development. Gonadal steroid hormones can both promote and hinder the survival of neurons, induce and suppress gene expression, or increase and decrease the density of neural connections (see Chapter 17 by Toran-Allerand, this volume). Whether steroids utilize distinct signal transduction pathways to exert these divergent effects on neurons remains to be determined, but it is likely that complex interactions between steroid hormone receptors and other transcription factors, or hormonal regulation of second-messenger systems, underlie at least some of the observed events. Equally important is the definition of the cellular mechanisms regulating the temporal organization of developmental processes, for it is possible that transient changes in the expression of transcription factors, growth factor receptors, or neurotransmitter receptors define windows of enhanced or diminished sensitivity to steroid hormone action. Together with a detailed dissection of the neural basis of sexually dimorphic functions, answers to these questions will contribute to our appreciation of sex steroid hormones as powerful trophic substances with profound influences on neural structure and function.

Acknowledgments

I thank Dr. L.W. Swanson for many illuminating discussions and support during much of the work described here. I am also grateful for expert technical assistance provided by L. McCall, B. Young, and A. Carr. Work in the author's laboratory is supported by NIH Grants NS26723, MH49234, and RR00163 and by the Alfred P. Sloan Foundation. This is ORPRC publication number 1948.

References

Akema, T., Tadakoro, Y., and Kimura, F. (1984). Regional specificity in the effect of estrogen implantation within the forebrain on the frequency of pulsatile luteinizing hormone secretion in the ovariectomized rat. *Neuroendocrinology* 39: 517–523.

Akesson, T. R. (1993). Androgen concentration by a sexually dimorphic population of tachykinin-immunoreactive neurons in the rat ventral premammillary nucleus. *Brain Res.* 608: 319–323.

Akesson, T. R., Simerly, R. B., and Micevych, P. E. (1988). Estrogen concentrating hypothalamic and limbic neurons project to the medial preoptic nucleus. *Brain Res.* 451: 381–385.

Alexander, M. J., Dobner, P. R., Miller, M. A., Bullock, B. P., Dorsa, D. M., and Leeman, S. E. (1989). Estrogen induces neurotensin/neuromedin N messenger ribonucleic acid in a preoptic nucleus essential for the preovulatory surge of luteinizing hormone in the rat. *Endocrinology* 125: 2111–2117.

Alexander, M. J., Kiraly, Z. J., and Leeman, S. E. (1991). Sexually dimorphic distribution of neurotensin/neuromedin N mRNA in the rat preoptic area. *J. Comp. Neurol.* 311: 84–96.

Arai, Y. (1981). Synaptic correlates of sexual differentiation. *Trends Neurosci.* 4: 291–293.

Arai, Y., Matsumoto, A., and Nishizuka, M. (1986). Synaptogenesis and neuronal plasticity to gonadal steroids: Implications for the development of sexual dimorphism in the neuroendocrine brain. *Curr. Top. Neuroendocrinol.* 7: 291–307.

Armstrong, R. C., and Montminy, M. R. (1993). Transsynaptic control of gene expression. *Annu. Rev. Neurosci.* 16: 17–29.

Arnold, A. P., and Breedlove, S. M. (1985). Organizational and activational effects of sex steroids on brain and behavior: A reanalysis. *Horm. Behav.* 19: 469–498.

Arnold, A. P., and Gorski, R. A. (1984). Gonadal steroid induction of structural sex differences in the central nervous system. *Annu. Rev. Neurosci.* 7: 413–442.

Arnold, A. P., and Jordan, C. L. (1988). Hormonal organization of neural circuits. In *Frontiers in Neuroendocrinology*, Vol. 10, ed. L. Martini and W. F. Ganong, pp. 185–214. New York: Raven Press.

Babcock, A. M., Bloch, G. J., and Micevych, P. E. (1988). Injections of cholecystokinin into the ventromedial hypothalamic nucleus inhibit lordosis behavior in the rat. *Physiol. Behav.* 43: 195–199.

Barraclough, C. A. (1979). Sex differentiation of cyclic gonadotropin secretion. In *Advances in the Biosciences*, Vol. 25, ed. A. M. Kaye and M. Kaye, pp. 433–450. Oxford: Pergamon Press.

Beatty, W. W. (1979). Gonadal hormones and sex differences in nonreproductive behaviors in rodents: Organizational and activational influences. *Horm. Behav.* 12: 112–163.

Black, I. B., Adler, J. E., Dreyfus, C. F., Friedman, W. J., LaGamma, E. F., and Roach, A. H. (1987). Experience, neurotransmitter plasticity, and behavior. In *Psychopharmacology: The Third Generation of Progress*, ed. H. Y. Meltzer, pp. 463–469. New York: Raven Press.

Black, I. B., Adler, J. E., and LaGamma, E. F. (1986). Impulse activity differential regulates co-localized transmitters by altering messenger RNA levels. *Prog. Brain Res.* 68: 121–127.

Blaustein, J. D., King, J. C., Toft, D. O., and Turcotte, J. (1988). Immunocytochemical localization of estrogen-induced progestin receptors in guinea pig brain. *Brain Res.* 474: 1–15.

Blaustein, J. D., and Olster, D. H. (1989). Gonadal steroid hormone receptors and social behaviors. In *Advances in Comparative and Environmental Physiology*, Vol. 3, ed. J. Balthazart, pp. 31–104. Berlin: Springer.

Bleier, R., Byne, W., and Siggelkow, I. (1982). Cytoarchitectonic sexual dimorphisms of the medial preoptic and anterior hypothalamic areas in guinea pig, rat, hamster, and mouse. *J. Comp. Neurol.* 212: 118–130.

Bloch, G. J., Dornan, W. A., Babcock, A. M., Gorski, R. A., and Micevych, P. E. (1989). Effects of site-specific CNS microinjection of cholecystokinin on lordosis behavior in the male rat. *Physiol. Behav.* 46: 725–730.

Bloch, G. J., and Gorski, R. A. (1988a). Cytoarchitectonic analysis of the SDN–POA of the intact and gonadectomized rat. *J. Comp. Neurol.* 275: 604–612.

Bloch, G. J., and Gorski, R. A. (1988b). Estrogen/progesterone treatment in adulthood affects the size of several components of the medial preoptic area in the male rat. *J. Comp. Neurol.* 275: 613–622.

Breedlove, S. M. (1986). Cellular analyses of hormone influence on motoneuronal development and function. *J. Neurobiol.* 17: 157–176.

Breedlove, S. M. (1992). Sexual dimorphism in the vertebrate nervous system. *J. Neurosci.* 12: 4133–4142.

Brown, T. J., Clark, A. S., and MacLusky, N. J. (1987). Regional sex differences in progestin receptor induction in the rat hypothalamus: Effects of various doses of estradiol benzoate. *J. Neurosci.* 7: 2529–2536.

Brown, T. J., Hochberg, R. B., Zielinski, J. E., and MacLusky, N. J. (1988). Regional sex differences in cell nuclear estrogen-binding capacity in the rat hypothalamus and preoptic area. *Endocrinology* 123: 1761–1770.

Burgess, L. H., and Handa, R. J. (1993). Hormonal regulation of androgen receptor mRNA in the brain and anterior pituitary gland of the male rat. *Mol. Brain Res.* 19: 31–38.

Canteras, N. S., Simerly, R. B., and Swanson, L. W. (1992a). Connections of the posterior nucleus of the amygdala. *J. Comp. Neurol.* 324: 143–179.

Canteras, N. S., Simerly, R. B., and Swanson, L. W. (1992b). Projections of the ventral premammillary nucleus. *J. Comp. Neurol.* 324: 195–212.

Canteras, N. S., Simerly, R. B., and Swanson, L. W. (1994). Projections of the hypothalamic ventromedial nucleus and the adjacent ventromedial lateral hypothalamic area. *J. Comp. Neurol.* 348: 41–79.

Canteras, N. S., and Swanson, L. W. (1992). Projections of the ventral subiculum to the amygdala, septum, and hypothalamus: A PHAL anterograde tract-tracing study in the rat. *J. Comp. Neurol.* 324: 180–194.

Carrillo, A. J., and Sheridan, P. J. (1980). Estrogen receptors in the medial basal hypothalamus of the rat following complete hypothalamic deafferentation. *Brain Res.* 186: 157–164.

Clancy, A. N., Bonsall, R. W., and Michael, R. P. (1992). Immunohistochemical labeling of androgen receptors in the brain of rat and monkey. *Life Sci.* 50: 409–417.

Cottingham, S. L., and Pfaff, D. (1986). Interconnectedness of steroid hormone-binding neurons: Existence and implications. In *Current Topics in Neuroendocrinology*, Vol. 7, ed. D. Ganten and D. W. Pfaff, pp. 223–240. Berlin: Springer.

del Abril, A., Segovia, S., and Guillamón, A. (1987). The bed nucleus of the stria terminalis in the rat: Regional sex differences controlled by gonadal steroids early after birth. *Dev. Brain Res.* 32: 295–300.

DeVoogd, T. J. (1991). Endocrine modulation of the development and adult function of the avian song system. *Psychoneuroendocrinology* 16: 41–66.

De Vries, G. J. (1990). Sex differences in neurotransmitter systems. *J. Neuroendocrinol.* 2: 1–13.

De Vries, G. J., Buijs, R. M., Van Leeuwen, F. W., Caffé, A. R., and Swaab, D. F. (1985). The vasopressinergic innervation of the brain in normal and castrated rats. *J. Comp. Neurol.* 233: 236–254.

De Vries, G. J., Duetz, W., Buijs, R. M., Van Heerikhuize, J, and Vreeburg, J. T. M. (1986). Effects of androgens and estrogens on the vasopressin and oxytocin innervation of the adult rat brain. *Brain Res.* 399: 296–302.

Diamond, M. I., Miner, J. N., Yoshinaga, S. K., and Yamamoto, K. R. (1990). Transcription factor interactions: Selectors of positive or negative regulation from a single DNA element. *Science* 249: 1266–1272.

Dodson, R. E., Shryne, J. E., and Gorski, R. A. (1988). Hormonal modification of the number of total and late-arising neurons in the central part of the medial preoptic nucleus of the rat. *J. Comp. Neurol.* 275: 623–629.

Etgen, A. M. (1984). Progestin receptors and the activation of female reproductive behavior: A critical review. *Horm. Behav.* 18: 411–430.

Evans, R. M. (1988). The steroid and thyroid hormone receptor superfamily. *Science* 240: 889–895.

Fink, G. (1988). Gonadotropin secretion and its control. In *The Physiology of Reproduction*, ed. E. Knobil and J. Neill, pp. 1349–1376. New York: Raven Press.

Fitzsimmons, M. D., Olschowka, J. A., Wiegand, S. J., and Hoffman, G. E. (1992). Interaction of opioid peptide-containing terminals with dopaminergic perikarya in the rat hypothalamus. *Brain Res.* 581: 10–18.

Fuxe, K., Cintra, A., Agnati, L. F., Härfstrand, A., Wikström, A.-C., Okret, S., Zoli, M., Miller, L. S., Greene, J. L., and Gustafsson, J.-A. (1987). Studies on the cellular localization and distribution of glucocorticoid receptor and estrogen receptor immunoreactivity in the central nervous system of the rat and their relationship to the monoaminergic and peptidergic neurons of the brain. *J. Steroid Biochem.* 27: 159–170.

Gerall, A. A., and Givon, L. (1992). Early androgen and age-related modifications in female rat reproduction. In *Handbook of Behavioral Neurobiology*, ed. A. A. Gerall, H. Moltz, and I. L. Ward, pp. 313–354. New York: Plenum Press.

Gerall, A. A., Moltz, H., and Ward, I. L. (eds)(1992). Handbook of Behavioral Neurobiology, Vol. 11: *Sexual Differentiation*. New York: Plenum Press.

Gorski, R. A. (1985). The 13th J. A. F. Stevenson Memorial Lecture: Sexual differentiation of the brain – possible mechanisms and implications. *Can. J. Physiol. Pharmacol.* 63: 577–594.

Goy, R. W. and McEwen, B. S. (eds.)(1980). *Sexual Differentiation of the Brain*. Cambridge, Mass.: MIT Press.

Gu, G., and Simerly, R. (in press). *Hormones and Behavior*.

Halpern, M. (1987). The organization and function of the vomeronasal system. *Annu. Rev. Neurosci.* 10: 325–362.

Hammer, R. P., Bogic, L. and Handa, R. J. (1993). Estrogenic regulation of proenkephalin mRNA expression in the ventromedial hypothalamus of the adult male rat. *Mol. Brain Res.* 19:129–134.

Hansen, S., Köhler, C., Goldstein, M., and Steinbusch, H. V. M. (1982). Effects of ibotenic acid-induced neuronal degeneration in the medial preoptic area and the lateral hypothalamic area on sexual behavior in the male rat. *Brain Res.* 239: 213–232.

Harlan, R. E. (1988). Regulation of neuropeptide gene expression by steroid hormones. *Mol. Neurobiol.* 2: 183–200.

Harris, G. W., and Levine, S. (1965). Sexual differentiation of the brain and its experimental control. *J. Physiol.* 181: 379–400.

Harrison, N. L., Majewska, M. D., Meyers, D. E. R., and Barker, J. L. (1989). Rapid actions of steroids on CNS neurons. In *Neural Control on Reproductive Function*, ed. J. M. Lakoski, J. R. Perez-Polo, and D. K. Rassin, pp. 137–166. New York: Liss.

Hashimoto, R., and Kimura, F. (1986). Inhibition of gonadotropin secretion induced by cholecystokinin implants in the medial preoptic area by the dopamine receptor blocker, pimozide, in the rat. *Neuroendocrinology* 42: 32–37.

Herbison, A. E. (1992). Identification of a sexually dimorphic neural population immunoreactive for calcitonin gene-related peptide (CGRP) in the rat medial preoptic area. *Brain Res.* 591: 289–295.

Herbison, A. E., and Dye, S. (1993). Perinatal and adult factors responsible for the sexually dimorphic calcitonin gene-related peptide-containing cell population in the rat preoptic area. *Neuroscience* 54: 991–999.

Hovath, T. H., Naftolin, F., and Leranth, C. (1993). Luteinizing hormone-releasing hormone and gamma-aminobutyric acid neurons in the medial preoptic area are synaptic targets of dopamine axons originating in anterior periventricular areas. *J. Neuroendocrinol*. 5: 71–79.

Johns, M. A. (1986). The role of the vomeronasal organ in behavioral control of reproduction. *Ann. N.Y. Acad. Sci*. 474: 148–157.

Ju, G., and Swanson, L. W. (1989). Studies on the cellular architecture of the bed nuclei of the stria terminalis in the rat, I: Cytoarchitecture. *J. Comp. Neurol*. 280: 587–602.

Kelly, M. J. (1982). Electrical effects of steroids in neurons. In *Hormonally Active Brain Peptides*, ed. K. W. McKerns and V. Pantic, pp. 253–277. New York: Plenum Press.

Kimura, F., Mitsugi, N., Arita, J., Akema, T., and Yoshida, K. (1987). Effects of preoptic injections of gastrin, cholecystokinin, secretin, vasoactive intestinal peptide and PHI on the secretion of luteinizing hormone and prolactin in ovariectomized estrogen-primed rats. *Brain Res*. 410: 315–322.

Kubo, K., Gorski, R. A., and Kawakami, M. (1975). Effects of estrogen on neuronal excitability in the hippocampal–septal–hypothalamic system. *Neuroendocrinology* 18: 176–191.

Larsson, K. (1979). Features of the neuroendocrine regulation of masculine sexual behavior. In *Endocrine Control of Sexual Behavior*, ed. C. Beyer, pp. 77–163. New York: Raven Press.

Lauber, A. H., Mobbs, C. V., Muramatsu, M., and Pfaff, D. W. (1991a). Estrogen receptor messenger RNA expression in rat hypothalamus as a function of genetic sex and estrogen dose. *Endocrinology* 129: 3180–3186.

Lauber, A. H., Romano, G. J., and Pfaff, D. W. (1991b). Sex difference in estradiol regulation of progestin receptor mRNA in rat mediobasal hypothalamus as demonstrated by in situ hybridization. *Neuroendocrinology* 53: 608–613.

Li, C. S., Kaba, H., Saito, H., and Seto, K. (1992a). Oestrogen infusions into the amygdala potentiate excitatory transmission from the accessory olfactory bulb to tuberoinfundibular arcuate neurones in the mouse. *Neurosci. Lett*. 143: 48–50.

Li, C. S., Kaba, H., Saito, H., and Seto, K. (1992b). Cholecystokinin: Critical role in mediating olfactory influences on reproduction. *Neuroscience* 48: 707–713.

Lieberburg, I., MacLusky, N., and McEwen, B. S. (1980). Androgen receptors in the perinatal rat brain. *Brain Res*. 196: 125–138.

MacLusky, N. J., and Brown, T. J. (1989). Control of gonadal steroid receptor levels in the developing brain. In *Neural Control of Reproductive Function*, ed. J. M. Lakoski, J. R. Perez-Polo, and D. K. Rassin, pp. 45–59. New York: Liss.

Matsumoto, A. (1992). Hormonally induced synaptic plasticity in the adult neuroendocrine brain. *Zoo. Sci*. 9: 679–695.

McEwen, B. S. (1983). Gonadal steroid influences on brain development and sexual differentiation. In *Reproductive Physiology IV: International Review of Physiology*, Vol. 27, ed. R. O. Greep, pp. 99–145. Baltimore: University Park Press.

McEwen, B. S., Coirini, H., Westlind Danielsson, A., Frankfurt, M., Gould, E., Schumacher, M., and Woolley, C. (1991). Steroid hormones as mediators of neural plasticity. *J. Steroid. Biochem. Mol. Biol*. 39: 223–232.

McGinnis, M. Y., Phelps, C. P., Nance, D. M., and McEwen, B. S. (1982). Changes in estrogen and progestin receptor binding resulting from retrochiasmatic knife cuts. *Physiol. Behav*. 29: 225–229.

McGinnis, M. Y., Lumia, A. R., and McEwen, B. S. (1985). Increased estrogen receptor binding in amygdala correlates with facilitation of feminine sexual behavior induced by olfactory bulbectomy. *Brain Res*. 33419: 19–25.

Meaney, M. J., Aitken, D. H., Jensen, L. K., McGinnis, M. Y., and McEwen, B. S.

(1985). Nuclear and cystosolic androgen receptor levels in the limbic brain of neonatal male and female rats. *Dev. Brain Res.* 23: 179–185.

Micevych, P. (1991). Sexual differentiation of a neuropeptide circuit regulating female reproductive behavior. In *Behaviorial Biology: Neuroendocrine Axis*, ed. T Archer and S. Heusen, pp. 139–150. Hillsdale, NJ: Erlbaum.

Micevych, P., Akesson, T., and Elde, R. (1988). Distribution of cholecystokinin-immunoreactive cell bodies in the male and female rat, II: Bed nucleus of the stria terminalis and amygdala. *J. Comp. Neurol.* 269: 381–391.

Micevych, P., and Ulibarri, C. (1992). Development of the limbic–hypothalamic cholecystokinin circuit: A model of sexual differentiation. *Dev. Neurosci.* 14: 11–34.

Murakami, S., and Arai, Y. (1989). Neuronal death in the developing sexually dimorphic periventricular nucleus of the preoptic area in the female rat: Effect of neonatal androgen treatment. *Neurosci. Lett.* 102: 185–190.

Naftolin, F., Leranth, C., and Garcia-Segura, L. M. (1992). Ultrastructural changes in hypothalami cells during estrogen-induced gonadotrophin feedback. *Neuroprotocols: A Companion to Methods in Neurosciences* 1: 16–26.

Neill, J. D. (1980). Neuroendocrine regulation of prolactin secretion. In *Frontiers in Neuroendocrinology*, Vol. 6, ed. L. Martini and W. F. Ganong, pp. 129–155. New York: Raven Press.

Oro, T., Simerly, R. B., and Swanson, L. W. (1988). Estrous cycle variations in levels of CCK immunoreactivity within cells of three interconnected sexually dimorphic forebrain nuclei. Evidence for a regulatory role for estrogen. *Neuroendocrinology* 47: 225–235.

Pannzica, G. C., Viglietti-Panzica, C., Calcagni, M., Anselmetti, G. C., Schumacher, M., and Balthazart, J. (1987). Sexual differentiation and hormonal control of the sexually dimorphic medial preoptic nucleus in the quail. *Brain Res.* 416: 59–68.

Park, O.-K., Gugneja, S., and Mayo, K. (1990). Gonadotropin-releasing hormone gene expression during the rat estrous cycle: Effects of pentobarbital and ovarian steroids. *Endocrinology* 127: 365–372.

Park-Sarge, O. K., Mordacg, J., Rahal, J., and Mayo, K. E. (1992). Progesterone receptor mRNA expression in the female rat hypothalamus. *Soc. Neurosci. Abstr.* 18: 113.

Pearce, D., and Yamamoto, K. R. (1993). Mineralocorticoid and glucocorticoid receptor activities distinguished by nonreceptor factors at a composite response element. *Science* 259: 1161–1165.

Peterson, S. L., and Barraclough, C. A. (1989). Suppression of spontaneous LH surges in estrogen-treated ovariectomized rats by microimplants of antiestrogens into the preoptic brain. *Brain Res.* 484: 279–289.

Pfaff, D., and Keiner, M. (1973). Atlas of estradiol-concentrating cells in the central nervous system of the female rat. *J. Comp. Neurol.* 151: 121–158.

Pfaff, D. W., and Schwartz-Giblin, S. (1988). Cellular mechanisms of female reproductive behaviors. In *The Physiology of Reproduction*, ed. E. Knobil and J. Neill, pp. 1487–1568. New York: Raven Press.

Power, R. F., Mani, S. K., Codina, J., Conneely, O. M., and O'Malley, B. W. (1991). Dopaminergic and ligand-independent activation of steroid hormone receptors. *Science* 254: 1636–1639.

Priest, C. A., Eckersell, C., and Micevych, P. E. (in press). Temporal regulation by estrogen of preproenkephalin-A mRNA expression in rat ventromedial nucleus of the hypothalamus. *Mol. Brain Res.*

Rainbow, T. C., Parsons, B., and McEwen, B. S. (1982). Sex differences in rat brain oestrogen and progestin receptors. *Nature* 300: 648–649.

Raisman, G., and Field, P. M. (1973). Sexual dimorphism in the neuropil of the preoptic area of the rat and its dependence on neonatal androgen. *Brain Res.* 54: 1–29.

Reinoso, B. S., and Simerly, R. B. (1991). Hormone-sensitive sexually dimorphic neurons in the anteroventral periventricular nucleus project to the arcuate nucleus of the hypothalamus. *Soc. Neurosci. Abstr.* 17: 1229.

Rhees, R. W., Shryne, J. E., and Gorski, R. A. (1990). Termination of the hormone-sensitive period for differentiation of the sexually dimorphic nucleus of the preoptic area in male and female rats. *Dev. Brain Res.* 52: 17–23.

Roberts, J. L., Dutlow, C. M., Jakubowski, M., Blum, M., and Millar, R. P. (1989). Estradiol stimulates preoptic area-anterior hypothalamic proGnRH–GAP gene expression in ovariectomized rats. *Mol. Brain Res.* 6: 127–134.

Romano, G. J., Harlan, R. E., Shivers, B. D., Howells, R. D., and Pfaff, D. W. (1988). Estrogen increases proenkephalin messenger ribonucleic acid levels in the ventromedial hypothalamus of the rat. *Mol. Endocrinol.* 2: 1320–1328.

Romano, G. J., Krust, A., and Pfaff, D. W. (1989a). Expression and estrogen regulation of progesterone receptor mRNA in neurons of the mediobasal hypothalamus: An *in situ* hybridization study. *Mol. Endocrinol.* 3: 1295–1300.

Romano, G. J., Mobbs, C. V., Howells, R. D., and Pfaff, D. W. (1989b). Estrogen regulation of proenkephalin gene expression in the ventromedial hypothalamus of the rat: Temporal qualities and synergism with progesterone. *Mol. Brain Res.* 5: 51–58.

Romano, G. J., Shivers, B. D., Harlan, R. E., Howells, R. D., and Pfaff, D. W. (1987). Haloperidol increases proenkephalin mRNA levels in the caudate-putamen of the rat: A quantitative study at the cellular level using in situ hybridization. *Mol. Brain Res.* 2: 33–41.

Ronnekleiv, O. K., and Kelly, M. J. (1986). Luteinizing hormone-releasing hormone neuronal system during the estrous cycle of the female rat: Effects of surgically induced persistent estrus. *Neuroendocrinology* 43: 564–576.

Roselli, C. E. (1991). Sex differences in androgen receptors and aromatase activity in microdissected regions of the rat brain. *Endocrinology* 128: 1310–1316.

Roselli, C. E., and Resko, J. A. (1993). Aromatase activity in the rat brain: Hormonal regulation and sex differences. *J. Steroid. Biochem. Mol. Biol.* 44: 499–508.

Sachs, B. D., and Meisel, R. L. (1988). The physiology of male sexual behavior. In *The Physiology of Reproduction*, ed. E. Knobil and J. Neill, pp. 1393–1485. New York: Raven Press.

Sar, M., and Stumpf, W. E. (1975). Distribution of androgen-concentrating neurons in rat brain. In *Anatomical Neuroendocrinology*, ed. W. E. Stumpf and L. D. Grant, pp. 120–133. Basel: Karger.

Schachter, B. S., Durgerian, S., Harlan, R. E., Pfaff, D. W., and Shivers, B. D. (1984). Prolactin mRNA exists in the rat hypothalamus. *Endocrinology* 114: 1947–1949.

Segovia, S., and Guillamón, A. (1993). Sexual dimorphism in the vomeronasal pathway and sex differences in reproductive behaviors. *Brain Res. Rev.* 18: 51–74.

Sengelaub, D. R., Nordeen, E. J., Nordeen, K. W., and Arnold, A. P. (1989). Hormonal control of neuron number in sexually dimorphic spinal nuclei of the rat, III: Differential effects of the androgen dihydrotestosterone. *J. Comp. Neurol.* 280: 637–644.

Shivers, B. D., Harlan, R. E., Morrell, J. I., and Pfaff, D. W. (1983). Absence of oestradiol concentration in cell nuclei of LHRH-immunoreactive neurones. *Nature* 304: 345–347.

Shughrue, P. J., Bushnell, C. D., and Dorsa, D. M. (1992). Estrogen receptor messenger ribonucleic acid in female rat brain during the estrous cycle: A comparison with ovariectomized females and intact males. *Endocrinology* 131: 381–388.

Simerly, R. B. (1989). Hormonal control of the development and regulation of tyrosine hydroxylase expression within a sexually dimorphic population of dopaminergic cells in the hypothalamus. *Mol. Brain Res.* 6: 297–310.

Simerly, R. B. (1990). Hormonal control of neuropeptide gene expression in sexually dimorphic olfactory pathways. *Trends Neurosci.* 13: 104–110.

Simerly, R. B. (1991). Prodynorphin and proenkephalin gene expression in the anteroventral periventricular nucleus of the rat: Sexual differentiation and hormonal regulation. *Mol. Cell. Neurosci.* 2: 473–484.

Simerly, R. B. (1993). Distribution and regulation of steroid hormone receptor gene expression in the central nervous system. In *Advances in Neurology*, Vol. 59, ed. F. J. Seil, pp. 207–226. New York: Raven Press.

Simerly, R. B. (1995). Anatomical substrates of hypothalamic integration. In *The Rat Nervous System*, ed. G. Paxinos, pp. 353–376. San Diego, CA: Academic Press.

Simerly, R. B., Chang, C., Muramatsu, M., and Swanson, L. W. (1990). Distribution of androgen and estrogen receptor mRNA-containing cells in the rat brain: An in situ hybridization study. *J. Comp. Neurol.* 294: 76–95.

Simerly, R. B., Gorski, R. A., and Swanson, L. W. (1986). Neurotransmitter specificity of cells and fibers in the medial preoptic nucleus: An immunohistochemical study in the rat. *J. Comp. Neurol.* 246: 343–363.

Simerly, R. B., McCall, L. D., and Watson, S. J. (1988). The distribution of opioid peptides in the preoptic region: Immunohistochemical evidence for a steroid sensitive enkephalin sexual dimorphism. *J. Comp. Neurol.* 276: 442–459.

Simerly, R. B., and Swanson, L. W. (1986). The organization of neural inputs to the medial preoptic nucleus of the rat. *J. Comp. Neurol.* 246: 312–342.

Simerly, R. B., and Swanson, L. W. (1987a). Castration reversibly alters levels of CCK immunoreactivity within cells of three interconnected sexually dimorphic nuclei in the rat. *Proc. Natl. Acad. Sci. (USA)* 84: 2087–2091.

Simerly, R. B., and Swanson, L. W. (1987b). The distribution of neurotransmitter-specific cells and fibers in the anteroventral periventricular nucleus: Implications for the control of gonadotropin secretion in the rat. *Brain Res.* 400: 11–34.

Simerly, R. B., and Swanson, L. W. (1988). Projections of the medial preoptic nucleus: A *Phaseolus vulgaris* leucoagglutinin anterograde tract-tracing study in the rat. *J. Comp. Neurol.* 270: 209–242.

Simerly, R. B., Swanson, L. W., and Gorski, R. A. (1984). Demonstration of a sexual dimorphism in the distribution of serotonin-immunoreactive fibers in the medial preoptic nucleus of the rat. *J. Comp. Neurol.* 225: 151–166.

Simerly, R. B., Swanson, L. W., and Gorski, R. A. (1985a). The distribution of monoaminergic cells and fibers in a periventricular nucleus involved in the control of gonadotropin release: Immunohistochemical evidence for a dopaminergic sexual dimorphism. *Brain Res.* 330: 55–64.

Simerly, R. B., Swanson, L. W., Handa, R. J., and Gorski, R. A. (1985b). The influence of perinatal androgen on the sexually dimorphic distribution of tyrosine hydroxylase-immunoreactive cells and fibers in the anteroventral periventricular nucleus of the rat. *Neuroendocrinology* 40: 501–510.

Simerly, R. B. and Young, B. J. (1991). Regulation of estrogen receptor messenger ribonucleic acid in rat hypothalamus by sex steroid hormones. *Mol. Endocrinol.* 5: 424–432.

Simerly, R. B., Young, B. J., Capozza, M. A., and Swanson, L. W. (1989). Estrogen differentially regulates neuropeptide gene expression in a sexually dimorphic olfactory pathway. *Proc. Natl. Acad. Sci. (USA)* 86: 4766–4770.

Swanson, L. W. (1983). Neuropeptides: New vistas on synaptic transmission. *Trends Neurosci.* 6: 294–295.

Swanson, L. W. (1986). Organization of mammalian neuroendocrine system. In *Handbook of Physiology*, Vol. 4: *The Nervous System*, ed. F. E. Bloom, pp. 317–363. Baltimore: Waverly Press.

Swanson, L. W. (1987). The hypothalamus. In *Handbook of Chemical Neuroanatomy,* Vol. 5: *Integrated Systems of the CNS,* Part I, ed. A. Björklund, T. Hökfelt, and L. W. Swanson, pp. 1–124. Amsterdam: Elsevier.

Terasawa, E., and Davis, G. A. (1983). The LHRH neuronal system in female rats: Relation to the medial preoptic nucleus. *Endocrinol. Japon.* 30: 405–417.

Terasawa, E., Wiegand, S. J., and Bridson, W. E. (1980). A role for medial preoptic nucleus on afternoon of proestrus in female rats. *Am. J. Physiol.* 238: E533–E539.

Tischkau, S. A., and Ramirez, V. D. (1993). A specific membrane binding protein for progesterone in rat brain: Sex differences and induction by estrogen. *Proc. Natl. Acad. Sci. (USA)* 90: 1285–1289.

Tobet, S. A., and Fox, T. O. (1992). Sex differences in neuronal morphology influenced hormonally throughout life. In *Handbook of Behavioral Neurobiology,* Vol. 11: *Sexual Differentiation,* ed. A. A. Gerall, H. Moltz, and I. L. Ward, pp. 41–83. New York: Plenum Press.

Toyooka, K. R., Connolly, P. B., Handa, R. J., and Resko, J. A. (1989). Ontogeny of androgen receptors in fetal guinea pig brain. *Biol. Reprod.* 41: 204–212.

Tung, L., Mohamed, M. K., Hoeffler, J. P., Takimoto, G. S., and Horwitz, K. B. (1993). Antagonist-occupied human progesterone B-receptors activate transcription without binding to progesterone response elements and are dominantly inhibited by A-receptors. *Mol. Endocrinol.* 7: 1256–1265.

Vandenbergh, J. G. (1988). Pheromones and mammalian reproduction. In *The Physiology of Reproduction,* ed. E. Knobil and J. Neill, pp. 1678–1696. New York: Raven Press.

Vegeto, E., Shahbaz, M. M., Wen, D. X., Goldman, M. E., O'Malley, B. W., and McDonnell, D. P. (1993). Human progesterone receptor A form is a cell- and promoter-specific repressor of human progesterone receptor B function. *Mol. Endocrinol.* 7: 1224–1255.

Vito, C. C., and Fox, T. O. (1982). Androgen and estrogen receptors in embryonic and neonatal rat brain. *Dev. Brain Res.* 2: 97–110.

Warembourg, M. (1978). Uptake of ^{3}H-labeled synthetic progestin by rat brain and pituitary: A radioautographic study. *Neurosci. Lett.* 9: 329–332.

Warembourg, M., Logeat, F., and Milgrom, E. (1986). Immunocytochemical localization of progesterone receptor in the guinea pig central nervous system. *Brain Res.* 384: 121–131.

Watson, R. E., Jr., Hoffmann, G. E., and Wiegand, S. J. (1986). Sexually dimorphic opioid distribution in the preoptic area: Manipulation by gonadal steroids. *Brain Res.* 398: 157–163.

Weiner, R. I., Findell, P. R., and Kordon, C. (1988). Role of classic and peptide neuromediators in the neuroendocrine regulation of LH and prolactin. In *The Physiology of Reproduction,* ed. E. Knobil and J. Neill, pp. 1235–1281. New York: Raven Press.

Wiegand, S. J., and Terasawa, E. (1982). Discrete lesions reveal functional heterogeneity of suprachiasmatic structures in regulation of gonadotropin secretion in the female rat. *Neuroendocrinology* 34: 395–404.

Wray, S., and Hoffman, G. (1986). A developmental study of the quantitative distribution of LHRH neurons within the central nervous system of the postnatal male and female rats. *J. Comp. Neurol.* 252: 522–531.

Wray, S., Key, S., Bachus, S., and Gainer, H. (1993). Regulation of LHRH and oxytocin gene expression in CNS slice-explant cultures: Effects of second messengers. *Soc. Neurosci. Abstr.* 19: 1395.

Wysocki, C. J. (1979). Neurobiological evidence for the involvement of the vomeronasal system in mammalian reproduction. *Neurosci. Biobehav. Rev.* 3: 301–341.

Wysocki, C. J. (1987). The vomeronasal system. In *Neurobiology of Taste and Smell,* ed. T. Finger and W. Silver, pp. 125–150. New York: Wiley.

Part II

Sex steroid interactions with specific
neurochemical circuits

5

Ovarian steroid interactions with hypothalamic oxytocin circuits involved in reproductive behavior

LORETTA M. FLANAGAN AND BRUCE S. McEWEN

Introduction

Circulating gonadal and adrenal steroid hormones readily cross the blood–brain barrier and affect neural circuits that subserve various aspects of brain function and behavior. When investigating the interactions between hormones and neurotransmitter systems, it is particularly important to study a model system where there is a behavioral endpoint for which the neural control sites are discrete and accessible. Mating behavior in the female rat provides such a system (Pfaff 1980). The ovarian steroid hormones estradiol (E) and progesterone (P) synchronize the occurrence of mating and ovulation, and thereby increase the likelihood of reproductive success. Our goal is to take advantage of this useful model system in order to elucidate the mechanisms of steroid hormone action on neuronal activity and connectivity. In addition to enhancing our knowledge of the specific events that control this biologically important behavior, such studies may have broader implications for the analysis of hormone–neurotransmitter interactions.

It is well known that ovariectomized (OVX) female rats display very little sexual behavior. Estradiol replacement reverses this condition, however, by sensitizing specific neural circuits to stimuli that elicit sexual behavior. Approximately 48 hours after E is administered, OVX female rats display sexual behavior when they encounter a male. Progesterone treatment, given after E and about 4 hours before behavioral testing, further enhances female sexual behavior (Boling and Blandau 1939). Treatment with P is ineffective unless the animals have been primed with E, in part because E induces expression of a subset of P receptors (MacLusky and McEwen 1978). This paradigm of steroid treatment approximates the plasma profile of ovarian steroids in the intact cycling female rat (Butcher et al. 1974). The present chapter discusses how both the estrous cycle and this E + P treatment regimen facilitate sexual behavior by acting on a neural circuit involving the neuropeptide oxytocin, in which the ventromedial nucleus of the hypothalamus (VMH) is involved.

Actions of E and P that influence the VMH

Several studies have demonstrated that the VMH is the critical site of action for E-induced sexual behavior. For instance, lesions of the VMH prevent enhancement of sexual behavior after E treatment (Pfaff and Sakuma 1979a), and electrical stimulation of the VMH facilitates the lordosis reflex in E-treated rats (Pfaff and Sakuma 1979b). Moreover, discrete infusions of E and P into the VMH, rather than systemically, are sufficient for these steroids to synergistically cause the expression of female sexual behavior (Rubin and Barfield 1983). However, while VMH lesions block sexual behavior induced by E alone, the combination of E and P can still elicit sexual behavior (Mathews and Edwards 1977). Furthermore, intracranial infusion of P antagonists into the VMH is very effective in reducing sexual behavior, as is similar administration into the habenula, preoptic area, and interpeduncular region of the midbrain (Etgen and Barfield 1986). These data suggest that while the VMH is critical to E action, it is only one of several sites where P facilitates sexual behavior.

The mechanisms whereby E and P modulate neural systems to allow female sexual behavior clearly involve changes in gene expression, initiated by the steroids binding to intracellular receptors. These activated receptor–hormone complexes, in turn, bind to specific DNA sites, referred to as hormone response elements, to inhibit or promote gene transcription (Freedman 1992). Application of inhibitors of either gene transcription or protein synthesis into the VMH block the effects of E on mating behavior (Rainbow et al. 1982a; Meisel and Pfaff 1984), and hormone-induced facilitation of sexual behavior requires synthesis of polyadenylate (polyA) mRNA (Yahr and Ulibarri 1986). The temporal relation between the translocation of occupied P receptors to the nuclei of the VMH neurons and the appearance of mating behavior (McGinnis et al. 1981), as well as the ability of protein synthesis inhibitors to block P effects (Rainbow et al. 1982b; Glaser and Barfield 1984), support the hypothesis that at least some aspects of P action are genomic.

Recent evidence, however, also indicates nongenomic mechanisms of P effects. Specifically, midbrain infusions of P conjugated to albumin, presumably confined to the extracellular space, are capable of enhancing female sexual behavior in hamsters. This is in contrast to VMH infusions, where only free P, and not the albumin conjugate, is effective (Frye et al. 1992). These results suggest that while the actions of P are genomic in the VMH, they might involve a membrane action in the midbrain. These membrane effects of P might involve a γ-aminobutyric acid (GABA)-ergic mechanism, since P metabolites and the adrenal steroid deoxycorticosterone modulate GABA receptors (Gee et al. 1987; Purdy et al. 1991). Indeed, administration of two GABA-active steroid metabolites, 5α,3α-hydroxydihydroprogesterone and 5α-tetrahydrodeoxycorticosterone, into the ventral tegmental area promotes lordosis in hamsters after P priming in the VMH region (Frye and DeBold 1993).

Some of the sequelae of the genomic changes induced by E and P are known, although how specific protein changes contribute to sexual behavior is a matter of speculation. Using two-dimensional gel electrophoresis, Jones et al. (1986, 1987, 1988) showed that E and P have additive, counteractive, and independent effects on the array of proteins expressed in the VMH. The identity and function of these steroid-regulated proteins are still incompletely understood, although one might be the oxytocin receptor, as discussed in detail later. Another proposed E-induced protein in the VMH is an isoform of phospholipase C-α (Mobbs et al. 1991), which may have protease activity (Urade et al. 1992).

Estrogen treatment of female rats also evokes a more general morphological change in the VMH, which indicates an increased formation of proteins. In the first study of this kind, OVX female rats were treated with 10 μg estradiol benzoate (EB) for 15 days. In the VMH, increased stacking of rough endoplasmic reticulum, indicative of increased capacity for protein biosynthesis, was observed along with an increased number of dense core vesicles (Cohen and Pfaff 1981). A single dose of 100 μg EB in OVX rats increased protein-synthetic activity 48 hours later in the VMH examined by means of quantitative electron microscopy (Carrer and Aoki 1982). Although the number of neurons was not affected, VMH neurons were hypertrophied and contained increased stacking of rough endoplasmic reticulum, condensed nucleolar material, enlarged Golgi, and irregular nuclear envelopes and pleiomorphic mitochondria. In addition, the number of synaptic contacts in the ventrolateral region of the VMH increased, without changing the proportion of synapses on spines and shafts.

A subsequent study involving much lower doses of EB and shorter time intervals demonstrated a more rapid series of changes (i.e., within 6 hours) in the morphology of VMH neurons, including rounding and enlargement of the nucleus, increased size of the nucleolus and cell soma, and increased area of the endoplasmic reticulum (Jones et al. 1985). The E-induced changes in nucleolar volume and rough endoplasmic reticulum were accompanied by elevated levels of 28 S ribosomal RNA in VMH neurons within 30 minutes to 6 hours (Jones et al. 1987, 1990). Thus, at this early interval, an initial array of changes in cellular activity sets the stage for the increased metabolic capacity and altered pattern of gene expression induced by E observed at later times.

More recent studies have shown that OVX rats have lower dendritic spine density in the VMH, and E replacement reverses this effect within 72 hours (Frankfurt et al. 1990). More important, spine density and axodendritic synapses fluctuate during the estrous cycle, peaking on proestrus and falling during estrus and diestrus (Frankfurt and McEwen 1991a,b). In the hamster, EB treatment for 2 days increased dendritic length in the VMH by almost 50% (Meisel and Luttrell

1990). Thus, it appears that the VMH is subject to cyclical fluctuations of synaptic connectivity and that VMH neurons respond rapidly to circulating E by increasing their capacity to make proteins, some of which are undoubtedly used to generate new dendritic membrane and spines.

A host of neurochemical effects of E and P have been documented in the VMH, including alterations in monoamine turnover (Renner et al. 1987; Gereau et al. 1993), changes in preproenkephalin mRNA levels (Romano et al. 1989), and regulation of receptors including $GABA_A$ (Schumacher et al. 1989b), acetylcholine (Rainbow et al. 1984), cholecystokinin (Akesson et al. 1987; Chapter 7 by Popper et al., this volume; but see also Schumacher et al. 1991a), and serotonin (Fischette et al. 1983). The scope of this chapter does not permit full discussion of all the documented changes in the VMH that correlate with the expression of female sexual behavior. However, studies of steroid interactions with an oxytocin (OT) neurocircuit have led to many insights into the intricacies of steroid facilitation of sexual behavior and steroid–peptide interactions.

Steroid-dependent effect of OT on sexual behavior

A potential role of OT in sexual behavior was first suggested by Arletti and Bertolini (1985), who showed that intracerebroventricular (icv) OT (1 ng) given 15–60 minutes before testing increased the lordosis quotient in OVX female rats primed with EB (10 ng) and P (500 μg). This finding was corroborated by Caldwell et al. (1986), who observed that animals primed with three daily injections of a low dose of E (0.15–0.5 μg) increased their sexual behavior 20–90 minutes after receiving icv OT (0.6–50 μg). In contrast, Gorzalka and Lester (1987) found that animals treated with two daily injections of E (10 μg) displayed no increased sexual behavior after receiving a smaller dose of OT (~10 ng, icv); if this dose of OT was given to animals that received P (200 μg) after their E priming, however, sexual behavior was facilitated. The results of the two latter reports seem incongruous, and led Gorzalka and Lester to conclude that the effect of OT was P-dependent, unless a pharmacological dose of OT was used. For purposes of comparison, these parameters are listed in Table 5.1.

It is now believed that this discrepancy is due to the presence of P produced by adrenal secretion in OVX animals, which contributes to sexual receptivity (Bartosik et al. 1971; Mann and Barraclough 1973; Barfield and Lisk 1974; Witcher and Freeman 1985). In an elegant group of studies, M. Schumacher and H. Coirini (unpublished results) found that after 2 days of EB treatment, a low dose of OT facilitated sexual behavior only when the animals were also pretreated with P (in agreement with Gorzalka and Lester); a larger dose of OT was effective in the absence of exogenous P after 3 days of EB treatment (corroborating Caldwell), except when the animals were adrenalectomized (Fig. 5.1). These data

Table 5.1. *Steroid and OT doses used to elicit female sexual behavior in published reports*

EB (μg)	P (μg)	OT (μg)	Latency (min)[a]	Reference
0.010, 2 days	500	0.001	15	Arletti and Bertolini (1985)
0.5, 3 days	—	0.6–50	20	Caldwell et al. (1986)
10, 2 days	200	0.010	20	Gorzalka and Lester (1987)
2, 2 days	250	0.200	60	Schumacher et al. (1989a)

[a]Latency refers to the behavioral testing time after central administration of OT.

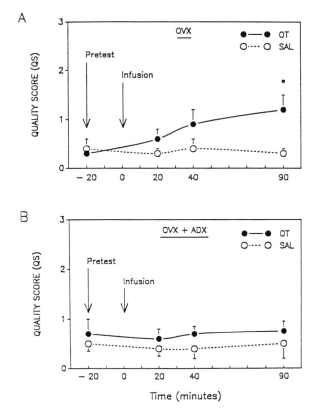

Figure 5.1. Effect of adrenalectomy (ADX) on the facilitation of lordosis behavior by OT in OVX rats treated with 0.5 μg EB for 3 days. Rats were infused with saline (SAL) or with OT dissolved in saline icv (OT; 9 animals per group). In OVX females (A), OT significantly increased lordosis behavior across successive tests (lordosis quality scores $p \leq .05$ by Kruskal–Wallis one-way ANOVA). However, after removal of the adrenal glands (B), OT no longer facilitated lordosis ($p > .05$ when compared with the corresponding SAL group by Mann–Whitney tests). Data from M. Schumacher and H. Coirini (unpublished results).

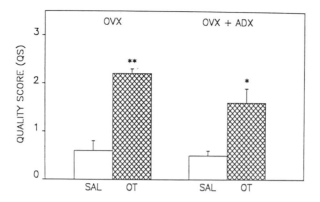

Figure 5.2. Effect of adrenalectomy (ADX) on the facilitation of lordosis behavior by OT in OVX females injected with 2 μg of EB followed 48 hours later by 250 μg P. Rats were infused with saline (SAL) or with 0.2 μg OT dissolved in saline (OT; 6 animals per group). Removal of the adrenal glands did not prevent the facilitation of lordosis behavior by OT. Data of M. Schumacher and H. Coirini (unpublished results).

suggest that prolonged E treatment stimulates adrenal secretion of P sufficient to permit OT to facilitate sexual behavior. Removal of the adrenal glands did not have a nonspecific inhibitory effect on sexual behavior, because these animals displayed lordosis when treated with EB and P (Fig. 5.2), justifying the claim that the effect of OT on sexual behavior is P-dependent.

Subsequent experiments have critically evaluated a host of variables, including the doses of EB, P, and OT and adrenal sources of steroid. The dose of EB and P must be low enough for a behavioral facilitation by OT to be apparent. Depending on the doses of EB, P, and OT, and the sexual experience of the animal, OT will differentially facilitate sexual behavior either in the dark only or in both dark and light periods (Schumacher et al. 1991b). Discrete intracranial administration of OT (100 ng) has localized at least one site of action for OT facilitation of sexual behavior in the caudal VMH (Schumacher et al. 1989a, 1990), and the dependence of this behavioral effect on OT receptors was confirmed using a uterotonic antagonist (Caldwell et al. 1990).

Effects of E on OT neuronal activity and biosynthesis

Regulation by the estrous cycle

Given the evidence for the facilitative role of OT in a steroid-dependent behavior, it is logical to ask whether these steroids affect oxytocinergic neurotransmission. Investigations of OT activity during the estrous cycle generally indicate that the release of OT occurs on the late afternoon or evening of proestrus (the time sexual behavior is most likely to occur). Levels of OT in portal blood peak on the after-

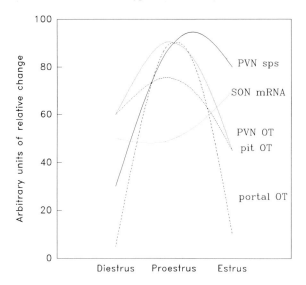

Figure 5.3. Summary of oxytocinergic activity during the estrous cycle. Values are not quantitative, but merely indicate relative changes. Each curve is based on PVN sps (spikes per second of presumed OT neurons; Negoro et al. 1979); PVN OT (Greer et al. 1986); SON mRNA (Van Tol et al. 1988); pit OT (pituitary; Crowley et al. 1978; Van Tol et al. 1988); portal blood OT (Sarkar and Gibbs 1984).

noon of proestrus (Sarkar and Gibbs 1986). The content of OT in the pituitary during the estrous cycle is highest at 1,000 hours on proestrus and declines again at 1,700 hours that day (Crowley et al. 1978; Van Tol et al. 1988), suggesting that OT is released on the afternoon of proestrus.

Most endogenous OT is produced by the paraventricular (PVN) and supraoptic (SON) nuclei. Neurons of the PVN are most active during proestrus and estrus (Negoro et al. 1979). In micropunches, OT levels in the PVN are lowest during diestrus, increase during metestrus, and remain high during proestrus and estrus (Greer et al. 1986). Whether the estrous cycle affects OT mRNA has been difficult to ascertain. In one study, levels of OT mRNA were elevated on the day of estrus in SON, measured by Northern and dot–blot analysis (Van Tol et al. 1988), but in other studies, there was no cyclical variation by *in situ* hybridization or Northern analysis (Zingg and Lefebvre 1988; Miller et al. 1989; Chibbar et al. 1990). Figure 5.3 summarizes the pattern of OT activity during the estrous cycle.

Effect of E on OT secretion and neural activity

Estrous cycle regulation of OT activity and release may be directly or indirectly related to E levels. Estradiol treatment increases plasma levels of OT–neurophysin in women (Robinson 1974). Similarly, E treatment for 2 days increased plasma levels of OT in OVX rats (Yamaguchi et al. 1979). This positive effect of

E on OT plasma levels was disrupted by deafferentation of the PVN, which suggests that afferents to OT neurons, rather than OT neurons themselves, are responsive to E. Electrophysiologically, 2 days of E treatment increased the excitability of oxytocinergic cells in the PVN (Negoro et al. 1979; Akaishi and Sakuma 1985). Collectively, these results support the idea that E treatment increases firing of oxytocinergic neurons and release of OT.

Effect of E on OT synthesis

In contrast to the consistently observed effects of E replacement on OT firing activity and release, effects on OT peptide and mRNA levels have been difficult to reconcile. For example, OT levels in the pituitary and in SON and PVN micropunches measured by radioimmunoassay were not significantly different in OVX animals treated with vehicle or E for 7 days (Burbach et al. 1990). Also, the optical density of OT immunocytochemical staining did not change after 1 month of E treatment in the PVN and SON (Rhodes et al. 1981). It is conceivable that constant hypothalamic and pituitary OT peptide levels in the face of elevated secretion in OVX animals treated with E might reflect an equilibrium of OT activity, based on coupled synthesis, transport, and release. Yet, although coupled synthesis and release might prevent detection of altered peptide levels, one might expect to see a change in mRNA. In this case, however, the data are mixed. In one study, 7 days of E treatment produced no change of OT mRNA in the SON or PVN compared with oil-treated controls (Burbach et al. 1990). On the other hand, OVX decreased OT mRNA, and E replacement for 40 days increased OT mRNA (Miller et al. 1989; Chibbar et al. 1990). Single-cell analyses using *in situ* hybridization revealed that OT mRNA levels were elevated by E treatment in a subpopulation of OT neurons in the SON and accessory nuclei (Chung et al. 1991). This study indicates that a failure to detect changes in hypothalamic OT mRNA could be due to the heterogeneity of the OT neuron population.

The reported effects of E on plasma OT (Robinson 1974; Yamaguchi et al. 1979) are consistent with the reports of increased OT neuronal activity (Negoro et al. 1979; Akaishi and Sakuma 1985). These findings, however, seem at odds with the inconsistent effects of E treatment on OT peptide and mRNA levels in the pituitary and hypothalamus. We suspect that E increases OT release and neural activity, but because synthesis and transport are coupled to release, only transient changes could be detected in OT peptide levels in the hypothalamus and posterior pituitary during the estrous cycle. Furthermore, it may be difficult to detect changes in OT mRNA due to its long half-life and great abundance. Finally, a change in the length of the polyA tail might improve the rate of translation and increase the rate of peptide biosynthesis without changing the number of mRNA molecules. Vasopressin polyA tail length is diurnally regulated (Robinson et al.

1988), and OT mRNA polyA tail length increases during lactation and pregnancy (Zingg and Lefebvre 1989).

Central to the issue of the mechanism of E action on OT neurons is the question of whether the effects of E are direct or indirect and, hence, whether these cells have E receptors. Richard and Zingg (1990) reported that the human OT gene contains a consensus sequence for an E response element in its promoter region. This promoter appears active and specific for the E receptor complex, because transfection of a rat OT promoter allows E to stimulate transcription of the reporter gene, but only in a cell line with E receptors (Burbach et al. 1990). It is noteworthy, therefore, that several studies indicate that most OT neurons do *not* have E receptors. A brain map of E-concentrating cells indicated that only a few such cells exist in the PVN, and virtually none are present in the SON (Pfaff and Keiner 1973). Rhodes et al. (1982) showed that the percentage of neurophysin-stained neurons in vasopressin-deficient Brattleboro rats (thus, presumably oxytocinergic) that concentrate E was very low in the magnocellular SON and PVN, but higher in the parvocellular lateral subnucleus (17%) and the posterior subnucleus (64%) of the PVN, which supports the view that E may have a direct effect on a functional subpopulation of OT neurons in the PVN. Alternatively, E could affect OT gene transcription indirectly via neural inputs, as suggested by knife cut studies (Yamaguchi et al. 1979). For example, E receptor–containing catacholaminergic cells in the brain stem may project to OT neurons (Heritage et al. 1980; Simerly et al. 1990).

Diurnal regulation

Estradiol also produces diurnal effects on regional OT levels throughout the day (Flanagan et al. 1991). Following treatment of OVX rats with EB (10 μg in 100 μl) on 2 consecutive days, levels of OT in the PVN, SON, and pituitary did not change compared with vehicle treatment; however, there was an effect of time of day in these regions (Fig. 5.4). In the PVN, OT levels were threefold higher ($p \leq$ .01) at 1400 hours than at all other times. In contrast, in the SON there was a biphasic diurnal effect on OT levels. These levels were significantly ($p \leq$ .02) higher at 1400 and 2000 hours compared with 1000 hours. The posterior pituitary displayed yet another pattern; OT levels were significantly ($p \leq$ .05) higher by approximately 50% at 1000 hours than at later times.

The results suggest that in the PVN, SON, and posterior pituitary there are no main effects of E regardless of the time of day; however, the level of OT in both oil-treated and EB-treated animals has a diurnal rhythm. Collectively, these results suggest that a diurnal signal increases OT levels in the PVN and SON at 1400 hours. Whether there is a diurnal signal to promote synthesis in anticipation of increased secretion or whether synthesis is induced by previous neural activity

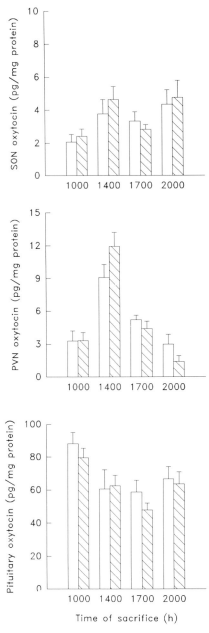

Figure 5.4. Levels of OT in the SON, PVN, and posterior pituitary as a function of time of day and steroid treatment (picograms per microgram protein; 7–16 animals per group). There was a significant effect of time, but not EB treatment, in all areas. Specifically, in the PVN, levels were significantly elevated at 1400 hours ($p \leq .01$). In the SON, levels at 1400 and 2000 hours were higher than those at 1000 hours ($p \leq .02$). Posterior pituitary levels of OT were significantly higher at 1000 hours than at all later times ($p \leq .05$). Striped bars represent estradiol treatment; open bars, vehicle.

is not clear. In light of the apparent depletion of posterior pituitary stores of OT at 1400 hours, it seems that this might occur in anticipation of, or concomitant with, increased release.

The key question is whether the fluctuation in OT levels is related to changes in OT neuronal activity. In this regard, fewer OT neurons of the PVN contain Fos protein at 1200 hours, while numerous cells contain Fos at 1700 hours in OVX rats (Arey and Freeman 1992). These data suggest the presence of a daily rhythm of OT activity that coincides with the changes in OT peptide levels in the PVN. Yet there is no daily rhythm of OT mRNA levels in males (Burbach et al. 1988). Either there is a sex difference in the daily fluctuation of OT levels or changes in OT mRNA do not underlie changes in peptide levels. Consistent with the lack of effect of E on the diurnal rhythm of hypothalamic OT levels, the daily rhythm of OT levels in the cerebrospinal fluid of female rhesus monkeys (Perlow et al. 1982) is not regulated by estrogen (Amico et al. 1990).

The physiological implications of a diurnal fluctuation of OT may relate to a variety of reproductive functions. Samson et al. (1986) showed that OT is involved in the circadian rhythm of prolactin release, which occurs between 1300 and 1600 hours. In their study, plasma OT levels increased from 16.4 to 51.0 pg/ml between 1100 and 1200 hours, whereas passive immunization to OT reduced the diurnal rise in prolactin. In addition, OT may play a role in the luteinizing hormone (LH) surge of intact cycling rats (Johnston et al. 1988; Robinson and Evans 1990), which also occurs between 1400 and 1600 hours. It is noteworthy that the effect of OT on LH secretion depends on the presence of E (Evans et al. 1992), given that hypothalamic OT levels rise in the afternoon even in the absence of E (Fig. 5.4) and that the elevated plasma OT levels reported by Samson and co-workers are also present in OVX rats. Taken together, these data suggest that the critical event in E regulation of OT effects on LH secretion might be upregulation of OT receptors in the anterior pituitary rather than alteration of OT.

Consistent with our results, Windle et al. (1992) observed that plasma OT rose during the light phase; however, the content of OT in the pituitary was inversely related to plasma levels (i.e., levels were lower late in the light period). In whole hypothalamus, OT levels paralleled the pattern in plasma, which suggests coupling of synthesis and release. Our observations that the pattern in the PVN was not identical to that in the SON were somewhat surprising. This might be attributed to either a dissociation of parvocellular and magnocellular OT activity or a differential innervation of SON and PVN magnocellular cells.

Of course, interpretation of the observed changes in peptide levels is limited because it is impossible to know whether these changes reflect regulation of mRNA transcription, protein synthesis, degradation, or transport. For example, vasopressin transport is coupled to synthesis (Roberts et al. 1991). Because no diurnal change in OT mRNA was detected (Burbach et al. 1988), increased rate and efficiency of translation and/or decreased degradation may account for the changes in peptide level. A possible explanation for the lack of correspondence between the change in peptide levels and steady OT mRNA levels might be a change in polyA tail lengths, which could thereby improve the efficiency of translation, as discussed earlier (Robinson et al. 1988).

Steroid effects on OT receptors in the VMH

While there is still uncertainty regarding the role of E in the regulation of OT synthesis, there is strong evidence for E induction of OT receptors. Estradiol treatment (20 μg) caused a fivefold increase in the binding density of OT receptors in the VMH (De Kloet et al. 1986). A comprehensive autoradiographic survey examined binding of OT receptors in the brain as a function of gender and steroid sensitivity (Tribollet et al. 1990). In addition to the VMH, OT binding was found in the anterior olfactory nuclei, islands of Calleja, lateral septum, bed nucleus of the stria terminalis (BNST), caudate putamen, central nucleus of the amygdala, and subiculum, with similar densities in males and females. Gonadectomy reduced OT binding in the islands of Calleja and VMH, and this effect was reversed by either E or testosterone in both males and females.

The caudal region of the VMH, which is implicated in sexual behavior, expresses a higher density of OT receptors, whereas the rostral portion requires a higher dose of E to achieve maximal OT binding (Johnson et al. 1989a). The time course of OT receptor binding in the VMH significantly increases within 24 hours of E treatment and continues to rise after 96 hours of continuous E administration (Johnson et al. 1989). Intact cycling females exhibit increased OT receptor binding in the VMH during behavioral estrus compared with females that are not in estrus (Insel 1990), suggesting a role for these receptors in reproductive behavior.

After E-induced OT receptor synthesis in the cell bodies of the VMH, receptors are transported laterally toward the region of high density of OT-immunoreactive fibers coursing in the neurohypophyseal tract (Coirini et al. 1991; Johnson et al. 1991b). In addition, P may cause further lateral extension of OT receptors (Schumacher et al. 1989a, 1990), which is apparently a nongenomic effect, since it occurs rapidly in brain sections *in vitro*. Rather than a lateral movement of OT receptors, P may promote a conformational change that increases the affinity of the receptor for OT ligands. Because P does not change the electrophysiological response of VMH neurons to OT (Kow et al. 1991), the functional significance of apparent receptor movement is unclear. One possible explanation involves a fiber tract ventral and lateral to the VMH, which contains OT axons traveling to the neurohypophysis. Oxytocin might be released at the level of the VMH from these fibers of passage, although there is presently no evidence for this. If the area containing dense OT receptors does not expand toward the endogenous ligand, the significance of this P-induced effect is difficult to imagine. It is not likely that the difference in area is an autoradiographic artifact due to density changes, because it was observed whether the density increased or decreased (Schumacher et al. 1989a, 1990). Thus, this novel nongenomic P action has not yet been completely incorporated into our understanding of oxytocinergic neurotransmission.

This nongenomic effect of P might be related to the fact that P is required for OT facilitation of sexual behavior (discussed earlier), but linking these phenomena has not been straightforward. First, it is not immediately obvious why the facilitation of lordosis by exogenous OT would necessitate expansion of the area containing OT receptors in the VMH, if such expansion normally brings these receptors into proximity with OT fibers. Second, *in vitro* intracellular recordings of VMH neurons have shown that P does not augment the excitatory action of OT produced by E pretreatment (Kow et al. 1991). Thus, an alternative explanation might be that P may sensitize the next link in the neural circuit for sexual behavior. In this regard it is noteworthy that OT increases extracellular norephinephrine levels in the VMH of animals pretreated with both E and P, but not in animals primed with only E (Vincent and Etgen 1993). Whatever the exact nature of the P dependency of OT actions, P does not markedly alter the changes induced by E on the excitatory postsynaptic effects of OT binding to its receptor.

Little is known about the cell types that contain OT receptors in the VMH. Oxytocin has an excitatory effect on VMH neurons (Inenaga et al. 1991), and this effect is enhanced in animals treated with E (Kow et al. 1991). Other agents known to facilitate sexual behavior are also excitatory to VMH neurons. Future histochemical and electrophysiological studies should identify the axonal targets of and neurotransmitters utilized by neurons containing OT receptors, as well as the interactions between OT and other neurotransmitter systems in the VMH.

To address other regulatory influences on OT receptors, OT binding in the VMH was observed in intact male rats after PVN lesions, daily icv bolus infusions of OT, or continuous icv infusions of OT (Insel et al. 1992). Acute infusions and PVN lesions did not alter OT receptor density; however, continuous OT infusion decreased binding significantly. These data suggest that OT receptors are downregulated in the presence of continuous, but not intermittent ligand, yet they are not conversely upregulated by the absence of OT (assuming that the PVN is an important endogenous source of OT in the VMH). As a comparison, it would have been interesting to include analysis of the dorsal vagal complex, where PVN oxytocinergic neurons are known to project and to release OT (Sawchenko and Swanson 1982).

Following the recent sequencing of the human uterine OT receptor (Kimura et al. 1992), *in situ* hybridization has been utilized to generate a map of OT receptor mRNA in the rat brain (Yoshimura et al. 1993) that is largely in agreement with the distribution of OT receptor based on autoradiography with radiolabeled OT or

OT analogues. Although there is a significant database for uterine receptors, there are many unknown details about brain OT receptor function. The human uterine OT receptor has a length of 388 amino acids with seven transmembrane domains typical of G-protein-coupled receptors. In a variety of species, the uterine OT receptor is coupled to hydrolysis of inositol phospholipid (e.g., Ruzycky and Triggle 1988). Further studies will bridge the gap in our understanding of sub-types, signal transduction, and regulation of the OT receptor in the brain.

The importance of OT receptors for sexual behavior

The E induction of OT receptors in the VMH, and particularly their regulation during the estrous cycle, further supports the role of this neurotransmitter in sexual behavior. In order to ascertain whether endogenous OT has the predicted facilitative effect on sexual behavior, Witt and Insel (1991) determined the effect of a specific OT antagonist on animals primed with EB alone or EB followed by P. When the antagonist (500 ng to 1 μg) was administered icv to animals pre-treated with EB alone (10 μg for 2 days), there was no reduction of sexual behavior. However, when the antagonist was given just before P (250 μg) treatment, 2, 4, or 6 hours before behavioral testing, in animals primed with either 1 or 10 μg EB, it reduced the lordosis quotient, decreased hopping and darting, and increased vocalizations, which may indicate rejection behavior. These authors noted that the antagonist was not effective unless given at the same time as P, which suggests that the effect of P (either direct or indirect) on OT must occur at some critical time before mating actually occurs. By occupying available sites, the antagonist decreased autoradiographic labeling of OT receptors in the VMH by 85% for at least 6 hours after the injection, and maximal OT receptor blockade occurred within 1 hour. The fact that the OT antagonist did not reduce sexual behavior in animals primed with EB alone is consistent with the notion that EB-induced sexual behavior is OT-independent and with the results of the exogenous OT studies discussed earlier which indicated that OT is ineffective in the absence of P.

McCarthy et al. (1994) observed that infusion of antisense oligonucleotides complementary to the OT receptor into the VMH reduced OT receptor binding in the VMH by 30% without affecting amygdaloid OT receptor binding. This anti-sense infusion decreased the lordosis quotient from 70 to 30% and increased rejection behavior in animals treated with EB (10 μg), but not P. However, the antisense infusions did not blunt the sexual behavior of animals primed with EB (1 μg) and P (1 mg). At first glance, these data seem at odds with the results obtained with the OT antagonist, which decreased behavior primed by EB and P, but not by E alone. These different mechanisms of blocking the OT receptor could lead to the discrepant results. The reduction in OT receptor binding was much greater after the antagonist was administered than after the antisense infusion. Also, the larger dose of P used in the antisense experiments may have had such a strong activating effect on other neural circuits, in addition to the OT component, that the partial reduction in OT receptor availability caused by antisense was inconsequential. The larger blockade of OT receptors produced by the antagonist may be required to markedly reduce (without abolishing) sexual behavior. Although further studies are required to reconcile these findings completely, it is noteworthy that the results with the antagonist are consistent with the hypothesis that P is important for the behavioral effects of OT.

Despite these differences, the common suggestion drawn from both studies is that endogenous OT is involved in sexual behavior. The relative importance of OT could depend on the extent to which the behavior is influenced by P. For example, in E-treated animals in the complete absence of P, OT is ineffective (Fig. 5.1); in the presence of moderate levels of P (from the adrenal), OT markedly augments sexual behavior, but only when a full complement of OT receptors is present in the VMH (Caldwell et al. 1986; McCarthy et al. 1994); when P levels are very high (supplemental injection), sexual behavior is facilitated by OT, even when OT receptor binding is reduced by 30%, but not if it is reduced by 80% (Witt and Insel 1991; McCarthy et al. 1994). Further substantiation of the quantitative effects of OT on sexual behavior would significantly advance our understanding of the nonlinear regulation of sexual behavior by steroids.

Oxytocinergic innervation of the VMH

Despite the wealth of data suggesting the enhancement of sexual behavior by OT in the VMH, there is presently no definitive evidence of an oxytocinergic projection to the VMH. In the absence of ultrastructural or retrograde tracing studies indicating such selective innervation, examining the induction of neuronal activation by sexual behavior would be useful for identifying a possible source of VMH oxytocinergic afferents. Fos, the early immediate gene protein product and transcription factor, might provide such a marker of neuroendocrine activity (Hoffman et al., in press). Oxytocin neurons express Fos-like immunoreactivity (Fos-ir) during a variety of conditions that are known to coincide with the release of OT (Giovannelli et al. 1990; Verbalis et al. 1991; but also see Fenelon et al. 1993). We have recently examined whether a subset of OT neurons are activated by sexual behavior indicated by the presence of Fos-ir in these neurons. Although steroid treatment itself did not cause Fos expression, Fos-ir appeared in OT neurons of the parvocellular PVN and BNST (Figs. 5.5 and 5.6) after sexual behavior in steroid-primed females (Flanagan et al. 1993). Although not definitive, these results suggest that OT neurons of the parvocellular PVN or BNST or both might project to the VMH.

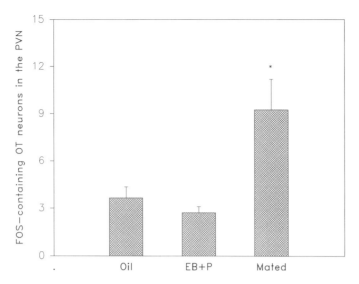

Figure 5.5. The percentage of OT-immunoreactive cells in the PVN with nuclear Fos immunostaining. Data are expressed as mean ± SEM. Asterisk indicates $p \leq .02$, compared with the other two groups. Reprinted from Flanagan et al. 1993, with permission.

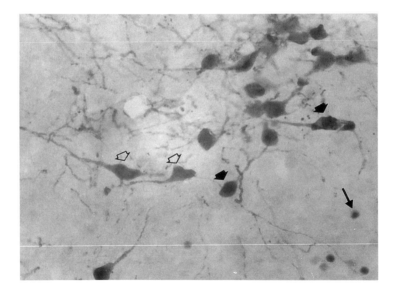

Figure 5.6. Representative PVN neurons immunocytochemically stained for oxytocin (cytoplasm) and Fos (nucleus). Cells stained for OT only are indicated by an open arrow, for Fos only by a narrow arrow, and for both OT and Fos by a thick filled arrow.

An induction of Fos-ir in cells of the ventrolateral VMH was also observed in the vicinity of OT fibers after mating behavior. Taken together with other evidence, these results support the notion that OT release contributes to neuronal excitability in the ventrolateral VMH and hence facilitates female sexual behavior in rats.

Additional preliminary studies have explored whether the pattern of mating-induced Fos expression is specific, or whether the expression of other oncogenes also exhibits this pattern. Brain sections from oil-treated, E- and P-treated, and E- and P-primed and mated animals were stained with antisera to Jun, JunB, and JunD. Each of these transcription factors showed a unique pattern of expression, which was not necessarily dependent on steroid treatment or behavior. For example, JunD was expressed in the same region-specific fashion in all treatment groups and was found in the suprachiasmatic nucleus, arcuate nucleus, PVN, preoptic area (POA), and cortex of the amygdala. In contrast, sexual behavior increased the expression of JunB in the POA, PVN, VMH, medial amygdala, amygdala cortex, and BNST; however, staining in the dorsal striatum and sensorimotor cortex occurred in all three groups, albeit more intensely in steroid-treated animals. Like Fos staining, Jun staining was observed only after mating in cell groups implicated in reproduction, including the POA, PVN, and medial amygdala. Unlike Fos staining, however, Jun staining was present in the arcuate nucleus after sexual behavior.

Sex differences

Male and female rats differ with regard to the type of sexual responses they display after hormonal priming. Males display mounting behavior, intromission, and ejaculation, whereas females exhibit lordosis in response to mating attempts by the male. Yet males and females have similar levels and distributions of receptors for estrogen, androgen, and progestin, as well as aromatizing enzymes and 5α-reductase, which convert testosterone to E and 5α-dihydrotestosterone, respectively (Lieberburg and McEwen 1977).

One of the reasons that males and females differ in their sexual responses is the developmental programming that occurs under the influence of testosterone secreted perinatally by the testes. The changes arising from this differential exposure to testosterone can be classified into two interacting components: (a) induction of sex differences in brain anatomy and connectivity (Arnold and Gorski 1984) and (b) programming of gender-dependent responses of brain cells to circulating E, P, and testosterone (McEwen et al. 1983; McEwen 1991). The developmental actions of testosterone produce male patterns of response to circulating hormones by determining the way the brain responds to circulating hormones, that is, which neural circuits are activated and what cellular responses are induced. Thus, the same levels of hormones in males and females produce different neural responses and different behavioral states.

In the VMH, developmental actions of testosterone increase the densities of axodendritic and axosomal synapses (Matsumoto and Arai 1986), thus producing a permanently different circuitry. Sex differences also occur in the induction of progestin receptors by E, with females showing greater induction than males across a range of E priming doses (Rainbow et al. 1982b; Parsons et al. 1984; Coirini and McEwen 1990; Brown et al. 1992). These differences are reversed by the presence in the female, or absence in the male, of testosterone during the early postnatal period (Parsons et al. 1984). In addition, E priming induces synapses on VMH neurons of adult OVX females, whereas it fails to do so in adult castrated males (Frankfurt et al. 1990; Frankfurt and McEwen 1991a,b). Thus, it appears likely that this sex difference, as well, is due to the presence or absence of testosterone during early postnatal development (Segarra and McEwen 1991), although the definitive experiment has not yet been performed.

One likely underlying reason for the reduced potency of E in the male VMH is the reduction of both cytosolic and nuclear forms of the E receptor in the male VMH compared with the female VMH (Brown et al. 1992). However, such quantitative sex differences do not fully account for the virtual absence of E-induced synapses in the male VMH, since E induces a significant number of progestin receptors in the male VMH compared with the female VMH. This point is underscored by the observation that E induces virtually the same number of OT receptors in the VMH of male and female rats (Coirini et al. 1989; Tribollet et al. 1990). Testosterone also induces OT receptors in the VMH, and this effect appears to involve the aromatization of testosterone to E (Johnson et al. 1989b, 1991a; Tribollet et al. 1990). The functional role of hormone-induced OT receptors in the male VMH is unknown, but it may be related to the role of OT in penile erection, stretching, and yawning. At least one study, however, has failed to demonstrate that the VMH is sensitive to the direct application of OT (Melis et al. 1986).

The contrast between the limited ability of E to increase synapses and induce progestin receptors in the VMH of males relative to females and its equipotent induction of OT receptors in males and females argues strongly in favor of sex differences in neuronal events mediated by circulating hormones; such differences might operate at the genomic level (McEwen et al. 1983; McEwen 1991). In addition, it is conceivable that these sexually dimorphic E effects might be related to sex differences in afferent innervation of the VMH, particularly in light of new evidence that excitatory amino acids play a critical role in E induction of synapses (Woolley and McEwen 1993).

Conclusions

Female sexual behavior provides a useful model system for studying the modulation of brain function by circulating steroid hormones. In the VMH, the genomic consequences of E administration alter neuronal metabolic activity and synaptic configuration, as well as the neurochemistry of an array of transmitter systems. Progesterone is also essential and has both genomic and membrane effects that can potentiate or antagonize actions of E.

Several lines of evidence support a facilitative role for OT in this behavior and indicate that ovarian steroids modulate the effects of OT in various ways. Estradiol administration causes a marked augmentation of OT binding in the VMH, and the area containing OT receptors expands laterally in the direction of VMH dendrites. It also increases the neuronal activity of OT neurons in the PVN, although the scarcity of E receptor–containing OT neurons suggests that it may do so indirectly. On the other hand, it appears that only a subset of OT neurons may be involved in sexual behavior, and it would be important to establish the presence or absence of E receptors in these particular neurons. The identification of behaviorally relevant OT neurons was accomplished by co-labeling OT with Fos, a marker of neuronal activation that is induced after rats display lordosis. The induction of Fos by sexual behavior suggests, but does not prove, that these OT-containing neurons release their peptide during mating. Like other neurotransmitters that enhance sexual behavior, the application of OT is excitatory to VMH neurons. Furthermore, interference with OT receptor either by blockade with selective receptor antagonists or by inhibition of E-induced receptor synthesis using antisense oligonucleotides attenuates lordosis. Finally, a third level of steroid regulation is represented by the requirement for P to produce the behavioral effects of OT. In contrast to the straightforward effect of E on OT receptor synthesis, however, the mechanism by which P permits OT effects on sexual behavior remains controversial. It is not certain whether P alters OT receptor binding directly or indirectly by regulating OT receptor coupling to second-messenger systems, or whether it modulates OT interactions with other cells or neurotransmitters.

In conclusion, these converging data strongly support the notion that OT is an important neuropeptide involved in the ovarian steroid modulation of female behavior. Studies of this model system have indicated several levels of interaction, including E effects on the induction of OT receptor, expansion of OT receptor territory in the VMH, and altered excitatory effects of OT on VMH neurons. As further studies clarify the cascade of steroid actions that modulate OT neurotransmission, novel interactions between these circulating hormones and neurotransmitters will almost certainly have broad implications for brain function and behavior.

Acknowledgments

This work was supported by USPHS Award NS07080. L. Flanagan was supported by USPHS Training Grant MH15125 and Postdoctoral Fellowship NS09026. Some of these data were presented in preliminary form at the Society for Neuroscience meeting in New Orleans (November 1991). The authors thank Dr. Nancy Weiland for her invaluable advice in technical and theoretical discussions of the experiments presented herein; Dr. John Fernstrom, at Western Psychiatric Institute and Clinic, for generously furnishing antisera to oxytocin; and Dr. Robert K. Cato, Dr. Daniel K. Yee, and Dr. Steven J. Fluharty, whose helpful comments improved earlier drafts of this chapter.

References

Akaishi, T., and Sakuma, Y. (1985). Estrogen excites oxytocinergic, but not vasopressinergic cells in the paraventricular nucleus of female rat hypothalamus. *Brain Res.* 335: 302–305.

Akesson, T. R., Mantyh, P. W., Mantyh, C. R., Matt, D. W., and Micevych, P. E. (1987). Estrous cyclicity of ^{125}I-cholecystokinin octapeptide binding in the ventromedial hypothalamic nucleus. *Neuroendocrinology* 45: 257–262.

Amico, J. A., Janosky, J. E., and Cameron, J. L. (1990). Effect of estradiol and progesterone upon the circadian rhythm of oxytocin in the cerebrospinal fluid of rhesus monkey. *Neuroendocrinology* 51: 543–551.

Arey, B. J., and Freeman, M. E. (1992). Activity of oxytocinergic neurons in the paraventricular nucleus mirrors the periodicity of the endogenous rhythm regulating prolactin secretion. *Endocrinology* 130: 126–132.

Arletti, R., and Bertolini, A. (1985). Oxytocin stimulates lordosis behavior in female rats. *Neuropeptides* 6: 247–253.

Arnold, A., and Gorski, R. (1984). Gonadal steroid induction of structural sex differences in the central nervous system. *Annu. Rev. Neurosci.* 7: 413–442.

Barfield, M. A., and Lisk, R. D. (1974). Relative contributions of ovarian and adrenal progesterone to the timing of heat in the 4 day cyclic rat. *Endocrinology* 94: 571–575.

Bartosik, D., Szarowski, D. H., and Watson, D. J. (1971). Influence of functioning ovarian tissue on the secretion of progesterone by the adrenal glands of female rats. *Endocrinology* 88: 1425–1427.

Boling, J. L., and Blandau, R. J. (1939). The estrogen–progesterone induction of mating responses in the spayed female rat. *Endocrinology* 25: 359–364.

Brown, T., Naftolin, F., and MacLusky, N. (1992). Sex differences in estrogen receptor binding in the rat hypothalamus: Effects of subsaturating pulses of estradiol. *Brain Res.* 578: 129–134.

Burbach, J. P. H, Adan, R. A. H., Van Tol, H. H. M., Verbeeck, M. A. E., Axelson, J. F., Van Leeuwen, F. W., Beekman, J. M., and Geert, A. (1990). Regulation of the rat oxytocin gene by estradiol. *J. Neuroendocrinol.* 2: 633–639.

Burbach, J. P. H., Liu, B., Voorhuis, T. A. M., and Van Tol, H. H. M. (1988). Diurnal variation in vasopressin and oxytocin messenger RNA in hypothalamic nuclei of the rat. *Mol. Brain Res.* 4: 157–160.

Butcher, R. L., Collins, W. E., and Fugo, N. W. (1974). Plasma concentration of LH,

FSH, prolactin, progesterone, and estradiol-17β throughout the 4-day estrous cycle of the rat. *Endocrinology* 94: 1704–1708.

Caldwell, J. D., Barakat, A. S., Smith, D. D., Hruby, V. J., and Pedersen, C. A. (1990). A uterotonic antagonist blocks the oxytocin-induced facilitation of female sexual receptivity. *Brain Res.* 512: 291–296.

Caldwell, J. D., Prange, A. J. Jr., and Pedersen, C. A. (1986). Oxytocin facilitates the sexual receptivity of estrogen-treated female rats. *Neuropeptides* 7: 175–189.

Carrer, H. F., and Aoki, A. (1982). Ultrastructural changes in the hypothalamic ventromedial nucleus of ovariectomized rats after estrogen treatment. *Brain Res.* 240: 221–233.

Chibbar, R., Toma, J. G., Mitchell, B. F., and Miller, F. D. (1990). Regulation of neural oxytocin gene expression by gonadal steroids in pubertal rats. *Mol. Endocrinol.* 4: 2030–2038.

Chung, S. K., McCabe, J. T., and Pfaff, D. W. (1991). Estrogen influences on oxytocin mRNA expression in preoptic and anterior hypothalamic regions studied by in situ hybridization. *J. Comp. Neurol.* 307: 281–295.

Cohen, R. S., and Pfaff, D. W. (1981). Ultrastructure of neurons in the ventromedial nucleus or the hypothalamus in ovariectomized rats with or without estrogen treatment. *Cell Tiss. Res.* 217: 451–470.

Coirini, H., Johnson, A. E., and McEwen, B. S. (1989). Estradiol modulation of oxytocin binding in the ventromedial hypothalamic nucleus of male and female rats. *Neuroendocrinology* 50: 193–198.

Coirini, H., and McEwen, B. S. (1990). Progestin receptor induction and sexual behavior by estradiol treatment in male and female rats. *J. Neuroendocrinol.* 2: 467–472.

Coirini, H., Schumacher, M., Flanagan, L. M., and McEwen, B. S. (1991). Transport of estradiol-induced oxytocin receptors in the ventromedial hypothalamus. *J. Neurosci.* 11: 3317–3324.

Crowley, W. R., O'Donohue, T. L., George, J. M., and Jacobowitz, D. M. (1978). Changes in pituitary oxytocin and vasopressin during the estrous cycle and after ovarian hormones: Evidence for mediation by norepinephrine. *Life Sci.* 23: 2579–2586.

De Kloet, E. R., Voorhuis, D. A. M., Boschma, Y., and Elands, J. (1986). Estradiol modulates density of putative "oxytocin receptors" in discrete rat brain regions. *Neuroendocrinology* 44: 415–421.

Etgen, A. E., and Barfield, R. J. (1986). Antagonism of female sexual behavior with intracerebral implants of antiprogestin RU 38486: Correlation with binding to neural progestin receptors. *Endocrinology* 119: 1610–1617.

Evans, J. J., Robinson, G., and Catt, K. J. (1992). Luteinizing hormone response to oxytocin is steroid-dependent. *Neuroendocrinology* 55: 538–543.

Fenelon, V. S., Poulain, D. A., and Theodosis, D. T. (1993). Oxytocin neuron activation and Fos expression: A quantitative immunocytochemical analysis of the effect of lactation, parturition, osmotic and cardiovascular stimulation. *Neuroscience* 53: 77–89.

Fischette, C., Biegon, A., and McEwen, B. S. (1983). Sex differences in serotonin 1 binding in rat brain. *Science* 222: 333–335.

Flanagan, L. M., Frankfurt, M., and McEwen, B. S. (1991). Estrogen regulation of oxytocinergic innervation of the ventromedial nucleus of the hypothalamus. *Soc. Neurosci. Abstr.* 17: 1461.

Flanagan, L. M., Pfaus, J. G., Pfaff, D. W., and McEwen, B. S. (1993). Induction of Fos immunoreactivity in oxytocin neurons after sexual activity in female rats. *Neuroendocrinology* 58: 352–358.

Frankfurt, M., Gould, E., Woolley, C. S., and McEwen, B. S. (1990). Gonadal steroids modify dendritic spine density in ventromedial hypothalamic neurons: A Golgi study in the adult rat. *Neuroendocrinology* 51: 530–535.

Frankfurt, M., and McEwen, B. S. (1991a). Estrogen increases axodendritic synapses in the ventromedial hypothalamus of rats after ovariectomy. *NeuroReport* 2: 380–382.

Frankfurt, M., and McEwen, B. S. (1991b). 5,7-Dihydroxytryptamine and gonadal steroid manipulation alter spine density in ventromedial hypothalamic neurons. *Neuroendocrinology* 54: 653–657.

Freedman, L. P. (1992). Anatomy of the steroid receptor zinc finger region. *Endocrine Rev.* 13: 129–145.

Frye, C. A., and DeBold, J. F. (1993). 3′-OH-DHP and 5′-THDOC implants to the ventral tegmental area facilitate sexual receptivity in hamsters after progesterone priming to the ventral medial hypothalamus. *Brain Res.* 612: 130–137.

Frye, C. A., Mermelstein, P. G., and DeBold, J. F. (1992). Evidence for a non-genomic action of progestins on sexual receptivity in hamster ventral tegmental area but not hypothalamus. *Brain Res.* 578: 87–93.

Gee, K. W., Chang, W. C., Brinton, R. E., and McEwen, B. S. (1987). GABA-dependent modulation of the Cl-ionophore by steroids in rat brain. *Eur. J. Pharmacol.* 136: 419–423.

Gereau, R. W., Kedzie, K. A., and Renner, K. J. (1993). Effect of progesterone on serotonin turnover in rats primed with estrogen implants into the ventromedial hypothalamus. *Brain Res. Bull.* 32: 293–300.

Giovannelli, L., Shiromani, P. J., Jirikowski, G. F., and Bloom, F. E. (1990). Oxytocin neurons in the rat hypothalamus exhibit c-fos immunoreactivity upon osmotic stress. *Brain Res.* 531: 299–303.

Glaser, J. H., and Barfield, R. J. (1984). Blockage of progesterone-activated estrous behavior in rats by intracerebral anisomycin is site specific. *Neuroendocrinology* 38: 337–343.

Gorzalka, B. B., and Lester, G. L. L. (1987). Oxytocin-induced facilitation of lordosis behavior in rats is progesterone-dependent. *Neuropeptides* 10: 55–65.

Greer, E. R., Caldwell, J. D., Johnson, M. F., Prange, A. J., and Pedersen, C. A. (1986). Variations in concentration of oxytocin and vasopressin in the paraventricular nucleus of the hypothalamus during the estrous cycle in rats. *Life Sci.* 38: 2311–2318.

Heritage, A. S., Stumpf, W. E., Sar, M., and Grant, L. D. (1980). Brainstem catecholamine neurons are target sites for sex steroid hormones. *Science* 207: 1377–1379.

Hoffman, G. E., Smith, M. S., and Verbalis, J. G. (in press). c-Fos and related immediate early gene products as markers of activity in neuroendocrine systems. *Front. Neuroendocrinol.*

Inenaga, K., Kannan, H., Yamashita, H., Tribollet, E., Raggenbass, M., and Dreifuss, J. J. (1991). Oxytocin excites neurons located in the ventromedial nucleus of the guinea-pig hypothalamus. *J. Neuroendocrinol.* 3: 569–573.

Insel, T. R. (1990). Regional induction of c-*fos*-like protein in rat brain following estradiol administration. *Endocrinology* 126: 1849–1853.

Insel, T. R., Winslow, J. T., and Witt, D. M. (1992). Homologous regulation of brain oxytocin receptors. *Endocrinology* 130: 2602–2608.

Johnson, A. E., Ball, G. F., Coirini, H., Harbaugh, C. R., McEwen, B. S., and Insel T. R. (1989). Time course of the estradiol-dependent induction of oxytocin receptor binding in the ventromedial hypothalamic nucleus of rat. *Endocrinology* 125: 1414–1419.

Johnson, A. E., Coirini, H. Ball, G. F., and McEwen, B. S. (1989a). Anatomical localization of the effects of 17β-estradiol on oxytocin receptor binding in the ventromedial hypothalamic nucleus. *Endocrinology* 124: 207–211.

Johnson, A., Coirini, H., Insel, T., and McEwen, B. S. (1991a). The regulation of oxytocin receptor binding in the ventromedial hypothalamic nucleus by testosterone and its metabolites. *Endocrinology* 128: 891–896.

Johnson, A., Coirini, H., McEwen, B. S., and Insel, T. (1989b). Testosterone modulates oxytocin binding in the hypothalamus of castrated male rats. *Neuroendocrinology* 50: 199–203.

Johnson, A. E., Harbaugh, C. R., and Gelhard, R. E. (1991b). Projections from the ventromedial nucleus of the hypothalamus contain oxytocin binding sites. *Brain Res.* 567: 332–336.

Johnston, C. A., Lopez, F., Samson, W. K., and Negro-Vilar, A. (1988). Physiologically important role of central oxytocin in the preovulatory release of luteinizing hormone. *Neurosci. Lett.* 120: 256–258.

Jones, K. J., Chikaraishi, D. M., Harrington, C. A., McEwen, B. S., and Pfaff, D. W. (1986). In situ hybridization detection of estradiol-induced changes in ribosomal RNA levels in rat brain. *Mol. Brain Res.* 1: 145–152.

Jones, K. J., Harrington, C. A., Chikaraishi, D. M., and Pfaff, D. W. (1990). Steroid hormone regulation of ribosomal RNA in rat hypothalamus: Early detection using in situ hybridization and precursor-product ribosomal DNA probes. *J. Neurosci.* 10: 1513–1521.

Jones, K. J., McEwen, B. S., and Pfaff, D. W. (1987). Quantitative assessment of the synergistic and independent effects of estradiol and progesterone on ventromedial hypothalamic and preoptic-area proteins in female rat brain. *Metab. Brain Dis.* 2: 271–281.

Jones, K. J., McEwen, B. S., and Pfaff, D. W. (1988). Quantitative assessment of early and discontinuous estradiol-induced effects on ventromedial hypothalamic and preoptic area proteins in female rat brain. *Neuroendocrinology* 48: 561–568.

Jones, K. J., Pfaff, D. W., and McEwen, B. S. (1985). Early estrogen-induced nuclear changes in rat hypothalamic ventromedial neurons: An ultrastructural and morphometric analysis. *J. Comp. Neurol.* 239: 255–266.

Kimura, T., Tanizawa, O., Mori, K., Brownstein, M. J., and Okayama, H. (1992). Structure and expression of a human oxytocin receptor. *Nature* 356: 526–529.

Kow, L.-M., Johnson, A. E., Ogawa, S., and Pfaff, D. W. (1991). Effects of ovarian steroid hormones on electrophysiological actions of oxytocin on hypothalamic neurons in vitro. *Neuroendocrinology* 54: 526–535.

Lieberburg, I., and McEwen, B. S. (1977). Brain cell nuclear retention of testosterone metabolites: 5a-Dihydrotestosterone and estradiol-17β in adult rats. *Endocrinology* 100: 588–597.

MacLusky, N. J., and McEwen, B. S. (1978). Oestrogen modulates progestin receptor concentrations in some rat brain regions but not in others. *Nature* 274: 276–278.

Mann, D. R., and Barraclough, C. A. (1973). Changes in peripheral plasma progesterone during the rat 4-day estrous cycle: An adrenal diurnal rhythm. *Proc. Soc. Exp. Biol. Med.* 142: 1226–1229.

Mathews, D., and Edwards, D. A. (1977). The ventromedial nucleus of the hypothalamus and the hormonal arousal of sexual behaviors in the female rat. *Horm. Behav.* 8: 40–51.

Matsumoto, A., and Arai, Y. (1986). Male–female difference in synaptic organization of the ventromedial nucleus of the hypothalamus in the rat. *Neuroendocrinology* 42: 232–246.

McCarthy, M. M., Kleopoulos, S. P., Mobbs, C. V., and Pfaff, D. W. (1994). Infusion of antisense oligonucleotides to the oxytocin receptor in the ventromedial hypothalamus reduces estrogen-induced sexual receptivity and oxytocin receptor binding in the female rat. *Neuroendocrinology* 59: 432–440.

McEwen, B. S. (1991). Our changing ideas about steroid effects on an ever-changing brain. *Sem. Neurosci.* 3: 497–507.

McEwen, B. S., Biegon, A., Fischette, C., Luine, V., Parsons, B., and Rainbow, T. (1983). Sex differences in programming of response to estradiol in the brain. In *Sexual Differentiation*, ed. M. Serio, M. Motta, M. Zanisi, and L. Martini, pp. 93–98. New York: Raven Press.

McGinnis, M. Y., Parsons, B., Rainbow, M. C., Krey, L. C., and McEwen, B. S. (1981). Temporal relationship between cell nuclear progestin receptor levels and sexual receptivity following intravenous progesterone administration. *Brain Res.* 218: 365–371.

Meisel, R. L., and Luttrell, V. R. (1990). Estradiol increases the dendritic length of ventromedial hypothalamic neurons in female Syrian hamsters. *Brain Res. Bull.* 25: 165–168.

Meisel, R. L., and Pfaff, D. W. (1984). RNA and protein synthesis inhibitors: Effects on sexual behavior in female rats. *Brain Res. Bull.* 12: 187–193.

Melis, M., Argiolas, A., and Gessa, G. (1986). Oxytocin-induced penile erection and yawning: Site of action in the brain. *Brain Res.* 398: 259–265.

Miller, F. D., Ozimek, G, Milner, R. J., and Bloom, F. E. (1989). Regulation of neuronal oxytocin mRNA by ovarian steroids in the mature and developing hypothalamus. *Proc. Natl. Acad. Sci. (USA)* 86: 2468–2472.

Mobbs, C. V., Kaplitt, M., Kow, L.-M., and Pfaff, D. W. (1991). PLC-α: A common mediator of the action of estrogen and other hormones? *Mol. Cell. Endocrinol.* 80: C187–C191.

Negoro, H., Visessuwan, S., and Holland, R. C. (1979). Unit activity in the paraventricular nucleus of female rats at different stages of the reproductive cycle and after ovariectomy, with or without oestrogen or progesterone treatment. *J. Endocrinol.* 59: 545–558.

Parsons, B., Rainbow, T., and McEwen, B.S. (1984). Organizational effects of testosterone via aromatization on feminine reproductive behavior and neural progestin receptors in rat brain. *Endocrinology* 115:1412–1417.

Perlow, M. J., Reppert, S. M., Artman, H. A., Fisher, D. A., Seif, S. M., and Robinson, A. G. (1982). Oxytocin, vasopressin and estrogen-stimulated neurophysin: Daily patterns of concentration in cerebrospinal fluid. *Science* 216: 1416–1418.

Pfaff, D. W. (1980). *Estrogens and Brain Function*. New York: Springer.

Pfaff, D., and Keiner, M. (1973). Atlas of estradiol-concentrating cells in the central nervous system of the female rat. *J. Comp. Neurol.* 151: 121–158.

Pfaff, D. W., and Sakuma, Y. (1979a). Deficit in the lordosis reflex of female rats caused by lesions in the ventromedial nucleus of the hypothalamus. *J. Physiol.* 288: 203–210.

Pfaff, D. W., and Sakuma, Y. (1979b). Facilitation of the lordosis reflex of female rats from the ventromedial nucleus of the hypothalamus. *J. Physiol.* 288: 189–202.

Purdy, R. H., Morrow, A. L., Moore, P. H., and Paul, S. M. (1991). Stress-induced elevations of gamma-aminobutyric acid type-A receptor active steroids in the rat brain. *Proc. Natl. Acad. Sci. (USA)* 88: 4553–4557.

Rainbow, T. C., McGinnis, M. Y., Davis, P. G., and McEwen, B. S. (1982a). Application of anisomycin to the lateral ventromedial nucleus of the hypothalamus inhibits the activation of sexual behavior by estradiol and progesterone. *Brain Res.* 233: 417–423.

Rainbow, T., Parsons, B., and McEwen, B. S. (1982b). Sex differences in rat brain oestrogen and progestin receptors. *Nature* 300: 648–649.

Rainbow, T., Snyder, L., Berck, D., and McEwen, B. S. (1984). Correlation of muscarinic receptor induction in the ventromedial hypothalamus nucleus with the activation of female sexual behavior by estradiol. *Neuroendocrinology* 39: 476–480.

Renner, K. J., Krey, L. C., and Luine, V. N. (1987). Effect of progesterone on mono-amine turnover in the brain of the estrogen-primed rat. *Brain Res. Bull.* 19: 195–202.

Rhodes, C. H., Morrell, J. I., and Pfaff, D. W. (1981). Changes in oxytocin content in the magnocellular neurons of the rat hypothalamus following water deprivation or estrogen treatment: Quantitative immunohistochemical studies. *Cell Tiss. Res.* 216: 47–55.

Rhodes, C. H., Morrell, J. I., and Pfaff, D. W. (1982). Estrogen-concentrating neuro-physin-containing hypothalamic magnocellular neurons in the vasopressin-deficient (Brattleboro) rat: A study combining autoradiography and immunocytochemistry. *J. Neurosci.* 2: 1718–1724.

Richard, S., and Zingg, H. H. (1990). The human oxytocin gene promoter is regulated by estrogens. *J. Biol. Chem.* 265: 1–6.

Roberts, M. M., Robinson, A. G., Hoffman, G. E., and Fitzsimmons, M. D. (1991). Vasopressin transport regulation is coupled to the synthesis rate. *Neuroendocrinology* 53: 416–422.

Robinson, A. G. (1974). Elevation of plasma neurophysin in women on oral contracep-tives. *J. Clin. Invest.* 54: 209–212.

Robinson, B. G., Frim, D. M., Schwartz, W. J., and Majzoub, J. A. (1988). Vaso-pressin mRNA in the suprachiasmatic nuclei: Daily regulation of polyadenylate tail length. *Science* 241: 342–344.

Robinson, G., and Evans, J. J. (1990). Oxytocin has a role in gonadotrophin regulation in rats. *J. Endocrinol.* 125: 425–432.

Romano, G. J., Harlan, R. E., Shivers, B. D., Howells, R. D., and Pfaff, D. W. (1989). Estrogen increases proenkephalin messenger ribonucleic acid levels in the ventromedial hypothalamus of the rat. *Mol. Endocrinol.* 2: 1320–1328.

Rubin, B. S., and Barfield, R. J. (1983). Induction of estrous behavior in ovariectomized rats by sequential replacement of estrogen and progesterone to the ventromedial hy-pothalamus. *Neuroendocrinology* 37: 218–224.

Ruzycky, A. L., and Triggle, D. J. (1988). Role of inositol phospholipid hydrolysis in the initiation of agonist-induced contractions of rat uterus: Effects of domination by 17-estradiol and progesterone. *Can. J. Physiol. Pharmacol.* 66: 10–17.

Samson, W. K., Lumpkin, M. D., and McCann, S. M. (1986). Evidence for a physi-ological role for oxytocin in the control of prolactin secretion. *Endocrinology* 119: 554–560.

Sarkar D. K., and Gibbs, D. M. (1984). Cyclic variation of oxytocin in the blood of pituitary portal vessels of rats. *Neuroendocrinology* 39: 481–483.

Sawchenko, P. E., and Swanson, L. W. (1982). Immunohistochemical identification of neurons in the paraventricular nucleus of the hypothalamus that project to the medulla and to the spinal cord in the rat. *J. Comp. Neurol.* 250: 260–272.

Schumacher, M., Coirini, H., Frankfurt, M., and McEwen, B. S. (1989a). Localized actions of progesterone in hypothalamus involve oxytocin. *Proc. Natl. Acad. Sci. (USA)* 86: 6798–6801.

Schumacher, M., Coirini, H., and McEwen, B. S. (1989b). Regulation of high-affinity GABAa receptors in specific brain regions by ovarian hormones. *Neuroendocrinol-ogy* 50: 315–320.

Schumacher, M., Coirini, H., McEwen, B. S., and Zaborsky, L. (1991a). Binding of [3H]cholecystokinin in the ventromedial hypothalamus is modulated by an afferent brainstem projection but not by ovarian steroids. *Brain Res.* 564: 102–108.

Schumacher, M., Coirini, H., Pfaff, D. W., and McEwen, B. S. (1990). Behavioral effects of progesterone associated with rapid modulation of oxytocin receptors. *Sci-ence* 250: 691–694.

Schumacher, M., Coirini, H., Pfaff, D. W., and McEwen, B. S. (1991b). Light–dark differences in behavioral sensitivity to oxytocin. *Behav. Neurosci.* 105: 487–492.

Segarra, A., and McEwen, B. S. (1991). Estrogen increases spine density in ventromedial hypothalamic neurons of peripubertal rats. *Neuroendocrinology* 54: 365–372.

Simerly, R. B., Chang, C., Muramatsu, M., and Swanson, L. W. (1990). Distribution of androgen and estrogen receptor mRNA-containing cells in rat brain: An in situ hybridization study. *J. Comp. Neurol.* 294: 76–95.

Tribollet, E., Audigier, S., Dubois-Dauphin, M., and Dreifuss, J. J. (1990). Gonadal steroids regulate oxytocin receptors but not vasopressin receptors in the brain of male and female rats: An autoradiographical study. *Brain Res.* 511: 129–140.

Urade, R., Nasu, M., Moriyama, T., Wada, K., and Kito, M. (1992). Protein degradation by the phosphoinositide-specific phospholipase C-α family from rat liver endoplasmic reticulum. *J. Biol. Chem.* 267: 15152–15159.

Van Tol, H. H. M., Bolwerk, E. L. M., Liu, B., and Burbach, J. P. H. (1988). Oxytocin and vasopressin gene expression in the hypothalamo-neurohypophyseal system of the rat during estrous cycle, pregnancy, and lactation. *Endocrinology* 122: 945–951.

Verbalis, J. G., Stricker, E. M., Robinson, A. G., and Hoffman, G. E. (1991). Cholecystokinin activates c-*fos* expression in hypothalamic oxytocin and corticotropin-releasing hormone neurons. *J. Neuroendocrinol.* 3: 205–213.

Vincent, P. A., and Etgen, A. M. (1993). Steroid priming promotes oxytocin-induced norepinephrine release in the ventromedial hypothalamus of female rats. *Brain Res.* 620: 189–194.

Windle, R. J., Forsling, M. L., and Guzek, J. W. (1992). Daily rhythms in the hormone content of the neurohypophyseal system and release of oxytocin and vasopressin in the male rat: Effect of constant light. *J. Endocrinol.* 133: 282–290.

Witcher, J. A., and Freeman, M. E. (1985). The proestrous surge of prolactin enhances sexual receptivity in the rat. *Biol. Reprod.* 32: 834–839.

Witt, D. M., and Insel, T. R. (1991). A selective oxytocin antagonist attenuates progesterone facilitation of female sexual behavior. *Endocrinology* 128: 3269–3276.

Woolley, C., and McEwen, B. S. (1993). Estradiol regulates hippocampal dendritic spine density via an NMDA receptor dependent mechanism. *Soc. Neurosci. Abstr.* 19: 379.15.

Yahr, P., and Ulibarri, C. (1986). Estrogen induction of sexual behavior in female rats and synthesis of polyadenylated messenger RNA in the ventromedial nucleus of the hypothalamus. *Mol. Brain Res.* 1: 153–165.

Yamaguchi, K., Akaishi, T., and Negoro, H. (1979). Effect of estrogen treatment on plasma oxytocin and vasopressin in ovariectomized rats. *Endocrinol. Jpn.* 26: 197–205.

Yoshimura, R., Kiyama, H., Kimura, T., Araki, T., Maeno, H., Tanizawa, O., and Tohyama, M. (1993). Localization of oxytocin receptor messenger ribonucleic acid in the rat brain. *Endocrinology* 133: 1239–1246.

Zingg, H., and Lefebvre, D. L. (1988). Oxytocin and vasopressin gene expression during gestation and lactation. *Mol. Brain Res.* 4: 1–6.

Zingg, H., and Lefebvre, D. L. (1989). Oxytocin mRNA: Increase of polyadenylate tail size during pregnancy and lactation. *Mol. Cell Endocrinol.* 65: 59–62.

6

Sex steroid regulation of hypothalamic opioid function

RONALD P. HAMMER, JR., AND SUN CHEUNG

The discovery that the hypothalamus is responsible for controlling both reproductive hormones and behavior suggested various mechanisms by which hormonal and behavioral cycles are inexorably linked, and even co-regulated. While our knowledge about this process has grown dramatically, our understanding of the essential control circuits that operate during normal reproduction, or fail in abnormal functioning, is still limited. Various neurotransmitter candidates have been proposed as essential elements of the systems that regulate reproduction. However, few are involved in so many aspects of reproduction as are the opioid peptides, which play a critical or supporting role in (a) controlling hormonal cycling in females (Akabori and Barraclough 1986; Kalra 1985; Wiesner et al. 1984), (b) regulating reproductive behavior in males (Hughes et al. 1988; Matuszewich and Dornan 1992; Myers and Baum 1979) and females (Pfaus and Pfaff 1992; Sirinathsinghji 1986; Wiesner and Moss 1986a), and even (c) modulating mesolimbic dopamine release mediated by reinforcing sexually relevant olfactory stimuli (Mitchell and Gratton 1991).

Gonadal steroid regulation of hypothalamic opioids represents an important feedback system by which to control reproduction. Hypothalamic (opioid) circuits regulate hormonal releasing hormones that control pituitary secretion. This regulation in turn affects gonadal steroid hormones, which act centrally to alter opioid function and facilitate reproductive behavior. Since such feedback is vitally important for the regulation of hormonal and behavioral events during the estrous cycle, most of this discussion will be limited to opioid action in females. Many of the experiments that we will describe utilized models of hormone manipulation to investigate natural regulation of hypothalamic opioid systems in animals, primarily rodents. Thus, the examination of gonadal steroid effects on hypothalamic opioids provides a means by which to describe both opioid control of reproductive hormones and behavior and feedback regulation of opioid control systems.

143

Heterogeneity of opioid systems

The diversity of opioid function derives, in part, from the multiplicity of endogenous opioid peptides and the variety of discrete opioid receptor types (and subtypes) that exist in the central nervous system. The rapid development of our knowledge in this area extends from the initial description of stereospecific binding of opiate analgesics to membrane receptors (Pert and Snyder 1973; Simon et al. 1973; Terenius 1973) and the characterization of several classes of endogenous opioid peptides (Cox et al. 1976; Goldstein et al. 1979; Hughes et al. 1975).

Opioid receptors may be described as belonging to one of three main classes or types: μ, δ, and κ. Each type has a unique ontogenetic pattern and adult distribution in the central nervous system, which have been described in detail elsewhere (reviewed by Leslie and Loughlin 1993). We now know that μ-, δ-, and κ-receptors are products of homologous but distinct genes (Chen et al. 1993; Evans et al. 1992; Fukuda et al. 1993; Wang et al. 1993; Yasuda et al. 1993). Only κ-receptors are localized to the hypothalamus in high density (Mansour et al. 1987), even though these receptors have rarely been implicated in reproductive function.

With regard to the endogenous ligands of these receptors, three families of opioid peptides have been identified: endorphins, enkephalins, and dynorphins. These compounds are produced by transcription of three distinct genes and subsequent translation of the resulting mRNAs to form precursor molecules that may be post-translationally processed into various end products (Evans et al. 1988). β-Endorphin (β-Endo) and its various processed forms are derived from the precursor proopiomelanocortin (POMC), which also contains the sequences of melanocyte-stimulating hormone and adrenocorticotropic hormone. The pentapeptides methionine- (met-) and leucine- (leu-) enkephalin (Enk), as well as several longer opioids, are contained within the precursor proenkephalin (PPE). Finally, the various forms of dynorphin (A and B) are contained within the precursor prodynorphin. The relationship of this family of endogenous peptides to the three opioid receptor types is not, however, exclusive. For example, β-Endo has approximately equal affinity for μ- and δ-receptors with little activity at κ-receptors; met- and leu-Enk have higher affinity for δ- than μ-receptors with little κ-receptor activity, although longer enkephalinergic peptides may have selective affinity for μ-receptors (Hurlbut et al. 1987); and dynorphin shows preferential activity at the κ-receptor (Paterson et al. 1983). Fortunately, various selective ligands with which to study the binding characteristics of and hormonal effects on μ-, δ-, and κ-receptors have been identified.

Gonadal steroid effects on opioid receptors

Despite the reported low density of μ- and δ-receptors in the hypothalamus (Mansour et al. 1987), significant sex differences of hypothalamic μ-receptor density have been observed (Hammer 1984; Ostrowski et al. 1987). In fact, brain opioid receptors were among the first neurotransmitter receptor systems reportedly altered by gonadal steroid hormones (Hahn and Fishman 1979). However, studies of hormonal effects on hypothalamic brain homogenates have yielded inconsistent results (Martini et al. 1989; Wilkinson et al. 1985), probably because certain hormone-induced functional changes in distinct hypothalamic opioid circuits might be obscured by the assessment of large portions of homogenized hypothalamus. Therefore, *in vitro* receptor autoradiography has been utilized to examine cyclical and hormone-induced changes of μ-receptor binding density in morphologically defined hypothalamic regions.

The medial preoptic nucleus (MPN) and its composite segments, the medial (MPNm), lateral (MPNl), and central (MPNc) subdivisions, are sexually dimorphic in rats, providing a morphological substrate for the neurochemical sex differences present in this region (Simerly et al. 1984). The MPNc, or sexually dimorphic nucleus of the preoptic area, is larger in male rats and contains a greater number of neurons (Gorski et al. 1978, 1980). Autoradiographic studies reveal that the binding density of [^{3}H]D-Ala^2-MePhe^4-Gly-ol^5-enkephalin (DAGO) to μ-receptors is elevated in the MPN of female rats and varies across the estrous cycle, with low density during proestrus and high density during met- and diestrus (Hammer 1990). In contrast, μ-receptor labeling is lower in the same region of males, and no sex difference of κ-receptor binding is observed. These receptors are initially expressed during an early postnatal critical period (Hammer 1985a), but only in the absence of androgen receptor activation (Hammer 1985b, 1988).

The cyclical expression of MPN μ-receptors is dependent on the presence of gonadal steroid hormones in a varying pattern which is similar to that of the estrous cycle (Mateo et al. 1992). Neither estradiol (E_2) alone nor E_2 followed by 3 hours of progesterone (P) exposure [E_2P(3)] increases MPN μ-receptor density in ovariectomized (OVX) female rats (Fig. 6.1). However, [^{3}H]DAGO binding is significantly elevated 27 hours after P treatment [E_2P(27)], and subsequently declines 24 hours later in the presence or absence of E_2. This hormone-induced alteration of [^{3}H]DAGO labeling in the MPN is due to an apparent increase in the binding capacity (B_{max}) or number of μ-receptors, since the density of labeling with a saturating concentration of ligand increased significantly following E_2P(27) treatment (Fig. 6.2; Zhou and Hammer 1993). Furthermore, the relative difference in binding density between E_2P(27) and vehicle treatment is greater when a saturating concentration of ligand is used than when a lower ligand concentration is used (compare Fig. 6.2 with Fig. 6.1), suggesting that a compensatory change of binding affinity might occur in concert with the hormonal induction of μ-receptor expression. Precise characterization of hormonal effects on μ-receptor binding in the MPN, however, would require autoradiographic analyses of Scatchard experiments conducted in regional tissue sections. These data

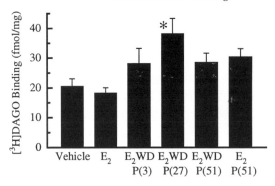

Figure 6.1. [³H]DAGO binding to μ-receptors in the medial preoptic area of OVX female rats treated with E_2 in a Silastic capsule implanted subcutaneously for 48 hours (E_2), E_2 capsule withdrawn after 48 hours followed by 3 hours P exposure [E_2WD P(3)], E_2 capsule withdrawn after 48 hours followed by 27 hours P exposure [E_2WD P(27)], E_2 capsule withdrawn after 48 hours followed by 51 hours P exposure [E_2WD P(51)], E_2 capsule priming followed by 51 hours P exposure [E_2 P(51)], or vehicle. Binding density is significantly ($p \leq .05$) increased following E_2WD P(27) treatment compared with vehicle or E_2 treatment. Taken with permission from Mateo et al. (1992; S. Karger AG, Basel).

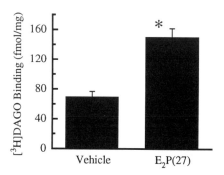

Figure 6.2. Binding density of 10 nM [³H]DAGO to μ-receptors in the medial preoptic area of OVX female rats treated with E_2 capsule for 48 hours followed by 27 hours P exposure [E_2P(27)] or vehicle. At this concentration, approximately 95% of receptors are occupied by the radioligand. Binding density is significantly ($p \leq .05$) increased following E_2WD P(27) treatment compared with vehicle treatment. Data from Zhou and Hammer (1993).

show that gonadal steroid hormones can upregulate MPN μ-receptors. However, the time course of this effect suggests either that extensive processing of the receptor protein occurs before emergence of functional binding sites or that these receptors are synthesized some distance away from the MPN and transported there along the axons of afferent neurons. The resulting receptors could be located either pre- or postsynaptically. This question could be elucidated by *in situ* hybridization histochemistry with nucleotide probes complementary to the cloned sequence of the receptor (Chen et al. 1993; Wang et al. 1993) to examine the localization of hypothalamic neurons containing μ-receptor mRNA and the effects of hormonal manipulation on μ-receptor expression in these neurons.

The observed pattern of hormonal effects suggests that MPN μ-receptors are physiologically regulated by gonadal hormone feedback. MPN μ-receptor density is low when the plasma E_2 level is elevated (e.g., during proestrus) or immediately after the administration of a hormonal priming regimen that facilitates lordosis in females receiving appropriate stimuli (Pfaff and Sakuma 1979). Such hormonal priming upregulates MPN μ-receptors 27 hours later, as occurs during the estrous cycle wherein μ-receptor density is elevated during metestrus (Hammer 1990). The subsequent decline of MPN μ-receptor density during estrous cycling could result from turnover of receptor proteins or from downregulation. This process appears not to be dependent on E_2 level, since μ-receptor density declines in the presence or absence of E_2 (Mateo et al. 1992). However, MPN μ-receptor density is positively correlated with declining P level at parturition (Hammer et al. 1992), suggesting that the process of receptor turnover could dominate in the absence of an active stimulus that normally promotes receptor expression. It should be noted that other factors present during the estrous cycle of intact females, but not in hormone-treated OVX animals, could also play a role.

An additional mechanism by which E_2 can affect μ-receptors has been described. μ-Opioid agonists act to hyperpolarize arcuate (Kelly et al. 1990) and medial preoptic neurons (M. Kelly, personal communication) by activating potassium conductance, without altering receptor binding. This effect is suppressed in the presence of E_2, probably by a decrease in the efficacy of G-protein coupling of the receptor to potassium channels (Kelly et al. 1992). Thus, opioid activity would be reduced further in the MPN and other hypothalamic regions in the presence of E_2. This mechanism certainly appears consistent with the outcome of cyclical hormone regulation of MPN μ-receptor density described earlier and provides a mechanism for "functional" downregulation of μ-receptor efficacy during proestrus.

Regulation of opioids by gonadal steroid hormones

Gonadal steroid hormones regulate the expression and distribution of Enk and β-Endo within hypothalamic nuclei. Although these peptides might affect the same opioid receptors in some circumstances, they represent distinct neuronal systems and their regulation should be considered separately.

Alteration of enkephalinergic circuits by gonadal hormones

Early reports showed that E_2 replacement increased preoptic met-Enk content (Dupont et al. 1980), which was observed to be elevated during proestrus (Kumar et al. 1979). Immunohistochemical studies reveal that the MPNl contains a larger dense plexus of Enk-like immunoreactive (IR) fibers in female than male rats, with enkephalinergic cell bodies present in the MPNm (Simerly et al. 1988). This sex difference, however, is not present until postnatal day 12 (Ge et al. 1993), which is after the critical period for sexual differentiation of the MPN (Dohler et al. 1986; Jacobson et al. 1985). The periventricular preoptic area also contains a sexually dimorphic met-Enk distribution that is denser in female rats (Watson et al. 1986).

In addition to the enkephalinergic cells in the MPNm, PPE mRNA is densely localized in the anterior hypothalamus and ventrolateral portion of the ventromedial nucleus of the hypothalamus (VMHvl) (Harlan et al. 1987). These regions probably contribute to the MPN enkephalinergic fiber field as neurons of both areas innervate the MPN (Simerly and Swanson 1986). As many as 20% of VMHvl PPE neurons project to the MPN (Hammer et al. 1991), and VMHvl PPE neurons are also known to innervate the mesencephalic central gray region (Yamano et al. 1986), which facilitates lordosis in female rats (Sakuma and Pfaff 1979). Most (80–90%) of the neurons located in the VMHvl of female rats contain PPE (Fig. 6.3), and chronic E_2 treatment further increases the number of VMHvl neurons expressing PPE and doubles the amount of PPE mRNA per cell (Romano et al. 1988). The time course of this effect has been studied by means of *in situ* hybridization histochemistry. Estrogenic induction of VMHvl PPE mRNA is biphasic and rapid; following the administration of a pulse of estradiol benzoate, the number of VMHvl neurons expressing PPE mRNA doubles within 1 hour, declines after 4 hours, and gradually increases again by 48 hours, returning to control level by 96 hours (Fig. 6.3; Priest et al., in press). This effect could be related to E_2 level, since PE expression falls rapidly after the removal of E_2-containing capsules unless P is present (Romano et al. 1989; also see Chapter 15 by Freidin and Pfaff, this volume, for further discussion of this regulatory mechanism). The extent of this effect is surprising, because only 30% of VMHvl Enk-IR neurons are thought to concentrate E_2 (Akesson and Micevych 1991). However, some VMHvl neurons might express low levels of E_2 receptor that are undetectable using E_2 labeling methods, thereby allowing induction of gene expression upon E_2 exposure.

Regulation of POMC products by gonadal hormones

Chronic E_2 treatment is thought to decrease hypothalamic β-Endo content (Wardlaw et al. 1982), and a similar effect might underlie the relative reduction of anterior hypothalamic β-Endo content observed during proestrus compared with diestrus (Knuth et al. 1983). Immunohistochemical analyses of β-Endo-like IR fibers in the preoptic area reveal a density gradient from medial to lateral, with the highest density in the periventricular region (PVN) and the lowest density in the lateral preoptic area (Fig. 6.4). Although no sex difference in fiber density was observed between males and OVX female rats treated with E_2 (Simerly et al.

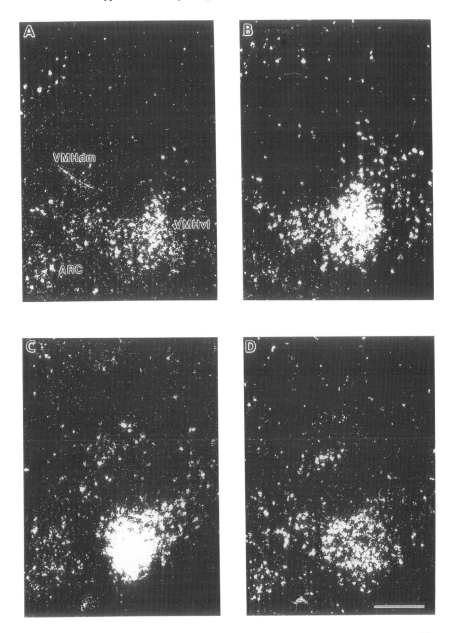

Figure 6.3. Dark-field microscopic images following *in situ* hybridization with an ^{35}S-labeled riboprobe complementary to the sequence of proenkephalin (PE) mRNA. Animals were ovariectomized for 2 weeks, then treated with a subcutaneous injection of 50 μg E_2 in safflower oil, and tissues were processed for histochemistry 0 (A), 1 (B), 48 (C), or 96 (D) hours after E_2 injection. The number of neurons labeled in the ventrolateral portion of the VMHvl increases significantly ($p \leq .05$) by 1 hour, declines slightly, then gradually increases to peak at 48 hours, and decreases to control level by 96 hours. The dorsomedial portion of the ventromedial hypothalamus (VMHdm) and the ARC are also visible. Scale bar, 150 μm. Original illustration prepared by Cathy Priest, after data from Priest et al. (in press).

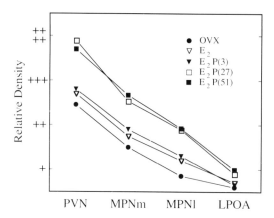

Figure 6.4. Relative density on a scale from 0 to 4 (+ + + +) of hormone effects on β-Endo-like IR fiber density in the preoptic area. Treatment groups are the same as those in Figure 6.1. A descending density gradient is present from the paraventricular nucleus (PVN) medially, through the medial portion of the medial preoptic nucleus (MPNm) and lateral portion of the medial preoptic nulceus (MPNl), to the lateral preoptic area (LPOA) laterally. E_2P treatment increased the density of fibers present in the PVN, MPNm, and MPNl only after longer P exposure times. Data from Cheung et al. (1992).

1988), gonadal steroid hormone treatment does alter preoptic β-Endo-like IR fibers in OVX rats (Cheung et al. 1992). The relative density of such fibers in the PVN, MPNm, and MPNl increases 27 hours after $E_2P(27)$ treatment, and remains high for at least 24 hours more (Fig. 6.4). It should be noted that E_2 alone or $E_2P(3)$ treatment has little effect on β-Endo-like IR fiber density. Thus, hormonal treatment does not produce an immediate effect on preoptic β-Endo content; instead, E_2P increases β-Endo-like IR fiber density roughly 1 day later. The pattern of hormone-induced changes is quite similar to that exhibited by μ-receptor density observed in the same brain region following similar steroid hormone treatment (Mateo et al. 1992)

In the cycling female rat, the time course of hormonal induction of preoptic β-Endo-like IR fibers appears to be similar (Ge and Hammer 1992). The density of fibers increases on the day of estrus, but is lower on all other days of the cycle (Fig. 6.5), and the medial-to-lateral gradient of decreasing fiber density is retained throughout the cycle. Thus, preoptic β-Endo-like IR fiber density increases within 1 day after the peak concentrations of E_2 and P are achieved (Freeman 1988), similar to the effect of E_2P administration in OVX rats. However, the rapid subsequent reduction of β-Endo-like IR staining observed during the estrous cycle suggests that some additional factor might either decrease expression or increase release of β-Endo during metestrus in the cycling animal.

POMC-Expressing neurons are ideally situated in the arcuate nucleus (ARC) to influence either hypothalamohypophyseal activity through short projections to the median eminence (ME) or neuronal activity and reproductive behavior via ascending projections. Chronic E_2 treatment decreases POMC mRNA in the ARC

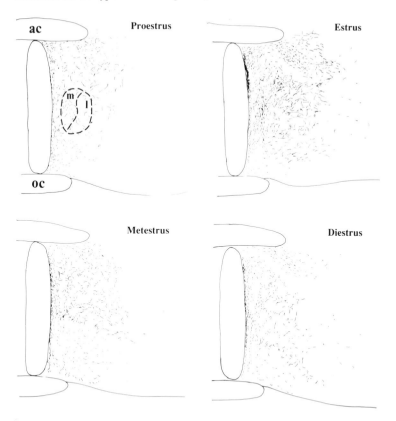

Figure 6.5. Camera lucida drawing of β-Endo-like IR fibers in the preoptic area of female rats in proestrus, estrus, metestrus, and diestrus. A higher density of labeling is present in all regions during estrus. The approximate location of the medial (m) and lateral (l) portions of the medial preoptic nucleus is shown by dashed lines. Abbreviations: ac, anterior commissure; oc, optic chiasm. Taken with permission from Ge and Hammer (1992).

(Wilcox and Roberts 1985), and the normal diurnal rhythm of POMC mRNA expression in ARC neurons is abolished in OVX animals, while POMC mRNA levels increase after E_2P treatment (Wise et al. 1990). Even in males, the regulation of ARC POMC mRNA by testosterone is possible only after aromatization to E_2 (Chowen et al. 1990). About 10% of ARC POMC neurons innervate the preoptic area, terminating more densely in the MPNm than the MPNl (Cheung and Hammer 1993). Thus, hormonal regulation of POMC mRNA expression could underlie the alterations of preoptic β-Endo content observed following hormone treatment or during the estrous cycle, even though the timing of hormone-induced mRNA induction might differ from that of protein expression at axon terminals.

Opioid regulation of reproduction

A major function of hypothalamic opioids in female reproduction is the control of cyclical sex hormones. Implantation of crystals containing the opioid antagonist naloxone into the medial preoptic area (MPOA) or ARC ME rapidly stimulates the release of luteinizing hormone (LH) (Kalra 1981). In contrast, β-Endo injected into the MPOA or VMH decreases plasma LH in OVX rats (Wiesner et al. 1984), suggesting that a reduction of hypothalamic opioid tone could precipitate LH release during proestrus (Kalra 1985). This process probably involves additional neurotransmitters such as norepinephrine (Akabori and Barraclough 1986; Kalra and Kalra 1984) and/or glutamate (Bonavera et al. 1993). Control of LH release is dependent on LH-releasing hormone (LHRH). Most LHRH neurons are located in the ventral portion of the preoptic area, and their axons reach the ME by passing caudally along either a medial or a lateral route (King et al. 1982). Thus, ARC POMC neurons could mediate LHRH neuronal activity through their β-Endo-containing axons, which contact LHRH neuron dendrites in the MPOA (Leranth et al. 1988), as shown in Figure 6.6. These POMC neurons could also modulate LHRH release via direct projections to the site of LHRH terminals in the ME, as has been demonstrated in the ewe (Conover et al. 1993). Since opioids have a tonic inhibitory effect on LHRH release (Kalra 1985), opioid tone must decrease at either or both ends of the LHRH neuron in order to facilitate the preovulatory surge of LHRH. This is precisely what occurs in the MPOA, wherein both β-Endo-like IR fiber density (Ge and Hammer 1992) and μ-receptor density (Hammer 1990) are lowest during proestrus. This simultaneous, cyclical downregulation of μ-receptors and their endogenous ligands probably occurs independently, since compensatory alteration is not observed. Moreover, the gradual rate of increasing opioid fiber and receptor density in this region suggests that preoptic opioids are not involved in the rapid postsurge *decrease* of LHRH release, but rather might promote a sustained inhibition of LHRH activity after the surge. However, other functions might also be subserved by this subsequent increase of preoptic opioid activity.

Another likely target for hypothalamic opioids is the modulation of lordosis behavior. Injection of β-Endo into the cerebral ventricle (Wiesner and Moss 1984, 1986b) of hormone-primed female rats suppresses lordosis behavior, and systemic pretreatment with μ-antagonists, but not δ-antagonists, prevents this effect (Wiesner and Moss 1986b). Direct infusion of β-Endo into the mesencephalic central gray matter (CG) also suppresses lordosis, perhaps via interaction with LHRH terminals (Sirinathsinghji et al. 1983). In addition, β-Endo injection into the MPOA suppresses lordosis (Sirinathsinghji 1986), which is facilitated by site-specific injection of a selective μ-receptor antagonist into the MPN (Hammer et al. 1989). Finally, intraventricular administration of a selective μ-receptor agonist inhibits, while a δ-agonist facilitates and a κ-agonist does not affect, lordosis (Pfaus and Pfaff 1992).

Even though few of these data were obtained by site-specific injection of opioid ligands and they cannot, therefore, resolve the locus of action for the observed

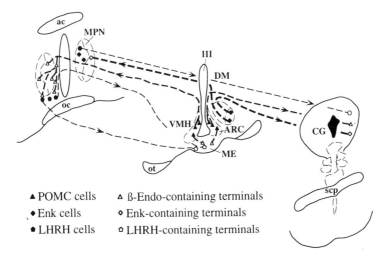

▲ POMC cells   △ ß-Endo-containing terminals
◆ Enk cells   ◇ Enk-containing terminals
● LHRH cells   ○ LHRH-containing terminals

Figure 6.6. Schematic illustration of hypothalamic opioid neurons and their fibers at the level of the MPOA, VMH, and CG. POMC-Containing neurons located in the ARC project to the MPOA and MPN, the median eminence (ME), and the CG. In the MPOA, β-Endo-containing terminals of these cells affect neurons which produce LHRH. LHRH neurons project by various routes to the ME. Neurons that express proenkephalin are located in the MPN and ventrolateral VMH. These VMH enkephalinergic neurons project rostrally to supply the MPN and caudally toward the CG. MPN neurons also project directly to the CG. See text for further explanation of functional circuitry. Abbreviations: ac, anterior commissure; oc, optic chiasm; ot, optic tract; DM, dorsomedial nucleus; III, third ventricle; scp, decussation of the superior cerebellar peduncle.

effects, the results suggest that (a) the MPOA, VMH, and CG are involved in the opioid regulation of lordosis, and (b) μ- and δ-receptors might have opposing effects on lordosis. Electrical stimulation of the CG facilitates, while lesions localized to this midbrain region reduce, lordosis (Sakuma and Pfaff 1979). The MPN also provides a rich innervation to the CG (Simerly and Swanson 1988). Therefore, the CG could represent the final common path for opioid control of lordosis behavior mediated either directly, via ARC β-Endo or VMHvl Enk projections to this region, or indirectly, via opioid interactions in the MPN, which then affect MPN projections to the CG. These pathways are illustrated schematically in Figure 6.6.

The increase of POMC transcription following E_2P treatment could inhibit lordosis by facilitating β-Endo release in CG projections. Alternatively, the gradual hormone-induced increase of β-Endo in the MPN, wherein μ-receptors are already "primed" to respond optimally, could decrease local neuronal activity, which would provide less input to the CG, thereby suppressing lordosis (Fig. 6.6). The neurotransmitter(s) contained within these MPN projection neurons are unknown; however, they might utilize Enk or some other sexually dimorphic neurotransmitter substance.

Although the influence of hormone levels on PPE expression is quite clear, the ultimate effect of Enk on lordosis behavior remains equivocal. If the E_2-induced rise of VMHvl PPE expression were rapidly translated into elevated Enk content and release from CG terminals, then Enk could facilitate lordosis by affecting δ-receptors there (Fig. 6.6). In fact, the dose-dependent estrogenic induction of VMHvl PPE expression reportedly correlates with increasing lordosis behavior (Lauber et al. 1990). However, the finite rate of mRNA translation, post-translational processing, and subsequent transport of the peptide product from VMHvl neurons to CG terminals precludes an immediate impact of altered PPE expression on Enk release (Priest et al., in press). Furthermore, site-specific injection of met-Enk into the CG in the presence of kelatorphan, an enkephalinase inhibitor, *suppresses* lordosis in E_2P-primed female rats (Bednar et al. 1987). Alternatively, VMHvl PPE neurons could act locally to affect activity in other inhibitory neurons that project to the CG, the net effect of which would be an indirect facilitation of lordosis. In any case, opioid peptides localized in specific brain circuits might be capable of regulating both the initiation and the termination of lordosis behavior.

It is clear, then, that endogenous opioid peptides acting at μ- and δ-receptors are important participants in the regulation of reproduction. Obviously, the complete mechanism is even more complicated than that proposed herein; we have not discussed the potential interaction of opioids with other peptide and/or nonpeptide neurotransmitters acting on the same postsynaptic neurons or synapses, nor have we addressed the putative interaction of additional neurotransmitters that might be co-released with opioids, but differentially regulated. Rather, we have described but one element of an elaborate process that permits the precise regulation of a complex behavioral and physiological process.

References

Akabori, A., and Barraclough, C. A. (1986). Gonadotropin responses to naloxone may depend upon spontaneous activity in noradrenergic neurons at the time of treatment. *Brain Res.* 362: 55–62.

Akesson, T. R., and Micevych, P. E. (1991). Endogenous opioid-immunoreactive neurons of the ventromedial hypothalamic nucleus concentrate estrogen in male and female rats. *J. Neurosci. Res.* 28: 359–366.

Bednar, I., Forsberg, G., and Södersten, P. (1987). Inhibition of sexual behavior in female rats by intracerebral injections of met-enkephalin in combination with an inhibitor of enkephalin degrading enzymes. *Neurosci. Lett.* 79: 341–345.

Bonavera, J. J., Kalra, S. P., and Kalra, P. S. (1993). Evidence that luteinizing hormone suppression in response to inhibitory neuropeptides, beta-endorphin, interleukin-1Beta, and neuropeptide-K, may involve excitatory amino acids. *Endocrinology* 133: 178–182.

Chen, Y., Mestek, A., Liu, J., Hurley, J. A., and Yu, L. (1993). Molecular cloning and functional expression of a mu-opioid receptor from rat brain. *Mol. Pharmacol.* 44: 8–12.

Cheung, S., and Hammer, R. P. (1993). Differential innervation of the rat medial preoptic nucleus by proopiomelanocortin neurons of the arcuate nucleus. *Anat. Rec.* S1: 43.

Cheung, S., Salinas, J., and Hammer, R. P. (1992). Gonadal steroid hormone-dependence of β-endorphin-like immunoreactive fibers in the medial preoptic area of the rat. *Anat. Rec.* 232: 26A.

Chowen, J. A., Argente, J., Vician, L., Clifton, D. K., and Steiner, R. A. (1990). Pro-opiomelanocortin messenger RNA in hypothalamic neurons is increased by testosterone through aromatization to estradiol. *Neuroendocrinology* 52: 581–588.

Conover, C. D., Kuljis, R. O., Rabii, R., and Advis, J.-P. (1993). Beta-endorphin regulation of luteinizing hormone-releasing hormone release at the median eminence in ewes: Immunocytochemical and physiological evidence. *Neuroendocrinology* 57: 1182–1195.

Cox, B. M., Goldstein, A., and Li, C. H. (1976). Opioid activity of a peptide [β-LPH-(61–91)], derived from β-lipotropin. *Proc. Natl. Acad. Sci. (USA)* 73: 1821–1826.

Dohler, K. D., Coquelin, A., Davis, F., Hines, M., Shryne, J. E., Sickmoller, P. M., Jarzab, B., and Gorski, R. A. (1986). Pre- and postnatal influence of an estrogen antagonist and an androgen antagonist on differentiation of the sexually dimorphic nucleus of the preoptic area in male and female rats. *Neuroendocrinology* 42: 443–448.

Dupont, A., Barden, N., Cusan, L., Merand, Y., Labrie, F., and Vaudry, H. (1980). β-Endorphin and met-enkephalin: Their distribution, modulation by estrogens and haloperidol, and role in neuroendorine control. *Fed. Proc.* 39: 2544–2550.

Evans, C. J., Keith, D. E., Morrison, H., Magendzo, K., and Edwards, R. H. (1992). Cloning of a delta opioid receptor by functional expression. *Science* 258: 1952–1955.

Evans, C. J., Hammond, D. L., and Fredrickson, R. C. A. (1988). The opioid peptides. In *The Opiate Receptors,* ed. G. Pasternak, pp. 23–71. New York: Humana Press.

Freeman, M. E. (1988). The ovarian cycle of the rat. In *The Physiology of Reproduction,* ed. E. Knobil and J. Neill, pp. 1893–1928. New York: Raven Press.

Fukuda, K., Kato, S., Mori, K., Nishi, M., and Takeshima, H. (1993). Primary structures and expression from cDNAs of rat opioid receptor delta- and mu-subtypes. *FEBS Lett.* 327: 311–314.

Ge, F., and Hammer, R. P. (1992). Alteration of β-endorphin-like immunoreactivity in the rat medial preoptic area across the estrous cycle. *Neuroendocrinol. (Life Sci. Adv.)* 11: 101–104.

Ge, F., Hammer, R. P., and Tobet, S. A. (1993). Ontogeny of leu-enkephalin and β-endorphin in the medial preotic area of the hypothalamus in rats. *Dev. Brain Res.* 73: 273–281.

Goldstein, A., Tachibana, S., Lowney, L. I., Hunkapiller, M., and Hood, L. (1979). Dynorphin-(1-13), an extraordinarily potent opioid peptide. *Proc. Natl. Acad. Sci. (USA)* 76: 6666–6670.

Gorski, R. A., Gordon, J. H., Shryne, J. E., and Southam, A. M. (1978). Evidence for a morphological sex difference within the medial preoptic area of the rat brain. *Brain Res.* 148: 333–346.

Gorski, R. A., Harlan, R. E., Jacobson, C. D., Shryne, J. E., and Southam, A. M. (1980). Evidence for the existence of a sexually dimorphic nucleus in the preoptic area of the rat. *J. Comp. Neurol.* 193: 529–539.

Hahn, E., and Fishman, J. (1979). Changes in rat brain opiate receptor content upon castration and testosterone replacement. *Biochem. Biophys. Res. Comm.* 90: 819–823.

Hammer, R. P. (1984). The sexually dimorphic region of the preoptic area in rats contains denser opiate receptor binding sites in females. *Brain Res.* 308: 172–176.

Hammer, R. P. (1985a). Ontogeny of opiate receptors in the rat medial preoptic area: Critical periods in regional development. *Intern. J. Dev. Neurosci.* 3: 541–548.

Hammer, R. P. (1985b). The sex hormone-dependent development of opiate receptors in the rat medial preoptic area. *Brain Res.* 360: 65–74.

Hammer, R. P. (1988). Opiate receptor ontogeny in the rat medial preoptic area is androgen-dependent. *Neuroendocrinology* 48: 336–341.

Hammer, R. P. (1990). μ-Opiate receptor binding in the medial preoptic area is cyclical and sexually dimorphic. *Brain Res.* 515: 187–192.

Hammer, R. P., Brady, L. S., Abelson, L., and Micevych, P. E. (1991). Differential localization of opioid neurons projecting to the medial preoptic nucleus. *Anat. Rec.* 229: 35A.

Hammer, R. P., Dornan, W. A., and Bloch, G. J. (1989). Sexual dimorphism and function of μ-opiate receptors in the rat medial preoptic area: Involvement in regulation of lordosis behavior. *Intern. Conf. Horm. Brain Behav. Abstr.* 77–78.

Hammer, R. P., Mateo, A. R., and Bridges, R. S. (1992). Hormonal regulation of medial preoptic μ-opiate receptor density before and after parturition. *Neuroendocrinology* 56: 38–45.

Harlan, R., Shivers, B., Romano, G., Howells, R., and Pfaff, D. (1987). Localization of preproenkephalin mRNA in the rat brain and spinal cord by *in situ* hybridization. *J. Comp. Neurol.* 258: 159–184.

Hughes, A. M., Everitt, B. J., and Herbert, J. (1988). Selective effects of β-endorphin infused into the hypothalamus, preoptic area and bed nucleus of the stria terminalis on the sexual and ingestive behavior of male rats. *Neuroscience* 27: 689–698.

Hughes, J., Smith, T. W., Kosterlitz, H. W., Fothergill, L. A., Morgan, B. A., and Morris, H. R. (1975). Identification of two related pentapeptides from the brain with potent opiate agonist activity. *Nature* 258: 577–579.

Hurlbut, D. E., Evans, C. J., Barchas, J. D., and Leslie, F. M. (1987). Pharmacological properties of a proenkephalin-derived opioid peptide: BAM 18. *Eur. J. Pharmacol.* 138: 359–366.

Jacobson, C. D., Davis, F. C., and Gorski, R. A. (1985). Formation of the sexually dimorphic nucleus of the preoptic area: Neuronal growth, migration and changes in cell number. *Dev. Brain Res.* 21: 7–18.

Kalra, S. P. (1981). Neural loci involved in naloxone-induced luteinizing hormone release. *Endocrinology* 109: 1805–1810.

Kalra, S. P. (1985). Neural circuits involved in the control of LHRH secretion: A model for estrous cycle regulation. *J. Steroid Biochem.* 23: 733–742.

Kalra, S. P., and Kalra, P. S. (1984). Opioid–adrenergic–steroid connection in regulation of luteinizing hormone secretion in the rat. *Neuroendocrinology* 38: 418–426.

Kelly, M. J., Loose, M. D., and Ronnekleiv, O. K. (1990). Opioids hyperpolarize β-endorphin neurons via μ-receptor activation of a potassium conductance. *Neuroendocrinology* 52: 268–275.

Kelly, M. J., Loose, M. D., and Ronnekleiv, O. K. (1992). Estrogen suppresses mu-opioid- and GABA_B-mediated hyperpolarization of hypothalamic arcuate neurons. *J. Neurosci.* 12: 2745–2750.

King, J. C., Tobet, S. A., Snavely, F. L., and Arimura, A. A. (1982). LHRH immunopositive cells and their projections to the median eminence and the organum vasculosum of the lamina terminalis. *J. Comp. Neurol.* 209: 287–300.

Knuth, U. A., Sikand, G. S., Casanueva, F. F., Havlicek, V., and Friesen, H. G.

(1983). Changes in beta-endorphin content in discrete areas of the hypothalamus throughout proestrus and diestrus of the rat. *Life Sci.* 33: 1443–1450.

Kumar, M. S. A., Chen, C. L., and Muther, T. F. (1979). Changes in the pituitary and hypothalamic content of methionine-enkephalin during the estrous cycle of rats. *Life Sci.* 25: 1687–1696.

Lauber, A. H., Romano, G. J., Mobbs, C. V., Howells, R. D., and Pfaff, D. W. (1990). Estradiol induction of proenkephalin messenger RNA in hypothalamus: Dose-response amd relation to reproductive behavior in the female rat. *Mol. Brain Res.* 8: 47–54.

Leranth, C., MacLusky, N. J., Shanabrough, M., and Naftolin, F. (1988). Immunohistochemical evidence for synaptic connections between pro-opiomelanocortin-immunoreactive axons and LH–RH neurons in the preoptic area of the rat. *Brain Res.* 449: 167–176.

Leslie, F. M., and Loughlin, S. E. (1993). Ontogeny and plasticity of opioid systems. In *The Neurobiology of Opiates,* ed. R. P. Hammer, pp. 85–123. Boca Raton, FL: CRC Press.

Mansour, A., Khachaturian, H., Lewis, M. E., Akil, H., and Watson, S. J. (1987). Autoradiographic differentiation of mu, delta, and kappa opioid receptors in the rat forebrain and midbrain. *J. Neurosci.* 7: 2445–2464.

Martini, L., Dondi, D., Limonta, P., Maggi, R., and Piva, F. (1989). Modulation by sex steroids of brain opioid receptors: Implications for the control of gonadotropins and prolactin secretion. *J. Steroid Biochem.* 33: 673–681.

Mateo, A. R., Hijazi, M., and Hammer, R. P. (1992). Dynamic patterns of medial preoptic μ-opiate receptor regulation by gonadal steroid hormones. *Neuroendocrinology* 55: 51–58.

Matuszewich, L., and Dornan, W. A. (1992). Bilateral injections of a selective μ-receptor agonist (morphiceptin) into the medial preoptic nucleus produces a marked delay in the initiation of sexual behavior in the male rat. *Psychopharmacology* 106: 391–396.

Mitchell, J. B., and Gratton, A. (1991). Opioid modulation and sensitization of dopamine release elicited by sexually relevant stimuli: A high speed chronoamperometric study in freely behaving rats. *Brain Res.* 551: 20–27.

Myers, B. M., and Baum, M. J. (1979). Facilitation by opiate antagonists of sexual performance in the male rat. *Pharmacol. Biochem. Behav.* 10: 615–618.

Ostrowski, N. L., Hill, J. M., Pert, C. B., and Pert, A. (1987). Autoradiographic visualization of sex differences in the pattern and density of opiate receptors in hamster hypothalamus. *Brain Res.* 421: 1–13.

Paterson, S. J., Robson, L. E., and Kosterlitz, H. W. (1983). Classification of opioid receptors. *Br. Med. Bull.* 39: 31–36.

Pert, C. B., and Snyder, S. H. (1973). Opiate receptor: Demonstration in nervous tissue. *Science* 179: 1011–1014.

Pfaff, D. W., and Sakuma, Y. (1979). Facilitation of the lordosis reflex of female rats from the ventromedial nucleus of the hypothalamus. *J. Physiol.* 288: 189–202.

Pfaus, J. G., and Pfaff, D. W. (1992). μ-, δ-, and κ-opioid receptor agonists selectively modulate sexual behaviors in the female rat: Differential dependence on progesterone. *Horm. Behav.* 26: 457–473.

Priest, C. A., Eckersell, C. B., and Micevych, P. E. (in press). Temporal regulation by estrogen of preproenkephalin-A mRNA in the rat ventromedial nucleus of the hypothalamus. *Mol. Brain Res.*

Romano, G. J., Harlan, R. E., Shivers, B. D., Howells, R. D., and Pfaff, D. W. (1988). Estrogen increases proenkephalin messenger ribonucleic acid levels in the ventromedial hypothalamus of the rat. *Mol. Endocrinol.* 2: 1320–1328.

Romano, G. J., Mobbs, C. V., Howells, R. D., and Pfaff, D. W. (1989). Estrogen regulation of proenkephalin gene expression in the ventromedial hypothalamus of the rat: Temporal qualities and synergism with progesterone. *Mol. Brain Res.* 5: 51–58.

Sakuma, Y., and Pfaff, D. (1979). Facilitation of female reproductive behavior from mesencephalic central gray in the rat. *Am. J. Physiol.* 237: R278–284.

Simerly, R. B., McCall, L. D., and Watson, S. J. (1988). Distribution of opioid peptides in the preoptic region: Immunohistochemical evidence for a steroid-sensitive enkephalin sexual dimorphism. *J. Comp. Neurol.* 276: 442–459.

Simerly, R. B., and Swanson, L. W. (1986). The organization of neural inputs to the medial preoptic nucleus of the rat. *J. Comp. Neurol.* 246: 312–342.

Simerly, R. B., and Swanson, L. W. (1988). Projections of the medial preoptic nucleus: A *Phaseolus vulgaris* leucoagglutinin anterograde tract tracing study in the rat. *J. Comp. Neurol.* 270: 209–242.

Simerly, R. B., Swanson, L. W., and Gorski, R. A. (1984). Demonstration of a sexual dimorphism in the distribution of serotonin-immunoreactive fibers in the medial preoptic nucleus of the rat. *J. Comp. Neurol.* 225: 151–166.

Simon, E. J., Hiler, J. M., and Edelman, I. (1973). Stereospecific binding of the potent narcotic analgesic [^{3}H]etorphine to rat brain homogenates. *Proc. Natl. Acad. Sci. (USA)* 70: 1947–1949.

Sirinathsinghji, D. J. S. (1986). Regulation of lordosis behaviour in the female rat by corticotropin-releasing factor, beta-endorphin/corticotropin and luteinizing hormone-releasing hormone neuronal systems in the medial preoptic area. *Brain Res.* 375: 49–56.

Sirinathsinghji, D. J. S., Whittington, P. E., Audsley, A., and Fraser, H. M. (1983). B-Endorphin regulates lordosis in female rats by modulating LH–RH release. *Nature* 301: 62–65.

Terenius, L. (1973). Stereospecific interaction between narcotic analgesics and a synaptic plasma membrane fraction of rat cerebral cortex. *Acta Pharmacol. Toxicol.* 32: 317–320.

Wang, J. B., Imai, Y., Eppler, C. M., Gregor, P., Spivak, C., and Uhl, G. R. (1993). μ opiate receptor: cDNA cloning and expression. *Proc. Natl. Acad. Sci. (USA)* 90: 10230–10234.

Wardlaw, S., Thoron, L., and Frantz, A. (1982). Effects of sex steroids on brain β-endorphin. *Brain Res.* 245: 327–331.

Watson, R. E., Hoffman, G. E., and Wiegand, S. J. (1986). Sexually dimorphic opioid distribution in the preoptic area: Manipulation by gonadal steroids. *Brain Res.* 398: 157–163.

Wiesner, J. B., Koenig, J. I., Krulich, L., and Moss, R. L. (1984). Site of action for β-endorphin-induced changes in plasma luteinizing hormone and prolactin in the ovariectomized rat. *Life Sci.* 34: 1463–1473.

Wiesner, J. B., and Moss, R. L. (1984). Beta-endorphin suppression of lordosis behavior in female rats: Lack of effect of peripherally-administered naloxone. *Life Sci.* 34: 1455–1462.

Wiesner, J. B., and Moss, R. L. (1986a). Behavioral specificity of β-endorphin suppression of sexual behavior: Differential receptor antagonism. *Pharmacol. Biochem. Behav.* 24: 1235–1239.

Wiesner, J. B., and Moss, R. L. (1986b). Suppression of receptive and proceptive behavior in ovariectomized, estrogen–progesterone-primed rats by intraventricular beta-endorphin: Studies of behavioral specificity. *Neuroendocrinology* 43: 57–62.

Wilcox, J. N., and Roberts, J. L. (1985). Estrogen decreases rat hypothalamic proopiomelanocortin messenger ribonucleic acid levels. *Endocrinology* 117: 2392–2396.

Wilkinson, M., Brawer, J. R., and Wilkinson, D. A. (1985). Gonadal steroid-induced modification of opiate binding sites in anterior hypothalamus of female rats. *Biol. Reprod.* 32: 501–506.

Wise, P. M., Scarbrough, K., Weiland, N. G., and Larson, G. H. (1990). Diurnal pattern of proopiomelanocortin gene expression in the arcuate nucleus of proestrous, ovariectomized, and steroid-treated rats: A possible role in cyclic luteinizing hormone secretion. *Mol. Endocrinol.* 4: 886–892.

Yamano, M., Inagaki, S., Kito, S., Matsuzaki, T., Shinohara, Y., and Tohyama, M. (1986). Enkephalinergic projections from the ventromedial hypothalamic nucleus to the midbrain central gray matter in the rat: An immunocytochemical analysis. *Brain Res.* 398: 337–346.

Yasuda, K., Raynor, K., Kong, H., Breder, C. D., Takeda, J., Reisine, T., and Bell, G. I. (1993). Cloning and functional comparison of kappa and delta opioid receptors from mouse brain. *Proc. Natl. Acad. Sci. (USA)* 90: 6736–6740.

Zhou, L., and Hammer, R. P. (1993). Gonadal steroid hormones upregulate rat medial preoptic μ-opiate receptors by rapid induction of protein synthesis. *Intern. Conf. Horm. Brain Behav. Abstr.* 205–206.

7

Effects of sex steroids on the cholecystokinin circuit modulating reproductive behavior

PAUL POPPER, CATHERINE A. PRIEST, AND
PAUL E. MICEVYCH

Introduction

During development and adulthood, the central nervous system (CNS) undergoes structural and neurochemical changes under the influence of endogenous stimuli. Among these stimuli are sex steroid hormones synthesized by the gonads, adrenal cortex, and CNS cells themselves (see Chapter 12 by Robel and Baulieu and Chapter 13 by Schlinger and Arnold, this volume). Sex steroid–sensitive neurons provide the anatomical substrate at which sex steroid hormones affect the structure and function of the CNS. Generally, these cells respond to the action of sex steroid hormones either directly, through hormone–receptor interaction, or indirectly, through trans-synaptic activation of membrane receptors and subsequently second messengers or immediate early genes.

In vertebrate species, most sex steroid–sensitive cells are part of the sexually differentiated limbic–hypothalamic circuit, which, within a more expansive neural network, controls reproductive events. To understand the functioning of the limbic–hypothalamic circuit, we must determine how sex steroids specify chemical connections within this circuit and how steroid hormone–induced changes in these connections are related to the display of sexual behavior. The phylogenetically conserved limbic–hypothalamic circuit consists of the posterior dorsal medial amygdala (MeA), encapsulated bed nucleus of the stria terminalis (BNST), medial preoptic nucleus (MPN), and ventromedial nucleus of the hypothalamus (VMH) (reviewed by Micevych and Ulibarri 1992). This circuit integrates a variety of neuronal inputs carrying somatic, olfactory, and hormonal information and sends projections to effector circuits to initiate neuroendocrine events (gonadotropin release) and sexual behavior (mounting or lordosis, reviewed by Pfaff and Schwartz-Giblin 1988).

We are interested in understanding the contribution of the limbic–hypothalamic

circuit to sexual behavior. Specifically, we have examined the expression, release, and receptor regulation of the neuropeptide cholecystokinin (CCK) within
the limbic–hypothalamic circuit and the effect of this peptide on lordosis behavior, the estrogen-stimulated female copulatory behavior. The investigation of the
sex steroid regulation of neuropeptide circuits in general, and CCK circuits in
particular, has been an extremely fruitful approach to understanding how steroids
regulate behaviorally relevant circuits in the CNS. Although certainly not the only
neuroactive peptide involved in the regulation of reproductive behavior (see
Chapter 5 by Flanagan and McEwen, Chapter 9 by Akesson and Micevych, Chapter 6 by Hammer and Cheung, and Chapter 11 by de Vries, this volume), the CCK
limbic–hypothalamic circuit has emerged as an excellent model for studying the
interactions of gonadal steroids with neural systems involved in reproduction (reviewed by Micevych and Ulibarri 1992).

CCK first was identified in the gastrointestinal tract (Ivy and Goldberg 1928)
and later detected in the CNS (Dockray 1980; Rehfeld 1978; Roberecht et al.
1978). Biologically active, carboxy-amidated CCK is formed by post-translational processing from a 115 amino acid precursor, prepro-CCK (pCCK; reviewed by Micevych and Ulibarri 1992). CCK is widely distributed in the CNS
(Abelson and Micevych 1991; Beinfeld et al. 1980; Crawley 1985; Emson et al.
1982; Innis et al. 1979; Larsson and Rehfeld 1979; Vanderhaegen 1985). Much of
the work on this neuropeptide has concentrated on its role in the physiology of the
striatum and hypothalamus. Initial studies focused on the relationship between
centrally distributed CCK and its role in the regulation of feeding (Della Ferra and
Baile 1979; Gibbs et al. 1973; Sturdevant and Goetz 1976). The role of CCK has
now been studied in analgesia (Faris et al. 1983) and exploratory behavior
(Crawley 1986; Crawley et al. 1984), as well as in the etiology of schizophrenia
(Wang et al. 1984), anxiety (reviewed by Woodruff et al. 1991) and reproductive
behavior (Bloch et al. 1987; Mendelson and Gorzalka 1984).

CCK and reproductive behavior

Adult behavioral responses to exogenous CCK-8

In the adult female rat, systemic injections of sulfated CCK octapeptide (sCCK-8)
facilitate lordosis behavior in animals that are minimally receptive (Bloch et al.
1987) and inhibit lordosis behavior in maximally receptive animals (Bloch et al.
1987; Mendelson and Gorzalka 1984). These conflicting behavioral effects have
been explained by the hypothesis that CCK has differential effects at different
nuclei and different CCK receptors (CCK-R). For example, injections of sCCK-8
into the MPN of estrogen-primed female rats facilitates lordosis behavior. Fur-

thermore, injection into the MPN of antiserum raised against sCCK-8 attenuates the frequency of lordosis behavior compared with that in control females (Dornan et al. 1989), indicating that endogenous CCK facilitates lordosis at the level of the MPN.

In contrast, a CCK-receptive cell population in the VMH mediates the attenuation of lordosis (Akesson et al. 1987; Babcock et al. 1988; Ulibarri et al. 1990). Injection of sCCK-8 into the VMH of minimally receptive rats has no effect on lordosis behavior; however, when the rats are receptive (lordosis quotient ≥ 60), sCCK-8 injection attenuates the behavior in a dose-dependent fashion. In addition, when rats are highly receptive, smaller amounts of sCCK-8 are required to attenuate lordosis, suggesting that sex steroids might sensitize cells to the effects of sCCK-8. Although lordosis is induced in ovariectomized female rats 24 hours after treatment with estrogen, sCCK-8 has no effect on lordosis until 48 hours after estrogen treatment (Fig. 11 in Micevych and Ulibarri 1992). This delay may be explained by the effect of estrogen on CCK-R binding. CCK-R binding in the VMH of ovariectomized female rats is low 24 hours and high 48 hours after estrogen treatment (Akesson et al. 1987). This suggests that the lack of effect of endogenous CCK on lordosis 24 hours after treatment of ovariectomized female rats with estrogen is due to the estrogen-induced reduction of CCK-R binding in the VMH. To characterize the steroid influence on the CCK circuit, we have examined the effect of sex steroid hormones on the expression of CCK mRNA, peptide, and CCK-R.

Sex steroid modulation of CCK expression

Steroid regulation of the CCK circuit occurs at two levels: presynaptic and postsynaptic. Presynaptic effects of sex steroids on the limbic–hypothalamic CCK circuit occur by regulating the mRNA level and subsequent peptide expression (Frankfurt et al. 1986; Micevych et al. 1988a; Oro et al. 1988) and release (Micevych et al. 1988b). One major advantage of studying neuropeptides is that their expression is regulated primarily at the transcriptional level (Comb et al. 1989), and thus changes in the amount of mRNA directly reflect subsequent quantitative changes in biologically active forms of the neuropeptide. In this context, we will refer to sex steroid effects on CCK mRNA and peptide levels as presynaptic; the effects of sex steroids on the CCK-R (Akesson et al. 1987; Popper et al. 1991; Ulibarri et al. 1990) will be described as postsynaptic.

The distribution of CCK-like immunoreactive (CCKir) cells in the limbic–hypothalamic circuit overlaps the distribution of sex steroid–receptive cells, is sexually dimorphic (Micevych et al. 1987, 1988a; Micevych and Bloch 1989), and is extremely sensitive to the adult gonadal steroid milieu (Frankfurt et al.

1986; Micevych et al. 1988a, b; Simerly and Swanson 1987). Castration dramatically decreases pCCK mRNA and the number of CCKir cells in the posterodorsal MeA, BNST, and MPN of male rats, which are restored by treatment with either testosterone or estrogen (Figs. 7.1A and 7.2A; Micevych and Bloch 1989; Micevych et al. 1992, 1994; Simerly 1990; Simerly and Swanson 1987). Similarly, in female rats, estrogen or testosterone significantly increases the number of CCKir and pCCK mRNA–containing cells, as well as the amount of pCCK mRNA per cell (Figs. 7.1B and 7.2B). These results are consistent with the hypothesis that estrogen is primarily responsible for increasing the expression of CCK in both adult male and female rats.

Ontogeny of the limbic--hypothalamic CCK circuit. Gonadal steroid regulation of gene expression underlies structural and functional sexual differentiation of the vertebrate CNS. Traditionally, sex steroid effects on neurons and behavior are categorized as organizational or activational (Phoenix et al. 1959). Organizational effects in the CNS are produced by gonadal steroids during fetal and neonatal development and determine the number, morphology, and connections of neurons devoted to specific tasks (Arnold and Breedlove 1985; Gorski 1985). The CCK circuit begins to develop as early as embryonic day 15, but the translated pro-CCK peptide is not post-translationally modified to a biologically active, α-amidated form until after birth (reviewed by Rehfeld 1989). The expression of the biologically active, adult form of CCK in the forebrain and brain stem precedes its expression in the limbic–hypothalamic circuit (Cho et al. 1983; Duchemin et al. 1987; Micevych and Ulibarri 1992). On postnatal day 1, high levels of pCCK mRNA are detected in the neocortex, hippocampus, amygdala, olfactory tubercle, and many brain stem nuclei, including the motor columns of rats of either sex; however, there is no detectable expression of pCCK mRNA in cells in the central subdivision of the MPN (MPNc), BNST, or MeA. By postnatal day 5, pCCK mRNA is detectable in the MPNc, BNST, and MeA, but only in the MPNc does the number of pCCK cells increase significantly before puberty (Micevych 1991). In the MPNc, there is a demonstrable sex difference favoring the male in the number of pCCK mRNA cells by postnatal day 10, which continues to develop into adulthood. In the BNST and the MeA, very few pCCK cells are observed before postnatal day 30, but by postnatal day 42, the number of pCCK cells in these nuclei reaches the adult level (Abelson et al. 1989; Micevych 1991; Micevych and Ulibarri 1992). This increase in the number of pCCK mRNA–expressing cells is most dramatic in male rats; before puberty, there are only 15–20% of the number of cells observed after puberty. These results suggest that the amount and pattern of gonadal steroid secretion that occurs at puberty regulate pCCK mRNA expression in the MPNc and BNST of adult rats.

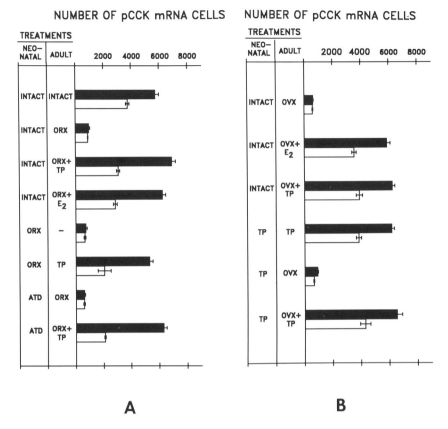

Figure 7.1. Histographs of cells expressing pCCK mRNA in the BNST (open bars) and the posterodorsal MeA (solid bars) in male (A) and female (B) rats. Neonatal and adult steroid manipulations were as follows. Within 1 hour of birth, male pups were cryoanesthetized and either gonadectomized or sham gonadectomized, or the aromatase inhibitor andros-1,6-tene-3,17-dione (ATD; in 4-mm Silastic capsules, 1.57 mm i.d. × 3.17 mm o.d.) or empty capsules were subcutaneously implanted. Female pups were similarly anesthetized, and testosterone propionate (TP; in 4-mm Silastic capsules) or empty capsules were implanted. On postnatal day (PND) 10, all capsules were removed. On PND 90, rats were anesthetized with Equithesin and castrated. In addition, some animals from each neonatal treatment group were implanted with 40-mm Silastic capsules containing TP, estradiol, or nothing. Four weeks later, rats were deeply anesthetized with Equithesin and perfused transcardially with 0.9% saline followed by 4% *para*-formaldehyde in 0.1 M Sörensen's phosphate buffer. Values are means ± SEM. All treatment groups include at least four animals. Abbreviations: ORX, orchidectomy; OVX, ovariectomy; E_2, estradiol-filled Silastic capsule; TP, testosterone propionate–filled Silastic capsule.

Sex steroids have organizational effects on the presynaptic elements of the limbic–hypothalamic CCK circuit. The early postnatal steroid environment determines the level of constitutive (non-estrogen-stimulated) expression of pCCK in the limbic–hypothalamic circuit (Micevych et al. 1994), since there are more

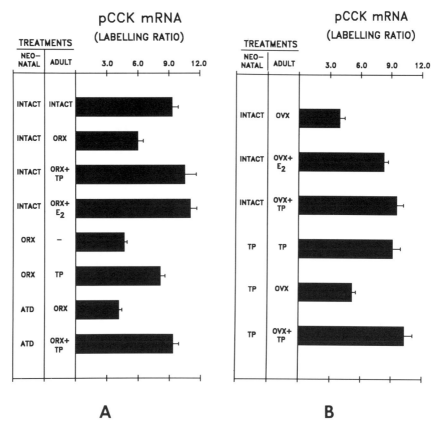

Figure 7.2. Histographs of pCCK mRNA labeling ratio in the BNST (open bars) and the MeA (solid bars) in male (A) and female (B) rats. Neonatal and adult steroid manipulations are described in the legend for Figure 7.1. Values are means ± SEM. All treatment groups include at least four animals. Statistical comparisons are discussed in the text. Abbreviations: E_2, estradiol-filled Silastic capsule; TP, testosterone propionate–filled Silastic capsule.

pCCK cells in the castrated adult male than in the adult ovariectomized female. This sex difference can be eliminated by neonatal treatment of the ovariectomized female with estrogen or testosterone, or by blocking aromatization of testosterone neonatally, or by neonatal castration of the male. This indicates that estrogen is the metabolite of testosterone responsible for establishing the sex difference in pCCK expression (Figs. 7.1 and 7.3). Similarly, in animals that are gonadectomized in adulthood, males have a higher level of pCCK mRNA expression in the MeA and BNST than females. Neonatal treatment of females with estrogen or testosterone, followed by ovariectomy as adults, eliminates this difference. Similarly, pCCK mRNA expression can be reduced in the male by neonatal treatment with the aromatase inhibitor ATD (Figs. 7.2A, 7.4, and 7.5). Although these

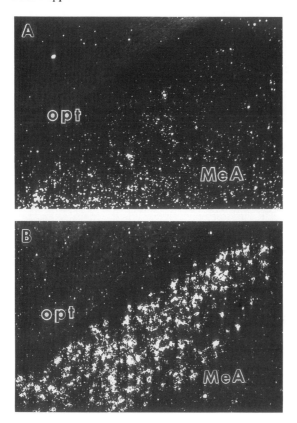

Figure 7.3. Dark-field photomicrographs of pCCK cRNA *in situ* hybridization to cells in the MeA in a prepubertal (A) and an adult rat (B) (opt, optic tract).

differences are interesting, it must be remembered that the constitutive levels of pCCK expression (the level of pCCK mRNA in the absence of sex steroids) in the adult MeA, BNST, and MPN are extremely low and therefore the overwhelming effect of sex steroids on the adult pattern of CCK distribution and the level of mRNA expression is activational.

CCK expression in adulthood. Activational effects are considered nonpermanent and are the result of sex steroids activating or inhibiting circuits that already exist [e.g., alteration of neurotransmitter or neuropeptide expression (Blanco et al. 1992; Micevych et al. 1987, 1988b; Oro et al. 1988; Sengelaub et al. 1989; Simerly and Swanson 1986; see also Chapter 5 by Flanagan and McEwen, Chapter 6 by Hammer and Cheung, Chapter 9 by Akesson and Micevych, and Chapter 11 by de Vries, this volume), release (Drouva et al. 1985; Micevych et al. 1988b), or distribution of their receptor populations (Akesson and Micevych 1986; Akesson et al. 1987; Biegon et al. 1982; Fischette et al. 1983; McEwen and

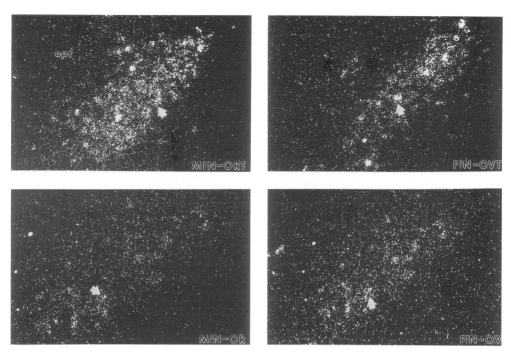

Figure 7.4. Dark-field photomicrographs of pCCK cRNA *in situ* hybridization to cells in the MeA in male and female rats. Arrows indicate specific hybridization to some cells in the MeA. Abbreviations: MIN-IN, male, neonatally intact – intact in adulthood; MIN-OR, male, neonatally intact – orchidectomized as an adult; MIN-ORT, male, neonatally intact – orchidectomized, treated with testosterone propionate (TP) as an adult; MAT-OR, male, neonatally treated with ATD – orchidectomized as an adult; MAT-ORT, male, neonatally treated with ATD – orchidectomized, TP-treated as an adult; FIN-OV, female, neonatally intact – ovariectomized as an adult; FIN-OVT, female, neonatally intact – ovariectomized, TP-treated as an adult; FT-OVT, female, neonatally TP-treated – ovariectomized, TP-treated as an adult; FIN-OVE, female, neonatally intact – ovariectomized, estradiol benzoate–treated as an adult.

Parson 1982; Sengelaub et al. 1989)]. Although subtle differences in pCCK mRNA expression can be observed between males and females, the primary influence on this expression is the adult steroid environment (Figs. 7.1 and 7.2). When this is equalized by gonadectomy and replacement with either testosterone or estrogen implants, the level of CCK expression in the limbic–hypothalamic circuit also becomes equal, regardless of the neonatal steroid environment. These data suggest the presence of a very weak organizational role of sex steroids on CCK expression in the limbic–hypothalamic circuit. Furthermore, they indicate that the pattern of pCCK mRNA expression observed in the limbic–hypothalamic circuit of adult animals is dependent on both the concentration and pattern of gonadal steroid exposure in adulthood.

In females, the expression of pCCK mRNA and CCKir cells in the limbic–

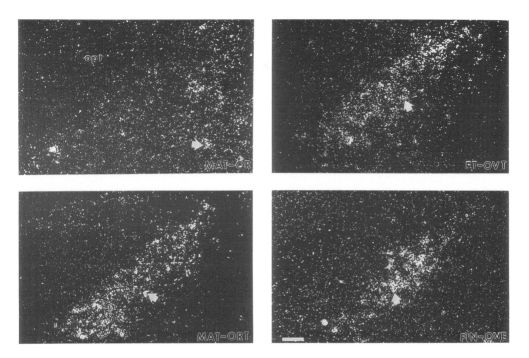

Figure 7.5. Dark-field photomicrographs of pCCK cRNA *in situ* hybridization to cells in the MeA in male and female rats. (See legend to Figure 7.4.) Scale bar, 100 μm.

hypothalamic circuit is determined by the cyclicity of estrogen secretion from the ovaries. Indeed, tissue levels of CCK vary during the estrous cycle; in the MeA, BNST, and VMH, tissue levels of CCK drop on the afternoon of proestrus and remain low through the morning of estrus (Frankfurt et al. 1986; Micevych et al. 1988b). *In vitro* superfusion of the mediobasal hypothalamus with estradiol-17β does not alter basal release of CCK. However, estradiol-17β facilitates potassium-stimulated release of CCK in a calcium-sensitive, dose-dependent manner (Micevych et al. 1988b), while estradiol-17α is ineffective. This stereospecific modulation of depolarization-induced CCK release suggests that estrogen itself is not responsible for initiating CCK release, but rather that estrogen plays a permissive role in CCK release. One way in which estrogen could affect CCK release is by altering neurotransmitter input to CCK neurons such that there is a net stimulatory influence. This alteration of input could be through estrogen alteration of transmitter release, alteration of transmitter binding to receptors on CCK cells, or alteration of signal transduction mechanisms in the CCK-releasing cells.

Thus, sex steroids have subtle organizational and dramatic activational effects on the presynaptic elements of the limbic–hypothalamic CCK circuit. The perinatal estrogen environment determines the estrogen-independent levels of pCCK mRNA expression and the number of pCCK mRNA–expressing cells. However, the adult pattern of CCK distribution and the level of expression are defined by the acute sex steroid hormonal state of the animal. In gonadally intact adult animals, the level of pCCK expression in the limbic–hypothalamic circuit is determined not by the neonatal steroid environment, but rather by the level and pattern of sex steroid secretion during adulthood. Males that have constantly high circulating testosterone levels also have high levels of pCCK mRNA and CCK immunoreactivity in the limbic–hypothalamic circuit, while the estrous cycle induces dramatic variations of pCCK mRNA and peptide expression in the limbic–hypothalamic circuit of female rats.

Mechanisms of CCK regulation by estrogen

The presynaptic elements of the CCK circuit are highly sensitive to the effects of estrogen; however, there is good evidence that estrogen does not act directly on CCK-expressing cells. Several conditions must be satisfied for estrogen to act on cells expressing CCK; these cells should contain estrogen receptors, and if CCK expression is regulated directly by estrogen the CCK gene would have to have an estrogen response element (ERE) through which the estrogen receptor could mediate trans-activation of the CCK gene. However, CCKir cells do not accumulate estrogen (Akesson and Micevych 1988), a necessary, but not sufficient, indication of direct steroid action in those cells, nor does estrogen receptor immunoreactivity co-localize in CCKir neurons (Herbison and Theodosis 1992). Furthermore, there is no evidence of an ERE in the 5′ promoter region of the pCCK gene (Haun and Dixon 1990). Thus, estrogen is thought to affect pCCK transcription indirectly by activating other cells, which, in turn, affect CCK neurons to alter their level of gene expression and activity trans-synaptically. This situation is similar to the proposed action of estrogen on luteinizing hormone–releasing hormone (LHRH) expression and release, since LHRH-expressing neurons neither accumulate estradiol (Shivers et al. 1983) nor contain estrogen receptor immunoreactivity (Herbison et al. 1992).

A potential trans-synaptic site for estrogenic induction of pCCK mRNA expression is the synaptic activation of second messengers. A number of neuropeptide-encoding genes contain elements that are regulated by both cAMP and phorbol esters, including pCCK (Haun and Dixon 1990; Monstein et al. 1990), prolactin (Cooke et al. 1981), corticotropin-releasing factor (Shibihara et al. 1983), pro-enkephalin (Horikawa et al. 1983), and somatostatin (Shen and Rutter 1984). In this scheme, G-protein-coupled receptors mediate the induction of inositol phospholipid hydrolysis, which, in turn, increases intracellular calcium and activates protein kinase C activity, or they may activate adenylate cyclase, leading to an increase in cAMP. The pCCK gene contains several cAMP-responsive elements that have been shown to be active in modulating transcription. This scheme of steroid-induced activation of pCCK gene expression has the advantage that it does not involve any special mechanism for gene regulation in pCCK cells of the limbic–hypothalamic circuit, but rather invokes signal transduction pathways common to most neurons. Estrogen increases neuronal activity to influence pCCK-expressing cells, producing an upregulation of pCCK mRNA. Thus, at the presynaptic level, estrogen appears to modulate trans-synaptically CCK expression and release by cells of the limbic-hypothalamic circuit, thereby regulating lordosis.

CCK binding

In both the CNS and periphery, CCK-R are classified into two subtypes, CCK_A-R and CCK_B-R. This distinction is based on the differential affinities of the CCK-R for various antagonists and for a family of neuropeptides that are functionally and structurally similar, sharing identical pentapeptide sequences at the carboxy terminal, but differing in the pattern of tyrosyl sulfation (Wank et al. 1992). The CCK_A-R has a 500- to 1,000-fold higher affinity for sulfated analogs of CCK than for nonsulfated analogs and is highly selective for the antagonist L-364,718, whereas the CCK_B-R has equally high affinity for the sulfated and nonsulfated peptide analogs and is selectively recognized by the antagonist L-365,260 (Chang and Lotti 1986; Lotti and Chang 1989; Moran et al. 1986; Saito et al. 1980). We and others have used several methods to study the distribution and function of CCK-R in the brain. First, cells that express specific CCK-R mRNA can be identified using *in situ* hybridization histochemistry. Second, one can use radiolabeled ligands to characterize the binding and determine whether CCK-R mRNA is transcribed to produce functional protein. Finally, site-specific microinjection of CCK followed by behavioral assessment allows the determination of whether the CCK-R mRNA and CCK binding sites are functionally coupled to behavior. We have applied these methods to the study of CCK-R in the VMH. Wank and colleagues (1992) cloned the rat CCK_A-R and the CCK_B-R and mapped the distribution of the CCK_A-R and CCK_B-R mRNA throughout the rat brain (Honda et al. 1993). The distribution of CCK-R mRNA was in good agreement with that determined using radiolabeled agonists and antagonists. One exception, however, was the distribution of CCK-R in the ventral striatum. Previous studies suggested the presence of CCK_A-R (Crawley 1991; Hill et al. 1987, 1990), but *in situ* hybridization histochemical studies indicate the presence of CCK_B-R mRNA. In addition, Honda and colleagues (1993) found that mRNA encoding CCK_B-R in the hypothalamus was prominent in the supraoptic nucleus, paraventricular nucleus, lateral hypothalamus, and VMH. The mRNA encoding CCK_A-R was localized in the arcuate nucleus, dorsomedial nucleus, and MPN. Using the same probes, we determined that the MPN, and especially the MPNc, is rich in cells expressing CCK_A-R mRNA. Thus, *in situ* hybridization histochemical studies with cRNA probes specific for CCK_A-R and CCK_B-R reveal that cells of the MPN and the VMH express different CCK receptor subtypes.

It is tempting to propose that CCK acts at different nuclei within the circuit and on different receptor subtypes to produce differential modulation of estrogen-induced lordosis behavior. We have demonstrated that CCK-R binding in the supraoptic nucleus and the VMH is regulated by estrogen (Akesson and Micevych 1986; Akesson et al. 1987) and that both these nuclei contain mRNA encoding CCK_B-R. We propose that estrogen differentially regulates these two receptor subtypes; estrogen could modulate CCK_B-R binding, but not CCK_A-R binding. The amount of [^{125}I]sCCK-8 binding in the VMH varies across the estrous cycle (Akesson et al. 1987); the highest binding occurs when the plasma level of estrogen is lowest. In addition, ovariectomy enhances binding, and the amount of [^{125}I]sCCK-8 binding is attenuated 24 hours after estrogen injection in ovariectomized females, matching the time course of binding changes that occur across the estrous cycle at the level of the VMH. CCK in the VMH inhibits the display of lordosis, and the variation of CCK-R binding across the estrous cycle suggests that CCK-R are modulated negatively by estrogen to allow for the display of lordosis. Indeed, a significant negative correlation ($r^2 = .75, p \leq .001$) is present between the amount of [^{125}I]sCCK-8 binding in the VMH and the lordosis quotient in ovariectomized, steroid-primed female rats (Ulibarri et al. 1990).

Steroid regulation of CCK-R binding in the VMH

Ontogeny of CCK-R binding. CCK-R development in the VMH proceeds in parallel in males and females, suggesting that these receptors develop independently of the sex steroid milieu. CCK-R binding increases postnatally, with approximately the same time course as the increase in pCCK mRNA (Abelson et al. 1989; Micevych and Ulibarri 1992). In the VMH, the ontogeny of CCK-R binding is more rapid than the appearance of immunoreactive CCK peptide. On postnatal day 1, low levels of binding are detected in the VMH. By postnatal day 10, the area containing binding is restricted to the core of the VMH, although binding reaches levels similar to those that can be measured in orchidectomized adult males and estradiol-treated ovariectomized females (~20 fmol [^{125}I]sCCK-8 per milligram protein). By postnatal day 15, binding is elevated significantly and covers the entire area of the VMH. These levels are similar to those seen in gonadally intact males and ovariectomized females treated chronically with estrogen. In the male, the CCK-R binding levels decrease somewhat at postnatal day 30 (Fig. 7.6), and then increase to adult levels. The ontogeny of CCK-R binding in the VMH shows a pattern similar to that of CCK-R binding in a number of areas (Pelaprat et al. 1988) as well as to other neurotransmitter receptors (e.g., Palacios et al. 1979;

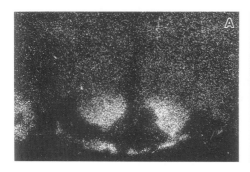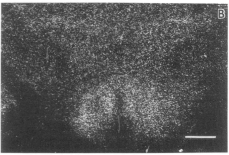

Figure 7.6. Autoradiographic images of [^{125}I]sCCK-8 binding (arrow) in the ventromedial nucleus of a 15-day-old (A) and 30-day-old rat (B). Scale bar, 416 μm.

Patey et al. 1980; Roth and Beinfeld 1985). CCK-R binding peaks in the first postnatal weeks, and decreases thereafter to reach adult levels. One possible explanation for this phenomenon is that there is an overgrowth of CCK innervation of the VMH. Since there are no CCK-containing somata in the VMH, tissue levels of CCK reflect the levels of the peptide in terminals. In fact, there is a good correlation between CCK levels and CCK-R binding in the VMH, both of which increase during the first postnatal weeks of development and then decrease to adult levels, suggesting that the hypothesis of axonal/terminal overgrowth and subsequent pruning may be correct.

Adult response of CCK-R binding to steroidal manipulations. In adulthood, the organization of synaptic inputs (Dodson et al. 1988; Matsumoto et al. 1988a,b; Micevych et al. 1988b) and the expression of neuroactive molecules (Micevych and Bloch 1989; Simerly et al. 1989) are dependent on the level of gonadal steroid hormones that first appear during the peripubertal period (Miller et al. 1989, Webb et al. 1990). The modification of neurochemical circuits, both during development and in adulthood, has profound behavioral consequences. For example, treatment of castrated adult male rats with estrogen increases their display of lordosis behavior.

The response of an animal to sCCK-8 microinjection into the VMH is correlated highly with the level of [^{125}I]sCCK-8 binding in the VMH, which appears to be regulated by the postnatal gonadal steroid environment (Ulibarri et al 1990). If perinatal steroids determine the response of CCK-R to steroids in adulthood, one would predict that manipulating steroid levels in adulthood would produce differential effects on binding in the VMH of males and females. Indeed, ovariectomy induces a high level of CCK-R binding, which is dramatically reduced 24 hours after estrogen treatment and is restored to initial levels 48 hours after the estrogen exposure (Figs. 7.7 and 7.8). The pattern of these data is similar to that observed

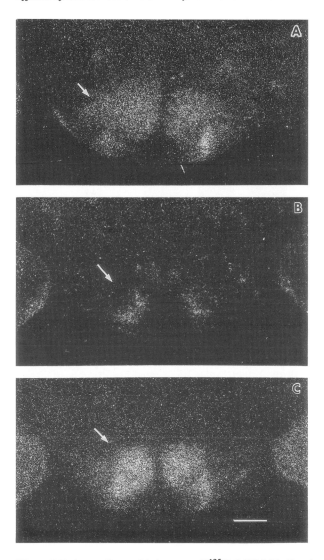

Figure 7.7. Autoradiographic images of [^{125}I]sCCK-8 binding (arrow) in the ventromedial nucleus of an ovariectomized female (A), an ovariectomized female that received 50 μg estradiol benzoate and was killed 24 hours later (B), and an ovariectomized female that received 300 μg testosterone propionate and was killed 24 hours later (C). Scale bar, 416 μm.

during the estrous cycle, in which the lowest level of CCK-R binding occurs on the morning of estrus, approximately 24 hours after the proestrus estrogen surge (Akesson et al. 1987). The results are consistent with the capacity of exogenous sCCK-8 microinjected into the VMH to block lordosis behavior; sCCK-8 is ineffective 24 hours after estrogen priming but produces a significant inhibition 48 hours after estrogen treatment (Fig. 11 in Micevych and Ulibarri 1992).

In sharp contrast to their actions in females, endogenous sex steroids do not

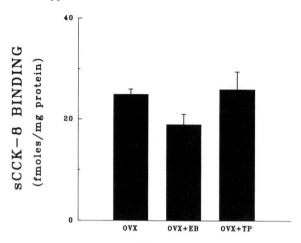

Figure 7.8. Histograph of [^{125}I]sCCK-8 binding in the ventromedial nucleus of ovariec-tomized female rats (OVX), ovariectomized female rats that received 50 μg estradiol benzoate and were killed 24 hours later (OVX + EB), and ovariectomized female rats that received 300 μg testosterone propionate and were killed 24 hours later (OVX + TP). Values are means ± SEM. All treatment groups include at least four animals.

downregulate CCK-R binding levels in males. In fact, testosterone may "protect" the levels of CCK-R binding. Intact males have very high levels of CCK-R bind-ing, which are not altered 2 weeks after orchidectomy. A bolus injection of tes-tosterone propionate (300 μg) into orchidectomized males does not change bind-ing levels to those measured in intact males (Figs. 7.9 and 7.10). It is interesting that estrogen affects CCK-R binding similarly in males and females. Twenty-four hours after treatment of orchidectomized males with 50 μg of estradiol benzoate, CCK-R binding is downregulated to levels observed in ovariectomized females 24 hours after estrogen treatment.

To test the idea that testosterone has a positive regulatory effect on CCK-R binding, 300 μg of testosterone propionate was injected into ovariectomized fe-males. Twenty-four hours after the injection, CCK-R binding levels were similar to those measured in ovariectomized females or intact males (Figs. 7.7 and 7.8). Since testosterone treatment did not alter CCK-R binding in gonadectomized male or female rats, one possibility is that testosterone has no effect on binding. Be-cause cells in the VMH contain high levels of aromatase in both sexes, we expect that estrogen, derived from testosterone, would act similarly on the VMH CCK-R in both sexes. Androgen, in the form of either testosterone or its 5α-reduced metabolite dihydrotestosterone (DHT), could counteract estrogenic downregula-tion, producing the same amount of binding as in gonadectomized females or males or intact males. While the expression of pCCK mRNA appears to be regu-lated in both sexes by estrogen, steroid regulation of CCK-R binding is regulated by both estrogen and androgen. Estrogen decreases CCK-R binding in orchidec-

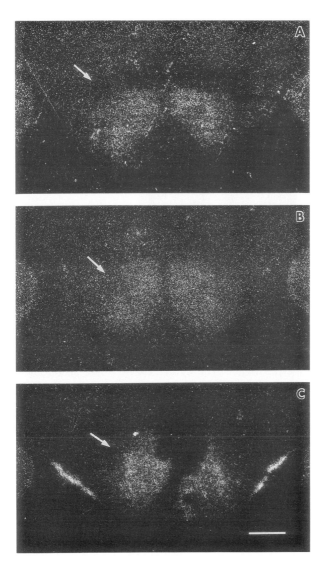

Figure 7.9. Autoradiographic images of [^{125}I]sCCK-8 binding (arrow) in the ventromedial nucleus of an orchidectomized male (A), an orchidectomized male that received 50 μg estradiol benzoate and was killed 24 hours later (B), and an orchidectomized male that received 300 μg testosterone propionate and was killed 24 hours later (C). Scale bar, 416 μm.

tomized males, while testosterone does not change CCK-R binding in ovariectomized females.

Several possible mechanisms could account for the results observed following gonadal steroid administration. In the VMH, the predominant form of CCK-R is CCK$_B$-R. Estrogen could downregulate CCK$_B$-R expression in the VMH by acting directly on cells that express CCK-R. Alternatively, the estrogen effects on

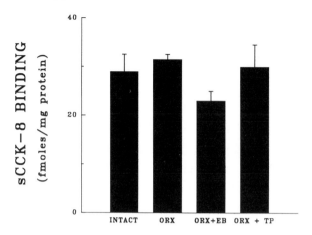

Figure 7.10. Histograph of [^{125}I]sCCK-8 binding in the ventromedial nucleus of go-nadally intact male rats, orchidectomized male rats (ORX), orchidectomized male rats that received 50 μg estradiol benzoate and were killed 24 hours later (ORX + EB), and orchidectomized male rats that received 300 μg testosterone propionate and were killed 24 hours later (ORX + TP). Values are means ± SEM. All treatment groups include at least four animals.

CCK-R binding could be the result of an estrogen-augmented release of CCK in the mediobasal hypothalamus (Micevych et al. 1988b) that produces a classic downregulation of receptors. The mechanism of this androgenic effect has not yet been elucidated; androgen may block the estrogen-facilitated release of CCK in the VMH, thus preventing a receptor downregulation. Alternatively, and perhaps additionally, androgen could act directly or trans-synaptically on CCK$_B$-R-expressing cells in the VMH. To test the possibility of effects of estrogen or testosterone treatment on the transcription of the CCK$_B$-R gene in the VMH we investigated CCK$_B$-R mRNA levels in the VMH from tissue sections from the same animals that were used for the receptor binding studies. *In situ* hybridization using an ^{35}S-labeled 40-mer oligonucleotide probe complementary to nucleotides 893–932 of the CCK$_B$-R mRNA showed no differences in CCK$_B$-R mRNA levels across treatment groups, indicating that gonadal steroids do not regulate CCK$_B$-R binding at the level of the CCK$_B$-R mRNA. However, a post-transcriptional effect on CCK$_B$-R of these steroids cannot be ruled out.

The most striking difference between the binding pattern in the male and the female rat occurs in their response to gonadectomy. Reducing circulating gonadal steroids increases CCK-R binding in females; however, CCK-R binding is un-affected in males. The absolute binding density in the intact male rat matches the binding density in the ovariectomized female rat. Indeed, CCK-R binding and behavioral responses to CCK are similar in both males and females after similar steroid priming. Microinjections of sCCK-8 into the VMH of estrogen-primed females or males attenuate the display of lordosis behavior (Fig. 11 in Micevych and Ulibarri 1992). However, the sensitivity of the lordosis behavior to exogenous sCCK-8 depends on the perinatal sex steroid milieu (Ulibarri et al. 1990; Ulibarri and Micevych 1993), suggesting that the lordosis circuit is organized by perinatal steroids.

Ontogeny of behavioral responses to exogenous CCK-8

Two behavioral experiments (Ulibarri et al. 1990; Ulibarri and Micevych 1993) have shown that the postnatal gonadal steroid environment alters the adult behavioral response to exogenous sCCK-8. Neonatally castrated males exhibit increased sensitivity to estrogen, are more behaviorally receptive, and are more sensitive to sCCK-8 microinfusions into the VMH than are intact adult males (Ulibarri et al. 1990). The CCK-regulated component of lordosis is very sensitive to the defeminizing effects of postnatal administration of testosterone propionate. Furthermore, the sexual differentiation of the response to sCCK-8 resembles the differentiation of female sexual behavior. The responses of animals to sCCK-8 were examined after manipulation of their neonatal steroid environments to determine whether androgens or estrogens act as the primary defeminizing steroid. Treatment of females with large doses of testosterone propionate within the first few hours of life substantially reduced the effect of microinjection of sCCK-8 into the VMH on lordosis during adulthood. Neonatal testosterone treatment of females does not reduce their estrogen-induced receptivity, but they are much less sensitive to the inhibitory effects of sCCK-8 microinjections. The results obtained are consistent with the hypothesis that estrogens are primarily responsible for sexual differentiation of the behavioral responses to exogenous sCCK-8. Males that are neonatally castrated or treated with capsules containing ATD show a dramatic inhibition of estrogen-induced lordosis behavior after sCCK-8 microinjection into the VMH; the magnitude of this inhibition is similar to that observed in control females. This demonstrates that estrogen, derived from testosterone, is primarily responsible for many of the defeminizing effects of testosterone on behavior, including animals' behavioral responses to exogenous sCCK-8. When behavioral data are compared with CCK-R binding data, it becomes clear that the behavioral response to exogenous CCK cannot be interpreted completely in terms of the effect of sex steroids on CCK-R binding in adulthood. Perinatal estrogen levels appear to determine the sensitivity of the response of CCK-R binding in the VMH to changes of sex steroid level during adulthood.

Summary

Recent evidence suggests that sex steroids act during development to determine the potential of specific circuits within the CNS to respond to changes in steroid level during adulthood. Perinatal steroid manipulation can alter an animal's display of sexual behavior during adulthood. Female rats are more sensitive to the effects of estrogen (Ulibarri et al. 1990; Ulibarri and Micevych 1993), since they require only one-third the amount of exogenous estrogen to induce lordosis behavior that male rats require. Males castrated on the day of birth or treated during the neonatal period with the aromatase inhibitor ATD are more likely to show lordosis behavior in response to estrogenic stimulation as adults than are intact males. Administration of testosterone or estrogen to neonatal female rats or castrated neonatal male rats inhibits their display of lordosis behavior during adulthood in response to estrogen priming. In contrast, neonatal treatment with DHT has no effect on adult female sexual behavior (Baum 1979; Clemens and Gladue 1978; Edwards and Thompson 1970; Parsons et al. 1984; Ulibarri and Micevych 1993), demonstrating that estrogen has an organizational effect on the differentiation of female sexual behavior.

The adult steroid environment appears to be the primary determining factor for the expression of CCK and its receptors. Our data do not support explicitly a role for perinatal steroids in the regulation of the CCK limbic–hypothalamic circuit, in terms of peptide expression. Rather, the data suggest that sex steroids act during development to determine the sensitivity of CCK-R in the limbic–hypothalamic circuit to changes in sex steroid levels during adulthood. Sex steroids apparently do not influence the development of CCK-R, but they do affect the behavioral response to microinjection of CCK into the VMH, which is mediated through CCK-R. Testosterone does not appear to regulate CCK-R binding; however, estrogen causes a transient downregulation of CCK-R binding. The mature limbic–hypothalamic circuit retains a great deal of plasticity during adulthood in terms of its capacity to express different levels of neuropeptides and receptors in response to changes in circulating levels of sex steroid hormones. This neurochemical plasticity allows alterations in the effectiveness of connections of limbic–hypothalamic circuits, strengthening some and weakening others, thereby eliciting a behavioral response. Thus, given the appropriate hormonal stimuli, behaviorally relevant circuits can be manipulated during adulthood, suggesting that sexual differentiation of the brain is as much a function of circulating steroid levels during adulthood as it is of the neonatal sex steroid environment.

Acknowledgments

The authors are grateful to Dr. Cesar Blanco and Ms. Krista Vink for helpful comments during the preparation of the manuscript. This work was supported by Grant NS 21220 to P.E.M.

References

Abelson, L., and Micevych, P. (1991). Distribution of preprocholecystokinin mRNA in motoneurons of the rat brainstem and spinal cord. *Mol. Brain Res.* 10: 327–335.

Abelson, L. A., Popper, P., Ulibarri, C., and Micevych, P. E. (1989). Ontogeny of cholecystokinin mRNA and receptors in male and female rats. *Soc. Neurosci. Abstr.* 15: 577.

Akesson, T. R., Mantyh, P. W., Mantyh, C. R., Matt, D. W., and Micevych, P. E. (1987). Estrous cyclicity of ^{125}I-cholecystokinin octapeptide binding in the ventromedial hypothalamic nucleus: Evidence for down modulation by estrogen. *Neuroendocrinology* 45: 257–262.

Akesson, T. R., and Micevych, P. E. (1986). Binding of ^{125}I-cholecystokinin-octapeptide in the paraventricular but not the supraoptic nucleus is increased by ovariectomy. *Brain Res.* 385: 165–168.

Akesson, T. R., and Micevych, P. E. (1988). Evidence for an absence of estrogen-concentration by CCK-immunoreactive neurons on the hypothalamus of the female rat. *J. Neurobiol.* 19: 3–16.

Arnold, A. P., and Breedlove, S. M. (1985). Organizational and activational effects of sex steroids on brain and behavior: A reanalysis. *Horm. Behav.* 19: 469–498.

Babcock, A. M., Bloch, G. J., and Micevych, P. E. (1988). Injections of cholecystokinin into the ventromedial hypothalamic nucleus inhibit lordosis behavior in the rat. *Phys. Behav.* 43: 195–199.

Baum, M. J. (1979). Differentiation of coital behavior in mammals: A comparative analysis. *Neurosci. Behav. Rev.* 3: 265–284.

Beinfeld, M. C., Meyer, B. K., and Brownstein, M. J. (1980). Cholecystokinin octapeptide in the rat hypothalamo-neurohypophysial system. *Nature* 288: 376–378.

Biegon, A., Fischette, C. T., Rainbow, T. C., and McEwen, B. S. (1982). Serotonin receptor modulation by estrogen in discrete brain nuclei. *Neuroendocrinology* 35: 287–291.

Blanco, C., Popper, P., and Micevych, P. E. (1992). α-CGRP mRNA expression in motoneurons is related to myosin heavy chain composition of specific rat hindlimb muscles. *Soc. Neurosci. Abstr.* 18: 858.

Bloch, G. J., Babcock, A. M., Gorski, R. A., and Micevych, P. E. (1987). Cholecystokinin stimulates and inhibits lordosis behavior in female rats. *Phys. Behav.* 39: 217–224.

Boden, P., and Hill, R. G. (1988). Effects of cholecystokinin and related peptides on neuronal activity in the ventromedial nucleus of the rat hypothalamus. *Br. J. Pharmacol.* 94: 246–252.

Chang, R. S. L., and Lotti, V. J. (1986). Biochemical and pharmacological characterization of an extremely potent and selective non-peptide cholecystokinin antagonist. *Proc. Natl. Acad. Sci. (USA)* 83: 4923–4926.

Cho, H. J., Shiotani, Y., Shiosaka, S., Inagaki, S., Kubota, Y., Kiyama, H., Umegaki, K., Tateishi, K., Hashimura, E., Hamaoka, T., and Tohyama, M. (1983). Ontogeny of cholecystokinin-8-containing neuron system of the rat: An immunohistochemical analysis: I. Forebrain and upper brainstem. *J. Comp. Neurol.* 218: 25–41.

Clemens, L. G., and Gladue, B. A. (1978). Feminine sexual behavior in rats is enhanced by prenatal inhibition of androgen aromatization. *Horm. Behav.* 30: 613–616.

Comb, M. J., Hyman, S. E., and Goodman, H. M. (1989). Mechanisms of trans-synaptic regulation of gene expression. *TINS* 10: 473–478.

Cooke, N., Coit, D., Shine, J., Baxter, J., and Martial, J. (1981). Human prolactin: cDNA structural analysis and evolutionary comparisons. *J. Biol. Chem.* 256: 4007–4016.

Crawley, J. N. (1986). Cholecystokinin potentiation of dopamine-mediated behaviors in the nucleus accumbens. *Ann. N.Y. Acad. Sci.* 448: 283–292.

Crawley, J. N. (1991). Cholecystokinin–dopamine interactions. *Trends Pharmacol. Sci.* 12: 232–236.

Crawley, J. N., St.-Pierre, S., and Gaudreau, P. (1984). Analysis of the behavioral activity of C- and N-terminal fragments of cholecystokinin octapeptide. *J. Pharmacol. Exp. Ther.* 230: 438–446.

Della Ferra, M. A., and Baile, C. A. (1979). Cholecystokinin octapeptide: Continuous picomolar injections into cerebral ventricles of sheep supress feeding. *Science* 206: 471–473.

Dockray, G. J. (1980). Cholecystokinin in rat cerebral cortex: Identification, purification and characterization by immunochemical methods. *Brain Res.* 118: 155–165.

Dodson, R. E., Shryne, J. E., and Gorski, R. A. (1988). Hormonal modification of the number of total and late-arising neurons in the central part of the medial preoptic nucleus of the rat. *J. Comp. Neurol.* 275: 623–629.

Dornan, W. A., Bloch, G. J., Priest, C. A., and Micevych, P. E. (1989). Microinjection of cholecystokinin into the medial preoptic nucleus facilitates lordosis behavior in the female rat. *Phys. Behav.* 45: 969–974.

Drouva, S. V., Laplante, E., Gautron, J. P., and Kordon, C. (1985). Effects of 17β-estradiol on LH-RH release from rat mediobasal hypothalamic slices. *Neuroendocrinology* 38: 152–157.

Duchemin, A. M., Quach, T. T., Iadarola, M. J., Deschenes, R. J., Schwartz, J. P., and Wyatt, R. J. (1987). Expression of the cholecystokinin gene in rat brain during development. *Dev. Neurosci.* 9: 61–67.

Edwards, D. A., and Thompson, M. L. (1970). Neonatal androgenization and estrogenization and the hormonal induction of sexual receptivity in rats. *Physiol. Behav.* 5: 1115–1119.

Emson, P. C., Rehfeld, J. F., and Roser, N. M. (1982). Distribution of cholecystokinin-like peptide in the human brain. *J. Neurochem.* 38: 117–119.

Faris, P. L., Kominsaruk, B. R., Watkins, L. R., and Mayer, J. D. (1983). Evidence of the neuropeptide cholecystokinin as an antagonist of opiate analgesia. *Science* 219: 310–312.

Fischette, C. T., Biegon, A., and McEwen, B. S. (1983). Sex differences in serotonin-1 receptor binding in rat brain. *Science* 222: 333–335.

Frankfurt, M., Siegel, R. A., Sim, I., and Wuttke, W. (1986). Estrous cycle variations in cholecystokinin and substance P concentration in discrete areas of the rat brain. *Neuroendocrinology* 42: 226–231.

Gibbs, J., Young, R. C., and Smith, G. P. (1973). Cholecystokinin elicits satiety in rats with open gastric fistulas. *Nature* 245: 323–325.

Gorski, R. A. (1985). The 13th J. A. Stevenson Memorial Lecture: Sexual differentiation of the brain–possible mechanisms and implications. *Can. J. Physiol. Pharmacol.* 63: 577–594.

Haun, R. S., and Dixon, J. E. (1990). A transcriptional enhancer essential for the expression of the rat cholecystokinin gene contains a sequence identical to the −296 element of the human c-*fos* gene. *J. Biol. Chem.* 265: 15455–15463.

Herbison, A. E., Robinson, J. E., and Skinner, D. C. (1992). Distribution of estrogen receptor-immunoreactive cells in the preoptic area of the ewe: Co-localization with glutamic acid decarboxylase but not luteinizing hormone-releasing hormone. *Neuroendocrinology* 57: 751–759.

Herbison, A. E., and Theodosis, D. T. (1992). Localization of oestrogen receptors in preoptic neurons containing neurotensin but not tyrosine hydroxylase, cholecystokinin or luteinizing hormone-releasing hormone in the male and female rat. *Neuroscience* 50(2): 283–298.

Hill, D. R., Campbell, N. J., Shaw, T. M., and Woodruff, G. N. (1987). Autoradiographic localization and characterization of peripheral type CCK receptors in rat CNS using highly selective nonpeptide CCK antagonists. *J. Neurosci.* 7: 2967–2976.

Hill, D. R., and Woodruff, G. N. (1990). Differentiation of central cholecystokinin receptor binding sites using the non-peptide antagonists MK-329 and L-365,260. *Brain Res.* 526: 276–283.

Honda, T., Wada, E., Battey, J. F., and Wank, S. A. (1993). Differential expression of CCK_A and CCK_B receptors in the rat brain. *Mol. Cell. Neurosci.* 4: 143–154.

Horikawa, S., Takai, T., Toyosato, M., Takahashi, H., Noda, M., Kakidani, H., Kubo, T., Hirose, T., Inayama, S., Hayashida, H., Miyata, T., and Numa, S. (1983). Isolation and structural organization of the human preproenkephalin B gene. *Nature* 306: 611–614.

Innis, R. B., Correa, F. M. A., Uhl, G. R., Schneider, B., and Snyder, S. H. (1979). Cholecystokinin octapeptide-like immunoreactivity: Histochemical localization in the brain. *Proc. Natl. Acad. Sci. (USA)* 76: 521–525.

Ivy, A. C., and Goldberg, E. (1928). A hormone mechanism for gallbladder contraction and evacuation. *Am. J. Physiol.* 86: 599–613.

Larsson, L.-I., and Rehfeld, J. F. (1979). Localization of molecular heterogeneity of cholecystokinin in the central and peripheral nervous system. *Brain Res.* 165: 210–218.

Lotti, V. J., and Chang, R. S. L. (1989). A potent and selective non-peptide gastrin antagonist and CCK_B receptor ligand: L-365,260. *Eur. J. Pharmacol.* 162: 273–280.

Matsumoto, A., Arnold, A. P., Zampighi, G. A., and Micevych, P. E. (1988a). Androgenic regulation of gap junctions between motoneurons in the rat spinal cord. *J. Neurosci.* 8: 4177–4183.

Matsumoto, A., Micevych, P. E., and Arnold, A. P. (1988b). Androgen regulates synaptic input to motoneurons of the adult rat spinal cord. *J. Neurosci.* 8: 4168–4176.

McEwen, B. S., and Parsons, B. (1982). Gonadal steroid action on the brain: Neurochemistry and neuropharmacology. *Annu. Rev. Pharmacol. Toxicol.* 22: 555–598.

Mendelson, S. D., and Gorzalka, B. B. (1984). Cholecystokinin-octapeptide produces inhibition of lordosis in the female rat. *Pharmacol. Biochem. Behav.* 21: 755–759.

Micevych, P. E. (1991). Sexual differentiation of a neuropeptide circuit regulating female reproductive behavior. In *Behavioral Biology: Neuroendocrine Axis*, ed. T. Archer and S. Hansen, pp. 139–150. Hillsdale, NJ: Erlbaum.

Micevych, P. E., Abelson, L., Fok, H., Ulibarri, C., and Priest, C. A. (1994). Gonadal steroid control of pCCK mRNA in the limbic–hypothalamic circuit: Comparison of adult with neonatal steroid treatment. *J. Neurosci. Res.* 38: 386–398.

Micevych, P. E, Akesson, T., and Elde, R. (1988a). Distribution of cholecystokinin-immunoreactive cell bodies in the male and female rat: II. Bed nucleus of the stria terminalis and amygdala. *J. Comp. Neurol.* 269: 381–391.

Micevych, P. E., and Bloch, G. J. (1989). Estrogen regulation of a reproductively relevant cholecystokinin circuit in the hypothalamus and limbic system of the rat. In *The Neuropeptide Cholecystokinin (CCK)*, ed. J. Huges, G. Dockray, and G. Woodruff, pp. 68–73. New York: Wiley.

Micevych, P. E., Matt, D. W., and Go, V. L. W. (1988b). Concentrations of cholecystokinin, substance P, and bombesin in discrete regions of male and female rat brain: Sex differences and estrogen effects. *Exp. Neurol.* 100: 416–426.

Micevych, P. E., Park, S. S., and Akesson, T. R. (1987). Distribution of cholecystokinin-immunoreactive cell bodies in the male and female rat: I. Hypothalamus. *J. Comp. Neurol.* 255: 124–136.

Micevych, P. E., and Ulibarri, C. (1992). Development of the limbic–hypothalamic cholecystokinin circuit: A model of sexual differentiation. *Dev. Neurosci.* 14: 11–34.

Micevych, P. E., Ulibarri, C., Abelson, L., and Fok, H. (1992). Expression of pCCK mRNA is dependent on adult gonadal steroids, not neonatal steroid environment. *Soc. Neurosci. Abstr.* 18: 234.

Miller, F. D., Ozimek, G., Milner, R. J., and Bloom, F. E. (1989). Regulation of neuronal oxytocin mRNA by ovarian steroids in the mature and developing hypothalamus. *Proc. Natl. Acad. Sci. (USA)* 86: 2468–2472.

Monstein, H. J., Folkesson, R., and Geijer, T. (1990). Procholecystokinin and pro-enkephalin A mRNA expression is modulated by cyclic AMP and noradrenaline. *J. Mol. Endocrinol.* 4: 37–41.

Moran, T. H., Robinson, P. H., Goldrich, M. S., and McHugh, P. R. (1986). Two brain cholecystokinin receptors: Implications for behavioral actions. *Brain Res.* 362: 175–179.

Oro, A. E., Simerly, R. B., and Swanson, L. W. (1988). Estrous cycle variations in levels of cholecystokinin immunoreactivity within cells of three interconnected sexually dimorphic forebrain nuclei. *Neuroendocrinology* 47: 225–235.

Palacios, J. M., Nichoff, D. L., and Kuhar, J. M. (1979). Ontogeny of GABA and benzodiazepine receptors: Effects of Triton X-100, bromide and muscimol. *Brain Res.* 179: 390–395.

Parson, B., Rainbow, T. C., and McEwen, B. S. (1984). Organizational effects of testosterone via aromatization on feminine reproductive behavior and neural progestin receptors in rat brain. *Endocrinology* 115: 1412–1417.

Patey, G., De la Baume, S., Gros, C., and Schwartz, J. (1980). Ontogenesis of enkephalinergic systems in rat brain: Postnatal changes in enkephalin levels, receptors and degrading enzyme activities. *Life Sci.* 27: 245–252.

Pelaprat, D., Dusart, I., and Peschanski, M. (1988). Postnatal development of cholecystokinin (CCK) binding sites in the rat forebrain and midbrain: An autoradiographic study. *Dev. Brain Res.* 44: 119–132.

Pfaff, D. W., and Schwartz-Giblin, S. (1988). Cellular mechanisms of female reproductive behaviors. In *The Physiology of Reproduction,* Vol. 2, ed. E. Knobil, J. D. Neill, L. L. Ewing, G. S. Greenwald, C. L. Market, and D. W. Pfaff, pp. 1487–1568. New York: Raven Press.

Phoenix, C. H., Goy, R. W., Gerall, A. A., and Young, C. W. (1959). Organizing action of prenatally administered testosterone propionate on the tissues mediating mating behavior in the female guinea pig. *Endocrinology* 65: 369–382.

Popper, P., Priest, C., and Micevych, P. E. (1991). The effects of sex hormones on cholecystokinin receptor binding in the rat brain. *Soc. Neurosci. Abstr.* 17: 1062.

Rehfeld, J. F. (1989). Post-translational attenuation of CCK gene transcription. In *The Neuropeptide Cholecystokinin (CCK),* ed. J. Hughes, G. Dockray, and G. Woodruff, pp. 58–67. New York: Wiley.

Roberecht, P., Descodt-Lanckerman, M., and Vanderhaegen, J.-J. (1978). Demonstration of histological activity of brain gastrin-like peptide material in the human: Its relationship with COOH-terminal octapeptide of cholecystokinin. *Proc. Natl. Acad. Sci. (USA)* 75: 524–528.

Roth, B. L., and Beinfeld, M. C. (1985). The postnatal development of VIP binding sites in rat forebrain and hindbrain. *Peptides* 6: 27–30.

Saito, A., Sankaran, H., Goldfine, I. D., and Williams, J. A. (1980). Cholecystokinin receptors in the brain: Characterization and distribution. *Science* 208: 115–116.

Sengelaub, D. R., Nordeen, E. J., Nordeen, K. W., and Arnold, A. P. (1989). Hormonal control of neuron number in sexually dimorphic spinal nuclei of the rat. III: Differential effects of dihydrotestosterone on spinal nucleus of the bulbocavernosus and dorsolateral nucleus. *J. Comp. Neurol.* 280: 637–644.

Shen, L. P., and Rutter, W. J. (1984). Sequence of the human somatostatin gene. *Science* 224: 168–171.

Shibihara, S., Morimoto, Y., Furutani, Y., Notake, M., Takahashi, H., Shimizu, S., Horikawa, S., and Numa, S. (1983). Isolation and response analysis of the human corticotropin-releasing factor precursor gene. *EMBO J.* 2: 775–779.

Shivers, B. D., Harlan, R. E., Morrell, J. I., and Pfaff, D. W. (1983). Absences of estradiol concentration in cell nuclei of LHRH-immunoreactive neurons. *Nature* 304: 345–347.

Simerly, R. B. (1990). Hormonal control of neuropeptide gene expression in sexually dimorphic olfactory pathways. *TINS* 13: 104–109.

Simerly, R. B., and Swanson, L. W. (1986). The organization of neural inputs to the medial preoptic nucleus of the rat. *J. Comp. Neurol.* 246: 312–342.

Simerly, R. B., and Swanson, L. W. (1987). Castration reversibly alters levels of cholecystokinin immunoreactivity within cells of three interconnected sexually dimorphic forebrain nuclei in the rat. *Proc. Natl. Acad. Sci. (USA)* 84: 2087–2091.

Simerly, R. B., Young, B. J., Capozza, M. A., and Swanson, L. W. (1989). Estrogen differentially regulates neuropeptide gene expression in a sexually dimorphic olfactory pathway. *Proc. Natl. Acad. Sci. (USA)* 86: 4766–4770.

Sturdevant, R. A., and Goetz, H. (1976). Cholecystokinin both stimulates and inhibits human food intake. *Nature* 261: 714–715.

Ulibarri, C., and Micevych, P. E. (1993). Role of perinatal estrogens in sexual differentiation of the inhibition of lordosis by exogenous cholecystokinin. *Phys. Behav.* 54: 95–100.

Ulibarri, C., Popper, P., and Micevych, P. E. (1990). Role of postnatal androgens in sexual differentiation of the lordosis-inhibiting effect of central injections of cholecystokinin. *J. Neurobiol.* 21: 796–807.

Vanderhaegen, J. J. (1985). Neuronal cholecystokinin. In *Handbook of Chemical Neuroanatomy*, Vol. 4: *GABA and Neuropeptides in the CNS, Part 1*, ed. A. Bjorklund and T. Hokfelt, pp. 406–435. Amsterdam: Elsevier.

Wang, R. Y., White, F. J., and Voigt, M. M. (1984). Cholecystokinin, dopamine and schizophrenia. *Trends Pharmacol. Sci.* 5: 436–438.

Wank, S. A., Harkins, R., Jensen, R. T., Shapira, H., Weerth, A. D., and Saltery, T. (1992). Purification, molecular cloning and functional expression of the cholecystokinin receptor from rat pancreas. *Proc. Natl. Acad Sci. (USA)* 89: 8691–8695.

Webb, D. K., Moulton, B. C., and Khan, S. A. (1990). Estrogen induced expression of the C-jun proto-oncogene in the immature and mature rat uterus. *Biochem. Biophys. Res. Commun.* 168: 721–726.

Woodruff, G. N., Hill, D. R., Boden, P., Singh, L., and Hughes, J. (1991). Functional role of brain CCK receptors. *Neuropeptides* 19(Suppl): 45–56.

8

Cholinergic regulation of female sexual behavior

GARY DOHANICH

Acetylcholine was the first endogenous chemical to be identified as a neuro-transmitter. In addition to its vital role in physiology, acetylcholine has been implicated in the regulation of mammalian behaviors that range from reflexive (Potter et al. 1990) to regulatory (Hagan et al. 1987) to cognitive (Levin et al. 1990). The control of these heterogeneous behaviors appears to be possible because of diffuse cholinergic circuits distributed throughout the mammalian brain (Mesulam et al. 1983). The organization of these systems, particularly in the forebrain and midbrain, places cholinergic neurons in regions involved in sensory, motor, and motivational processes.

Our work during the past decade has indicated that cholinergic systems also play an important role in the regulation of certain behaviors exhibited by mammalian females during mating. It is well known that the sexual behaviors of female rodents are controlled closely by steroid hormones, primarily estrogen and progesterone secreted by the ovaries. However, the sequence of neural events initiated by these hormones to cause complex behavioral responses has not been described fully. The capacity of cholinergic mechanisms to affect rodent sexual behavior suggests a key interface between endocrine activity and cholinergic function. Our primary hypothesis states that ovarian steroids regulate brain function, and consequently behavior, by altering the activity of cholinergic systems within neural target structures. According to current theories of hormone action, steroids could access nuclear genomes (O'Malley and Means 1974) and surface membranes (Schumacher 1990) to alter the nature and number of various proteins associated with cholinergic neurotransmission.

Investigations of the role of acetylcholine in the regulation of female sexual behavior in rodents have advanced along several interrelated lines. First, the techniques of behavioral pharmacology have been employed to determine the effects of manipulations of central cholinergic neurochemistry on female sexual behavior. Second, the locations of central sites at which cholinergic systems act to

184

control female sexual behavior in rodents have been explored with anatomical methods. Finally, the effects of ovarian steroids on cholinergic neurochemistry have been examined using biochemical techniques.

Pharmacological regulation of female sexual behavior

Female sexual behavior

Mating behaviors are exhibited by female rats only during an 8- to 12-hour period of the 4- to 5-day estrous cycle. During the period of sexual receptivity, female rats display a series of proceptive behaviors that include hops, darts, and ear wiggles before the mount of the male and receptive behavior that includes lordosis during the mount of the male. Lordosis is a distinctive posture assumed by females of several mammalian species during the period of sexual receptivity. It is defined as a flexion of the lumbar epaxial muscles that causes a ventral arching of the spine and elevation of the perineum (Pfaff 1980). Despite the simplicity of this behavior, lordosis is a critical component of mating for rodents. In species such as the rat, in which the male achieves multiple intromissions before ejaculating, lordosis promotes genital stimulation of both sexes and facilitates sperm deposition during ejaculation (Clemens and Christensen 1975). The lordosis response is elicited in the female rat by the mount of a male and is dependent on somatosensory stimulation to the flanks, rump, and tail base (Pfaff 1980).

Removal of the ovaries abolishes the capacity for proceptive behavior and lordosis in female rats, while systemic administration of estrogen (estradiol) for several days followed by progesterone restores these behaviors in ovariectomized females several hours after progesterone injection. This treatment regimen loosely simulates the natural hormonal profile of the intact cyclic female such that maximum receptivity is preceded by several days of rising estrogen titers followed by a brief increase in progesterone levels (Clemens and Christensen 1975). In addition to being hormone dependent and biologically relevant, lordosis is elicited readily in the laboratory and recorded easily by simple observation. Consequently, lordosis provides an ideal behavior for studying the interaction of hormonal and neurochemical factors that regulate mammalian functions.

Facilitation of lordosis by cholinergic agonists

Agents that enhance cholinergic neurotransmission facilitate the occurrence of lordosis. We demonstrated that intracerebral administration of cholinergic agonists rapidly activated lordosis in ovariectomized rats primed with doses of estrogen that are behaviorally ineffective (Table 8.1) (Clemens et al. 1980, 1981, 1989; Dohanich et al. 1984). Lordosis could be elicited by stimulus male rats

Table 8.1. *Effects of cholinergic agents on female sexual behavior*

Agent	Dose (μg)[a]	Behavioral effect	Reversal
Cholinergic agonist			
Carbachol	0.5	Facilitation	Atropine
Muscarinic agonists			
Oxotremorine	1	Facilitation	Scopolamine
Pilocarpine	30	Facilitation	Scopolamine
Bethanechol	Crystals	Facilitation	Atropine
Acetylcholine	5	Facilitation	Atropine
Oxotremorine-M (M2)	0.1	Facilitation	—
McN-A-343 (M1)	20	No effect	—
Cholinesterase inhibitors			
Physostigmine	10	Facilitation	Atropine
Neostigmine	1	Facilitation	—
Choline uptake inhibitor			
Hemicholinium-3	1.25	Inhibition	Choline, carbachol
Muscarinic antagonists			
Atropine	20	Inhibition	—
Scopolamine	10	Inhibition	—
Amitriptyline	20	No effect	—
Pirenzepine (M1)	80	No effect	—

[a]Lowest intracerebral dose that produced the full behavioral response. Doses were delivered bilaterally in 0.5-μl volumes of buffered vehicle.

within 15 minutes after bilateral infusion of cholinergic agonists delivered in 0.5-μl volumes of a buffered vehicle into the lateral ventricles or specific brain regions (Fig. 8.1). This facilitation of behavior usually dissipated within about 90 minutes. Increases in the incidence of lordosis elicited by males were observed following intracerebral administration of the cholinergic receptor agonist carbachol (0.125–1 μg bilaterally) and the muscarinic receptor agonists oxotremorine (0.5–2 μg bilaterally), pilocarpine (30–60 μg bilaterally), and bethanechol (crystals). The acetylcholinesterase inhibitor physostigmine (eserine), which elevates synaptic levels of endogenous acetylcholine, also facilitated lordosis in female rats following intracerebral infusion (2.5–10 μg bilaterally). In addition, exogenous acetylcholine (2.5–5 μg bilaterally) facilitated lordosis following intracerebral infusion in combination with low concentrations of physostigmine (0.5 μg bilaterally), which prevented degradation of the acetylcholine. The effects of these cholinergic agonists were mediated solely by muscarinic receptors, because their facilitation of lordosis was blocked completely by systemic pretreatment with the muscarinic receptor blockers atropine and scopolamine (Fig. 8.2; Clemens et al. 1980, 1989; Dohanich et al. 1984). Nicotinic receptors

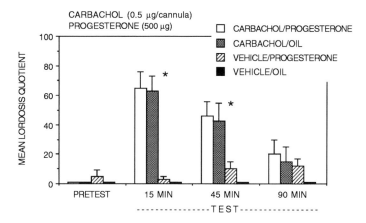

Figure 8.1. Facilitation of lordosis by the cholinergic agonist carbachol. Ovariectomized rats were injected intramuscularly with estradiol benzoate (0.5 μg/kg) at 72, 48, and 24 hours before behavioral testing. Progesterone (500 μg) or oil vehicle was injected intramuscularly 5 hours before testing. Each female was mounted 10 times by a stimulus male rat, the incidence of lordosis was recorded, and a lordosis quotient was calculated [lordosis quotient = (number of lordoses/10 mounts) × 100]. Following this behavioral pretest, each female received bilateral intracerebral infusions of carbachol or vehicle solution. Behavioral effects were tested 15, 45, and 90 minutes after infusion. *$p \leq .01$, carbachol vs. vehicle.

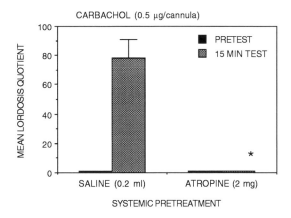

Figure 8.2. Prevention of carbachol facilitation of lordosis by pretreatment with the muscarinic receptor blocker atropine. Ovariectomized rats were injected intramuscularly with estradiol benzoate (0.5 μg/kg) at 72, 48, and 24 hours before behavioral testing. Each female was injected intraperitoneally with atropine or saline 45 minutes before bilateral intracerebral infusion of carbachol. *$p \leq .01$, atropine vs. saline.

do not appear to play a significant role in the facilitation of lordosis by cholinergic systems (Weaver and Clemens 1987).

In order to facilitate lordosis reliably with cholinergic agonists, it was necessary to prime ovariectomized females with low doses of estrogen injected intramuscularly as estradiol benzoate. In the absence of estrogen pretreatment, carbachol facilitated lordosis only weakly at high intracerebral doses (2.5 μg bilaterally) that also disrupted general activity (Clemens et al. 1983). Therefore, a cholinergic agonist could not simply substitute for estrogen, but rather depended on the presence of adequate estrogen stimulation to achieve its behavioral effect.

The level of estrogen required to support the action of a cholinergic agonist was quite low, however. In the absence of cholinergic treatment, our typical estrogen priming regimens (e.g., 0.5 μg/kg estradiol benzoate for 3 days) consistently failed to facilitate lordosis in ovariectomized rats (Fig. 8.1). Furthermore, the doses of estradiol benzoate utilized in our experiments failed to induce lordosis even when combined with an active dose (500 μg) of systemic progesterone (Fig. 8.1). Consequently, intracerebral carbachol facilitated lordosis at lower doses of estrogen priming than did systemic progesterone. The effects of carbachol and progesterone also did not appear to be additive, because infusions of carbachol were equally effective in females treated with estradiol benzoate and in females treated with estradiol benzoate and progesterone (Fig. 8.1). These results indicated that carbachol did not simply substitute for progesterone following administration to ovariectomized female rats primed with estradiol benzoate. Indeed, cholinergic agonists and progesterone appeared to exert their effects on female sexual behavior by different mechanisms.

In an earlier series of experiments, Lindstrom and Meyerson (1967) found that systemic administration of muscarinic receptor agonists reduced the incidence of lordosis in ovariectomized female rats that had been made sexually receptive by treatment with estrogen and progesterone. Because pretreatment with a monoamine synthesis inhibitor prevented this inhibition of lordosis by pilocarpine, Lindstrom (1971) concluded that cholinergic agonists inhibited lordosis in female rats via a serotonergic mechanism. Our results conflicted with those of Lindstrom. However, these discrepant outcomes might be explained by differences in the routes and doses of drug administration. Although Lindstrom reported that intraperitoneal injection of oxotremorine (0.5–2 mg/kg) or pilocarpine (10–50 mg/kg) inhibited lordosis, we found that intracerebral administration of these same muscarinic agonists in microgram doses produced consistent increases in the incidence of lordosis. While it is difficult to assess the effects of these agonists delivered by different routes and doses of administration, the report of Lindstrom and Meyerson (1967) indicated that varied degrees of tremor and inactivity were induced by systemic administration of cholinergic agents despite pretreatment with methylatropine to reduce the peripheral effects of the agonists. Such peripheral side effects might account for the inhibition of lordosis observed in these earlier experiments.

In a later study, muscarinic agonists increased the incidence of lordosis 4 hours after systemic administration to ovariectomized females that had been primed with estrogen, an effect attributed to pituitary-adrenal mechanisms (Lindstrom

1973). It is unlikely that the facilitation of lordosis following intracerebral infusion of cholinergic agonists in our experiments could be mediated by adrenal activity. In our studies, intracerebral administration of cholinergic agonists activated lordosis within 15 minutes after infusion and the facilitation of behavior faded within 90 minutes. Lordosis activated by the release of adrenal steroids would persist at high levels for more than 90 minutes after release. For example, following exposure to ether, which induces the release of adrenal secretions, the incidence of lordosis in female rats increased linearly from the time of exposure before peaking at 120–180 minutes (Franck 1977). In addition, the facilitation of lordosis by muscarinic agonists persisted in the absence of the adrenal glands (Clemens et al. 1980). Finally, as indicated earlier, the administration of progesterone proved to be behaviorally ineffective at the levels of estrogen necessary to promote facilitation of lordosis by cholinergic agonists (Fig. 8.1).

Inhibition of lordosis by cholinergic antagonists

In our experiments, intracerebral infusion of agents that stimulate cholinergic receptors or enhance acetylcholine action facilitated lordosis in ovariectomized rats primed with estrogen. In another series of experiments, we assessed the effects on lordosis of agents that reduce cholinergic transmission in ovariectomized female rats treated with estrogen and progesterone. Intracerebral infusion of hemicholinium-3 (1.25–7.5 µg bilaterally), an agent that retards the synthesis of acetylcholine by inhibiting the transport of choline across neuronal membranes, reduced the incidence of lordosis in female rats made sexually receptive by intramuscular injection of estradiol benzoate and progesterone (Clemens and Dohanich 1980). In addition, when either choline (120 µg bilaterally) or carbachol (0.5 µg bilaterally) was infused along with hemicholinium-3, the inhibition of lordosis was reduced or prevented (Clemens and Dohanich 1980; Clemens et al. 1981). In another series of experiments, the cholinergic muscarinic receptor blockers scopolamine and atropine also reduced the incidence of lordosis following systemic or intracerebral administration to ovariectomized female rats primed with estradiol benzoate and progesterone (Fig. 8.3; Clemens and Dohanich 1980; Clemens et al. 1989). The inhibition of lordosis by muscarinic receptor antagonists was demonstrated first by Singer (1968) following systemic administration of atropine. Lindstrom (1973) also reported decreases in lordosis following systemic administration of scopolamine to receptive female rats. The capacity of an acetylcholine synthesis inhibitor, as well as muscarinic receptor blockers, to reduce the occurrence of lordosis in receptive female rats indicated that cholinergic activity is critical to the display of this sexual behavior. Cholinergic systems appear to play an important role in other aspects of female mating, as indicated by recent evidence that systemic or intraventricular administration of scopolamine

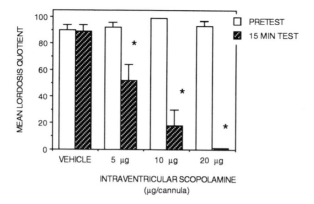

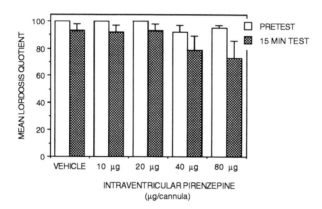

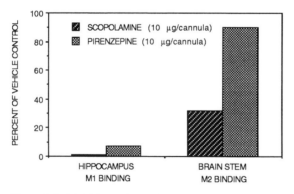

Figure 8.3. Inhibition of lordosis by the muscarinic receptor blocker scopolamine, but not by the M1 receptor blocker pirenzepine. Ovariectomized rats were injected intramuscularly with estradiol benzoate (1 µg/kg) at 72, 48, and 24 hours before behavioral testing. Progesterone (500 µg) was injected intramuscularly 5 hours before testing. Lordosis was measured before and 15 minutes after bilateral intraventricular infusion of scopolamine (*upper*) or pirenzepine (*middle*). *$p \leq .01$, scopolamine vs. vehicle. M1 receptor binding in hippocampus and M2 receptor binding in brain stem were determined *in vitro* in tissues dissected within 20 minutes after bilateral intraventricular infusion of scopolamine or pirenzepine (*lower*).

interfered with the expression of solicitation behaviors, coital pacing, and mate preference normally displayed by receptive female rats (Dohanich et al. 1993).

Muscarinic receptor subtypes and lordosis

While earlier experiments indicated that muscarinic cholinergic receptors contribute to the regulation of lordosis, these receptors are now known to exist in multiple forms. Although five distinct forms of the muscarinic receptor have been identified (Buckley et al. 1989), traditional pharmacological methods have been used to study and manipulate primarily two subtypes, M1 and M2 (Hammer and Giachetti 1982). In our laboratory, the contributions of M1 and M2 receptors to the regulation of lordosis were investigated in a series of biochemical and behavioral experiments (Dohanich et al. 1991). Using *in vitro* competition assays, we confirmed that the M1-selective antagonist pirenzepine bound with moderately high affinity to M1 binding sites in the rat brain ($IC_{50} = 13.7$ nM) and with very weak affinity to M2 binding sites ($IC_{50} = 635.8$ nM). In behavioral experiments, intraventricular infusion of pirenzepine (10, 20, 40, or 80 μg bilaterally) failed to reduce the incidence of lordosis in ovariectomized rats primed with estrogen and progesterone (Fig. 8.3). *Ex vivo* binding analyses indicated that intraventricular infusion of pirenzepine, even at the lowest dose of 10 μg bilaterally, inhibited M1 binding almost completely without affecting M2 binding (Fig. 8.3). Consequently, female rats displayed a high incidence of lordosis despite complete blockade of M1 binding sites. These results suggested that the M1 receptor subtype is not critical to the display of lordosis.

Cholinergic agonists also display different affinities for M1 and M2 receptor subtypes. In experiments with muscarinic agonists (Dohanich et al. 1991), we found that oxotremorine-M and McN-A-343 bound with higher affinities to M1 sites than did carbachol. However, oxotremorine-M and carbachol displayed higher affinities for M2 binding sites than did McN-A-343. In behavioral tests, intraventricular infusion of oxotremorine-M (0.1 μg bilaterally) or carbachol (1 μg bilaterally) activated lordosis in ovariectomized females primed with low doses of estrogen (Fig. 8.4). In contrast, McN-A-343 failed to increase lordosis significantly following infusion (1, 10, or 20 μg bilaterally). Consequently, the capacity of these agents to activate lordosis in female rats was related to their affinities for M2 binding sites (oxotremorine-M > carbachol > McN-A-343) but not M1 binding sites (oxotremorine-M > McN-A-343 > carbachol). These results support a putative role of the M2 receptor subtype in the control of lordosis, although differences in the intrinsic activities of these cholinergic agonists must be considered (Dohanich et al. 1991). The effects of various cholinergic agents on lordosis behavior displayed by female rats are summarized in Table 8.1.

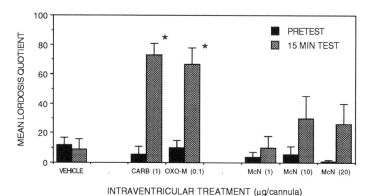

Figure 8.4. Facilitation of lordosis as related to the affinities of cholinergic agonists for the M2 receptor subtype oxotremorine-M > carbachol > McN-A-343. Ovariectomized rats were injected intramuscularly with estradiol benzoate (1 μg/kg) at 72, 48, and 24 hours before behavioral testing. Lordosis was measured before and 15 minutes after bilateral intraventricular infusion of each agonist. *$p \leq .01$, agonist vs. vehicle.

Cholinergic manipulations in other preparations

Cholinergic manipulations have been found to alter sexual behaviors under several other conditions. These results suggest that the regulation of mating by acetylcholine may be a general phenonemon. In ovariectomized hamsters made sexually receptive by treatment with estrogen and progesterone, scopolamine reduced total lordosis duration following systemic (1 mg/kg) or intraventricular (10 and 20 μg bilaterally) administration (Dohanich et al. 1990). In addition, the acetylcholinesterase inhibitor physostigmine facilitated lordosis of short duration following intraventricular infusion (10 μg bilaterally) in ovariectomized hamsters primed only with estradiol benzoate. It has also been reported that intracerebral administration of acetylcholine increased sexual motivation and induced spontaneous orgasm in human females (Heath 1972). Consequently, central cholinergic systems may be an important neural component involved in the control of female sexual behavior in a number of mammalian species.

Earlier studies of cholinergic regulation of female sexual behavior were limited to ovariectomized rats primed with exogenous hormones. The use of this traditional model provided an experimental preparation in which hormone factors could be tightly controlled. Because this hormone replacement model may not precisely reflect conditions in gonadally intact females, we also studied the effects of cholinergic manipulations in naturally cycling female rats. In gonadally intact females (Menard and Dohanich 1989), we found that scopolamine reduced the incidence of lordosis displayed by intact females during proestrus within 15 minutes after intraventricular infusion (10 or 20 μg bilaterally) without interrupting cyclicity. In a second set of experiments (Menard and Dohanich 1990), lordosis was activated in intact, unreceptive females within 15 minutes after intraventricular infusion of physostigmine (10 μg bilaterally). However, this facilitation of lordosis in intact females was dependent on the stage of the estrous cycle. Physostigmine activated lordosis only at proestrus before the onset of natural receptivity and not at other stages of the cycle. If intact females were primed with very low doses of free estradiol (e.g., 0.1 μg) at 24 and 36 hours before mid-diestrus or diestrus II, then intraventricular infusion of physostigmine activated lordosis at mid-diestrus and diestrus II (C. Menard and G. Dohanich, unpublished data). Consequently, cholinergic facilitation of lordosis may be possible only after sufficient estrogen exposure from either endogenous or exogenous sources.

Anatomical sites in the cholinergic regulation of lordosis

Although the administration of cholinergic agents has predictable and reliable effects on the occurrence of lordosis in female rodents, the sites of action of these agents in the central nervous system have not been identified definitively. Several brain regions are appropriate candidates, based on three criteria: (1) the capacity to concentrate estrogen from the circulation, (2) a presumed involvement in the regulation of lordosis as evidenced by lesion, stimulation, and hormone implant studies, and (3) a capacity to respond to estrogen with changes in cholinergic neurochemistry. The medial preoptic area, ventromedial hypothalamus, and midbrain central gray matter satisfy each of these criteria, and the contributions of these regions to the cholinergic control of lordosis have been examined in various experiments. In addition, the locations of cholinergic cell bodies that might provide innervation to cholinergic projection sites involved in lordosis have been investigated.

Bilateral infusions of cholinergic agonists in small volumes (0.5 μl) into the medial preoptic area or ventromedial hypothalamus activated lordosis in a dose-related manner in ovariectomized rats primed with estrogen (Dohanich et al. 1984). Dose–response relationships indicated that the agonists carbachol and oxotremorine were more effective in facilitating lordosis when infused into the medial preoptic area than when infused into the ventromedial hypothalamus. However, when agonists were infused into the medial preoptic area through cannulae angled to avoid traversing the lateral ventricles, these agents failed to facilitate lordosis (Fig. 8.5). Infusions directly into the lateral ventricles of the same females, in contrast, effectively facilitated behavior. From these experiments, it was concluded that cholinergic agonists infused directly into the medial preoptic area might facilitate lordosis by stimulating cholinergic receptors outside the medial preoptic area following diffusion through the ventricular system to distant sites. Agents infused though cannulae that cross brain ventricles can pass easily up the outer barrel of the cannula and gain access to the ventricular system (Rout-

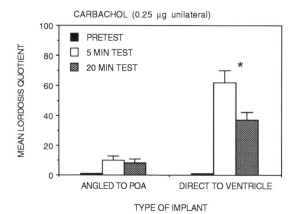

Figure 8.5. Failure of the cholinergic agonist carbachol to activate lordosis when infused into the medial preoptic area (POA) through cannulae angled to avoid traversing the lateral ventricles. Ovariectomized rats were injected intramuscularly with estradiol benzoate (0.5 µg/kg) at 72, 48, and 24 hours before behavioral testing. Carbachol was infused unilaterally into the POA or the contralateral lateral ventricle on separate tests. *$p \leq .01$, angled preoptic infusions vs. lateral ventricle infusions.

tenberg 1972). A similar conclusion could not be drawn about infusions into the ventromedial hypothalamus, because the structure of the ventricular system prevented the placement of angled cannulae into the ventromedial hypothalamus by conventional techniques. Consequently, our infusion experiments did not provide definitive support for either the medial preoptic area or the ventromedial hypothalamus as a critical site at which cholinergic agents exert their effects on lordosis.

Taking another approach, Kaufman et al. (1988) found that crystalline implants of the antagonist scopolamine in the ventromedial hypothalamus inhibited lordosis in ovariectomized rats primed with estrogen and progesterone. Autoradiographic analysis of the spread of [N-^{3}H]methylscopolamine indicated that diffusion of the drug was confined to a limited area surrounding the implant site. However, this study did not completely resolve the localization issue. First, the concentration of scopolamine required to inhibit lordosis following intracerebral administration was several orders of magnitude greater than the concentration of [N-^{3}H]methylscopolamine used to assess diffusion. Consequently, the diffusion of the radioactive compound undoubtedly underestimated the actual area of scopolamine diffusion following infusion of a behaviorally effective dose. Second, the diffusion pattern of the hydrophilic ligand [N-^{3}H]methylscopolamine may be substantially different from the diffusion pattern of the lipophilic agent scopolamine, the compound used in the behavioral portion of their experiments. Additional support for the role of the ventromedial hypothalamus as a site of cholinergic action was provided by evidence that intraventricular infusion of the agonist oxotremorine failed to activate lordosis in female rats following electrolytic ablation of the ventromedial hypothalamus (Richmond and Clemens 1988). This experiment also suffered from interpretational limitations. Although females with lesions of the ventromedial hypothalamus failed to respond to oxotremorine, these females also failed to exhibit sexual behavior after treatment with estrogen and progesterone. Therefore, these results must be interpreted cautiously, because females with lesions of the ventromedial hypothalamus in this experiment may simply have been unable to exhibit lordosis under any condition.

Other experiments indicated that the infusion of cholinergic agonists into the midbrain central gray matter activated lordosis in ovariectomized rats primed with estrogen (Richmond and Clemens 1986). Furthermore, lateral ventricular infusion of cholinergic agonists was ineffective in females with electrolytic lesions of the central gray matter. In contrast to lesions of the ventromedial hypothalamus (Richmond and Clemens 1988), females with lesions of the central gray matter failed to respond to cholinergic agonists but did exhibit sexual behavior following treatment with estrogen and progesterone. Unfortunately, in a separate experiment, tritiated compounds infused into the central gray matter were found to diffuse as far away as the ventromedial hypothalamus (Meyers et al. 1985). Therefore, as in the cases of the medial preoptic area and the ventromedial hypothalamus, the area stimulated by infusion of cholinergic agonists directly into the central gray matter might not be limited to the site of infusion. Nevertheless, intact central gray matter appeared to be necessary for cholinergic agonists to facilitate lordosis, which suggested that the midbrain central gray matter is an anatomical locus implicated in cholinergic control of lordosis.

Although most attempts to localize the sites of cholinergic control of lordosis have focused on axon terminal fields, some efforts have been directed at cholinergic cell bodies that might provide innervation of relevant areas. Cholinergic neurons located in the basal forebrain regions of the rat send axons to several telencephalic projection sites, including the cortex, hippocampus, and amygdala (Mesulam et al. 1983). Electrolytic lesions that eliminated various nuclei of this basal forebrain system, including the horizontal limb of the diagonal band of Broca and portions of the substantia innominata and magnocellular preoptic nucleus, did not alter the occurrence of lordosis in ovariectomized rats treated with estrogen (Dohanich and McEwen 1986). However, females with basal forebrain lesions displayed an extremely high incidence of rejection behavior when mounted by male rats. Although estrogen treatment is known to increase the activity of cholinergic enzymes in this region (Luine and McEwen 1983), these neurons apparently do not play a significant role in cholinergic regulation of lordosis.

Another major group of cholinergic cell bodies found in the midbrain of the rat send ascending projections to an assortment of regions, including the thalamus, hypothalamus, septum, and central gray matter (Mesulam et al. 1983). Ovari-ectomized rats treated with estrogen and progesterone failed to exhibit lordosis following the infliction of electrolytic lesions that destroyed the laterodorsal tegmental nucleus or the pedunculopontine nucleus, midbrain regions containing cholinergic cell bodies (Clemens and Brigham 1988). Furthermore, lordosis was displayed by females with lesions after intraventricular infusion of cholinergic agonists. These results suggested that cholinergic cell bodies located in both midbrain regions provide cholinergic innervation that is critical to the display of lordosis in female rats.

These anatomical experiments have not produced a clear characterization of the neuronal circuitry involved in cholinergic control of lordosis. Nevertheless, the evidence seems to indicate that the source of cholinergic innervation is more likely to arise from cholinergic cell bodies located in midbrain than in forebrain regions. In addition, various data weakly suggest that the ventromedial hypothalamus and the midbrain central gray matter play a role in this circuit. The contributions of the medial preoptic area and septum, as well as other brain regions, as components of this system remain to be determined.

Regulation of cholinergic neurochemistry by steroids

Types of manipulation

As indicated previously, the supposition that cholinergic activity contributes to the regulation of hormone-dependent behaviors, such as lordosis, is based on the hypothesis that steroids alter neurotransmitter function through genomic and membrane actions. A direct test of this hypothesis can be accomplished by determining the effects of different steroid conditions on various components of cholinergic neurochemistry (Dohanich et al. 1985). A number of experiments have examined *in vivo* changes in cholinergic muscarinic binding sites, cholinergic synthetic and degradative enzymes, and acetylcholine content in ovariectomized rats after steroid replacement or in gonadally intact females over the estrous cycle. Membrane effects of gonadal steroids on muscarinic binding sites have also been investigated following *in vitro* exposure of brain tissues to estrogen or progesterone.

Effects of estrogen and progesterone on muscarinic receptors

Several laboratories have reported changes in muscarinic receptor binding in some brain regions of ovariectomized rats following systemic administration of estrogen. Increases in the binding of [^{3}H]quinuclidinyl benzilate, a specific muscarinic ligand, were demonstrated in the medial basal hypothalamus and the midbrain central gray matter of ovariectomized rats treated with estradiol or estradiol benzoate (Rainbow et al. 1980; Dohanich et al. 1982, 1985; Rainbow et al. 1984; Meyers and Clemens 1985; Olsen et al. 1988). Micropunch dissections of hypothalamic subregions revealed that estrogen treatment increased muscarinic binding in the anterior hypothalamus, the periventricular nucleus of the anterior hypothalamus, and the ventromedial hypothalamus (Rainbow et al. 1980, 1984). These increases in binding were not altered by the administration of progesterone in conjunction with estrogen (Rainbow et al. 1984). Saturation experiments further indicated that changes in muscarinic binding following estrogen treatment reflected increases in receptor number rather than receptor affinity (Rainbow et al. 1980; Dohanich et al. 1982). In addition, estrogen treatments that were effective in female rats failed to increase muscarinic binding in the hypothalamus of gonadectomized male rats (Dohanich et al. 1982, 1985; Rainbow et al. 1984; Olsen et al. 1988).

Despite independent demonstrations of this phenomenon, changes in muscarinic binding induced by estrogen treatment in female rats have been modest in magnitude (20–30%) and occasionally not reproducible. For example, an extremely high dose of estradiol (80 mg/kg) administered in two treatments to ovariectomized rats was reported to decrease, not increase, binding of the muscarinic antagonist [N-^{3}H]methylscopolamine in the hypothalamus and the amygdala (Al-Dahan and Thomas 1987). In addition, the decrease in muscarinic binding in the amygdala was significantly greater when progesterone (1 mg/kg) was administered along with estrogen. Even when administered alone, progesterone reduced binding in the amygdala within 30 minutes after subcutaneous injection, possibly as the result of a direct membrane action. Previous studies revealed no changes in muscarinic binding in the amygdala of female rats after the administration of estrogen or estrogen and progesterone (Dohanich et al. 1982; Rainbow et al. 1984). In another study, ovariectomy increased muscarinic binding of the novel antagonist [N-^{3}H]methyl-4-piperidyl benzilate in the median hypothalamus and estrogen treatment reduced binding to estrous levels (Egozi et al. 1982). Similar contradictory results have been reported in other brain regions such as the medial preoptic area, in which muscarinic binding was found to decrease, increase, or not change following estrogen treatment of ovariectomized rats (Dohanich et al. 1982, 1985; Rainbow et al. 1984; Olsen et al. 1988). Analysis of muscarinic receptor subtype binding has not been illuminating. Estrogen treatment of ovariectomized rats failed to alter binding of the M1-selective antagonist [^{3}H]pirenzepine or the M2-selective agonist [^{3}H]oxotremorine-M in the medial preoptic area, medial basal hypothalamus, central gray matter, or septum (Dohanich et al. 1991).

Various changes in muscarinic binding have also been detected when brain homogenates were incubated *in vitro* with steroids before and during the binding reaction. For example, 60% of the muscarinic binding sites measured with the antagonist [N-^{3}H]methyl-4-piperidyl benzilate were converted from high-affinity agonist sites to low-affinity agonist sites when exposed to physiological levels of estradiol during preincubation and incubation (Egozi and Kloog 1985). This conversion of muscarinic binding sites from a high-affinity to a low-affinity state was observed only in homogenates of the preoptic area from female rats killed on the morning of proestrus. Klangkalya and Chan (1988) failed to find any *in vitro* effects of various forms of estrogen on muscarinic binding in hypothalamic and pituitary homogenates from ovariectomized rats. Alternatively, in the same study, *in vitro* progestins inhibited muscarinic binding of [^{3}H]quinuclidinyl benzilate in hypothalamic and pituitary homogenates. However, we found that this *in vitro* inhibition of muscarinic binding by progesterone may be a nonspecific effect on binding that occurs only at micromolar concentrations of progesterone that are well above physiological levels (Dohanich et al. 1989).

Effects of the estrous cycle on muscarinic receptors

Studies that measured muscarinic binding over the estrous cycle of female rats have not clarified the effects of gonadal steroids on cholinergic receptors. Muscarinic binding as determined by binding of [^{3}H]quinuclidinyl benzilate was reported to be elevated significantly at proestrus compared with other stages of the estrous cycle (Olsen et al. 1988). This increase in binding was found only in homogenates of the medial preoptic area and not in the medial basal hypothalamus. In contrast, another report demonstrated decreased binding of [N-^{3}H]methylscopolamine in the hypothalamus and amygdala of female rats at proestrus compared with metestrus (Al-Dahan and Thomas 1987). More recently in our laboratory (C. Menard, T. Hebert, and G. Dohanich, unpublished data), we found that muscarinic binding as measured by [^{3}H]quinuclidinyl benzilate levels was elevated significantly at proestrus/estrus in the medial basal hypothalamus and central gray matter compared with diestrous stages (Fig. 8.6). In addition, muscarinic binding was significantly lower in the medial preoptic area at proestrus/estrus compared with diestrous stages. The outcomes of these studies appear to be in direct conflict. In the medial basal hypothalamus, there is evidence of an increase (C. Menard et al., unpublished data), a decrease (Al-Dahan and Thomas 1987), and no change (Olsen et al. 1988) in muscarinic binding at proestrus. In the medial preoptic area, there is evidence of an increase (Olsen et al. 1988) and a decrease (Menard et al., unpublished data) in binding at proestrus.

Steroid effects on other cholinergic endpoints

Steroid treatments and estrous stages exert effects on other components of the cholinergic system, including the enzymes choline acetyltransferase and

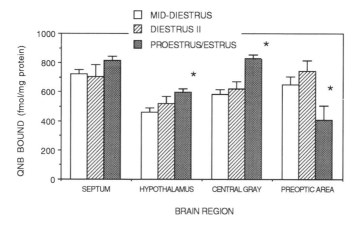

Figure 8.6. Changes in muscarinic receptor binding in intact female rats over the estrous cycle as measured by the antagonist [^{3}H]quinuclidinyl benzilate (QNB) (1 nM). *$p \leq .01$, proestrus/estrus vs. mid-diestrus.

acetylcholinesterase, and the endogenous neurotransmitter acetylcholine. The administration of estradiol to ovariectomized rats increased the activities of the synthesizing enzyme choline acetyltransferase and the degradative enzyme acetylcholinesterase in the horizontal limb of the diagonal band of Broca (Luine and McEwen 1983). Indeed, some acetylcholinesterase-containing cell bodies in this region were found to concentrate estradiol (Fallon et al. 1983). Choline acetyltransferase activity was also elevated following estradiol treatment in the basal amygdala, a region that receives projections from the horizontal limb of the diagonal band (Luine et al. 1975). Similar changes in cholinergic enzyme activities after estrogen treatment were not found in other regions normally implicated in sexual behavior such as the medial preoptic area and the medial basal hypothalamus (Luine et al. 1975; Luine and McEwen 1983). However, other reports described decreased activity of cholinergic enzymes in the anterior hypothalamus (Kobayashi et al. 1963) and the preoptic area (Libertun et al. 1973) at estrus compared with other stages of the estrous cycle.

Few studies have documented the levels or turnover of acetylcholine following steroid manipulations. However, at least two reports described changes induced by estrogen treatment in brain regions with reproductive significance. In one study employing a radioenzymatic technique, the concentration of acetylcholine was reduced in the periventricular nucleus and the ventral tegmental area in ovariectomized rats treated with estradiol benzoate (Muth et al. 1980). A later report using a chemiluminescence method demonstrated a 30% reduction of acetylcholine content in the preoptic area of female rats on the morning of proestrus compared with diestrus levels, followed several hours later by a precipitous rise to approximately 175% of diestrus levels (Egozi et al. 1986).

Implications of steroid effects on cholinergic neurochemistry

While it is clear that estrogen is capable of affecting various components of cholinergic neurochemistry in the female rat brain, documentation of these effects has proved elusive. A variety of factors may underlie the lack of reliability that is apparent in the literature. First, there appear to be significant methodological differences among many studies that could easily account for some discrepancies. These procedural issues include the duration of time after ovariectomy, the dose of estradiol, the time the animals were killed, the precise area sampled by dissection, the type of muscarinic radioligand, the assay technique, and the criteria used to define stages of the estrous cycle. Second, the effects of steroids such as estrogen on cholinergic activity, and specifically on muscarinic binding activity, were usually limited in magnitude. Although even modest changes in cholinergic activity could have significant consequences for brain function, reliable demonstrations of these effects may be beyond conventional assay techniques. Third, neurotransmission is a dynamic activity. Therefore, it is reasonable to argue that the components of any neurotransmitter system, including receptors, may be altered dramatically over narrow time frames. Evidence that rapid changes in cholinergic neurochemistry can occur is provided by the fivefold change in acetylcholine concentration observed in the preoptic area over a 3-hour period of proestrus (Egozi et al. 1986) and by the twofold change in unoccupied muscarinic binding sites reported in rat brain over a 6-hour period of the diurnal cycle (Mash et al. 1985). Furthermore, an additional seasonal rhythm appears to underlie the circadian variations in muscarinic binding sites (Kafka et al. 1981). Consequently, changes in components of the cholinergic system under varying hormone conditions may be obscured or complicated by sudden and transient fluctuations in endogenous cholinergic neurochemistry. Finally, the variability in steroid effects on cholinergic measures across studies is not unique to this system. Similar degrees of variability are reported in the literature describing steroid-induced changes in other neurotransmitter systems, including catecholaminergic, serotonergic, and peptidergic systems.

Despite discrepancies in the reported effects of steroids on cholinergic endpoints, a variety of findings suggest that estrogenic regulation of the cholinergic system is implicated in the control of lordosis. Specifically, many of the changes in muscarinic binding and acetylcholine content following estrogen manipulations occur in areas traditionally associated with the regulation of lordosis, including the medial basal hypothalamus, the medial preoptic area, and the midbrain central gray matter. In addition, fluctuations in muscarinic binding and acetylcholine content are often reported to occur at proestrus in some of these areas, at precisely the time when sexual receptivity is displayed. Furthermore, increases in muscarinic binding sites following estrogen treatment are associated with protein synthesis, since these increases were prevented by treatment with a protein synthesis inhibitor (Rainbow et al. 1984). In fact, administration of a protein synthesis inhibitor to ovariectomized rats during estrogen priming prevented the occurrence of lordosis, as well as the induction of muscarinic binding sites. Finally, the minimum dose and time of estrogen exposure necessary for the activation of lordosis and the induction of muscarinic binding sites are similar (Rainbow et al. 1984).

Summary

Cholinergic regulation of female sexual behavior

Steroid hormones regulate a diverse set of mammalian functions and behaviors that must often occur as cyclic, coordinated events. It is now clear that neurotransmitters play a primary role in mediating the effects of steroids on neuronal activity and consequent function. The study of the endocrine and pharmacological factors that control sexual behavior has advanced our knowledge of steroid action significantly. Using this system, investigators are able to study simple, measurable behaviors that are affected readily by hormonal and pharmacological manipulations.

Our experiments suggest that the interaction of gonadal steroids and acetylcholine is especially amenable to this type of analysis. Results from a variety of studies clearly indicate that the effects of gonadal steroids on sexual behaviors displayed by rodents are mediated, at least in part, by the activity of central acetylcholine neurons. More specifically, stimulation of cholinergic muscarinic receptors by exogenous agonists or by the endogenous neurotransmitter facilitates lordosis, an essential behavior displayed by receptive female rats. Inhibition of cholinergic function by exogenous cholinergic antagonists or by muscarinic receptor blockers reduced the occurrence of lordosis displayed by female rats. Various biochemical findings indicate that cholinergic neurochemistry can be altered by estrogen and progesterone, gonadal steroids that exert powerful control over the sexual behavior of female rats. Therefore, we propose that these steroids control the onset and occurrence of female sexual behavior by altering cholinergic systems within the brain.

The effects of cholinergic agents on lordosis have also been demonstrated in ovariectomized, hormone-primed hamsters (Dohanich et al. 1990) and gonadally intact female rats (Menard and Dohanich 1989, 1990). Furthermore, recent evidence implicates cholinergic systems in the display of other components of female mating behavior that are regulated by steroids, including solicitation behaviors, pacing of coital contacts, and mate preference (Dohanich et al. 1993). Together, these studies suggest that acetylcholine might be a general mediator of various steroid-dependent sexual behaviors displayed by different species.

The reliability, magnitude, and generality of cholinergic effects on female sexual behavior indicate that acetylcholine is a critical neurochemical component in the regulation of sexual behavior. However, a diverse group of neurotransmitter systems contribute to the control of this vital set of behaviors. To date, a vast literature has implicated monoamines, peptides, and amino acid neurotransmitters as participants in the regulation of steroid-dependent sexual behavior. As a primary goal, future research must integrate these various components into a model that will extend our understanding beyond the limitations imposed by the insular analyses that have dominated previous research.

Integration of systems

Can we begin to study functional interactions between the neurotransmitters that have been implicated independently in female sexual behavior? Evidence from a variety of biochemical, anatomical, and behavioral studies suggests the existence of intriguing interactions between neurotransmitter systems. For example, opiates are often reported to inhibit acetylcholine release in the brain (Rada et al. 1991), an action that might mediate the inhibitory effect of some opiates on lordosis (Pfaus and Gorzalka 1987; also see Chapter 6 by Hammer and Cheung, this volume). Alternatively, acetylcholine has been found to increase norepinephrine turnover in the hypothalamus (Coudray-Lucas et al. 1983), an action that might mediate the facilitative effects of acetylcholine on lordosis. At a cellular level, the same neurons in the ventromedial hypothalamus that displayed increased electrical activity following the application of oxytocin also responded to acetylcholine and norepinephrine (Kow et al. 1991). Furthermore, this correlated response of ventromedial hypothalamic neurons occurred only in tissue slices prepared from ovariectomized rats that had been primed with estrogen. Oxytocin, norepinephrine, and acetylcholine each have been reported to facilitate lordosis in female rats (Schumacher et al. 1990; Etgen et al. 1992; also see Chapter 5 by Flanagan and McEwen, this volume).

Thus, there appears to be a neuroanatomical substrate that fosters interactions between various neurotransmitter systems with potentially important functional consequences. Understanding the interconnection between different neurotransmitter systems will become critical as we attempt to interpret how these different systems interact to facilitate and terminate behavioral receptivity. At a molecular level, the existence of several receptor subtypes for each neurotransmitter represents an exciting, if not intimidating, development for behavioral neuroscience. These receptor subtypes clearly play important roles in neurotransmitter action. Selective manipulation of receptor subtypes should resolve discrepant results that arise from less specific pharmacological manipulations of some systems. However, the most important feature of receptor subtypes may be their specific couplings with signal transduction systems in the neuronal membrane. The significance of these effector systems for behavior lies in the possibility that different transmitters and their associated subtypes might act through a common signal transduction route to affect neuronal activity and consequent behavior.

Steroid hormone regulation of neurotransmitter systems is clearly a mechanism universal to mammals. Therefore, this avenue of research has important implications for understanding the influence of hormones on many mammalian behaviors. While advances in the study of receptors, molecular biology, and anatomy have been dramatic, the task of applying this information to the study of behavior remains. Clarifying the complexities of those neural events that control behavior continues to provide the greatest challenge to the field of neuroscience.

Acknowledgments

Research reviewed in this chapter was conducted with the support of various grants, including awards to the author from USPHS (HD-22235, HD-06258), NSF (BNS-9021447), and the Louisiana Educational Quality Support Fund [86-TUU (16)-133-11]. The author is grateful to Dr. Ann Clark, Dartmouth College, and Dr. Beth Wee, Tulane University, for valuable editorial comments.

References

Al-Dahan, M. I. M., and Thomas, P. J. (1987). Contrasting effects of testicular and ovarian steroids upon muscarinic binding sites in the brain. *Pharmacology* 34: 250–258.

Buckley, N. J., Bonner, T. I., Buckley, C. M., and Brann, M. R. (1989). Antagonist binding properties of five cloned muscarinic receptors expressed in CHO-K1 cells. *Mol. Pharmacol.* 35: 469–476.

Clemens, L. G., Barr, P. J., and Dohanich, G. P. (1989). Cholinergic regulation of female sexual behavior in rats demonstrated by manipulation of endogenous acetylcholine. *Physiol. Behav.* 45: 437–442.

Clemens, L. G., and Brigham, D. A. (1988). Identification of cholinergic cell groups that facilitate sexual behavior in female rats. *Soc. Neurosci. Abstr.* 14: 274.

Clemens, L. G., and Christensen, L. W. (1975). Sexual behavior. In *The Behaviour of Domestic Animals*, ed. E. S. E. Hafez, pp. 108–145. London: Bailliere Tindall.

Clemens, L. G., and Dohanich, G. P. (1980). Inhibition of lordotic behavior in female rats following intracerebral infusion of anticholinergic agents. *Pharmacol. Biochem. Behav.* 13: 89–95.

Clemens, L. G., Dohanich, G. P., and Barr, P. J. (1983). Cholinergic regulation of feminine sexual behaviour in laboratory rats. In *Hormones and Behaviour in Higher Vertebrates,* ed. J. Balthazart, E. Prove, and R. Gilles, pp. 56–68. Berlin: Springer-Verlag.

Clemens, L. G., Dohanich, G. P., and Witcher, J. A. (1981). Cholinergic influences on estrogen-independent sexual behavior in female rats. *J. Comp. Physiol. Psychol.* 95: 763–770.

Clemens, L. G., Humphrys, R. R., and Dohanich, G. P. (1980). Cholinergic brain mechanisms and the hormonal regulation of female sexual behavior in the rat. *Pharmacol. Biochem. Behav.* 13: 81–88.

Coudray-Lucas, C., Prioux-Guyonneau, M., Sentenac, H., Cohen, Y., and Wepierre, J. (1983). Brain catecholamine metabolism changes and hypothermia in intoxication by anticholinesterase agents. *Acta Pharmacol. Toxicol.* 52: 224–229.

Dohanich, G. P., Barr, P. J., Witcher, J. A., and Clemens, L. G. (1984). Pharmacological and anatomical aspects of cholinergic activation of female sexual behavior. *Physiol. Behav.* 32: 1021–1026.

Dohanich, G. P., Cada, D. A., and Wee, B. E. F. (1989). *In vitro* inhibition of muscarinic binding by steroid hormones. *Soc. Neurosci. Abstr.* 15: 628.

Dohanich, G. P., and McEwen, B. S. (1986). Cholinergic limbic projections and behavioral role of basal forebrain nuclei in the rat. *Brain Res. Bull.* 16: 477–482.

Dohanich, G. P., McMullan, D. M., and Brazier, M. M. (1990). Cholinergic regulation of sexual behavior in female hamsters. *Physiol. Behav.* 47: 127–131.

Dohanich, G. P., McMullan, D. M., Cada, D. A., and Mangum, K. A. (1991). Muscarinic receptor subtypes and sexual behavior in female rats. *Pharmacol. Biochem. Behav.* 38: 115–124.

Dohanich, G., Nock, B., and McEwen, B. S. (1985). Steroid hormones, receptors, and neurotransmitters. In *Molecular Mechanism of Steroid Hormone Action,* ed. V. K. Moudgil, pp. 701–732. New York: de Gruyter.

Dohanich, G. P., Ross, S. M., Francis, T. J., Fader, A. J., Wee, B. E. F., Brazier, M. M., and Menard, C. S. (1993). The effects of a muscarinic antagonist on various components of female sexual behavior. *Behav. Neurosci.* 107: 819–826.

Dohanich, G. P., Witcher, J. A., and Clemens, L. G. (1985). Prenatal antiandrogen feminizes behavioral but not neurochemical response to estrogen. *Pharmacol. Biochem. Behav.* 23: 397–400.

Dohanich, G. P., Witcher, J. A., Weaver, D. R., and Clemens, L. G. (1982). Alteration of muscarinic binding in specific brain areas following estrogen treatment. *Brain Res.* 241: 347–350.

Egozi, Y., Avissar, S., and Sokolovsky, M. (1982). Muscarinic mechanisms and sex hormone secretion in rat adenohypophysis and preoptic area. *Neuroendocrinology* 35: 93–97.

Egozi, Y., and Kloog, Y. (1985). Muscarinic receptors in the preoptic area are sensitive to 17β-estradiol during the critical period. *Neuroendocrinology* 40: 385–392.

Egozi, Y., Kloog, Y., and Sokolovsky, M. (1986). Acetylcholine rhythm in the preoptic area of the rat hypothalamus is synchronized with estrous cycle. *Brain Res.* 383: 310–313.

Etgen, A. M., Ungar, S., and Petitti, N. (1992). Estradiol and progesterone modulation of norepinephrine neurotransmission: Implications for the regulation of female reproductive behavior. *J. Neuroendocrinol.* 4: 255–271.

Fallon J. H., Loughlin, S. E., and Ribak, C. E. (1983). The islands of Calleja complex of rat basal forebrain. III: Histochemical evidence of a striatopallidal system. *J. Comp. Neurol.* 218: 91–120.

Franck, J. E. (1977). Short latency induction of lordosis by ether in estrogen-primed female rats. *Physiol. Behav.* 19: 245–247.

Hagan, J. J., Tonnaer, J. A. D. M., and Broekkamp, C. L. E. (1987). Cholinergic stimulation of drinking from the lateral hypothalamus: Indications for M2 muscarinic receptor mediation. *Pharmacol. Biochem. Behav.* 26: 771–779.

Hammer, R., and Giachetti, A. (1982). Muscarinic receptor subtypes: M1 and M2 biochemical and functional characterization. *Life Sci.* 31: 2991–2998.

Heath, R. G. (1972). Pleasure and brain activity in man. *J. Nerv. Ment. Dis.* 154: 3–18.

Kafka, M. S., Wirz-Justice, A., Naber, D., and Wehr, T. A. (1981). Circadian acetylcholine receptor rhythm in rat brain and its modification by imipramine. *Neuropharmacology* 20: 421–425.

Kaufman, L. S., McEwen, B. S., and Pfaff, D. W. (1988). Cholinergic mechanisms of lordotic behavior in rats. *Physiol. Behav.* 43: 507–514.

Klangkalya, B., and Chan, A. (1988). Inhibition of hypothalamic and pituitary muscarinic receptor binding by progesterone. *Neuroendocrinology* 47: 294–302.

Kobayashi, T., Kobayashi, T., Kato, J., and Minaguchi, H. (1963). Fluctuations in choline acetylase activity in hypothalamus of rat during estrous cycle and after castration. *Endocrinol. Jap.* 10: 175–182.

Kow, L.-M., Johnson, A. E., Ogawa, S., and Pfaff, D. W. (1991). Electrophysiological actions of oxytocin on hypothalamic neurons *in vitro*: Neuropharmacological characterization and effects of ovarian steroids. *Neuroendocrinology* 54: 526–535.

Levin, E. D., Rose, J. E., McGurk, S. R., and Butcher, L. L. (1990). Characterization of the cognitive effects of combined muscarinic and nicotinic blockade. *Behav. Neur. Biol.* 53: 103–112.

Libertun, C., Timiras, P. S., and Kragt, C. L. (1973). Sexual differences in the hypothalamic cholinergic system before and after puberty: Inductory effect of testosterone. *Neuroendocrinology* 12: 73–85.

Lindstrom, L. H. (1971). The effect of pilocarpine and oxotremorine on oestrous behavior in female rats after treatment with monoamine depletors or monoamine synthesis inhibitors. *Eur. J. Pharmacol.* 15: 60–65.

Lindstrom, L. H. (1973). Further studies on cholinergic mechanisms and hormone-activated copulatory behavior in the female rat. *J. Endocrinol.* 56: 275–283.

Lindstrom, L. H., and Meyerson, B. J. (1967). The effect of pilocarpine, oxotremorine and arecoline in combination with methylatropine or atropine on hormone activated oestrous behavior in ovariectomized rats. *Psychopharmacology* 11: 405–413.

Luine, V. N., Khylchevskaya, R. I., and McEwen, B. S. (1975). Effect of gonadal steroids on activities of monoamine oxidase and choline acetylase in rat brain. *Brain Res.* 86: 293–306.

Luine, V. N., and McEwen, B. S. (1983). Sex differences in cholinergic enzymes of diagonal band nuclei in the rat preoptic area. *Neuroendocrinology* 36: 475–482.

Mash, D. C., Flynn, D. D., Kalinoski, L., and Potter, L. T. (1985). Circadian variations in radioligand binding to muscarine receptors in rat brain dependent upon endogenous agonist occupation. *Brain Res.* 331: 35–38.

Menard, C. S., and Dohanich, G. P. (1989). Scopolamine inhibition of lordosis in naturally cycling female rats. *Physiol. Behav.* 45: 819–823.

Menard, C. S., and Dohanich, G. P. (1990). Physostigmine facilitation of lordosis in naturally cycling female rats. *Pharmacol. Biochem. Behav.* 36: 853–858.

Mesulam, M. M., Mufson, E. J., Wainer, B. H., and Levey, A. I. (1983). Central cholinergic pathways in the rat: An overview based on an alternative nomenclature (Ch1–Ch6). *Neuroscience* 10: 1185–1201.

Meyers, T. C., and Clemens, L. G. (1985). Dissociation of cholinergic facilitation of feminine sexual behavior and estrogen induction of muscarinic cholinergic receptors. *Conf. Reprod. Behav. Abstr.* 17: 68.

Meyers, T. C., Richmond, G., and Clemens, L. G. (1985). Diffusion pattern of cholinergic compounds infused into the ventral medial hypothalamus or midbrain central gray. *Soc. Neurosci. Abstr.* 11: 737.

Muth, E. A., Crowley, W. R., and Jacobowitz, D. M. (1980). Effect of gonadal hormones on luteinizing hormone in plasma and on choline acetyltransferase activity and acetylcholine levels in discrete nuclei of the rat brain. *Neuroendocrinology* 30: 329–336.

Olsen, K. L., Edwards, E, Schechter, N., and Whalen, R. E. (1988). Muscarinic receptors in preoptic area and hypothalamus: Effects of cyclicity, sex and estrogen treatment. *Brain Res.* 448: 223–229.

O'Malley, B., and Means, A. (1974). Female steroid hormones and target cell nuclei. *Science* 183: 610–620.

Pfaff, D. W. (1980). *Estrogens and Brain Function: Neural Analysis of a Hormone-Controlled Mammalian Reproductive Behavior*. New York: Springer.

Pfaus, J. G., and Gorzalka, B. B. (1987). Opioids and sexual behavior. *Neurosci. Biobehav. Rev.* 11: 1–31.

Potter, T. J., Cottrell, G. A., and Van Hartesveldt, C. (1990). Effects of cholinergic agonists on the dorsal immobility response. *Pharmacol. Biochem. Behav.* 36: 77–80.

Rada, P., Mark, G. P., Pothos, E., and Hoebel, B. G. (1991). Systemic morphine simultaneously decreases extracellular acetylcholine and increases dopamine in the nucleus accumbens of freely moving rats. *Neuropharmacology* 30: 1133–1136.

Rainbow, T. C., DeGroff, V., Luine, V. N., and McEwen, B. S. (1980). Estradiol 17-β increases the number of muscarinic receptors in hypothalamic nuclei. *Brain Res.* 198: 239–243.

Rainbow, T. C., Snyder, L., Berck, D. J., and McEwen, B. S. (1984). Correlation of muscarinic receptor induction in the ventromedial hypothalamic nucleus with activation of feminine sexual behavior by estradiol. *Neuroendocrinology* 39: 476–480.

Richmond, G., and Clemens, L. G. (1986). Evidence for involvement of midbrain central gray in cholinergic mediation of female sexual receptivity in rats. *Behav. Neurosci.* 100: 376–380.

Richmond, G, and Clemens, L. G. (1988). Ventromedial hypothalamic lesions and cholinergic control of female sexual behavior. *Physiol. Behav.* 42: 179–182.

Routtenberg, A. (1972). Intracranial chemical injection and behavior: A critical review. *Behav. Biol.* 7: 601–641.

Schumacher, M. (1990). Rapid membrane effects of steroid hormones: An emerging concept in neuroendocrinology. *Trends Neurosci.* 13: 359–362.

Schumacher, M., Coirini, H., Pfaff, D. W., and McEwen, B. S. (1990). Behavioral effects of progesterone associated with rapid modulation of oxytocin receptors. *Science* 250: 691–694.

Singer, J. J. (1968). The effects of atropine upon the female and male sexual behavior of female rats. *Physiol. Behav.* 3: 377–378.

Weaver, D. R., and Clemens, L. G. (1987). Nicotinic cholinergic influences on sexual receptivity in female rats. *Pharmacol. Biochem. Behav.* 26: 393–400.

9

Sex steroid regulation of tachykinin peptides in neuronal circuitry mediating reproductive functions

THOMAS R. AKESSON AND PAUL E. MICEVYCH

Substance P is the best-known member of the family of tachykinin (TAC) peptides. Its distribution in spinal ganglia and substantia gelatinosa of the dorsal horn initially suggested a role in the transmission of sensory information from the periphery to the central nervous system (Hökfelt et al. 1975). Subsequently, an extensive distribution in the limbic system and brain stem (Ljungdahl et al. 1978) implicated these peptides in integrative and other processes as well. Mammalian TAC peptides are derived from two separate genes, usually designated preprotachykinins A and B. Preprotachykinin A mRNA codes for substance P, neurokinin A (NKA), neuropeptide K (NPK), and neuropeptide γ (NKγ; Krause et al. 1990), and preprotachykinin B mRNA codes for neurokinin B (NKB). Thus, the rat central nervous system contains at least five TAC peptides (Arai and Emson, 1986; Takano et al. 1986; Takeda et al. 1990; Tatemoto et al. 1985). Although substance P has a well-established role in the regulation of luteinizing hormone–releasing hormone (LHRH) secretion and sexual behavior, reproductively relevant roles for other TAC peptides are only now being explored. Much of what we can infer about the functional roles of TAC peptides in the central nervous system is derived from immunohistochemical studies, but antibodies to substance P often cross-react with other TAC peptides. Therefore, we have used the term "TAC-immunoreactive" (TACir) to describe the labeling of cells using antibodies directed against these peptides.

Steroids, tachykinins, and lordosis

Successful fertilization of ova depends on the appropriate display of proceptive and receptive behaviors and on neuroendocrine mechanisms that time these behaviors with ovulation. Little is known about the central mechanisms mediating proceptive behavior, but circuitry regulating lordosis has been investigated extensively by means of a multidisciplinary approach (Micevych and Ulibarri 1992;

Pfaff and Schwartz-Giblin 1988). The ventromedial nucleus of the hypothalamus (VMH) plays a particularly important role in the mediation of lordosis by integrating predominantly inhibitory limbic signals (Law and Meagher 1958; Powers and Valenstein 1972; Yamanouchi and Arai 1983) with cyclic variation of ovarian steroids, resulting in a periodic stimulatory output to the midbrain periaqueductal gray (PAG) and the surrounding reticular formation (Pfaff 1980). Estrogen is essential for lordosis, and estrogen implants in the VMH restore the behavior in females made nonreceptive by ovariectomy (Rubin and Barfield 1980). Consistent with the possibility that the midbrain also has a role in the integration of sensory and endocrine information needed for lordosis, a convergence of ascending somatosensory and descending hypothalamic signals has been described in this region (Kow and Pfaff 1982). Lesions of the dorsal PAG and the region lateral to it cause deficits in lordosis (Sakuma and Pfaff 1979a), and electrical stimulation of the PAG facilitates lordosis (Sakuma and Pfaff 1979b).

TACir projections from the VMH to the PAG

Estrogen-binding cells in the VMH are located primarily in the ventrolateral subdivision of the VMH (VMHvl), and many of these project to the midbrain (Morrell and Pfaff 1982). TACir cells are found throughout the VMH, and a high proportion of these concentrate estrogen in the VMHvl (Fig. 9.1). Binding of ^{125}I-labeled substance P is low in the VMH (Shults et al. 1984), suggesting that much of the TAC produced in the VMH is transported away from the nucleus before release. Indeed, TACir projections from the VMH to the preoptic area (Yamano et al. 1986) and to the PAG have been demonstrated (Dornan et al. 1990).

The VMH sends projections to all levels of the PAG, but VMHvl projections have been shown to terminate preferentially in the PAG at intercollicular levels (Morrell et al. 1981). In the PAG, estrogen-concentrating cells are found throughout the rostrocaudal extent of the PAG but are much more numerous in the caudal aspect, especially at the intercollicular level and the level of the inferior colliculus (Akesson et al. 1993). The PAG contains a large population of TACir cells (Ljungdahl et al. 1978; Shults et al. 1984) located in areas that are dorsolateral, lateral, and ventrolateral to the cerebral aqueduct throughout the rostrocaudal extent of the nucleus, and many of the TACir cells in the ventrolateral PAG concentrate estrogen (Fig. 9.2). Thus, in addition to the TAC-containing projection from the VMH, the PAG itself produces TAC peptides. The possibility that estrogen stimulates TAC synthesis in the PAG and the destination of PAG-derived TAC are untested.

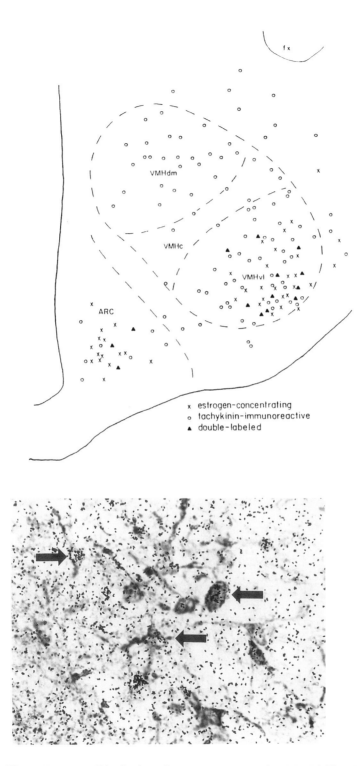

Figure 9.1. *Top:* Distribution of estrogen-concentrating (**x**), TACir (O), and double-labeled cells (▲) in the mediobasal hypothalamus of a female rat. Abbreviations: ARC, arcuate nucleus; fx, fornix; VMHdm, c, and vl, dorsomedial, central, and ventrolateral divisions of the ventromedial nucleus. *Bottom:* Photomicrograph of neurons in the VMHvl that bind estrogen (accumulation of silver grains) or contain TACir (DAB reaction product), or both. Arrows indicate coincidence of label.

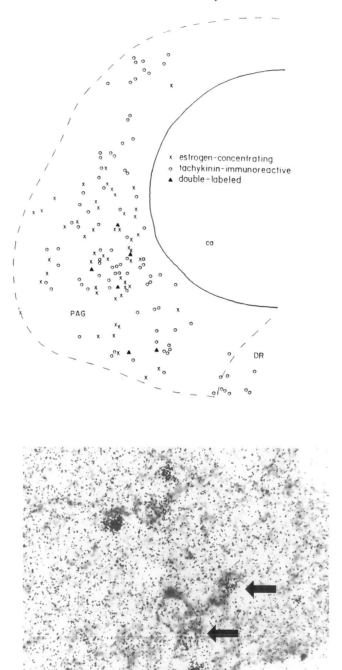

Figure 9.2. *Top*: Distribution of estrogen-concentrating (**x**), TACir (○), and double-labeled cells (▲) in the periaqueductal gray (PAG) of a female rat. Abbreviations: ca, cerebral aqueduct; DR, dorsal raphe. *Bottom*: Photomicrograph of neurons in the PAG that bind estrogen or contain TACir, or both. Arrows indicate coincidence of label.

Steroid regulation of TACs in the VMH projecting to the PAG

To test the effects of altering circulating steroid levels on the number of TACir cells in the VMH, we compared females that were ovariectomized for a period of 2 weeks, ovariectomized, and given an injection of 25 μg estradiol benzoate (EB) 2 days before death, or ovariectomized and given a capsule containing estradiol (E$_2$) for a period of 2 weeks (Fig. 9.3). Compared with ovariectomized females, those receiving estrogen replacement had an increased number of TACir cells in the VMHvl and in the anterior VMH (Fig. 9.4; Akesson 1994).

Since many of the TACir cells in the VMHvl concentrate estrogen (Fig. 9.1), the increase in labeling could be under the direct influence of estrogen (Akesson and Micevych 1988). In the anterior subdivision of the VMH, in contrast, there are few estrogen-concentrating cells, suggesting that an indirect mechanism might be involved. Specific mechanisms underlying the effect of estrogen on TAC production are unclear, because there is evidence both in favor of and against steroid

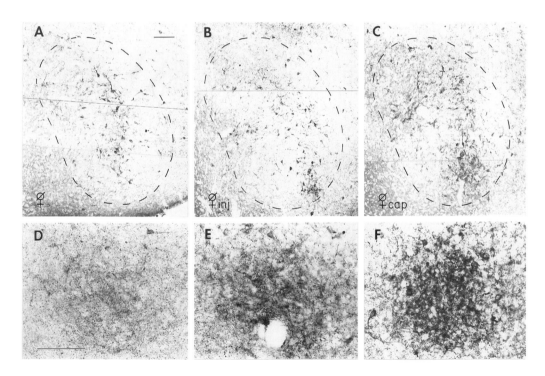

Figure 9.3. Photomicrographs of TACir cells in the VMH of a female that had been ovariectomized for a period of 2 weeks (A), ovariectomized, and given an injection of 25 μg EB 2 days before death (B), or ovariectomized and given a 5-mm capsule of E$_2$ for a period of 2 weeks (C). The number of TACir cells was increased by estrogen treatments. In the posterior pole of the VMH, the density of TACir fibers was increased by the same estrogen treatments (E and F) compared with the fiber density of ovariectomized females (D). Scale bars in A and D, 100 μm.

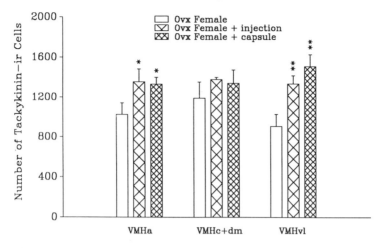

Figure 9.4. The effect of estrogen on the number of TACir cells in the anterior VMH (VMHa) and, at mid to posterior levels, number of TACir cells in the central and dorsomedial VMH (VMHc + dm) and ventrolateral VMH (VMHvl). Estrogen replacement significantly increased the number of TACir cells in the VMHa (*$p \leq .05$) and VMHvl (*$p \leq .01$) compared with ovariectomized females.

regulation of preprotachykinin (PPT) mRNA in the hypothalamus (Brown et al. 1990) and in the VMH (Priest et al., in press; Romano et al. 1989). However, there is a demonstrable influence of estrogen on immunoreactive TAC. In the posterior pole of the VMH there is a steroid-responsive aggregation of TACir fibers (Fig. 9.3). In ovariectomized females, these TACir fibers were lightly stained; in ovariectomized females that were exposed to EB for 2 days, the fiber staining was of moderate intensity; and in ovariectomized females that were exposed to E_2 for 2 weeks, TACir fibers formed a darkly stained plexus (Akesson 1994). Thus, the production of TAC peptides and transport from the VMH appear to vary with the steroid milieu.

Functional implications of TAC regulation by estrogen in the female

Estrogenic induction of protein synthesis is a prerequisite for lordosis (Meisel and Pfaff 1985), and VMH projections to the PAG are essential for this behavior (Akaishi and Sakuma 1986). Thus, regulation of lordosis presumably involves periodic and prolonged (12 or more hours) stimulation of certain gene products in the VMH and the transport and release of proteins in the PAG, which produces a behavioral response. A large number of peptides are found in the VMH, while TAC peptides (Ljungdahl et al. 1978) and endogenous opioids (Khachaturian et al. 1985) are synthesized in the VMH.

The number of TACir cells in the VMH and the density of TACir fibers in the posterior VMH are increased by estrogen. The paucity of TAC receptors in the

VMH, together with the findings that VMH levels of TAC are reduced at estrus (Micevych et al. 1988; but see Frankfurt et al. 1986), suggest that increased TAC immunoreactivity in the VMH is accompanied by transport of the peptide away from the VMH in axons and that these events normally occur at proestrus. A TACir projection from the VMH to the PAG has been demonstrated (Dornan et al. 1990), and substance P injections throughout much of the rostrocaudal extent of the dorsolateral PAG facilitate lordosis in minimally receptive females (Dornan et al. 1987). Thus, the induction of TAC expression in the VMH may be an important mechanism in the central regulation of lordosis.

Steroids, tachykinins, and male sexual behavior

Olfactory information needed to initiate and sustain male sexual behavior in rodents is conveyed from the main and accessory olfactory bulbs to the medial (MeA) and cortical (CoA) nuclei of the amygdala. Additional sensory information from the thalamus, hypothalamus, and brain stem is processed in the amygdala and subsequently conveyed back to the hypothalamus (Kostarczyk 1986). The MeA and CoA bind gonadal steroids and form a limbic circuit that allows integration of sensory and endocrine inputs through connections with other steroid-concentrating nuclei, including the bed nucleus of the stria terminalis (BNST), medial preoptic area (MPOA), and ventral premammillary nucleus (PMv). Projections from the MPOA travel in the medial forebrain bundle to terminate in the midbrain tegmentum, and tegmental projections to the striatum link these limbic circuits to circuitry that coordinates motor and autonomic functions essential for copulatory behavior (Sachs and Meisel 1988).

Several of the integrative components of this system exhibit sexual dimorphisms that result from a differential exposure to gonadal steroids during a critical period in development (Goy and McEwen 1980). In adulthood these dimorphisms function as neural substrates of sexually differentiated responses to gonadal steroids, some of which may be important for processes that activate male sexual behavior (Arnold and Breedlove 1985; Arnold and Gorski 1984). Portions of the preoptic area (Gorski et al. 1980), BNST, and amygdala are larger in males (Hines et al. 1992), and these nuclei participate in gonadal steroid–dependent, male-typical copulatory behaviors (Davidson 1966; Davis and Barfield 1979; Lisk 1967).

Lesions in the MeA and CoA have been associated with decreased arousal. Harris and Sachs (1975) observed an increased number of intromissions before ejaculation and increased ejaculation latencies after lesions were inflicted in the MeA and CoA. In another study, males with MeA and CoA lesions responded to stimulus females when levels of soliciting were high but not when they were low (Perkins et al. 1980). A similar decreased ejaculation potential or arousability has been shown to result from lesions in the BNST (Emery and Sachs 1976; Valcourt and Sachs 1979). In contrast, lesions of the MPOA (Larsson and Heimer 1964) or its connection with the midbrain tegmentum (Brackett and Edwards 1984) commonly result in a complete loss of male sexual behavior. In addition, tract tracing in combination with estrogen autoradiography has demonstrated steroid-specific projections from the MeA and BNST to the medial preoptic nucleus (MPN) in the male rat (Akesson et al. 1988). Thus, the MeA, BNST, and MPN are anatomically and functionally linked and the MPOA projection to the ventral tegmental area (VTA) is vital for the regulation of male copulatory behavior.

These sexually dimorphic structures that mediate male sexual behavior are targets of both estrogen and androgen. Over the past 20 years, a great deal of evidence has suggested the importance of estrogen in central nervous system mechanisms that motivate copulation in male rodents. We will first examine interactions between TAC peptides and estrogen as they relate to male sexual behavior, and then consider the behavioral significance of an androgen-specific TAC circuit.

Distribution of estrogen-receptive and TACir cells in the MeA–BNST–MPN–VTA pathway

The distribution of TACir cells in nuclei that participate in the mediation of male sexual behavior strongly suggests a regulatory role for TAC peptides. A large population of TACir cells is present in the rat MeA (Ljungdahl et al. 1978), and since knife cuts that isolated the amygdala did not cause a reduction in TAC immunoreactivity, it has been suggested that TAC contained in the MeA is intrinsic (Emson et al. 1978). The MeA gives rise to a TACir projection that travels in the stria terminalis and terminates in the BNST (Sakanaka et al. 1981; but see Emson et al. 1978) in a pattern that overlaps a high density of substance P binding sites (Shults et al. 1984). The BNST also contains numerous TACir cells (Ju et al. 1989), and this population gives rise to a TACir-specific projection that terminates, at least in part, in the MPOA (Paxinos et al. 1978). The posterodorsal MeA (MeApd) and encapsulated BNST (BNSTe) contain dense plexuses of TACir fibers that are sexually dimorphic in extent. The sex differences in size of the MeA TACir plexus (Malsbury and McKay 1989) and BNSTe TACir plexus (Malsbury and McKay 1987) exceed cytoarchitectural differences reported for these nuclei (Hines et al. 1992).

Medial nucleus of the amygdala. We have divided the MeA into anterior (MeAa) and posterior portions and further subdivided the posterior MeA into dorsal and ventral (MeApv) parts (MePD and MePV of Paxinos and Watson 1986). Estrogen-concentrating cells are scattered in the MeAa, are numerous in the MeApv, and are especially dense in the MeApd. TACir cells in the MeA are similarly distributed, and a high proportion of these concentrate estrogen (Fig. 9.5). Co-localization was infrequent in the MeAa and MeApv.

Bed nucleus of the stria terminalis. We have divided the BNST into anterior (BNSTa), posteromedial (BNSTpm), and posterolateral (BNSTpl) parts. The BNSTa represents approximately the anterior half of the rostrocaudal extent of the BNST. The BNSTpm division is composed mainly of the encapsulated BNST (McDonald 1983) (BNSTMPM of Paxinos and Watson 1986). The BNSTpl division corresponds to the BNSTMPL and BNSTMPI of Paxinos and Watson (1986).

The majority of estrogen-concentrating cells in the BNST are located in the BNSTpm, although substantial populations also exist in the BNSTa, especially along its anterodorsal aspect, below the ventral division of the lateral septum. A moderate number of TACir cells were present in the BNSTa and BNSTpl, and in agreement with Ju et al. (1989), a greater number of TACir cells were found in the BNSTpm. A high proportion of TACir cells in the BNSTpm concentrate estrogen, and these double-labeled cells were restricted almost entirely to the encapsulated portion of the BNSTpm (Fig. 9.6).

Preoptic area. The MPOA is a major target of estrogen, but TACir cell bodies are scattered and few in number, and co-localization of TACir within cells that concentrate estrogen was rare (Fig. 9.6). The MPOA, including the anteroventral periventricular nucleus, the medial subdivision of the medial preoptic nucleus, and preoptic periventricular nucleus, contains a particularly dense plexus of TACir fibers and terminals, suggesting that this is a major site of TAC release. There is some disagreement, however, as to the origin of these fibers. The stria terminalis (Paxinos et al. 1978), lateral part of the anterior hypothalamus (Takatsuki et al. 1983), and VMH (Yamano et al. 1986) have all been suggested as sources.

Steroid regulation of TAC immunoreactivity in the MeA–BNST–MPN pathway

To test the effects of altering circulating levels of gonadal steroids on the number of TACir cells in the MeA and BNST, intact males were compared with males that had been castrated for a period of 3 weeks (Figs. 9.7 and 9.8). Castration resulted in an 85% loss of TACir cells in the MeApd, a 49% loss in the MeApv, and a 73% loss in the BNSTpm (Fig. 9.9). TACir populations in the MeAa, BNSTa, and BNSTpl were not decreased by castration. These TACir losses in the MeApd and BNSTe greatly exceed the sexual dimorphisms in nuclear volume (Hines et al. 1992), as well as the reduction in nuclear volume that occurs

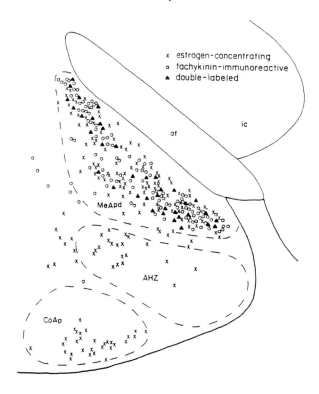

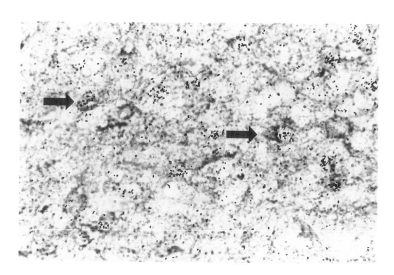

Figure 9.5. *Top*: Distribution of estrogen-concentrating (**x**), TACir (○), and double-labeled cells (▲) in the posterodorsal subdivision of the medial nucleus of the amygdala (MeApd) of a male rat. Abbreviations: AHZ, amygdalohippocampal transition zone; CoAp, posterior part of the cortical nucleus of the amygdala; ic, internal capsule; ot, optic tract. *Bottom*: Photomicrograph of neurons in the MeApd that bind estrogen or contain TACir, or both. Arrows indicate coincidence of label.

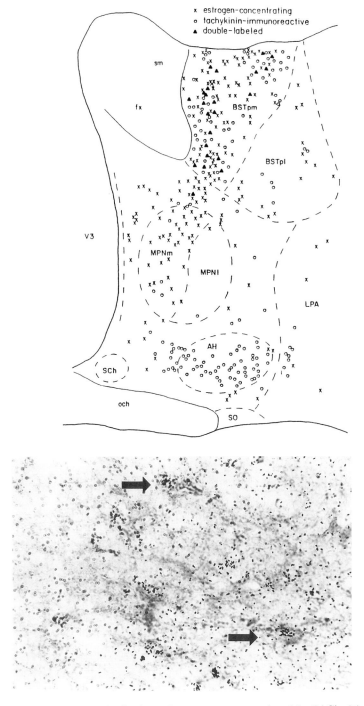

Figure 9.6. *Top*: Distribution of estrogen-concentrating (**x**), TACir (○), and double-labeled cells (▲) in the posteromedial subdivision of the bed nucleus of the stria terminalis (BSTpm) of a male rat. Abbreviations: AH, anterior hypothalamus; BSTpl, post-erolateral subdivision of the BNST; fx, fornix; LPA, lateral preoptic area; MPNm, medial division of the medial preoptic nucleus; MPNl, lateral division of the MPN; och, optic chiasm; SCh, suprachiasmatic nucleus; sm, stria medularis; SO, supraoptic nucleus; V3, third ventricle. *Bottom*: Photomicrograph of neurons in the BNSTpm that bind estrogen or contain TACir, or both. Arrows indicate coincidence of label.

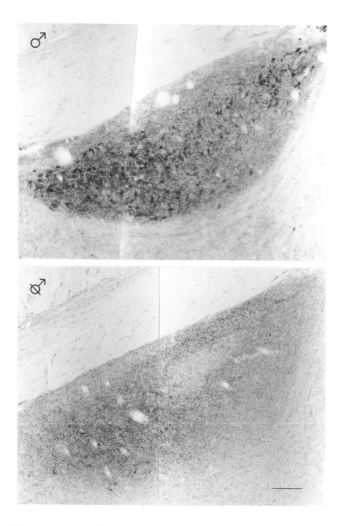

Figure 9.7. The effect of castration on the number of TACir cells in the MeApd. The number of TACir cells is greatly reduced in a male that had been castrated 3 weeks before death compared with an intact male. Scale bar, 100 μm.

as a result of gonadectomy (Malsbury and McKay 1994), suggesting a specific steroid–TAC interaction.

Functional implications of TAC regulation by estrogen in the male

Our data documenting the loss of TACir cells in the MeApd and BNSTe following castration (Figs. 9.7, 9.8, and 9.9) underscore the importance of gonadal steroids in the normal maintenance of TAC production in these regions. As was the case in the VMH, many of the TACir cells in the MeApd and BNSTe concentrate

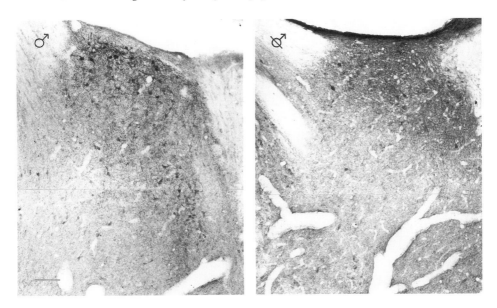

Figure 9.8. The effect of castration on the number of TACir cells in the BNSTpm. The number of TACir cells is greatly reduced in a male that had been castrated 3 weeks before death compared with an intact male. Scale bar, 100 μm.

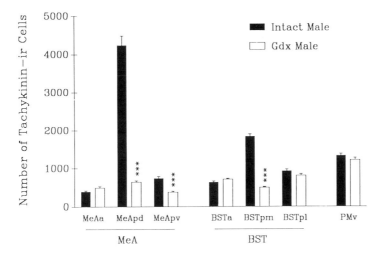

Figure 9.9. The effect of castration on the number of TACir cells in the MeA, BNST, and ventral premammillary nucleus (PMv). Castration caused a significant loss of TACir cells in the posterodorsal and posteroventral divisions of the MeA and in the posteromedial division of the BNST. ***$p \leq .001$.

estrogen (Figs. 9.5 and 9.6), and estrogen could, therefore, act directly on the genome responsible for the production of TAC peptides. However, hybridization with a PPT cRNA complementary to the α, β, and γ forms revealed no differences in the levels of PPT mRNA in the MeApd between animals that did or did not receive replacement steroid after gonadectomy (Simerly et al. 1989). Thus, TACir in the MeApd responds to alterations in circulating levels of gonadal steroids through a potential post-translational mechanism.

Our results confirm and extend the finding that the TACir fiber plexus in the MeApd is highly sensitive to alterations in circulating steroids (Dees and Kozlowski 1984). Indeed, 8 weeks after castration, this plexus was 42% smaller than in castrates given replacement testosterone (Malsbury and McKay 1994), and this loss could not be explained by a change in nuclear volume, which was reduced by 27%. In contrast, the reduction in size of the BNSTe plexus in castrated compared with intact males was of the same magnitude (26%) as the reduction in nuclear volume in the BNST (28%) that occurs after castration. The dramatic decline in the number of TACir cell bodies in the MeApd (85%), MeApv (49%), and BNSTe (73%), however, greatly exceeds reductions in nuclear volume and shrinkage of TACir fiber plexuses that occur after castration, suggesting a specific and sensitive steroid–TAC interaction.

It remains to be determined, however, whether the loss of TACir elements in the MeA and BNST contributes to the decline in male sexual behavior that follows castration, but observations that TAC peptides affect motor and motivational aspects of male sexual behavior strongly suggest a functional relevance of the steroid-dependent maintenance of TACir in these circuits. Bilateral injections of substance P in the MPOA caused a decrease in mount frequency and a reduction in intromission and ejaculation latencies (Dornan and Malsbury 1989), thus increasing the pace and the likelihood of copulation to ejaculation. Effective sites were located mainly in the MPN but extended into the anterior hypothalamus as well. In addition, antisera raised against substance P strongly inhibited male sexual behavior by increasing mount, intromission, and ejaculation latencies and intercopulatory interval (Dornan and Malsbury 1989). Few studies have tested the possibility that other TAC peptides participate in the mediation of copulatory behavior. Intracerebroventricular injections of NPK, an extended form of neurokinin A, disrupt male sexual behavior without causing a generalized increase in locomotor activity (Dornan et al. 1993; Kalra et al. 1991). NPK has been identified in neurons of the MeA, BNST, and PMv (Valentino et al. 1986).

Release of TAC peptides in the midbrain also appears to contribute to the mediation of male sexual behavior. In the VTA, substance P injections increase several locomotor and exploratory behaviors, presumably by exciting dopaminergic cells of the A10 group (Stinus et al. 1978; Eison et al. 1982; Kelly et al. 1985). The origins of TAC peptides in the VTA are not well known, but in addition to the habenula (Cuello et al. 1978), the septum and BNST are possible sources. The chemical identity of VTA TAC peptides is also uncertain, but several binding studies have shown that NK_3 receptors (which preferentially bind neurokinin B) and not NK_1 receptors (which preferentially bind substance P) predominate in this region (Bergström et al. 1987; Saffroy et al. 1988).

The excitatory influence TAC peptides might have on dopaminergic cells in the VTA is of interest because the A10 group could play a regulatory role in the execution of male sexual behavior. The dopamine agonist apomorphine has an inhibitory effect on dopaminergic neurons in the VTA. Reducing the activity of the mesolimbic dopamine pathway with apomorphine slowed copulatory rates by increasing the number of misdirected or incomplete behaviors without affecting the number of female-directed behaviors, which suggests disruption of motor rather than motivational aspects of male sexual behavior (Hull et al. 1991; see Chapter 10 by Hull, this volume). Thus, in addition to influencing motivational-limbic components of male copulatory behavior (Dornan and Malsbury 1989), TAC peptides appear to participate in mechanisms that relate to the performance of copulatory behavior.

Androgens, TACs, and male sexual behavior

Most testosterone that enters the brain is aromatized to estrogen or reduced to dihydrotestosterone (DHT; Lieberburg and McEwen 1977). Studies examining the behavioral effects of steroid replacement in castrated males have led to the general conclusion that DHT, which is not aromatized to estrogen, acts primarily in the periphery to complement the action of estrogen in the brain (Christensen and Clemens 1974; Feder et al. 1974). DHT alone does not restore male sexual behavior in castrated rats (McDonald et al. 1970), and the inhibition of aromatization of testosterone in the preoptic area blocks male sexual behavior (Christensen and Clemens 1975).

A role for DHT in circuitry that regulates copulatory behavior is supported by studies of the distribution of DHT-concentrating neurons and the connections that are made between nuclei that are targets of DHT. For instance, testosterone-derived androgen is concentrated in the BNSTe and MeApd (Sheridan, 1979), and there is a field of DHT-accumulating neurons in the lateral septum (Sar and Stumpf 1977) that does not overlap the distribution of estrogen-binding neurons. The lateral septum, BNSTe, and MeApd are among the strongest projections of the PMv (Canteras et al. 1992). Thus, the hypothalamus and limbic system might contain an androgen-specific circuit. Indeed, the possibility that androgens stimulate male copulatory behavior by acting in the brain has long been recognized (McGinnis and Dreifuss 1989; Paup et al. 1975; Whalen and Luttge 1971), particularly with respect to ejaculation (Baum and Vreeburg 1973). DHT implanted in the lateral septum or MeA of rats given systemic estrogen augmented male sexual behavior (Baum et al. 1982), whereas implants in the MPOA and VTA were ineffective.

However, elucidation of the relationship between brain androgen receptors and male mating behavior is made difficult by the overlapping distribution with estrogen receptors. The ventral premammillary nucleus (PMv) in the caudal hypothalamus is unusual in that it contains a population of DHT-binding cells that are well separated from estrogen-binding cells throughout much of the nucleus. This nucleus receives a massive projection from the MeA, BNST, and MPN (Heimer and Nauta 1969; Krettek and Price 1978; Simerly and Swanson 1986), and these connections have recently been shown to be reciprocal (Canteras et al. 1992). Cells that bind DHT in the BNST, MPOA, and VMH have been shown to project to the midbrain (Lisciotto and Morrell 1990), but other DHT-specific projections have not been identified.

About a quarter of the DHT-binding cells in the PMv are TACir (Figs. 9.10A and C), and although the volume of the PMv is approximately the same in males and females, males have about twice as many TACir cells (Fig. 9.11). To test for activational effects of gonadal steroids, ovariectomized females with or without estrogen replacement were compared with intact or castrated males (Akesson 1993). Castration had no effect on the number of TACir cells in the PMv (Fig. 9.9). Females had about half the number of TACir cells contained in the male PMv regardless of their steroid milieu (Akesson 1993).

To address the question of what behaviors are influenced by the activity of the PMv, bilateral injections of 2 μl of 1% *N*-methyl D-aspartate (NMDA) were used to lesion the PMv of young adult males that had been castrated and implanted with a 20-mm Silastic capsule filled with testosterone. Rats with PMv lesions displayed greatly exaggerated intromission and ejaculation latencies and postejaculatory intervals with no generally debilitative effects. Histological examination of effective lesion sites revealed an almost complete loss of cells in all but the caudal end of the PMv without discernible damage to nearby fiber tracts (T. Akesson, unpublished data).

To test the behavioral relevance of DHT in the rat PMv, castrated males were given 5-mm capsules of E_2 and crystalline DHT, which was delivered into the PMv via cannulae 24 and 48 hours before behavioral testing. Estrogen levels were insufficient to activate any sexual behaviors other than low rates of mounting in control animals ($n = 6$) that received E_2 capsules and intracranial cholesterol. However, four of seven rats receiving DHT in the PMv displayed sexual behavior and three of these copulated to ejaculation (T. Akesson, unpublished data). DHT implants in the lateral septum and MeA of castrated males given systemic estrogen increased the intromission rate and number of ejaculations per test (Baum et al. 1982). These findings are consistent with the conclusion that, within the brain, estrogen alone is capable of stimulating most measures of male sexual behavior, but nonaromatizable androgen, in very specific locations, is also stimulatory.

Steroids, TACs, and luteinizing hormone

Secretion of luteinizing hormone (LH) from the anterior pituitary is regulated by cyclic release of hypothalamic LHRH from the median eminence (Goodman and

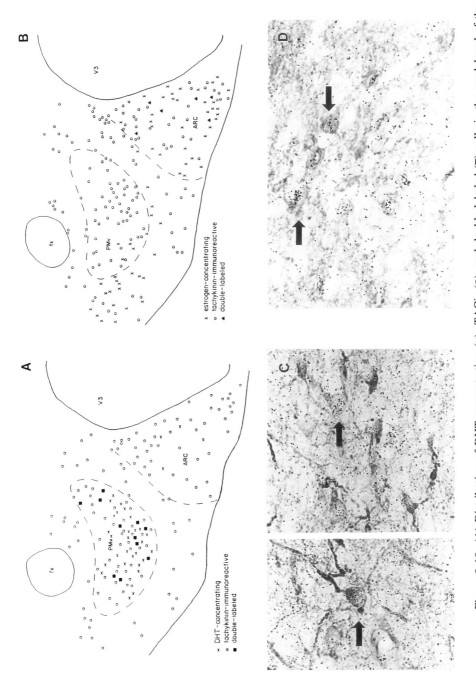

Figure 9.10. (A) Distribution of DHT-concentrating (x), TACir (○), and double-labeled (■) cells in the caudal end of the mediobasal hypothalamus of a male rat. (B) Photomicrograph of neurons in the ventral premammillary nucleus (PMv) that bind DHT or contain TACir, or both. Arrows indicate coincidence of label. (C) Distribution of estrogen-concentrating (x), TACir (○), and double-labeled (▲) cells in a male rat. (D) Photomicrograph of neurons in the arcuate nucleus (ARC) that bind estrogen or contain TACir, or both. Abbreviations: fx, fornix; V3, third ventricle.

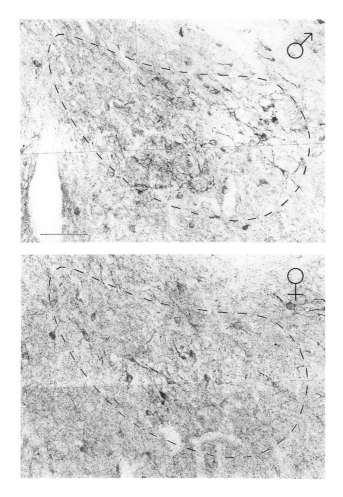

Figure 9.11. The number of TACir cells in the PMv is sexually dimorphic, with males having about two times more than females. Scale bar, 100 μm.

Knobil 1981; Sawyer 1975). Since the timing of ovulation is regulated by ovarian steroids and since LHRH cells do not, themselves, bind estrogen (Shivers et al. 1983), there has been considerable interest in identifying neural elements that convey estrogen-activated signals to LHRH cell bodies and axons. A substantial body of literature now supports the possibility that TAC-synthesizing cells form a subset of neurons that serve this function.

Vijayan and McCann (1979) observed that intracerebroventricular injections of substance P elevated circulating levels of LH in ovariectomized rats, and a stimulatory effect of substance P on LH secretion has been substantiated (Arisawa et al. 1990; Ohtsuka et al. 1987). In contrast, Kerdelhué et al. (1978) found that antisera directed against substance P stimulated LH release in cycling rats during proestrus, suggesting an inhibitory role for TAC peptides. Sahu and Kalra (1992) reported that intracerebroventricular injections of substance P, NKA, NKγ, and NKB had little effect on serum LH levels in the presence or absence of gonadal steroid priming, but that injections of NPK produced a striking decrease in circulating LH.

The results of studies of substance P and LHRH at the level of the median eminence (ME) suggest a functional relationship between the two peptides, but do not resolve the nature of the interaction. Substance P levels contained in the ME varied over the estrous cycle, with elevated levels at diestrus day 1 (Micevych et al. 1988) or diestrus day 2 (Antonowicz et al. 1982; Jakubowska-Naziemblo et al. 1985). The observation that substance P levels were elevated in the ME at proestrus led Parnet et al. (1990) to suggest that release of stored TAC peptides inhibits LHRH secretion.

In addition to hypothalamic sites, TAC peptides affect LH secretion through binding sites in the pituitary itself (Kerdelhué et al. 1985; Mikkelsen et al. 1989). Substance P inhibits (Kerdelhué et al. 1978) or stimulates (Shamgochian and Leeman 1992) release of LH from cultured anterior pituitary cells. Indeed, the anterior pituitary contains more immunodetectable substance P in males than females. In gonadectomized rats, pituitary levels of substance P are decreased by estrogen and increased by DHT (DePalatis et al. 1985). This evidence suggests that TAC peptides are involved in the regulation of LH secretion from the anterior pituitary, but that determining the functional nature of the interaction will require a more complete understanding of the site specificity, endocrine status, and perhaps the specific identity of the TAC involved.

The findings that the number of TACir cells in the arcuate nucleus (ARC) increased at proestrus (Tsuruo et al. 1984) and that TACir cells in the ARC concentrate estrogen (Figs. 9.10B and D; Akesson and Micevych 1988) suggest the possibility of a direct stimulatory regulation by estrogen. TAC peptides synthesized in the ARC could influence LHRH through projections to the ME (Palkovits et al. 1989). TACir fibers commonly make synaptic contact with LHRH cell bodies (Hoffman 1985), and some of these projections might originate in the ARC (Tsuruo et al. 1991). For these reasons, the effects of estrogen on TACir neurons in the ARC are of particular interest.

We found that most TACir cells in the ARC contained NKB mRNA and, in contrast to the stimulatory effects of gonadal steroids on TAC peptides in the VMH (Fig. 9.3), the number of ARC TACir cells was substantially reduced after 2 weeks of exposure to estrogen (Fig. 9.12; Akesson et al. 1991). Thus, the production of TAC peptides in the ARC appeared to correlate positively with cyclic variation in ovarian steroids (Tsuruo et al. 1984), but continuous exposure to estrogen strongly inhibited TAC synthesis.

This sensitive response by NKB mRNA/TACir cells in the ARC may have relevance to reproductive aging. As female rats age, ovulatory cycles become

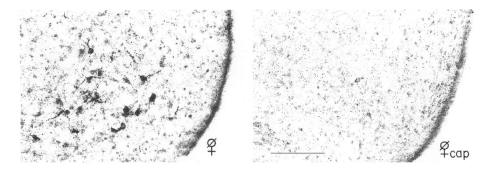

Figure 9.12. The number of TACir cells in the arcuate nucleus of an ovariectomized female rat compared with a female that had been ovariectomized and given a 5-mm capsule of E_2 for a period of 2 weeks. A continuous exposure to estrogen resulted in a loss of TACir cells. Scale bar, 100 μm.

irregular and this state is followed by a condition termed "persistent estrus," which is characterized by chronically high levels of estrogen and feedback inhibition of LH secretion (Lu et al. 1981). Our finding that continuous high levels of estrogen resulted in a substantial loss of ARC TACir cells suggests that TAC peptides might participate in steroidal feedback that is directed to LHRH-containing fibers or cells. Support for this hypothesis comes from studies in postmenopausal women in which loss of ovarian hormones produces a compensatory rise in plasma gonadotropins. Consistent with the data from rats, the loss of estrogen in humans was accompanied by hypertrophy and a greatly increased number of infundibular neurons expressing mRNA coding for substance P and NKB (Rance and Young 1991). Thus, TAC expression in the ARC could constitute one of the mechanisms that mediates feedback necessary for the periodic surge of LH that triggers ovulation, and continuous high (rat) or low (human) levels of estrogen may disrupt this feedback.

Conclusions

Tachykinins, first recognized in primary afferent fibers for their role in the perception of pain, are also produced in steroid-binding cells, which represent first-order neurons of neuroendocrine systems. These peptides appear to figure prominently in the sequence of events that follows cellular activation by gonadal steroids and leads to the stimulation or inhibition of sexual behavior and the release of LH from the anterior pituitary. Estrogen induction of TAC peptides in the VMH and their subsequent release in the PAG appear to represent a facilitatory mechanism underlying the lordosis reflex. In addition, the interactions of estrogen with TAC peptides in the MeA and BNST seem to be involved in the activation or suppression of male sexual behavior. We have presented evidence which suggests that androgen binding in a sexually dimorphic population of TACir cells in the PMv might also be relevant to processes that mediate male sexual behavior. Finally, events that follow estrogen accumulation by TACir cells in the ARC probably contribute to the regulation of circulating LH at the levels of the MPOA, ME, and anterior pituitary. Since there are multiple forms of TAC peptides and their receptors in the rodent brain, future studies are needed to resolve issues concerning the specific identities and sites of action of the TAC peptides in neural processes that regulate reproductive function.

References

Akaishi, T., and Sakuma Y. (1986). Projections of oestrogen-sensitive neurones from the ventromedial hypothalamic nucleus of the female rat. *J. Physiol.* 372: 207–220.

Akesson, T. R. (1993). Androgen concentration by a sexually dimorphic population of tachykinin-immunoreactive neurons in the rat ventral premammillary nucleus. *Brain Res.* 608: 319–323.

Akesson, T. R. (1994). Gonadal steroids regulate immunoreactive tachykinin in the ventromedial nucleus of the rat hypothalamus. *J. Comp. Neurol.* 341: 351–356.

Akesson, T. R., and Micevych, P. E. (1988). Estrogen concentration by substance P-immunoreactive cells in the medial basal hypothalamus of the female rat. *J. Neurosci. Res.* 19: 412–419.

Akesson, T. R., Simerly, R. B., and Micevych, P. E. (1988). Estrogen-concentrating hypothalamic and limbic neurons project to the medial preoptic nucleus. *Brain Res.* 451: 381–385.

Akesson, T. R., Sternini, C., and Micevych, P. E. (1991). Continuous estrogen decreases neurokinin B expression in the rat arcuate nucleus. *Mol. Cell. Neurosci.* 2: 299–304.

Akesson, T. R., Ulibarri, C. and Cao, Z. (1993). Synthesis of substance P in periaqueductal central gray: Cellular colocalization with receptors for estrogen and glutamate. *Soc. Neurosci. Abstr.* 19: 819.

Antonowicz, U., Jakubowska-Naziemblo, B., Cannon, D., and Powell, D. (1982). Immunoreactive substance P content in median eminence and pituitary gland during oestrus, dioestrus and after anterior hypothalamic deafferentation. *Endokrinologie* 79: 25–34.

Arai, H., and Emson, P. C. (1986). Regional distribution of neuropeptide K and other tachykinins (neurokinin A, neurokinin B and substance P) in rat central nervous system. *Brain Res.* 399: 240–249.

Arisawa, M., De Palatis, L., Ho, R., Snyder, G. D., Yu, W. H., Pan, G., and McCann, S. M. (1990). Stimulatory role of substance P on gonadotropin release in ovariectomized rats. *Neuroendocrinology* 51: 523–529.

Arnold, A. P., and Breedlove, S. M. (1985). Organizational and activational effects of sex steroids on brain and behavior: A reanalysis. *Horm. Behav.* 19: 469–498.

Arnold, A. P., and Gorski, R. A. (1984). Gonadal steroid induction of structural sex differences in the central nervous system. *Annu. Rev. Neurosci.* 7: 413–442.

Baum, M. J., Tobet, S. A., Starr, M. S., and Bradshaw, W. G. (1982). Implantation of dihydrotestosterone propionate into the lateral septum or amygdala facilitates copulation in castrated male rats given estradiol systemically. *Horm. Behav.* 16: 208–223.

Baum, M. J., and Vreeburg, J. T. M. (1973). Copulation in castrated male rats following combined treatment with estradiol and dihydrotestosterone. *Science* 182: 283–285.

Bergström, L., Torrens, Y., Saffroy, M., Beaujouan, J. C., Lavielle, S., Chassaing, G., Morgat, J. L., Glowinski, J., and Marquet, A. (1987). [^{3}H]Neurokinin B and ^{125}I-Bolton Hunter eledoisin label identical tachykinin binding sites in the rat brain. *J. Neurochem.* 48: 125–133.

Brackett, N. L., and Edwards, D. A. (1984). Medial preoptic connections with the midbrain tegmentum are essential for male sexual behavior. *Physiol. Behav.* 32: 79–84.

Brown, E. R., Harlan, R. E., and Krause, J. K. (1990). Gonadal steroid regulation of substance P (SP) and SP-encoding messenger ribonucleic acids in the rat anterior pituitary and hypothalamus. *Endocrinology* 126: 330–340.

Canteras, N. W., Simerly, R. B., and Swanson, L. W. (1992). Projections of the ventral premammillary nucleus. *J. Comp. Neurol.* 324: 195–212.

Christensen, L. W., and Clemens, L. G. (1974). Intrahypothalamic implants of testosterone or estradiol and resumption of masculine sexual behavior in long-term castrated male rats. *Endocrinology* 95: 984–990.

Christensen, L. W., and Clemens, L. G. (1975). Blockade of testosterone-induced mounting behavior in the male rat with intracranial application of the aromatization inhibitor, androst-1,4,6-triene-3,17-dione. *Endocrinology* 97: 1545–1551.

Cuello, A. C., Emson, P. C., Paxinos, G., and Jessell, T. (1978). Substance P containing and cholinergic projections from the habenula. *Brain Res.* 149: 413–429.

Davidson, J. M. (1966). Activation of the male rat's sexual behavior by intracerebral implantation of androgen. *Endocrinology* 79: 783–794.

Davis, P. G., and Barfield, R. J. (1979). Activation of masculine sexual behavior by intracranial estradiol benzoate implants in male rats. *Neuroendocrinology* 28: 217–227.

Dees, W. L., and Kozlowski, G. P. (1984). Effects of castration and ethanol on amygdaloid substance P immunoreactivity. *Neuroendocrinology* 39: 231–235.

DePalatis, L. R., Khorram, O., and McCann, S. M. (1985). Age-, sex-, and gonadal steroid-related changes in immunoreactive substance P in the rat anterior pituitary gland. *Endocrinology* 117: 1368–1373.

Dornan, W. A., Akesson, T. R., and Micevych, P. E. (1990). A substance P projection from the VMH to the dorsal midbrain central gray: Implication for lordosis. *Brain Res. Bull.* 25: 791–796.

Dornan, W. A., and Malsbury, C. W. (1989). Peptidergic control of male rat sexual behavior: The effects of intracerebral injections of substance P and cholecystokinin. *Physiol. Behav.* 46: 547–556.

Dornan, W. A., Malsbury, C. W., and Penny, R. B. (1987). Facilitation of lordosis by injection of substance P into the midbrain central gray. *Neuroendocrinology* 45: 498–506.

Dornan, W. A., Vink K. L., Malen P., Short, K., Struthers, W., and Barrett, C. (1993). Site-specific effects of intercerebral injections of three neurokinins (neurokinin A, neurokinin K, and neurokinin γ) on the expression of male rat sexual behavior. *Physiol. Behav.* 54: 249–258.

Eison, A. S., Eison, M. S., and Iversen, S. D. (1982). The behavioral effects of a novel substance P analogue following infusion into the ventral tegmental area or substantia nigra of the brain. *Brain Res.* 238: 137–152.

Emery, D. E., and Sachs, B. D. (1976). Copulatory behavior in male rats with lesions in the bed nucleus of the stria terminalis. *Physiol. Behav.* 17: 803–806.

Emson, P. C., Jessell, T., Paxinos, G., and Cuello, A. C. (1978). Substance P in the amygdaloid complex, bed nucleus and stria terminalis of the rat brain. *Brain Res.* 149: 97–105.

Feder, H. H., Naftolin, F., and Ryan, K. J. (1974). Male and female sexual responses in male rats given estradiol benzoate and 5α-androstan-17β-ol-3-one propionate. *Endocrinology* 94: 136–141.

Frankfurt, M., Siegel, R. A., Sim, I., and Wuttke, W. (1986). Estrous cycle variations in cholecystokinin and substance P concentrations in discrete areas of the rat brain. *Neuroendocrinology* 42: 226–231.

Goodman, R. L., and Knobil, E. (1981). The sites of action of ovarian steroids in the regulation of LH secretion. *Neuroendocrinology* 32: 57–63.

Gorski, R. A., Harlan, R. E., Jacobson, C. D., Shryne, J. E., and Southam, A. M. (1980). Evidence for the existence of a sexually dimorphic nucleus in the preoptic area of the rat. *J. Comp. Neurol.* 193: 529–539.

Goy, R. W., and McEwen, B. S. (1980). *Sexual Differentiation of the Brain.* Cambridge, MA: MIT Press.

Harris, V. H., and Sachs, B. D. (1975). Copulatory behavior in male rats following amygdaloid lesions. *Brain Res.* 86: 514–518.

Heimer, L., and Nauta, W. J. H. (1969). The hypothalamic distribution of the stria terminalis in the rat. *Brain Res.* 13: 284–297.

Hines, M., Allen, L. S., and Gorski, R. A. (1992). Sex differences in subregions of the medial nucleus of the amygdala and the bed nucleus of the stria terminalis of the rat. *Brain Res.* 579: 321–326.

Hoffman, G. E. (1985). Organization of LHRH cells: Differential apposition of neurotensin, substance P and catecholamine axons. *Peptides* 6: 439–461.

Hökfelt, T., Kellerth, J. O., Nilsson, G., and Pernow, B. (1975). Substance P: Localization in the central nervous system and in some primary sensory neurons. *Science* 190: 889–890.

Hull, E. M., Weber, M. S., Eaton, R. C., Dua, R., Markowski, V. P., Lumley, L., and Moses, J. (1991). Dopamine receptors in the ventral tegmental area affect motor, but not motivational or reflexive, components of copulation in male rats. *Brain Res.* 554: 72–76.

Jakubowska-Naziemblo, B., Antonowicz, U., Cannon, D., Powell, D., and Rohde, W. (1985). Immunoreactive substance P and LH-RH content in median eminence and pituitary gland during proestrus, oestrus, lactation and after anterior hypothalamic deafferentation. *Exp. Clin. Endocrinol.* 85: 155–166.

Ju, G., Simerly, R. B., and Swanson, L. W. (1989). Studies on the cellular architecture of the bed nuclei of the stria terminalis in the rat: II. Chemoarchitecture. *J. Comp. Neurol.* 280: 603–621.

Kalra, S., Sahu, A., Dube, M. G., and Kalra, P. S. (1991). Effects of various tachykinins on pituitary LH secretion, feeding, and sexual behavior in the rat. In *Substance P and Related Peptides: Cellular and Molecular Physiology*, ed. S. E. Leeman, J. E. Krause, and F. Lembeck. *Ann. N.Y. Acad. Sci.* 632: 332–338.

Kelly, A. E., Cador, M., and Stinus, L. (1985). Behavioral analysis of the effect of substance P injected into the ventral mesencephalon on investigatory and spontaneous motor behavior in the rat. *Psychopharmacology* 85: 37–46.

Kerdelhué, B., Tartar, A., Lenoir, V., El Abed, A., Hublau, P., and Millar, R. P. (1985) Binding studies of substance P anterior pituitary binding sites: Changes in substance P binding sites during the rat estrous cycle. *Regul. Pept.* 10: 133–143.

Kerdelhué, B., Valens, M., and Langlois, Y. (1978). Stimulation de la sécretion de la LH et de la FSH hypophysaire après immunoneutralisation de la Substance P endogene chez la ratte cyclique. *C.R. Acad. Sci. Paris D* 286: 977–979.

Khachaturian, H., Lewis, M. E., Schäfer, M. K.-H., and Watson, S. J. (1985). Anatomy of the CNS opioid systems. *TINS* 8: 111–119.

Kostarczyk, E. M. (1986). The amygdala and male reproductive functions: I. Anatomical and endocrine bases. *Neurosci. Biobehav. Rev.* 10: 67–77.

Kow, L.-M., and Pfaff, D. W. (1982). Responses of medullary reticulospinal and other reticular neurons to somatosensory and brain stem stimulation in anesthetized or freely moving ovariectomized rats with or without estrogen treatment. *Exp. Brain Res.* 47: 191–202.

Krause, J. E., Hershey, A. D., Dykema, P. E., and Takeda, Y. (1990). Molecular biological studies on the diversity of chemical signalling in tachykinin peptidergic neurons. In *A Decade of Neuropeptides: Past, Present, and Future*, ed. G. F. Koob, C. A. Sandman, and F. L. Strand. *Annu. N.Y. Acad Sci.* 579: 254–272.

Krettek, J. E., and Price, J. L. (1978). Amygdaloid projections to subcortical structures within the basal forebrain and brainstem in the rat and cat. *J. Comp. Neurol.* 178: 225–254.

Larsson, K., and Heimer, L. (1964). Mating behaviour of male rats after lesions in the preoptic area. *Nature* 202: 413–414.

Law, T., and Meagher, W. (1958). Hypothalamic lesions and sexual behavior in the female rat. *Science* 128: 1626–1627.

Lieberburg, I., and McEwen, B. S. (1977). Brain cell nuclear retention of testosterone metabolites, 5α-dihydrotestosterone and estradiol-17β, in adult rats. *Endocrinology* 100: 588–597.

Lisciotto, C. A., and Morrell, J. I. (1990). Androgen-concentrating neurons of the forebrain project to the midbrain in rats. *Brain Res.* 516: 107–112.

Lisk, R. D. (1967). Neural localization for androgen activation of copulatory behavior in the male rat. *Endocrinology* 80: 754–761.

Ljungdahl, Å., Hökfelt, T., and Nilsson, G. (1978). Distribution of substance P-like immunoreactivity in the central nervous system of the rat. I: Cell bodies and nerve terminals. *Neuroscience* 3: 861–943.

Lu, J. K. H., Gilman, D. P., Meldrum, D. R., Judd, H. L., and Sawyer, C. H. (1981). Relationship between circulating estrogens and the central mechanisms by which ovarian steroids stimulate luteinizing hormone secretion in aged and young female rats. *Endocrinology* 108: 836–841.

Malsbury, C. W., and McKay, K. (1987). A sex difference in the pattern of substance P-like immunoreactivity in the bed nucleus of the stria terminalis. *Brain Res.* 420: 365–370.

Malsbury, C. W., and McKay, K. (1989). Sex difference in the substance P-immunoreactive innervation of the medial nucleus of the amygdala. *Brain Res. Bull.* 23: 561–567.

Malsbury, C. W., and McKay, K. (1994). Neurotrophic effects of testosterone on the medial nucleus of the amygdala in adult male rats. *J. Neuroendocrinol.* 6: 57–69.

McDonald, A. J. (1983). Neurons of the bed nucleus of the stria terminalis: A Golgi study in the rat. *Brain Res. Bull.* 10: 111–120.

McDonald, P., Beyer, C., Newton, F., Brien, B., Baker, R., Tan, H. S., Campson, C., Kitching, P., Greenhill, R. and Pritchard, D. (1970). Failure of 5α-dihydrotestosterone to initiate sexual behavior in the castrated male rat. *Nature* 227: 964–965.

McGinnis, M. Y., and Dreifuss, R. M. (1989). Evidence for a role of testosterone–androgen receptor interactions in mediating masculine sexual behavior in male rats. *Endocrinology* 124: 618–626.

Meisel, R. L., and Pfaff, D. W. (1985). Specificity and neural sites of action of anisomycin in the reduction or facilitation of female sexual behavior in rats. *Horm. Behav.* 19: 237–251.

Micevych, P. E., Matt, D. W., and Go, V. L. W. (1988). Concentrations of cholecystokinin, substance P, and bombesin in discrete regions of male and female rat brain: Sex differences and estrogen effects. *Exp. Neurol.* 100: 416–425.

Micevych, P., and Ulibarri, C. (1992). Development of the limbic–hypothalamic cholecystokinin circuit: A model of sexual differentiation. *Dev. Neurosci.* 14: 11–34.

Mikkelsen, J. D., Larsen, P. J., Møller, M., Vilhardt, H., and Soermark, T. (1989).

Substance P in the median eminence and pituitary of the rat: Demonstration of immunoreactive fibers and specific binding sites. *Neuroendocrinology* 50: 100–108.

Morrell, J. I., Greenberger, L. M., and Pfaff, D. W. (1981). Hypothalamic, other diencephalic, and telencephalic neurons that project to the dorsal midbrain. *J. Comp. Neurol.* 201: 589–620.

Morrell, J. I., and Pfaff, D. W. (1982). Characterization of estrogen-concentrating hypothalamic neurons by their axonal projections. *Science* 217: 1273–1276.

Ohtsuka, S., Miyake, A., Nishizaki, T., Tasaka, K., Aono, T., and Tanizawa, O. (1987). Substance P stimulates gonadotropin-releasing hormone release from rat hypothalamus in vitro with involvement of oestrogen. *Acta Endocrinol.* 115: 247–252.

Palkovits, M., Kakucska, I., and Makara, G. B. (1989). Substance P-like immunoreactive neurons in the arcuate nucleus project to the median eminence in rat. *Brain Res.* 486: 364–368.

Parnet, P., Lenoir, V., Palkovits, M., and Kerdelhué, B. (1990). Estrous cycle variations in gonadotropin-releasing hormone, substance P and beta-endorphin contents in the median eminence, the arcuate nucleus and the medial preoptic nucleus in the rat: A detailed analysis of proestrus changes. *J. Neuroendocrinol.* 2: 291–296.

Paup, D. C., Mennin, S. P., and Gorski, R. A. (1975). Androgen- and estrogen-induced copulatory behavior and inhibition of luteinizing hormone (LH) secretion in the male rat. *Horm. Behav.* 6: 35–46.

Paxinos, G., Emson, P. C., and Cuello, A. C. (1978). Substance P projections to the entopeduncular nucleus, the medial preoptic area and lateral septum. *Neurosci. Lett.* 7: 133–136.

Paxinos, G., and Watson, C. (1986). *The Rat Brain in Stereotaxic Coordinates.* San Diego, CA: Academic Press.

Perkins, M. S., Perkins, M. N., and Hitt, J. C. (1980). Effect of stimulus female on sexual behavior of male rats given olfactory tubercle and corticomedial amygdaloid lesions. *Physiol. Behav.* 25: 495–500.

Pfaff, D. W. (1980). *Estrogens and Brain Function.* New York: Springer.

Pfaff, D. W., and Schwartz-Giblin, S. (1988). Cellular mechanisms of female reproductive behaviors. In *The Physiology of Reproduction*, Vol. 2, ed. E. Knobil, J. D. Neill, L. L. Ewing, G. S. Greenwalk, C. L. Market, and D. W. Pfaff, pp. 1487–1569. New York: Raven Press.

Powers, B., and Valenstein, E. W. (1972). Sexual receptivity: Facilitation by medial preoptic lesions in female rats. *Science* 175: 1003–1005.

Priest, C. A., Vink, K., and Micevych, P. E. (in press). Temporal sequence of estrogen regulation of beta-preprotachykinin mRNA expression in the ventromedial nucleus of the rat. *Mol. Brain Res.*

Rance, N. E., and Young, W. S. (1991). Hypertrophy and increased gene expression of neurons containing neurokinin-B and substance-P messenger ribonucleic acids in the hypothalami of postmenopausal women. *Endocrinology* 128: 2239–2247.

Romano, G. J., Bonner, T. E., and Pfaff, D. W. (1989). Preprotachykinin gene expression in the mediobasal hypothalamus of estrogen-treated and ovariectomized control rats. *Exp. Brain Res.* 76: 21–26.

Rubin, B. S., and Barfield, R. J. (1980). Priming of estrous responsiveness by implants of 17β-estradiol in the ventromedial hypothalamic nucleus of female rats. *Endocrinology* 106: 504–509.

Sachs, B. D., and Meisel, R. L. (1988). The physiology of male sexual behavior. In: *The Physiology of Reproduction*, Vol. 2, ed. E. Knobil, J. Neill, L. L. Ewing, G. S. Greenwald, C. L. Markert, and D. W. Pfaff, pp. 1393–1487. New York: Raven Press.

Saffroy, M., Beaujouan, J.-C., Torrens, Y., Besseyre, J., Bergström, L., and Glowinski, J. (1988). Localization of tachykinin binding sites (NK$_1$, NK$_2$, NK$_3$ ligands) in the rat brain. *Peptides* 9: 227–241.

Sahu, A., and Kalra, S. P. (1992). Effects of tachykinins on luteinizing hormone release in female rats: Potent inhibitory action of neuropeptide K. *Endocrinology* 130: 1571–1577.

Sakanaka, M., Shiosaka, S., Takatsuki, K., Inagaki, S., Takagi, H., Senba, E., Kawai, Y., Matsuzaki, T., and Tohyama, M. (1981). Experimental immunohistochemical studies on the amygdalofugal peptidergic (substance P and somatostatin) fibers in the stria terminalis of the rat. *Brain Res.* 221: 231–242.

Sakuma, Y., and Pfaff, D. W. (1979a). Mesencephalic mechanisms for integration of female reproductive behavior in the rat. *Am. J. Physiol.* 237: R285–290.

Sakuma, Y., and Pfaff, D. W. (1979b). Facilitation of female reproductive behavior from mesencephalic central gray in the rat. *Am. J. Physiol.* 237: R278–284.

Sar, M., and Stumpf, W. E. (1977). Distribution of androgen target cells in rat forebrain and pituitary after [^{3}H]-dihydrotestosterone administration. *J. Steroid Biochem.* 8: 1131–1135.

Sawyer, C. H. (1975). Some recent developments in brain–pituitary–ovarian physiology. *Neuroendocrinology* 17: 97–124.

Shamgochian, M. D., and Leeman, S. E. (1992). Substance P stimulates luteinizing hormone secretion from anterior pituitary cells in culture. *Endocrinology* 131: 871–875.

Sheridan, P. J. (1979). The nucleus interstitialis striae terminalis and the nucleus amygdaloideus medialis: Prime targets for androgen in the rat forebrain. *Endocrinology* 104: 130–136.

Shivers, B. D., Harlan, R. E., Morrell, J. I., and Pfaff, D. W. (1983). Absence of oestradiol concentration in cell nuclei of LHRH-immunoreactive neurons. *Nature* 304: 345–347.

Shults, C. W., Quirion, R., Chronwall, B., Chase, T. N., and O'Donohue, T. L. (1984). A comparison of the anatomical distribution of substance P and substance P receptors in the rat central nervous system. *Peptides* 5: 1097–1128.

Simerly, R. B., and Swanson, L. W. (1986). The organization of neural inputs to the medial preoptic nucleus of the rat. *J. Comp. Neurol.* 246: 312–342.

Simerly, R. B., Young, B. J., Capozza, M. A., and Swanson, L. W. (1989). Estrogen differentially regulates neuropeptide gene expression in a sexually dimorphic olfactory pathway. *Proc. Natl. Acad. Sci. (USA)* 86: 4766–4770.

Stinus, L., Kelley, A., and Iversen, S. D. (1978). Increased spontaneous activity following substance P infusion into A10 dopaminergic area. *Nature* 276: 616–618.

Takano, Y., Nagashima, A., Masui, H., Kuromizu, K., and Kamiya, H. (1986). Distribution of substance K (neurokinin A) in the brain and peripheral tissues of rats. *Brain Res.* 369: 400–404.

Takatsuki, K., Sakanaka, M., Takagi, H., Tohyama, M., and Shiotani, Y. (1983). Experimental immunohistochemical studies on the distribution and origins of substance P in the medial preoptic area of the rat. *Exp. Brain Res.* 53: 183–192.

Takeda, Y., Takeda, J., Smart, B. M., and Krause, J. E. (1990). Regional distribution of neuropeptide γ and other tachykinin peptides derived from the substance P gene in the rat. *Regul. Pept.* 28: 323–333.

Tatemoto, K., Lundberg, J. M., Jörvall, H., and Mutt, V. (1985). Neuropeptide K: Isolation, structure and biological activities of a novel brain tachykinin. *Biochem. Biophys. Res. Commun.* 128: 947–953.

Tsuruo, Y., Hisano, S., Okamura, Y., Tsukamoto, N., and Daikoku, D. (1984). Hypothalamic substance P-containing neurons: Sex-dependent topographical differences and ultrastructural transformations associated with stages of the estrous cycle. *Brain Res.* 305: 331–341.

Tsuruo, Y., Kawano, H., Hisano, S., Kagotani, Y., Daikoku, S., Zhang, T., and Yanaihara, N. (1991). Substance P-containing neurons innervating LHRH-containing neurons in the septo-preoptic area of rats. *Neuroendocrinology* 53: 236–245.

Valcourt, R. J., and Sachs, B. D. (1979). Penile reflexes and copulatory behavior in male rats following lesions in the bed nucleus of the stria terminalis. *Brain Res. Bull.* 4: 131–133.

Valentino, K. L., Tatemoto, K., Hunter, J., and Barchas, J. D. (1986). Distribution of neuropeptide K-immunoreactivity in the rat central nervous system. *Peptides* 7: 1043–1059.

Vijayan, E., and McCann, S. M. (1979). *In vivo* and *in vitro* effects of substance P and neurotensin on gonadotropin and prolactin release. *Endocrinology* 105: 64–68.

Whalen, R. E., and Luttge, W. G. (1971). Testosterone, androstenedione and dihydrotestosterone: Effects on mating behavior of male rats. *Horm. Behav.* 2: 117–125.

Yamano, M., Inagaki, S., Kito, S., and Tohyama, M. (1986). A substance P-containing pathway from the hypothalamic ventromedial nucleus to the medial preoptic area of the rat: An immunohistochemical analysis. *Neuroscience* 18: 395–402.

Yamanouchi, K., and Arai, Y. (1983). Forebrain and lower brainstem participation in facilitatory and inhibitory regulation of the display of lordosis in female rats. *Physiol. Behav.* 30: 155–159.

10

Dopaminergic influences on male rat sexual behavior

ELAINE M. HULL

Sexual behavior of male rodents depends heavily on the actions of testosterone and its metabolites. One means by which testosterone may facilitate copulation is by altering the release and/or effectiveness of neurotransmitters. This chapter integrates information and speculation concerning the multiple roles of one such neurotransmitter, dopamine, in the regulation of male rat sexual behavior.

The three major dopamine systems that are important for male sexual behavior, including sexual motivation and genital reflexes, will be reviewed. Then, evidence suggesting that testosterone influences dopamine activity in the medial pre-optic area (MPOA) and nucleus accumbens (NAc) will be discussed. Recent results suggesting that a novel messenger molecule, nitric oxide, might affect both dopamine release and sexual behavior will be presented, as will preliminary evidence suggesting that stimulation of the D_1 dopamine receptor promotes copulation-induced expression of the immediate early gene c-*fos* in the MPOA. Finally, a model summarizing some of the events related to central dopamine release will be described.

The problem of neural coordination of behavior

The execution of a behavior as complex as copulation requires exquisite coordination of neural activity in numerous sites. Olfactory, visual, auditory, and somatosensory stimuli elicit a precisely timed and coordinated motor sequence that includes locomotor pursuit, mounting and pelvic thrusting, penile erection and insertion, and ultimately ejaculation and postejaculatory grooming and quiescence. Steroid hormones facilitate this process by biasing sensorimotor integration, so that a sexually relevant stimulus is more likely to elicit a sexual response. Most effects of steroid hormones on sexual behavior are relatively long term. We are left with the problem of how this long-term biasing is translated into moment-to-moment control of behavior. One step in this translation is probably a change in the release or effectiveness of one or more neurotransmitters. One candidate for a central role is dopamine, since dopaminergic drugs have long been known to facilitate masculine, and possibly feminine, sexual behavior (reviewed by Bitran and Hull 1987). Furthermore, increased dopamine activity associated with copulation has been observed in several key integrative sites, as described later.

A common feature of dopaminergic action is enhancement of sensorimotor integration, which is achieved by removing tonic inhibition (Chevalier and Deniau 1990). Thus, dopamine does not elicit behavior directly, but allows stimuli easier access to output pathways. This short-term biasing is similar to the longer-term biasing of steroid hormones.

Three integrative systems

Three major systems control sexual motivation and genital and somatomotor responses in male rats (Fig. 10.1). A key factor in this model is that sensory input from a receptive female and/or the act of copulation elicits the release of dopamine in each of the three main integrative systems. Output from these systems controls the expression of behaviors, including sexual motivation, genital reflexes, and stereotyped somatomotor patterns of mounting and thrusting. According to this model, the MPOA facilitates all of these behavioral factors. The mesolimbic system regulates general activation and responsiveness to motivational stimuli. It also interfaces with the nigrostriatal system, which disinhibits appropriate somatomotor patterns, including locomotion, mounting, and pelvic thrusting.

A simplified anatomical model is presented in Figure 10.2. Each of the three integrative systems is described in in the following subsections, with major atten-

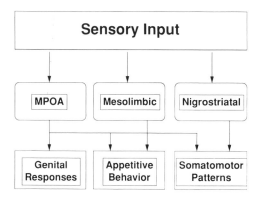

Figure 10.1. Conceptual model of the dopamine systems that control male sexual behavior. Sensory input from the female and from the act of copulation elicits dopamine release in the MPOA and the mesolimbic and nigrostriatal dopamine systems. MPOA dopamine influences genital responses, appetitive behavior, and somatomotor patterns; mesolimbic dopamine influences appetitive behavior and somatomotor patterns; and nigrostriatal dopamine facilitates somatomotor patterns of copulation.

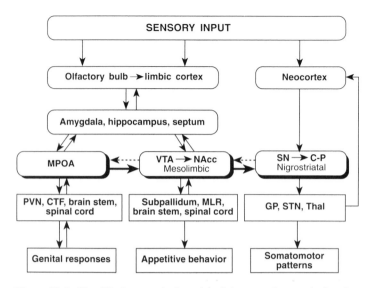

Figure 10.2. Simplified anatomical model of the central control of male sexual behavior. Sensory input from the female gains access to the MPOA and nucleus accumbens (NAcc) via the limbic system. Sensory input from the genitals arrives at the MPOA via the spinal cord, brain stem, and central tegmental field (CTF). Output from the MPOA influences genital reflexes via the PVN, brain stem, and spinal cord; it also influences appetitive behavior, possibly via input to the mesolimbic system. The NAcc, a major target of VTA dopamine neurons, serves as an interface between the limbic and motor systems to regulate appetitive behavior. The caudate-putamen (C-P), the major target of dopamine cells in the substantia nigra (SN), is part of a "loop" that facilitates somatomotor patterns and "readiness to respond." It receives input from the neocortex and sends output via the globus pallidus (GP) to the thalamus (Thal) and subthalamic nucleus (STN); the thalamus in turn projects back to the neocortex. A similar loop involving the NAcc and prefrontal cortex is not shown. MLR denotes mesencephalic locomotor region, including pedunculo-pontine nucleus.

tion devoted to the MPOA because of its critical role in masculine sexual behavior (reviewed in Sachs and Meisel 1988).

The MPOA integrative system

Sensory input to the MPOA. The MPOA receives indirect sensory input from virtually every sensory modality (reviewed by Simerly and Swanson 1986). Olfactory input, which is especially important for copulation in most male rodents, is relayed via the medial amygdala and bed nucleus of the stria terminalis. Auditory, visual, and somatosensory information, processed by the neocortex, is relayed via the ventral subiculum and lateral septum. Visceral and genital input arrives at the MPOA from the central tegmental field, the nucleus of the solitary tract, and the A1 noradrenergic region of the brain stem (Simerly and Swanson 1986). In summary, sensory input from virtually all sensory modalities has indirect access to the MPOA, via limbic and brain stem structures.

Two aspects of the sensory input to the MPOA are noteworthy. First, reciprocal connections with each source of input provide a means for the MPOA to modulate sensory processing, thereby enhancing input relevant to sexual behavior. Second, the MPOA and all of its major afferents concentrate gonadal hormones. Therefore, those hormones can promote the processing of sexually relevant stimuli. The importance of gonadal hormones for the processing of sensory input to the MPOA was first demonstrated in a classic electrophysiological recording study by Pfaff and Pfaffman (1969). In that experiment, castration rendered MPOA neurons unresponsive to the odor of an estrous female, although primary sensory processing in the olfactory bulbs remained intact. Therefore, neurons in a site critical for the integration of male sexual behavior were no longer responsive to a sexually relevant stimulus.

Major efferent projections of the MPOA. The major efferent projections from the MPOA are to hypothalamic, midbrain, and brain stem nuclei that regulate autonomic or somatomotor patterns and motivational states (reviewed in Simerly and Swanson 1988). The actual motor patterns are stored in the midbrain, brain stem, and spinal cord (Mogenson et al. 1980). The role of the MPOA is probably to remove the tonic inhibition on these patterns and thereby to enhance the ability of sensory stimuli to elicit a motor response. A similar role for nigrostriatal dopamine has been proposed in the context of somatomotor patterns (e.g., Albin et al. 1989).

Efferent projections from the MPOA are critical for the initiation of copulation. Lesions destroying most of the MPOA permanently abolish copulation, although the effects of smaller lesions depend on the size and location of the lesion and previous sexual experience of the animal (reviewed by Sachs and Meisel 1988). Males with MPOA lesions may still exhibit appetitive behavior to be with a female, but they appear to be unable to trigger the stereotypic mounting and thrusting pattern (Hansen et al. 1982; Everitt 1990).

Activation of dopamine release in the MPOA. Sensory, especially olfactory, input from a receptive female increases the release of catecholamines in the MPOA (Blackburn et al. 1992). The odor of an estrous female was almost as effective as copulation in increasing extracellular catecholamines in the MPOA, measured by *in vivo* voltammetry. The odor of an estrous female was less effective in eliciting catecholamine release in the NAc and was ineffective in the caudate-putamen. The relative contributions of dopamine and norepinephrine to the MPOA response could not be resolved with voltammetry; however, a subsequent microdialysis study (using high-performance liquid chromatography with electrochemical detection) confirmed that dopamine contributed to the catecholamine increase during copulation (Hull et al. 1993, unpublished data). Thus, sensory stimuli can activate dopamine release in the MPOA, which may in turn enhance the capacity of those stimuli to elicit a behavioral response.

The release of dopamine in the MPOA shows at least some behavioral specificity. Copulation increased MPOA dopamine measured by microdialysis, while eating a highly palatable food (a peanut-butter-filled cookie) did not alter MPOA dopamine level in the same animals (Hull et al. 1993). This is in contrast to the elicitation of dopamine release in the NAc by numerous types of stimuli (reviewed later).

Dopamine release may be affected by direct influences on the axon terminals, as well as by changes in the firing rate of dopamine cell bodies. For example, the recently discovered messenger molecule nitric oxide appears to enhance dopamine release in the MPOA. This effect will be discussed more fully later.

Source of MPOA dopamine. Dopamine input to the MPOA arises primarily or entirely from the periventricular system, including cell bodies in the medial portion of the MPOA (Simerly et al. 1986) and the anterior cell bodies of the incertohypothalamic tract (Bjorklund et al. 1975). A small number of axons from the ventral tegmental area (VTA) appear to reach the MPOA (Simerly and Swanson 1986), but they are probably nondopaminergic (Bjorklund et al. 1975; Lookingland and Moore 1984).

Consequences of dopamine release in the MPOA: facilitation of copulation. Microinjections of a dopamine agonist into the MPOA increased the rate and efficiency of copulation (Hull et al. 1986) and also increased the number of *ex copula* erections (Pehek et al. 1989). In contrast, blocking the access of endogenous dopamine to those receptors impaired both copulation (Pehek et al. 1988) and genital reflexes (Warner et al. 1991). It also decreased sexual motivation, measured as percent choice of the female in an X-maze, but it did not affect locomotion (Warner et al. 1991). Similar injections of a dopamine agonist (apomorphine) or antagonist (*cis*-flupenthixol) had no effect on eating or drinking in the home cage (Pehek et al. 1988). Thus, endogenous dopamine in the MPOA facilitates copulation and enhances genital reflexes and sexual motivation, without affecting locomotion or two other motivated behaviors. The decrease in sexual motivation was somewhat surprising, since most theories regard the MPOA as important for the consummatory aspects of copulation, but not for the appetitive aspects (e.g., Everitt 1990; Blackburn et al. 1992).

Additional evidence for the importance of dopamine in regulating copulation was obtained by destroying MPOA dopamine terminals with 6-hydroxydopamine. Such lesions impaired copulation if tests were made within a few hours after the lesion was inflicted (Bazzett et al. 1992) or after inhibition of local dopamine synthesis (Bitran et al. 1988; Bazzett et al. 1992). Rapid functional recovery within 24 hours was probably due to increased synthesis of dopamine in the remaining terminals (Bitran et al. 1988; Bazzett et al. 1992). Thus, either blocking dopamine receptors or destroying dopamine terminals in the MPOA can impair copulation, confirming the importance of endogenous dopamine for sexual behavior.

Differential influence of D_1 and D_2 receptors on genital reflexes. Dopamine D_1 and D_2 receptors have different roles in the regulation of male sexual behavior. D_1 receptors were originally defined as those positively coupled to adenylyl cyclase, whereas D_2 receptors were negatively or not coupled to that effector (Kebabian and Calne 1979). Additional subtypes have been cloned; however, they all fall into the same two families (Civelli et al. 1991). As in most of the pharmacological literature, references here to D_1 and D_2 receptors include any of the subtypes within the two families.

There appear to be three separate dopaminergic mechanisms in the MPOA that regulate genital reflexes; each has a different threshold of activation (Fig. 10.3). The low-threshold mechanism is activated by D_2 receptors, probably with the cooperation of D_1 receptors; this results in the removal of inhibition on genital reflexes. The removal of inhibition was inferred from a decrease in latency to the first penile reflex in tests using restrained supine male rats. Very low doses of a D_2 agonist (quinelorane, LY-163502) decreased reflex latency but had no effect on the number of penile reflexes (erections and anteroflexions, or "flips") or seminal emissions (Bazzett et al. 1991). The D_1 agonist dihydroxyphenyltetrahydrothienopyridine (THP) did not affect reflex latency (Markowski et al. 1994). However, the decreased latency observed with the mixed D_1/D_2 agonist apomorphine in the MPOA was partially blocked by both a D_1 (SCH 23390) and a D_2 (raclopride) antagonist (Hull et al. 1992). Therefore, in the presence of normal

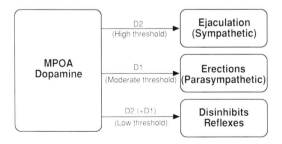

Figure 10.3. Model of MPOA dopamine influences on genital reflexes. A low-threshold mechanism mediated by D_2 receptors disinhibits genital reflexes; some stimulation of D_1 receptors may be permissive. A moderate-threshold mechanism, activated by stimulation of D_1 receptors, facilitates penile erection. A high-threshold D_2 mechanism facilitates seminal emission and inhibits erection. These mechanisms may be activated successively by increasing levels or longer durations of dopamine activity.

background D_1 activity, stimulation of D_2 receptors is sufficient to decrease reflex latency. However, blocking the D_1 background activity renders the D_2 mechanism less effective.

The moderate threshold mechanism is mediated by D_1 receptors. This mechanism increases the number of erections and anteroflexions, possibly in part by increasing the activity of the parasympathetic system (Hull et al. 1992). It also increases the rate of copulation (Markowski et al. 1994). Erections require the actions of both the parasympathetic system and the striated penile muscles (reviewed by Hart and Leedy 1985; Sachs and Meisel 1988). Thus, stimulation of D_1 receptors may disinhibit the parasympathetic system, and thereby direct blood flow to the genitals, and may also contribute to the disinhibition of the penile motor neurons. This presumed redirection of blood flow could have two effects: (a) It could provide the blood necessary for an erection; and (b) it could warm the genital area. Certain sensory receptors in the penis do not respond to warming per se, but they are much more responsive to tactile stimuli when they are passively warmed (Johnson and Kitchell 1987). Therefore, warming could result in greater sensory input to all levels of the system, including the MPOA. Because tonic inhibition is removed by the low-threshold D_2 mechanism, the increased sensory input would result in greater motor output, hence more penile erections and anteroflexions. Stimulation of D_1 receptors also appears to inhibit seminal emission, perhaps thereby preventing "premature ejaculation" (Hull et al. 1992).

The high-threshold mechanism is mediated by D_2 receptors, but opposed by D_1 receptors. This mechanism appears to shift autonomic balance to favor ejaculation. A D_2 agonist (quinelorane) injected into the MPOA facilitated ejaculation *in copula* (Hull et al. 1989) and increased seminal emissions *ex copula* (Bazzett et al. 1991). Ejaculation is mediated by the sympathetic nervous system and striated muscles of the penis and pelvic floor (reviewed by Hart and Leedy 1985; Sachs and Meisel 1988). Therefore, stimulation of D_2 receptors in the MPOA may disinhibit the sympathetic nervous system and possibly also the motoneurons innervating the penis and pelvic floor. In support of this interpretation, the increase in erections elicited by a low dose of the mixed D_1/D_2 agonist apomorphine was blocked only by a D_1 antagonist (Hull et al. 1992). The increase in seminal emissions evoked by a high dose of apomorphine was blocked only by the D_2 antagonist. Thus, either lower levels of endogenous dopamine or a shorter duration of dopamine release may disinhibit erectile mechanisms, whereas higher levels or longer durations of dopamine release appear to favor ejaculation. In addition to facilitating ejaculation, D_2 mechanisms might also contribute to impotence during the immediate postejaculatory period. Consistent with this interpretation, the D_2 agonist quinelorane delayed and slowed copulation (Hull et al. 1989), and decreased the number of *ex copula* erections and anteroflexions (Bazzett et al. 1991).

One might think that the antagonistic effects of D_1 and D_2 receptors would simply cancel each other out and make copulation impossible. However, the differential thresholds of the three mechanisms might contribute to the appropriate sequencing of behavioral patterns. Therefore, dopamine released in the MPOA before and during copulation could first disinhibit reflexes, then facilitate erections and anteroflexions, and finally facilitate ejaculation. It might also regulate the number of intromissions required to elicit ejaculation and thereby affect the male's ability to trigger a progestational state in the female (reviewed in Adler and Toner 1986). It is possible that the low-threshold and high-threshold D_2 mechanisms are mediated by different members of the D_2 receptor "family," having somewhat different affinities for dopamine and/or utilizing either different second messengers or different neuronal circuits.

The nigrostriatal integrative system

The nigrostriatal system enhances the readiness to respond to external stimuli (Albin et al. 1989; Salamone 1991; Robbins and Everitt 1992). Specifically, dopamine released in the caudatoputamen (CP) disinhibits pathways through which the cortex elicits movements (Chevalier and Deniau 1990). For example, the loss of dopamine in Parkinson's disease results in difficulty in initiating responses to environmental stimuli. Dopamine neurons in both the substantia nigra and the VTA respond with short latencies to a variety of stimuli that have "alerting, arousing and attention-grabbing properties" (Schultz 1992). Thus, they do not provide specific information about the stimuli that elicit them. Their major input may be from the reticular formation (Schultz 1992). The nigrostriatal system may be especially critical for the expression of "consummatory" movements (Robbins and Everitt 1992), such as the somatomotor patterns required for pursuit of the female, mounting, and thrusting. This system will not be discussed in detail, because it has been reviewed extensively elsewhere and because its specific contribution to masculine sexual behavior has not been clarified.

The mesolimbic integrative system

The mesolimbic system is critical for appetitive behavior and reinforcement. It is activated by a variety of motivational behaviors, including eating, drinking, copulating, drug self-administration, and intracranial self-stimulation (Mogenson et al. 1980; Spyraki et al. 1983; Stellar et al. 1983; Wise and Bozarth 1984; Mucha and Iversen 1986; Fibiger and Phillips 1988; Mark et al. 1991). There is disagreement as to whether mesolimbic dopamine is more important for reward processes per se (e.g., Wise and Bozarth 1984) or for the behavioral activation elicited by reinforcers – positive or negative, anticipated or attained (e.g., Koob 1992). However, there is virtual unanimity that the mesolimbic system is crucial for one or more aspects of appetitive behavior.

Dopamine release in the NAc is regulated by influences both on dopamine cell bodies in the VTA and on dopamine terminals in the NAc (reviewed by White 1991). Dopamine neurons in the VTA, as well as those in the substantia nigra, respond with short latencies to either natural or learned reinforcers (Schultz 1992). Their function is parallel to that of the nigrostriatal system, in that motor responses to motivational stimuli are disinhibited. Several limbic structures influence dopamine release presynaptically at the terminals in the NAc (reviewed by Simon and LeMoal 1988; Mogenson and Yim 1991). Input from the amygdala seems to be especially important for the responsiveness to conditioned reinforcers (Cador et al. 1989). In return, dopamine modulates the responsiveness of NAc neurons to limbic inputs (Mogenson and Yim 1991). This system exhibits considerable plasticity, both in the acquisition of secondary reinforcers and in the types of appetitive behavior utilized to attain this goal. Mogenson et al. (1980) proposed that the NAc serves as an interface between motivation and action via its connections with the CP, pallidum, and mesencephalic locomotor region.

Roles of the MPOA and mesolimbic system in sexual behavior

Everitt (1990) suggested that the MPOA (as well as the nigrostriatal system) is important for the consummatory aspects of copulation, whereas the mesolimbic system provides the motivational impetus. For example, manipulations of the NAc affected responding for a secondary reinforcer (a light that had been associated with a receptive female), but did not affect copulatory performance (Everitt et al. 1989). In contrast, lesions of the MPOA abolished copulation but did not affect lever pressing for the secondary reinforcer. Thus, there appears to be a double dissociation between mesolimbic and MPOA influences on sexual behavior, with the mesolimbic system contributing appetitive responses and the MPOA initiating the consummatory act.

However, such a strict dichotomy may be overly simplistic. There is evidence both that the MPOA can influence sexual motivation and that the mesolimbic system can affect copulatory performance. Injections of dopamine antagonists into the MPOA decreased sexual motivation as measured in an X-maze (Warner et al. 1991) or in a bilevel apparatus in which the male changes levels in search of a receptive female (Pfaus and Phillips 1991). In addition, electrolytic lesions of the MPOA decreased the preference of a male for an estrous female (Edwards and Einhorn 1986) and decreased the male's pursuit of an estrous female (Paredes et al. 1993). One interpretation of these data is that dopamine released in the MPOA enhances the processing of genital sensation and possibly also of olfactory cues from the estrous female. These stimuli, then, might channel the male's attention and behavior toward sexual pursuits. Manipulations of dopamine activity in the

MPOA did not affect general locomotion or responsiveness to food or water. Thus, dopamine activity in the MPOA may serve less to increase general activation than to redirect or bias behavior to favor sexual activities.

Decreased activity of the mesolimbic dopamine tract can also impair copulatory performance and general sensorimotor responsiveness. The firing rate of mesolimbic dopamine neurons is decreased by stimulating autoreceptors on their cell bodies (White et al. 1988). This treatment delayed and slowed copulation (Hull et al. 1990) and also slowed locomotion in an X-maze and in a videotaped study of copulation (Hull et al. 1991). In addition, motor patterns were poorly organized, with some males mounting from the female's head or side or performing numerous incomplete mounts. In contrast, specifically sexual motivation seemed unaffected, as neither the percent choice of the female in the X-maze nor the number of female-directed behaviors in the videotaped copulation study was affected. *Ex copula* genital reflexes were also not affected (Hull et al. 1991). Even though males were more sluggish, the specific channeling of activity into sexual behavior was not impaired. Thus, slowing of mesolimbic dopamine activity resulted in general hypoactivity and poorly organized copulatory patterns, but did not affect several measures of sexual motivation (Hull et al. 1991). Therefore, general sensorimotor responsiveness, rather than sexual motivation, was inhibited by the decrease of mesolimbic dopamine activity.

Summary of the integrative systems

Sexually relevant stimuli gain access to the MPOA and the mesolimbic dopamine system by way of several limbic structures, all of which contain neurons that concentrate gonadal steroids. Exteroceptive stimuli reach the nigrostriatal dopamine system via the neocortex. These stimuli increase the release of dopamine in the MPOA, NAc, and CP. Increases of dopamine cell firing in the mesolimbic and nigrostriatal systems may be mediated by the reticular formation, whereas limbic and cortical structures might regulate dopamine release in the axon terminals. Dopamine in turn enhances stimulus–response coupling in each of the three integrative systems and may also influence sensory processing by means of reciprocal connections with the sources of input. Output from the MPOA disinhibits genital responses and channels behavior toward sexual output. It may also help to disinhibit somatomotor patterns. The mesolimbic system promotes general activation and responsiveness to motivational stimuli, and interfaces with the nigrostriatal system, which disinhibits somatomotor patterns.

Hormone transmitter interactions

Testosterone and its metabolites are critical for copulatory behavior in male rodents. However, whereas testosterone levels decline rapidly after castration, copulatory behavior decreases gradually over a 2- to 10-week period (Davidson 1966b). Copulation can be restored by implants of either testosterone (Davidson 1966a) or estradiol (Christensen and Clemens 1974) into the MPOA. The prolonged period of behavioral decline after castration might be related to gradual changes in biochemical processes, cell size or number, or synaptic connectivity.

Restoration of copulation in castrated rats

Sexual behavior can be enhanced by the dopamine agonist apomorphine during the prolonged copulatory decline after castration (Malmnas 1977) or in animals with suboptimal injections of testosterone (Malmnas 1973). Even long-term (3 to 8 months) castrates, which had failed to mount on at least two weekly screening tests, responded to apomorphine with a partial restoration of copulation (Scaletta and Hull 1990). Systemic administration restored both mounting and intromitting, with a few animals ejaculating. Microinjections into the MPOA significantly restored only mounting, though a few animals intromitted. The significant increase in mounting but not intromitting is consistent with an influence of MPOA dopamine on sexual motivation. The effectiveness of apomorphine suggests that one result of castration may be a decrease of dopamine activity in one or more sites regulating copulation. The greater effectiveness of systemic injections than intra-MPOA injections suggests that dopamine receptors in sites other than the MPOA also contribute to the facilitative effects of dopamine agonists. On the other hand, the inability of apomorphine to restore copulation completely suggests that castration impairs factors other than dopamine release.

Effects of castration and hormone replacement on dopamine activity

Testosterone may facilitate dopamine activity, and thereby copulation, in the NAc. Castration decreased *ex vivo* tissue levels of both dopamine and its major metabolite, dihydroxyphenylacetic acid (DOPAC), in the NAc and decreased the incidence of female-directed behaviors over the same time course (Mitchell and Stewart 1989). Systemically administered testosterone or estradiol restored both biochemical and behavioral measures to normal. Tissue levels of dopamine in the MPOA actually increased after castration. However, DOPAC levels were below the threshold of detectability in all groups, so it was not clear whether the increased dopamine levels in the MPOA resulted from increased synthesis or decreased release.

In other studies, castration increased (Simpkins et al. 1983), decreased (Gunnett et al. 1986), or did not affect (Baum et al. 1986) *ex vivo* tissue levels or turnover of dopamine in the MPOA. Since none of these studies tested the ability to copulate shortly before the assays, the behavioral relevance of these findings is not clear. Also, tissue levels of neurotransmitters reflect primarily intracellular stores, so the ease of release of the transmitter in appropriate circumstances is difficult to assess.

In order to measure directly the influence of hormones on dopamine release and copulatory ability, we have begun a microdialysis study of castrates with and without testosterone replacement. Preliminary data suggest that castration interferes with the dopamine release usually elicited by the presence of a female (J. Du and E. Hull, unpublished results). Eight males received guide cannulae located above the MPOA at the time of castration, and were injected daily with 200 µg testosterone propionate or oil vehicle for 7 days thereafter. On the eighth day a microdialysis probe was lowered into the MPOA and baseline samples were collected 4 hours later. Baseline dopamine and DOPAC levels were determined from three consecutive samples, which showed less than 10% variability. A wire mesh cage containing a receptive female was then placed into the microdialysis arena. Two 20-minute samples were collected, after which the cage was removed and the two animals were allowed to copulate. Two additional samples were collected while the male had free access to the female, and two final samples were collected after the female was removed from the arena. For the next 7 days, the male received the alternate treatment, and the test was repeated on the eighth day.

Males with testosterone replacement showed a significant increase of extracellular dopamine in the first sample collected in the presence of the caged female and a slightly greater increase of dopamine during copulation. This effect was not observed in the absence of testosterone treatment. Seven of the eight animals copulated after testosterone replacement, while only three copulated after oil injections. Those that did copulate after oil injections had numerous mounts, few intromissions, and a long interval before ejaculation, and they did not show increased dopamine release.

In summary, these results suggest that one reason for the decline of sexual activity in castrates is a decrease of female-induced dopamine release in the MPOA. This observation may be related to the finding of Pfaff and Pfaffman (1969) that castration blocked the increase in MPOA firing rate usually elicited by the odor of an estrous female. Thus, castration abolishes the responsiveness of two measures of neural activity in the MPOA: firing rate and dopamine release. Therefore, stimuli that would normally elicit copulation cannot activate some measures of neural processing in an area critical for the initiation of copulation. It is equally important to determine whether castration blocks female-induced dopamine release in the NAc as well.

Influence of nitric oxide on dopamine release and copulation

Nitric oxide is a highly reactive gas given off when L-arginine is converted to citrulline by nitric oxide synthase. Nitric oxide has been implicated in the control of dopamine release in striatal slices (Hanbauer et al. 1992; Zhu and Luo 1992;

Lonart et al. 1993) and in a variety of other neural and nonneural effects (reviewed by Lancaster 1992; Snyder and Bredt 1992).

Large fibers containing nitric oxide synthase are scattered through the MPOA (Vincent and Kimura 1992). We recently reported that the nitric oxide precursor L-arginine, but not its inactive isomer D-arginine, increased dopamine release measured by microdialysis in the MPOA (Lorrain and Hull 1993). This increase was blocked by co-administration of the nitric oxide synthesis inhibitor N-monomethyl-L-arginine (NMMA), which decreased basal dopamine release when administered alone. This suggests that nitric oxide might contribute to the normal regulation of dopamine release in this region, including perhaps the prolonged increase of MPOA dopamine activity that occurs during copulation.

In addition to its enhancement of dopamine release in the MPOA, nitric oxide may mediate the capacity of dopamine and oxytocin to induce the penile erection and yawning syndrome in freely moving rats (Melis and Argiolas 1993). Systemic administration of the dopamine agonist apomorphine or intraventricular injection of oxytocin elicited an average of 3 to 4 penile erections and 15 to 20 yawns during the 60-minute observation period. These behaviors were decreased by either systemic or intraventricular injections of nitric oxide synthase inhibitors. These authors suggested that the paraventricular nucleus (PVN) is a likely site for this action of nitric oxide. Therefore, in addition to increasing dopamine release in the MPOA, nitric oxide may serve as an effector of dopamine activity in the PVN.

Nitric oxide also promotes erection by relaxing the smooth muscle of the penile corpora cavernosa (Ignarro et al. 1990; Burnett et al. 1992). We tested the effects of N-nitro-L-arginine methyl ester (NAME), an inhibitor of nitric oxide synthase, administered systemically before copulation testing. NAME dose-dependently impaired copulation and *ex copula* penile erections (Hull et al. 1994). Unexpectedly, it also increased the number of *ex copula* seminal emissions. It is likely that the decrease in erections resulted at least in part from impairment of vasodilation in the penile corpora. However, the stimulus for the increase in seminal emissions must have been neural, and we are currently investigating possible central mechanisms underlying this effect.

Dopamine regulation of copulation-induced c-*fos* expression

Several studies have shown that stimuli which elicit, and result from, copulation also increase the expression of Fos, the protein product of the immediate early gene c-*fos*, in several sites (Robertson et al. 1991; Baum and Everitt 1992; Kollack and Newman 1992). Fos is believed to promote the translation of synaptic activity into further alterations of gene expression, and therefore might lead to long-term changes in neural activity (Morgan and Curran 1989).

Baum and Everitt (1992) reported that increasing copulatory activity induced Fos-immunoreactive (Fos-IR) labeling. Furthermore, combined lesions of the medial amygdala and the central tegmental field (CTF) abolished the MPOA Fos response to copulation. The authors suggested that olfactory input from the medial amygdala and genital input via the CTF are essential for the copulation-induced expression of Fos in the MPOA.

One step in the induction of Fos by sexual stimuli might be stimulation of D_1 receptors (Robertson et al. 1989). The activation of c-*fos* in the striatum by amphetamine and cocaine is dependent on the stimulation of D_1 receptors (Graybiel et al. 1990; Young et al. 1991). We have examined whether stimulation of D_1 receptors in the MPOA is required for copulation-induced Fos expression (Lumley et al. 1993). Male rats were treated with the D_1 antagonist SCH 23390, the D_2 antagonist raclopride, or vehicle before copulating to ejaculation. Control animals received vehicle injections but were not allowed to copulate. Animals that copulated were killed 90 minutes after ejaculation; control animals were killed 120 minutes after injection. Preliminary results indicate that copulation increased Fos-IR labeling in vehicle-treated animals. This response was decreased by pretreatment with the D_1 antagonist, while the D_2 antagonist appeared to increase Fos-IR labeling. Thus, D_1 receptor stimulation might promote copulation-induced c-*fos* expression in the MPOA (Lumley et al. 1993). This suggests that a common biochemical mechanism might be involved in the regulation of behavioral- and drug-induced c-*fos* expression in the MPOA and the striatum, respectively. The observed opposing effects of D_1 and D_2 receptor stimulation also fit with our previous results showing similar effects of these receptor subtypes on genital reflexes. Thus, stimulation of D_1 receptors, which promotes erections (Hull et al. 1992a) and speeds copulatory rate (Markowski et al. 1994), also appears to induce the expression of c-*fos*. In contrast, stimulation of D_2 receptors, which promote ejaculation and inhibit erection (Hull et al. 1992a) and slow copulatory rate (Hull et al. 1989), may inhibit c-*fos* expression.

Summary

The smell, sound, and sight of an estrous female elicit dopamine release in both the MPOA and NAc (Fig. 10.4). Dopamine release in the MPOA may depend on the presence of both testosterone and nitric oxide. Stimulation of dopamine receptors in the MPOA enhances the rate and efficiency of copulation and facilitates genital reflexes. A low-threshold D_2 mechanism (perhaps with a D_1 contribution) appears to disinhibit genital reflexes (decreases reflex latency without affecting the number of reflexes). Stimulation of a moderate-threshold D_1 receptor mechanism facilitates penile erections and anteroflexions and may thereby speed copulatory rate. An additional consequence of the stimulation of D_1 receptors may be the

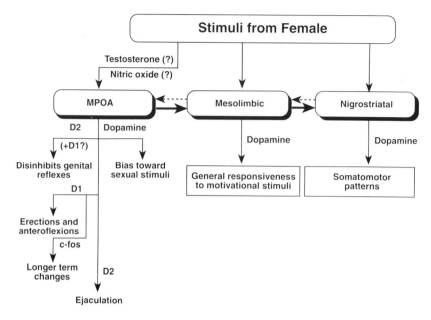

Figure 10.4. Summary of dopaminergic influences on male sexual behavior. Stimuli from a receptive female elicit dopamine release in the MPOA. Testosterone and nitric oxide might facilitate this release. Low levels or short durations of dopamine activity disinhibit genital reflexes via stimulation of D_2 receptors. Higher levels or longer durations of extracellular dopamine stimulate D_1 receptors, which may facilitate erection. Still higher levels or longer durations activate a high-threshold D_2 mechanism that shifts autonomic balance to favor ejaculation. Dopamine in the MPOA may also enhance sexual motivation by enhancing the processing of sexually relevant stimuli. Stimuli from the receptive female also increase dopamine release by mesolimbic terminals in the nucleus accumbens, thereby enhancing general responsiveness to motivational stimuli, including sexual stimuli. As copulation begins, dopamine is released by nigrostriatal terminals in the CP; this disinhibits somatomotor patterns and promotes "readiness to respond."

induction of the immediate early gene c-*fos*, which might in turn produce long-term changes in the control of copulation. Stimulation of a high-threshold D_2 mechanism facilitates ejaculation, but delays and slows copulation. Dopamine in the MPOA may also introduce a motivational bias toward sexual stimuli and responses. Dopamine released in the NAc appears to enhance general responsiveness to motivational stimuli, while dopamine released in the CP with the onset of copulation disinhibits somatomotor patterns of copulation.

References

Adler, N. T., and Toner, J. P. (1986). The effects of copulatory behavior on sperm transport and fertility in rats. In *Reproduction: A Behavioral and Neuroendocrine Perspective*, Ann. N.Y. Acad. Sci. Vol. 474, ed. B. R. Komisaruk, H. I. Siegel, M. F. Cheng, and H. H. Feder, pp. 21–32. New York: New York Academy of Science.

Albin, R. L., Young, A. B., and Penney, J. B. (1989). The functional anatomy of basal ganglia disorders. *Trends Neurosci.* 12: 366–376.

Baum, M. J., and Everitt, B. J. (1992). Increased expression of c-*fos* in the medial preoptic area after mating in male rats: Role of afferent inputs from the medial amygdala and midbrain central tegmental field. *Neuroscience* 50: 627–646.

Baum, M. J., Melamed, E., and Globus, M. (1986). Dissociation of the effects of castration and testosterone replacement on sexual behavior and neural metabolism of dopamine in the male rat. *Brain Res. Bull.* 16: 145–148.

Bazzett, T. J., Lumley, L. A., Bitran, D., Markowski, V. P., Warner, R. K., and Hull, E. M. (1992). 6-OHDA lesions of MPOA impair copulation in male rats. *Brain Res.* 580: 164–171.

Bazzett, T. J., Eaton, R. C., Thompson, J. T., Markowski, V. P., Lumley, L. A., and Hull, E. M. (1991). Dose dependent D_2 effects on genital reflexes after MPOA injections of quinelorane and apomorphine. *Life Sci.* 48: 2309–2315.

Bitran, D., and Hull, E. M. (1987). Pharmacological analysis of male rat sexual behavior. *Neurosci. Biobehav. Rev.* 11: 365–389.

Bitran, D., Hull, E. M., Holmes, G. M., and Lookingland, K. J. (1988). Regulation of male rat copulatory behavior by preoptic incertohypothalamic dopamine neurons. *Brain Res. Bull.* 20: 323–331.

Bjorklund, A., Lindvall, O., and Nobin, A. (1975). Evidence of an incertohypothalamic dopamine neuron system in the rat. *Brain Res.* 89: 29–42.

Blackburn, J. G., Pfaus, J. G., and Phillips, A. G. (1992). Dopamine functions in appetitive and defensive behaviours. *Prog. Neurobiol.* 39: 247–279.

Burnett, A. L., Lowenstein, C. J., Bredt, D. S., Chang, T. S. K., and Snyder, S. H. (1992). Nitric oxide: A physiologic mediator of penile erection. *Science* 257: 401–403.

Cador, M., Robbins, T. W., and Everitt, B. J. (1989). Involvement of the amygdala in stimulus-reward associations: Interactions with ventral striatum. *Neuroscience* 30: 77–86.

Chevalier, G., and Deniau, J. M. (1990). Disinhibition as a basic process in the expression of striatal functions. *Trends Neurosci.* 13: 277–280.

Christensen, L. W., and Clemens, L. G. (1974). Intrahypothalamic implants of testosterone or estradiol and resumption of masculine sexual behavior in long-term castrated male rats. *Endocrinology* 95: 984–990.

Civelli, O., Bunzow, J. R., Grandy, D. K., Zhou, Q. Y., and Vantol, H. H. M. (1991). Molecular biology of the dopamine receptors. *Eur. J. Pharmacol.–Mol. Pharmacol.* 207: 277–286.

Davidson, J. M. (1966a). Activation of the male rat's sexual behavior by intracerebral implantation of androgen. *Endocrinology* 79: 783–794.

Davidson, J. M. (1966b). Characteristics of sex behaviour in male rats following castration. *Anim. Behav.* 14: 266–272.

Edwards, D. A., and Einhorn, L. C. (1986). Preoptic and midbrain control of sexual motivation. *Physiol. Behav.* 37: 329–335.

Everitt, B. J. (1990). Sexual motivation: A neural and behavioural analysis of the mechanisms underlying appetitive and copulatory responses of male rats. *Neurosci. Biobehav. Rev.* 14: 217–232.

Everitt, B. J., Cador, M., and Robbins, T. W. (1989). Interactions between the amygdala and ventral striatum in stimulus-reward associations: Studies using a second-order schedule of sexual reinforcement. *Neuroscience* 30: 63–75.

Fibiger, H. C., and Phillips, A. G. (1988). Mesocorticolimbic dopamine systems and reward. In *The Mesocorticolimbic Dopamine System*, Ann. N.Y. Acad. Sci. Vol. 537, ed. P. W. Kalivas and C. B. Nemeroff, pp. 206–215. New York: New York Academy of Science.

Graybiel, A. M., Moratalla, R., and Robertson, H. A. (1990). Amphetamine and cocaine induce drug-specific activation of the c-*fos* gene in striosome-matrix compartments and limbic subdivisions of the striatum. *Neurobiology* 87: 6912–6916.

Gunnett J. W., Lookingland, K. J., and Moore, K. E. (1986). Comparison of the effects of castration and steroid replacement on incertohypothalamic dopaminergic neurons in male and female rats. *Neuroendocrinology* 44: 269–275.

Hanbauer, I., Wink, D., Osawa, Y., Edelman, G. M., and Gally, J. A. (1992). Role of nitric oxide in NMDA-evoked release of [^{3}H]-dopamine from striatal slices. *Neuroreport* 3: 409–412.

Hansen, S., Kohler, C., Goldstein, M., and Steinbusch, H. V. M. (1982). Effects of ibotenic acid-induced neuronal degeneration in the medial preoptic area and the lateral hypothalamic area on sexual behavior in the male rat. *Brain Res.* 239: 213–232.

Hart, B. L., and Leedy, M. G. (1985). Neurological bases of male sexual behavior: A comparative analysis. In *Reproduction, Handbook of Behavioral Neurobiology*, Vol. 7, ed. N. Adler, D. Pfaff, and R. W. Goy, pp. 373–422. New York: Plenum Press.

Hull, E. M., Bazzett, T. J., Warner, R. K., Eaton, R. C., and Thompson, J. T. (1990). Dopamine receptors in the ventral tegmental area modulate male sexual behavior in rats. *Brain Res.* 512: 1–6.

Hull, E. M., Bitran, D., Pehek, E. A., Warner, R. K., Band, L. C., and Holmes, G. M. (1986). Dopaminergic control of male sex behavior in rats: Effects of an intracerebrally infused agonist. *Brain Res.* 370: 73–81.

Hull, E. M., Eaton, R. C., Markowski, V. P., Moses, J., Lumley L. A., and Loucks J. A. (1992). Opposite influence of medial preoptic D_1 and D_2 receptors on genital reflexes: Implications for copulation. *Life Sci.* 51: 1705–1713.

Hull, E. M., Eaton, R. C., Moses, J., and Lorrain, D. (1993). Copulation increases dopamine activity in the medial preoptic area of male rats. *Life Sci.* 52: 935–940.

Hull, E. M., Lumley, L. A., Matuszewich, L., Dominguez, J., Moses, J., and Lorrain, D. S. (1994). The roles of nitric oxide in sexual function of male rats. *Neuropharmacology* 33: 1499–1504.

Hull, E. M., Warner, R. K., Bazzett, T. J., Eaton, R. C., Thompson, J. T., and Scaletta, L. L. (1989). D_2/D_1 ratio in the medial preoptic area affects copulation of male rats. *J. Pharmacol. Exp. Ther.* 251: 422–427.

Hull, E. M., Weber, M. S., Eaton, R. C., Dua, R., Markowski, V. P., Lumley, L., and Moses, J. (1991). Dopamine receptors in the ventral tegmental area affect motor, but not motivational or reflexive, components of copulation in male rats. *Brain Res.* 554: 72–76.

Ignarro, L. J., Bush, P. A., Buga, G. M., Wood, K. S., Fukuto, J. M., and Rajfer, J. (1990). NO and cyclic GMP formation upon electrical field stimulation cause relaxation of corpus cavernosum smooth muscle. *Biochem. Biophys. Res. Commun.* 170: 843–850.

Johnson, R. D., and Kitchell, R. L. (1987). Mechanoreceptor response to mechanical and thermal stimuli in the glans penis of the dog. *J. Neurophysiol.* 57: 1813–1836.

Kebabian, J. W., and Calne, D. B. (1979). Multiple receptors for dopamine. *Nature* 277: 93–96.

Kollack, S. S., and Newman, S. W. (1992). Mating behavior induces selective expression of protein within the chemosensory pathways of the male Syrian hamster brain. *Neurosci. Lett.* 143: 223–228.

Koob, G. F. (1992). Dopamine, addiction and reward. *Semin. Neurosci.* 4: 139–148.

Lancaster, Jr., J. R. (1992). Nitric oxide in cells. *Am. Scientist* 80: 248–259.

Lonart, G., Cassels, K. L., and Johnson, K. M. (1993). Nitric oxide induces calcium-dependent [^{3}H]dopamine release from striatal slices. *J. Neurosci. Res.* 35: 192–198.

Lookingland, K. J., and Moore, K. E. (1984). Dopamine receptor-mediated regulation of incertohypothalamic dopaminergic neurons in the rat. *Brain Res.* 304: 329–338.

Lorrain, D. S., and Hull, E. M. (1993). Nitric oxide increases dopamine and serotonin release in the medial preoptic area. *Neuroreport* 5: 87–89.

Lumley, L. A., Bazzett, T. J., and Hull, E. M. (1993). A D$_1$ antagonist decreases copulation-induced c-*fos* expression in MPOA. *Soc. Neurosci. Abstr.* 19: 1019.

Malmnas, C. O. (1973). Monoaminergic influence on testosterone-activated copulatory behavior in the castrated male rat. *Acta Physiol Scand.* (Suppl.) 395: 1–128.

Malmnas, C. O. (1977). Dopaminergic reversal of the decline after castration of rat copulatory behaviour. *Endocrinology* 73: 187–188.

Mark, G. P., Schwartz, D. H., Hernandez, H. L., West, H. L., and Hoebel, B. G. (1991). Application of microdialysis to the study of motivation and conditioning: Measurements of dopamine and serotonin in freely behaving rats. In *Techniques in the Behavioral and Neural Sciences*, Vol. 7: *Microdialysis in the Neurosciences*, ed. T. E. Robinson and J. B. Justice, Jr., pp. 369–388. New York: Elsevier.

Markowski, V. P., Lumley, L. A., Moses, J., and Hull, E. M. (1994). A D$_1$ agonist in the MPOA facilitates copulation in male rats. *Pharmacol. Biochem. Behav.* 47: 483–486.

Melis, M. R., and Argiolas, A. (1993). Nitric oxide synthase inhibitors prevent apomorphine-induced and oxytocin-induced penile erection and yawning in male rats. *Brain Res. Bull.* 32: 71–74.

Mitchell, J. B., and Stewart, J. (1989). Effects of castration, steroid replacement, and sexual experience on mesolimbic dopamine and sexual behaviors in the male rat. *Brain Res.* 491: 116–127.

Mogenson, G. J., and Yim, C. C. (1991). Neuromodulatory functions of the mesolimbic dopamine system: Electrophysiological and behavioural studies. In *The Mesolimbic Dopamine System: From Motivation to Action*, ed. P. Willner and J. Scheel-Kruger, pp. 105–130. New York: Wiley.

Mogenson, G. J., Jones, D. L., and Yim, C. Y. (1980). From motivation to action: Functional interface between the limbic system and the motor system. *Prog. Neurobiol.* 14: 69–97.

Morgan, I., and Curran, T. (1989). Stimulus–transcription coupling in neurons: Role of cellular immediate–early genes. *Trends Neurosci.* 12: 459–462.

Mucha, R. F., and Iversen, S. D. (1986). Increased food intake after opioid microinjections into nucleus accumbens and ventral tegmental area of rat. *Brain Res.* 397: 214–224.

Paredes, R. G., Highland, L., and Karam, P. (1993). Socio-sexual behavior in male rats after lesions of the medial preoptic area: Evidence for reduced sexual motivation. *Brain Res.* 618: 271–276.

Pehek, E. A., Thompson, J. T., and Hull, E. M. (1989). The effects of intracranial administration of the dopamine agonist apomorphine on penile reflexes and seminal emission in the rat. *Brain Res.* 500: 325–332.

Pehek, E. A., Warner, R. K., Bazzett, T. J., Bitran, D., Band, L. C., Eaton, R. C., and Hull, E. M. (1988). Microinjections of *cis*-flupenthixol, a dopamine antagonist, into the medial preoptic area impairs sexual behavior of male rats. *Brain Res.* 443: 70–76.

Pfaff, D. W., and Pfaffman, C. (1969). Olfactory and hormonal influences on the basal forebrain of the male rat. *Brain Res.* 15: 137–158.

Pfaus, J. G., and Phillips, A. G. (1991). Role of dopamine in anticipatory and consummatory aspects of sexual behavior in the male rat. *Behav. Neurosci.* 105: 727–743.

Robbins, T. W., and Everitt, B. J. (1992). Functions of dopamine in the dorsal and ventral striatum. *Sem. Neurosci.* 4: 119–128.

Robertson, H. A., Peterson, M. R., Murphy, K., and Robertson, G. S. (1989). D_1-Dopamine receptor agonists selectively activate striatal c-*fos* independent of rotational behavior. *Brain Res.* 503: 346–349.

Robertson, G. S., Pfaus, J. G., Atkinson, L. J., Matsumura, H., Phillips, A. G., and Fibiger, H. C. (1991). Sexual behavior increases c-*fos* expression in the forebrain of the male rat. *Brain Res.* 564: 352–357.

Sachs, B. D., and Meisel, R. L. (1988). The physiology of male sexual behavior. In *The Physiology of Reproduction*, Vol. 2, ed. E. Knobil and J. Neill, pp. 1393–1485. New York: Raven Press.

Salamone, J. D. (1991). Behavioral pharmacology of dopamine systems: A new synthesis. In *The Mesolimbic Dopamine System: From Motivation to Action*, ed. P. Willner and J. Scheel-Kruger, pp. 599–613. New York: Wiley.

Scaletta L. L., and Hull, E. M. (1990). Systemic or intracranial apomorphine increases copulation in long-term castrated male rats. *Pharmacol. Biochem. Behav.* 37: 471–475.

Schultz, W. (1992). Activity of dopamine neurons in the behaving primate. *Semin. Neurosci.* 4: 129–138.

Simerly, R. B., Gorski, R. A., and Swanson, L. W. (1986). Neurotransmitter specificity of cells and fibers in the medial preoptic nucleus: An immunohistochemical study in the rat. *J. Comp. Neurol.* 246: 343–363.

Simerly, R. B., and Swanson, L. W. (1986). The organization of neural inputs to the medial preoptic nucleus of the rat. *J. Comp. Neurol.* 246: 312–342.

Simerly, R. B., and Swanson, L. W. (1988). Projections of the medial preoptic nucleus: A *Phaseolus vulgaris* leucoagglutinin anterograde tract-tracing study in the rat. *J. Comp. Neurol.* 270: 209–242.

Simon, H., and LeMoal, M. (1988). Mesencephalic dopaminergic neurons: Role in the general economy of the brain. In *The Mesocorticolimbic Dopamine System*, Ann. N.Y. Acad. Sci. Vol. 537, ed P. W. Kalivas and C. B. Nemeroff, pp. 235–253. New York: New York Academy of Science.

Simpkins, J. W., Kalra, S. P., and Kalra, P. S. (1983). Variable effects of testosterone on dopamine activity in several microdissected regions in the preoptic area and medial basal hypothalamus. *Endocrinology* 112: 665–669.

Snyder, S. H., and Bredt D. S. (1992). Biological roles of nitric oxide. *Scientific Am.* 267: 68–77.

Spyraki, C., Fibiger, H. C., and Phillips, A. G. (1983). Attenuation of heroin reward in rats by disruption of the mesolimbic dopamine system. *Psychopharmacology* 79: 278–283.

Stellar, J. R., Kelley, A., and Corbett, D. (1983). Effects of peripheral and central dopamine blockade on lateral hypothalamic self-stimulation: Evidence for both reward and motor deficits. *Pharmacol. Biochem. Behav.* 18: 433–442.

Vincent, S. R., and Kimura, H. (1992). Histochemical mapping of nitric oxide synthase in the rat brain. *Neuroscience* 46: 755–784.

Warner, R. K., Thompson, J. T., Markowski, V. P., Loucks, J. A., Bazzett, T. J., Eaton, R. C., and Hull, E. M. (1991). Microinjection of the dopamine antagonist *cis*-flupenthixol into the MPOA impairs copulation, penile reflexes and sexual motivation in male rats. *Brain Res.* 540: 177–182.

White, F. J., Bednarz, L. A., Wachtel, S. R., Hjorth, S., and Brooderson, R. J. (1988). Is stimulation of both D_1 and D_2 receptors necessary for the expression of dopamine-mediated behaviors? *Pharmacol. Biochem. Behav.* 30: 189–193.

White, F. J. (1991). Neurotransmission in the mesoaccumbens dopamine system. In *The Mesolimbic Dopamine System: From Motivation to Action*, ed. P. Willner and J. Scheel-Kruger, pp. 61–103. New York: Wiley.

Wise, R. A., and Bozarth, M. A. (1984). Brain reward circuitry: Four elements "wired" in apparent series. *Brain Res. Bull.* 12: 203–208.

Young, S. T., Porrino, L. J., and Iadarola, M. J. (1991). Cocaine induces striatal Fos-immunoreactive proteins via dopaminergic D_1 receptors. *Proc. Natl. Acad. Sci (USA)* 88: 1291–1295.

Zhu, X. Z., and Luo, L-G. (1992). Effect of nitroprusside (nitric oxide) on endogenous dopamine release from rat striatal slices. *J. Neurochem.* 59: 932–935.

11

Studying neurotransmitter systems to understand the development and function of sex differences in the brain: the case of vasopressin

GEERT J. DE VRIES

Introduction

For neuroscientists, the study of sex differences in the brain promises at least two benefits. Investigations of their development can elucidate the processes that form brain structure during ontogeny that generates specific functions and behaviors, while investigations of the functional significance of these sex differences can reveal how brain morphology and function are related. Except for the fact that sex-related differences in the number of spinal motoneurons have been linked to sex-related differences in the number of specific muscle cells (Kelley 1988; Breedlove 1992), these benefits have been difficult to achieve, however. The complexity of the neuroanatomical connections to and from the brain regions where these differences are found and technical difficulties in manipulating specific sexually dimorphic elements in these areas have delayed the desired result.

This complexity, however, can be exploited. Given that all brain areas contain heterogeneous populations of cells and inputs, focusing on the neurotransmitter content of cells and inputs could reveal whether sexual differentiation selectively affects particular cell populations. This, in turn, could facilitate our understanding of the cellular processes underlying differentiation. Focusing on the neurotransmitter content may also help to reveal the anatomical connections of sexually dimorphic areas, and therefore to assess the impact of a particular dimorphism on other brain areas. Finally, knowing the neurotransmitter systems involved would allow specific manipulation of sexually dimorphic elements by applying specific agonists and antagonists (De Vries 1990). To illustrate some of these advantages, this chapter will discuss a particularly well characterized, sexually dimorphic system, the vasopressin projections of the bed nucleus of the stria terminalis (BNST) and the medial amygdaloid nucleus (MeA).

Sexually dimorphic vasopressin projections in the brain

Discovery of the sexually dimorphic vasopressin innervation of the brain

Both the BNST and MeA show some marked morphological sex differences. In rats, subregions of the BNST are bigger in males than in females (Del Abril et al. 1987; Hines et al. 1992) and contain more neurons in males than in females (Guillamon et al. 1988). Similar differences are found in the MeA (Mizukami et al. 1983; Hines et al. 1992). There are also differences in the number of synapses in the MeA (Nishizuka and Arai 1983) and in specific neurotransmitters other than vasopressin. For example, the posterodorsal area of the MeA and the encapsulated part of the BNST contain more cholecystokinin-immunoreactive cells and receive a denser substance P–immunoreactive innervation in males than in females (Malsbury and McKay 1987, 1989; Micevych et al. 1988).

Although vasopressin-immunoreactive (vasopressin-ir) cells in the BNST and MeA are present in the posterodorsal area of the MeA and the encapsulated area of the BNST, which are the subregions that show the most extreme sex-related differences in size (Hines et al. 1992), these cells are mostly found lateral and ventral to these areas (Caffé and Van Leeuwen 1983; Van Leeuwen and Caffé 1983). Therefore, they appear to contribute only partly to the sex differences in these areas. The sex differences in the projections of these cells were discovered serendipitously, before any other sex difference in a specific neurotransmitter system had been detected anatomically and, in fact, before it was known that these projections came from the BNST and MeA. While studying the development of what was then seen as the projections of the suprachiasmatic nucleus in rats, we stumbled on a large variation in the density of the vasopressin innervation of the lateral septum and lateral habenular nucleus in rats 12 days of age and older. A repeat experiment – with subjects separated according to sex – revealed that males have a vasopressin-ir fiber plexus from the twelfth postnatal day onward, while females show such a plexus only after the twenty-first postnatal day. In adulthood, males still have a denser vasopressin-ir fiber network than females (Fig.11.1; De Vries et al. 1981).

Vasopressin-ir projections in the brain have now been extensively traced. Vasopressin is synthesized by several cell groups, each projecting to distinct areas in the brain (Fig. 11.2). In addition to hypothalamic vasopressin-ir cells, there are several cell groups that are not readily stained for vasopressin unless animals are pretreated with the axonal transport blocker colchicine, most notably in the BNST and the MeA (Caffé and Van Leeuwen 1983; Van Leeuwen and Caffé 1983). The BNST and MeA cells project to several limbic structures, such as the lateral septum and lateral habenular nucleus, and to several midbrain structures, such as

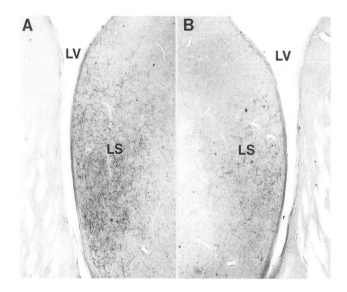

Figure 11.1. Vasopressin-ir fibers in the lateral septum (LS) of a male (A) and female.(B) rat; LV, lateral ventricle.

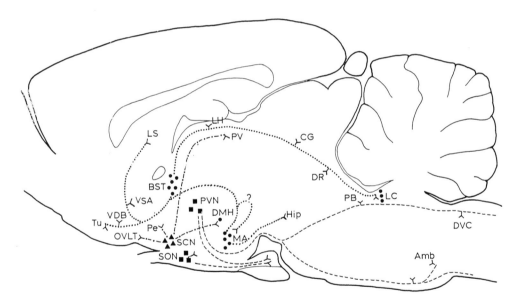

Figure 11.2. Scheme of the most prominent vasopressin-ir pathways. (·····) BNST and MeA projections to the lateral septum (LS), ventral septal area (VSA), perimeter of the diagonal band of Broca (VDB), olfactory tubercle (Tu), lateral habenular nucleus (LH), midbrain central gray (CG), dorsal raphe nucleus (DR), locus coeruleus (CR), and ventral hippocampus (Hip). (•—•) suprachiasmatic nucleus (SCN) projections to the perimeter of the organum vasculosum laminae terminalis (OVLT), the periventricular (Pe) and dorsomedial nucleus of the hypothalamus (DMH). (- - -) paraventricular nucleus (PVN) projections to the parabrachial nucleus (PB), dorsal vagal complex (DVC), and ambiguus nucleus (Amb); (- - -) and (- - -) PVN and supraoptic nucleus (SON) projections to the neurohypophysis. (Adapted from De Vries et al. 1985.)

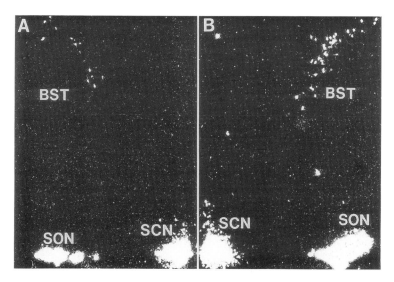

Figure 11.3. Dark-field-illuminated sections displaying cells labeled for vasopressin mRNA in the BNST of a female (A) and a male (B) rat treated with similar levels of testosterone. Labeled cells also show up in the supraoptic nucleus (SON) and the suprachiasmatic nucleus (SCN).

midbrain central gray, dorsal raphe nucleus, pontine peripeduncular nucleus, and the locus coeruleus (De Vries and Buijs 1983; Caffé et al. 1987, 1989).

In rats, all vasopressin-ir projections of the BNST and MeA appear to be denser in males than in females (De Vries and Al-Shamma 1990). Consistent with these differences, males have about two or three times more vasopressin-ir cells in the BNST than females (Van Leeuwen et al. 1985; De Vries and Al-Shamma 1990; Wang et al. 1993) and a similar difference in the number of cells that can be labeled for vasopressin mRNA (Fig. 11.3; Miller et al. 1989b; De Vries et al. 1994). Although a similar trend was found in the number of vasopressin-ir cells in the MeA in two studies (De Vries and Al-Shamma 1990; Wang et al. 1993), these differences were not significant, possibly because of the larger variation in the staining of MeA cells than in BNST cells. With *in situ* hybridization, nearly twice as many MeA cells labeled for vasopressin mRNA are present in males than in females (Szot and Dorsa 1993).

Such widespread sex-related differences in cell number and density of the projections have not been found in other vasopressin-ir projections, although some of these have sexually dimorphic portions. For example, in gerbils, the vasopressin-ir projections to the sexually dimorphic area of the preoptic/anterior hypothalamic area (SDA) are much denser in males than in females, but the projections to the periventricular nucleus of the hypothalamus do not differ (Crenshaw et al. 1992), even though both appear to come from the suprachiasmatic nucleus (Crenshaw and De Vries 1991). The widespread sex-related differences in the BNST and MeA projections might be related to the difference in the number of vasopressin-ir cells in the male and female BNST and MeA, whereas the sex differences in the presumed projections of the suprachiasmatic nucleus may be related to the presence or absence of sex differences in the areas they innervate; in this case, they may be related to the presence of a difference between the male and female SDA and the absence of such an obvious difference in the periventricular nucleus of the hypothalamus (Commins and Yahr 1984).

Gonadal hormone effects on vasopressin cells in the BNST and MeA

The vasopressin-ir projections of the BNST and MeA are extremely sensitive to steroid. After gonadectomy, BNST and MeA cells and their projections lose their vasopressin immunoreactivity and can no longer be labeled for vasopressin mRNA; treatment with gonadal steroids prevents these changes from occurring (Fig. 11.4; De Vries et al. 1984, 1985; Van Leeuwen et al. 1985; Miller et al. 1989a). In males, vasopressin immunoreactivity disappears gradually from BNST and MeA projections in about 2–3 months (De Vries et al. 1984); in mice, a similarly gradual decline was observed in females as well as males (Mayes et al. 1988). Biosynthesis of vasopressin declines much faster than vasopressin immunoreactivity. Only 1 day after castration, vasopressin mRNA levels in individual cells were significantly decreased, while the number of cells that could be labeled for vasopressin mRNA were reduced by 90% after 1 week (Miller et al. 1992). These findings suggest that although vasopressin biosynthesis is almost instantly reduced by gonadectomy, vasopressin remains present in projections for several weeks, possibly still capable of influencing brain function. The dramatic reduction of vasopressin biosynthesis, however, suggests that castration also dramatically reduces vasopressin release.

In this regard, there is indirect evidence that vasopressin release is reduced after castration. Following intracerebroventricular (icv) injection of vasopressin, rats showed motor disturbances following a second vasopressin injection given 2 days later (Poulin and Pittman 1991). Injections of hypertonic saline, which stimulates septal vasopressin release (Demotes-Mainard et al. 1986; Landgraf et al. 1988), sensitize rats just as well to the motor effects of vasopressin as does an icv vasopressin injection. In castrated rats, however, vasopressin injections can still sensitize rats to the motor effects of subsequent injections, though intraperitoneal injections of hypertonic saline cannot do so, suggesting that castration eliminates endogenous vasopressin release in the septal area (Poulin and Pittman 1991).

Although the sex-related differences in, and hormonal effects on, the vasopressin-ir projections of the BNST and MeA most likely influence vasopressin release, they do not seem to influence receptor density or sensitivity as do changes

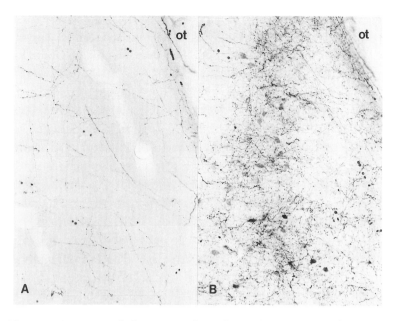

Figure 11.4. Vasopressin-immunoreactive cells and fibers in the medial amygdaloid nucleus of a castrated (A) and an intact male (B) rat. Note that after castration only a few nonbranching, thick-caliber fibers remain (ot, optic tract).

in other peptide systems (Catt et al. 1979). Long-term castration does not affect the distribution of vasopressin binding sites (Tribollet et al. 1988), nor does it affect the number and affinity of vasopressin receptors, vasopressin-stimulated phosphoinositol hydrolysis in septal tissue, or the ability of vasopressin to sensitize septal tissue to the motor effects of a subsequent vasopressin injection (Poulin and Pittman 1991).

Homologous sex differences in other vertebrates

Sexual dimorphism and hormone responsiveness appear to be conserved features of the vasopressin-ir projections of the BNST and MeA. They have been found not only in various rodent species [European (Buijs et al. 1986) and Djungarian hamsters (Bittman et al. 1991), Mongolian gerbils (Crenshaw et al. 1992), prairie and meadow voles (Bamshad et al. 1993), and mice (Mayes et al. 1988)] but also in other mammalian species [ferrets (G. De Vries and M. Baum, unpublished results)]. Similar dimorphism and hormonal responsiveness have been found in homologous, vasotocin-ir projections in nonmammalian vertebrates [amphibians: rough-skinned newts (Moore 1992) and bullfrogs (Boyd et al. 1992); reptiles: the lizards *Gekko gecko* (Stoll and Voorn 1985) and *Anolis carolinensis* (Propper et al. 1992), the turtle *Pseudemys scripta elegans*, and the snake *Python regius* (Smeets et al. 1990); birds: Japanese quails (Viglietti-Panzica et al. 1992) and canaries (Voorhuis et al. 1988)]. There are, however, animals with homologous vasotocin-ir or vasopressin-ir projections that are not notably sexually dimorphic [rainbow trouts (Van Den Dungen et al. 1982), the amphibians *Rana ridibunda, Xenopus-laevis,* and *Pleurodeles waltii* (Gonzales and Smeets 1992a,b), and guinea pigs (Dubois-Dauphin et al. 1989)]. Since these studies did not quantify immunostaining, it is not known whether sex differences are nonexistent or just subtle. The BNST of humans and the BNST and MeA of macaques have vaso-pressin-ir cells, and although the lateral septum of neither species showed dense vasopressin-ir projections, other areas that receive hormone-responsive vaso-pressin innervation in rats (i.e., the ventral tegmental area and midbrain central gray of macaques, and the locus coeruleus of humans and macaques) did show such projections, although without obvious sex differences (Fliers et al. 1986; Caffé et al. 1989). Syrian hamsters stand apart in that no trace of vasopressin-ir projections of the BNST and MeA can be found (Dubois-Dauphin et al. 1989; Albers et al. 1991). Ironically, the lateral septum of Syrian hamsters has vaso-pressin binding sites (Ferris et al. 1993) and is sensitive to behavioral effects of vasopressin (Irvin et al. 1990). These variations in vasopressin/vasotocin pro-jections may be exploited to reveal the processes underlying their sexual differen-tiation and their functional significance.

Sexual differentiation of the vasopressin projections of the BNST and MeA

Most research on the sexual differentiation of these projections has utilized rats. In these animals, gonadal hormones determine sexual differentiation of centrally regulated functions and behaviors presumably by influencing the development of specific neuronal systems during a restricted – often referred to as critical – pe-riod around birth (Gorski 1984; Yahr 1988). A first attempt to test whether go-nadal hormones also influence the sexual differentiation of the vasopressin-ir pro-jections indicated that gonadal hormones influence these projections not only in the first but also in the third postnatal week (De Vries et al. 1983), which is later than predicted from the critical periods of other sexually dimorphic neural sys-tems (Gorski 1984; Yahr 1988). In retrospect, this first study did not examine whether perinatal levels of gonadal hormones permanently influence the sexual differentiation of vasopressin-ir projections, since the subjects were killed within 4 weeks after birth. This is considerably less than the time required for vaso-pressin immunoreactivity to disappear completely from these fibers after gonadec-tomy in adulthood, which takes more than 8 weeks (De Vries et al. 1984). Conse-quently, it was impossible from these studies to distinguish between permanent and temporary effects of gonadal steroids.

We reconsidered whether perinatal levels of gonadal hormones permanently

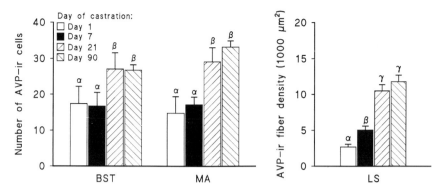

Figure 11.5. Effects of castration at 1, 7, 21, or 90 days after birth on the number (± SEM) of vasopressin-ir (AVP-ir) cells in the BNST and MeA and the vasopressin-ir fiber density in the lateral septum (LS). Greek letters indicate significant differences by ANOVA.

influence the sexual differentiation of vasopressin-ir projections, by comparing the effects of neonatal manipulations on the appearance of these projections in 3-month-old rats that had been treated with similar testosterone levels for 4 weeks before death. Male rats that were castrated at 3 months of age (control males) showed more vasopressin-ir cells in the BNST and a higher density of vaso-pressin-ir fibers in the lateral septum than neonatally castrated male rats, whose cell numbers and fiber density did not differ from those of female rats that were ovariectomized neonatally or at 3 months of age (control females). This suggested that testicular secretions after birth permanently influence the development of the vasopressin-ir projections of the BNST. A second experiment showed that male rats castrated on the day of birth or at postnatal day 7 had fewer vasopressin-ir cells in the BNST and MeA and a lower vasopressin-ir fiber density in the lateral septum than did either male rats castrated at postnatal day 21 or control males (Fig. 11.5), suggesting that testicular secretions influenced the differentiation of vasopressin-ir projections around day 7. A third experiment confirmed that tes-tosterone propionate treatment on the seventh postnatal day significantly in-creased vasopressin-ir fiber density in the lateral septum of neonatally gonadec-tomized male and female rats and fully restored the number of vasopressin-ir cells in the BNST of neonatally castrated males, but not of females (Wang et al. 1993).

Some discrepancies in the effects of hormonal manipulation on the vasopressin-ir cell number and fiber density may provide clues to the mechanism by which differentiation of this system occurs. There were, for example, no differences of vasopressin-ir cell number in the BNST and MeA of males castrated on postnatal day 1 or 7, or of control females. However, differences were found in the vaso-pressin-ir fiber density in the lateral septum, which was lower in neonatally cas-trated males than in males castrated on postnatal day 7 (Fig. 11.5). In addition, testosterone propionate injections on the seventh postnatal day increased the number of vasopressin-ir cells in the BNST to the level present in control males, whereas such treatment increased the vasopressin-ir fiber density in the lateral septum to a level intermediate between those of control males and females. Testosterone during development might influence the number of cells that produce vasopressin independently of the level at which individual vasopressin-ir cells can produce vasopressin. A similar discrepancy has been found in the sexually dimorphic nucleus of the bulbocavernosus (Lee et al. 1989). Testosterone determines the number of motoneurons in this nucleus in the last week of pregnancy and the first week after birth, while it determines the size of these motoneurons in adulthood during the first two weeks after birth. Recent studies of the effects of testosterone metabolites on vasopressin mRNA expression suggest that BNST and MeA cells display different sexually dimorphic features, which are determined by separate critical periods.

Cellular basis of sex-related differences in vasopressin expression

Testosterone influences vasopressin production by androgen as well as by estrogen receptor–mediated mechanisms. In castrated male rats, estradiol, which is metabolite of testosterone generated by aromatization (Naftolin et al. 1975), partially restores vasopressin immunostaining in castrated male rats, while dihydrotestosterone, which is a nonaromatizable, androgenic metabolite of testosterone generated by reduction (Lieberburg and McEwen 1975), does not by itself restore vasopressin immunostaining. However, if dihydrotestosterone is given in combination with estradiol, it enhances vasopressin immunostaining (De Vries et al. 1986). Since virtually all vasopressin-ir cells in the BNST and MeA in males are immunoreactive for estrogen as well as androgen receptors (Axelson and Van Leeuwen 1990; Zhou et al. 1994), androgens and estrogens may influence vasopressin production by directly acting on vasopressin ir cells.

Because there are no clearly recognizable androgen- and estrogen-responsive elements in the promoter region of the vasopressin gene (Young 1992; Adan and Burbach 1992), it is not clear whether androgens and estrogens can directly influence the transcription of vasopressin messenger RNA. It is even questionable whether they influence arginine vasopressin (AVP) mRNA transcription at all, since a nuclear run-on assay of BNST tissue did not show any effect of castration on AVP gene transcription. Since the same study showed that castration decreased, while testosterone increased, the length of the polyadenylate tail of AVP mRNA, which may enhance its stability, gonadal hormones might influence AVP mRNA levels at a post-transcriptional level (Carter and Murphy 1993).

Androgens and estrogens might also influence vasopressin production indirectly, each by influencing the effectiveness of the other. Estrogen might, for example, increase the responsiveness of individual BNST and MeA cells to dihydrotestosterone treatment by preventing the metabolic inactivation of dihydrotestosterone in the brain (cf. Södersten 1980). Estrogen might also increase the effectiveness of androgen receptors of vasopressin-producing cells by altering the duration of androgen receptor occupation, as has been observed in the preoptic area of male rats (Roselli and Fasasi 1992).

Sex-related differences in androgen levels cannot fully explain sex-related differences in vasopressin fiber staining and vasopressin mRNA levels observed in intact male and female rats (De Vries et al. 1981; Miller et al. 1989b). If male and female rats are treated with similar levels of testosterone, males show more vasopressin-ir and vasopressin mRNA–labeled cells in the BNST and MeA and denser vasopressin-ir fiber projections from the BNST and MeA than do females (De Vries and Al-Shamma 1990; Wang et al. 1993; De Vries et al. 1994). This sex difference, however, can be caused by differences in the metabolism of testosterone, since the BNST of male rats has a higher level of aromatase, the enzyme that catalyzes the aromatization of testosterone into estradiol, than the BNST of female rats (Roselli 1991). Testosterone could, therefore, stimulate BNST cells more effectively in male than in female rats.

Whether or not a sex-related difference in testosterone metabolism contributes to the sex-related differences in vasopressin cells, it cannot be the only factor, since there are still differences in the effects of estradiol and dihydrotestosterone on vasopressin mRNA expression in male and female gonadectomized rats. More BNST cells respond to estrogen stimulation in males than in females (Fig. 11.6). In addition, these BNST cells show more labeling per cell in males than females, suggesting that in addition to a difference in the number of cells that express vasopressin, there are also sex-related differences in estrogen responsiveness of individual vasopressin-producing cells. Sex-related differences in androgen responsiveness might contribute even more to the sex differences of vasopressin-producing cells in the BNST. Dihydrotestosterone, when given by itself, did not raise vasopressin mRNA levels over those of gonadectomized animals, but when given together with estradiol, it significantly increased the number of vasopressin mRNA labeled cells in males, but not in females (Fig. 11.6).

Since it is not known at which level androgens and estrogens act on vasopressin synthesis, it is not yet clear which factor contributes to the sex differences in androgen and estrogen responsiveness of vasopressin cells. One factor may be steroid receptor levels of individual cells. Although no sex-related differences of estrogen receptors in the BNST have been reported (Brown et al. 1992), the number of androgen receptors associated with the cell nuclear fraction in the BNST is higher in males than in females (Roselli 1991). Such a difference might

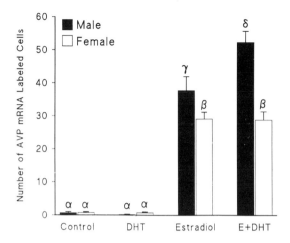

Figure 11.6. Differences in the number ($\pm$ SEM) of BNST cells labeled for vasopressin (AVP) mRNA in male and female rats that were gonadectomized (control) or gonadectomized and treated with dihydrotestosterone (DHT), estradiol, or a combination of estradiol and DHT (E + DHT). Greek letters indicate significant differences by ANOVA.

explain why dihydrotestosterone increases the number of vasopressin mRNA–labeled cells in males but not in females.

Ontogeny of the vasopressin-ir cells of the BNST and MeA

To understand which cellular features make the developing vasopressin-ir cells sensitive to the differentiating influences of gonadal hormones, it would be desirable to study the development of these cells. This is hampered, however, by the late onset of vasopressin mRNA expression in the BNST and MeA, which is detected at postnatal days 3 and 5 in males and at postnatal days 21 and 35 in females, respectively (Szot and Dorsa 1993). This makes it impossible to recognize vasopressin-ir cells during the period when gonadal steroids influence their sexual differentiation. However, these vasopressin cells can be distinguished from surrounding cells in the adult BNST using the thymidine analog bromo-2-deoxy-5-uridine (BrdU) as a "birth marker." Most vasopressin-ir cells appear to be born on embryonic days 12 and 13 (counting the day that sperm plugs were found as embryonic day 1), which places them in the earliest cohort of cells that will survive to form the adult forebrain (H. Al Shamma and G. De Vries, unpublished results). This was surprising, since the majority of the surrounding cells in the BNST and MeA are born on embryonic days 14–16 (Bayer 1980, 1987). This difference in cell birth may be exploited to study the sexual differentiation of the vasopressin-ir cells by following the development of cells labeled by BrdU injections on embryonic days 12 and 13.

Functional significance of sex differences in vasopressin systems

Sexual behavior

One can safely assume that the vasopressin-ir projections are involved in sexually dimorphic functions or at least in functions that are influenced by gonadal steroids, such as sexual behavior. Lesion studies have indeed implicated the BNST and MeA in male sexual behavior (Harris and Sachs 1975; Valcourt and Sachs 1979). Furthermore, the involvement of these vasopressin-ir projections in male sexual behavior fits with the gradual effects of castration on male sexual behavior (Davidson 1966) and with the similarity in the effects of testosterone metabolites on male sexual behavior and on the density of vasopressinergic innervation of the lateral septum (Baum and Vreeburg 1973; Larsson et al. 1973). Treating rats with a centrally acting vasopressin analog, in fact, reversed the decline of copulatory behavior following castration (Bohus 1977). However, injecting vasopressin directly into the lateral septum did not affect male sexual behavior (Koolhaas et al. 1991).

Lesion and stimulation studies have implicated areas innervated by the sexually dimorphic vasopressin fibers in female sexual behavior as well (Nance et al. 1974; Zasorin et al. 1975). In fact, intraventricular injections of vasopressin stimulate female sexual behavior, and similar injections of vasopressin antagonist inhibit it (Södersten et al. 1985). However, a direct link between the vasopressin innervation of the lateral septum and female sexual behavior has never been tested. See De Vries (1990) and De Vries et al. (1992) for more extensive reviews of a possible involvement of these fibers in sexual behavior. This review will concentrate on functions in which septal vasopressin has been directly implicated.

Nonreproductive functions

Vasopressin was one of the first peptides found to influence behavior and was, in fact, among the first to be called a neuropeptide (De Wied 1969). A great many papers have addressed the effects of vasopressin on centrally regulated functions such as learning and memory (De Wied 1969), cardiovascular regulation (Versteeg et al. 1983), thermoregulation (Kasting 1989), and motor behaviors (Kasting et al. 1980). Many of these studies involved injecting vasopressin or its analogs into the ventricles or into areas that do not necessarily receive vasopressin-ir fibers. Since there are several different vasopressin systems (Fig. 11.2), the results of such studies cannot always be related to any of these systems in particular. More recent studies, undertaken with the anatomy of the central vasopressin pathways in mind, have indicated a number of functions in which BNST and MeA projections could be involved. For example, the vasopressin innervation of the septal area has been implicated in thermoregulation, osmoregulation, social memory, motor disturbances, and aggressive behavior (see later). Competition studies using nonradioactive vasopressin analogs (Dorsa et al. 1988), electrophysiological studies (Raggenbass et al. 1988), *in situ* hybridization studies (Ostrowski et al. 1992), and functional studies suggest that vasopressin receptors in this area resemble vasopressor (V1) receptors and not the antidiuretic receptor or the vasopressin–oxytocin receptor that is found in the hippocampus (Audigier and Barberis 1985).

One of the first functions attributed to septal vasopressin was reduction of fever (Cooper et al. 1979). Since then, fever-reducing effects of vasopressin and fever-enhancing effects of $V1_a$ receptor antagonists injected into the ventral septal area have been demonstrated repeatedly for a variety of mammals including rats (Kasting 1989). The ventral septal area probably receives vasopressin-ir innervation from the BNST. This innervation is denser in males than in females (De Vries and Al-Shamma 1990) and disappears after castration (De Vries et al. 1985) as well as after lesioning of the BNST (De Vries and Buijs 1983). Similarly, electrical stimulation of the BNST alters the electrical activity of ventral septal neurons, as does exogenous application of vasopressin (Disturnal et al. 1985a,b). In addition, stimulation of the BNST attenuates pyrogen-induced fever, presumably by enhancing vasopressin release, since this effect can be prevented by administering a $V1_a$ receptor antagonist into the ventral septal area (Naylor et al. 1988). Castration, which apparently reduces vasopressin release in the lateral septum, lengthens pyrogen-induced fever (Pittman et al. 1988). Vasopressin treatment also reversed the effects of castration on the capacity of the pyrogen interleukin-1 to induce "sickness behavior" (increased sleepiness, lethargy, reduced social activities, reduced food intake). Using social investigation of juvenile conspecifics as an index of sickness behavior, castrated rats were observed to be more sensitive to the fever-inducing and behavioral effects of interleukin-1. Furthermore, vasopressin more effectively attenuated the effects of interleukin-1 in castrated than in intact rats, and, conversely, administration of a $V1_a$ receptor antagonist potentiated the effects of interleukin-1 in intact but not in castrated rats (Dantzer et al. 1991; Bluthe and Dantzer 1992a).

Although vasopressin release in the ventral septal area apparently reduces fever but does not affect body temperature in other ways (Kasting 1989), vasopressin release in the dorsal lateral septum may prevent hypothermia in European hamsters. In this species, the density of the vasopressin-ir fiber plexus in the lateral septum is high during the summer, when testosterone levels are elevated, and low during fall and winter, when testosterone levels are reduced and the animals hibernate (Buijs et al. 1986). Chronic infusions of vasopressin into the lateral septum prevented the bouts of hypothermia that are associated with hibernation

(Hermes et al. 1989). In addition, Silastic implants containing testosterone given at the beginning of the hibernation season kept the density of vasopressin-ir fibers in the lateral septum high and prevented hibernation, possibly by sustaining the release of vasopressin in the lateral septum.

Septal vasopressin could be involved in other homeostatic functions as well. Injections with hypertonic saline or hypovolemia in rats increased the release of vasopressin from the lateral septum (Demotes-Mainard et al. 1986; Landgraf et al. 1988), and dehydration reduced vasopressin immunoreactivity in the lateral septum (Epstein et al. 1983). Although the functional significance of these effects is unclear, they suggest that septal vasopressin may be involved in the regulation of body fluids, which is sexually dimorphic and influenced by gonadal steroids as well (Chow et al. 1992).

Septal vasopressin has been implicated in social behaviors that are influenced by gonadal steroids. Intermale aggression, for example, declines gradually after castration, as does male sexual behavior (DeBold and Miczek 1984). Injections of vasopressin into the lateral septum (or the MeA) reversed this decline (Koolhaas et al. 1990, 1991). Such studies also suggest that septal vasopressin enhances social recognition, defined as the ability of rats to recognize conspecifics they had previously investigated (Dantzer et al. 1988). However, social recognition is not impaired in long-term castrated rats and is even increased in female versus male rats. Social recognition of long-term castrated males and intact females appears to depend less, or not at all, on vasopressin, since it was not affected by $V1_a$ antagonists (Bluthe et al. 1990; Bluthe and Dantzer 1990).

Orchestration of steroid-responsive functions

Given the many actions of the sexually dimorphic vasopressin projections of the BNST and MeA [a list that is likely to grow, since regions innervated by these projections, such as the lateral septum, have multimodal inputs and equally diverse outputs (Jakab and Leranth 1991)], perhaps the most significant action of these projections is to change a set of functions in a coordinated fashion depending on the physiological condition. A good example of such coordinated changes are the physiological and behavioral transitions displayed by seasonal breeders, which also show dramatic changes in the density of the vasopressin-ir projections of the BNST and MeA (Buijs et al. 1986; Bittman et al. 1991).

Reproduction comprises another set of functions that in female rodents demand dramatic changes both in gonadal hormone secretions and in physiological processes such as the regulation of water and energy balance, and body temperature (Numan 1988). Vasopressin-ir projections of the BNST and MeA also exhibit reproduction-related changes in females. For example, vasopressin-ir fibers in the lateral septum are more intensely stained in pregnant than in nonpregnant guinea pigs (Merker et al. 1980). In addition, the lateral habenular nucleus contains higher levels of assayable vasopressin in lactating than in pregnant rats (Caldwell et al. 1987), and the release of vasopressin from the septum is markedly higher in pregnant and parturient, than in sexually naive, female rats (Landgraf et al. 1991). These results suggest that vasopressin-ir projections of the BNST and MeA are involved in behaviors or functions that change in pregnant and parturient rats, such as maternal behavior, which is altered by icv injections of vasopressin (Pedersen et al. 1982).

Parental behavior

To help differentiate between physiological and behavioral changes in which vasopressin-ir projections may be involved, we compared males and females that become parents. Specifically, we compared the vasopressin innervation of sexually naive with parental animals of two closely related species of voles that differ dramatically in their reproductive strategies: prairie voles (*Microtus ochrogaster*), a monogamous species in which fathers as well as mothers provide parental care, and meadow voles (*Microtus pennsylvanicus*), a promiscuous species in which only mothers provide parental care (Wilson 1982). In prairie vole males, the density of the vasopressin-ir innervation of the lateral septum and lateral habenular nucleus was lower in parental males than in sexually naive males. Since similar differences were not observed in meadow vole males, they might be related to the regulation of paternal behavior. Parental behavior did not affect the density of vasopressin-ir fibers in prairie vole females, which suggests that the involvement of vasopressin may be sexually dimorphic as well (Fig. 11.7, Bamshad et al. 1993). In a follow-up experiment designed to examine when the vasopressin-ir projections change, we showed that pairing male and female prairie voles dramatically influences the density of the vasopressin-ir projections in males but not in females. The density of these projections decreased after mating in males, then increased gradually during the gestation period only to decrease again after the birth of pups (Fig. 11.8; Bamshad et al. 1994). The initial reduction of vasopressin-ir fiber density in the lateral septum and lateral habenular nucleus could reflect an increase in vasopressin release, since it coincides with higher levels of vasopressin messenger RNA in the BNST and with higher levels of testosterone (Wang et al. 1994).

The changes in the pattern of vasopressin innervation of male prairie voles during the reproductive period might be related to changes in social behaviors as well. After mating, prairie voles form stable pair bonds and display increased aggression toward unfamiliar conspecifics and increased paternal responsiveness (Getz et al. 1981; Bamshad et al. 1994). Winslow et al. (1993) showed that icv injections of a $V1_a$ antagonist inhibit the mating-induced increase in aggression

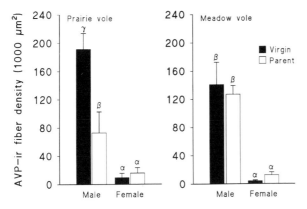

Figure 11.7. Vasopressin-ir fiber (AVP-ir) density ($\pm$ SEM) in the lateral septum of prairie and meadow voles that were either sexually naive or parental. Greek letters indicate significant differences by ANOVA.

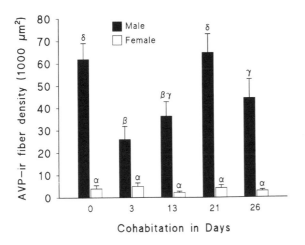

Figure 11.8. Vasopressin-ir (AVP-ir) fiber density ($\pm$ SEM) in the lateral septum before (0) and after 3, 13, 21, and 26 days of cohabitation in males and females. Greek letters indicate significant differences by ANOVA.

and pair bonding in males, suggesting that endogenous release of vasopressin stimulates the increase of aggression and pair bonding.

To test whether septal release of vasopressin also affects parental responsiveness, we injected saline, vasopressin, or the $V1_a$ receptor antagonist d(CH$_2$)$_5$-Tyr(Me) AVP into the lateral septum of sexually naive male prairie voles and recorded the four most prominent paternal activities displayed. Vasopressin stimulated grooming, crouching over, and contacting pups, while the $V1_a$ receptor antagonist blocked these behaviors (Fig. 11.9; Wang et al. 1994). We also observed that castrated voles spent less time grooming, contacting, and crouching over pups, and testosterone treatment reversed these changes (Fig. 11.10; Wang

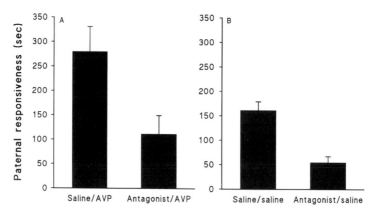

Figure 11.9. Time spent on paternal behavior during a 10-minute testing period by prairie vole males in which saline was injected into the septum followed by vasopressin (AVP) or in which a $V1_a$ antagonist was injected followed by vasopressin (t-test, $p \leq .05$), and by prairie voles injected twice with saline or with a $V1_a$ antagonist followed by saline (t-test, $p \leq .001$).

and De Vries 1993). Since castrated voles had a lower vasopressin-ir fiber density in the lateral septum than did the voles treated with endogenous or exogenous testosterone, castration may have reduced paternal responsiveness by changing the vasopressin innervation in the lateral septum. Although testosterone may also influence other systems involved in paternal activities, an involvement of the lateral septum in paternal behavior in prairie voles corresponds to the pattern of involvement of the septum in maternal behavior in rats and mice (Carlson and Thomas 1968; Fleischer and Slotnick 1978). In addition, the lateral habenular nucleus, wherein the vasopressin innervation exhibited effects of cohabitation similar to those observed in the lateral septum, might also be involved in paternal behavior, since lesioning of this area impairs the onset of maternal behavior in rats (Corodimas et al. 1983).

The role of the vasopressin-ir projections of the BNST and MeA in social behaviors could be related to vomeronasal influences on the brain. Pheromones play an important role in the changes of reproductive physiology induced by pairing male and female prairie voles (Dluzen et al. 1981; Lepri and Wysocki 1987). Although it is unknown whether pheromones play a similar role in male reproductive physiology, the aforementioned effects of septal vasopressin on social recognition in rats depend on an intact vomeronasal system. Lesioning of the vomeronasal system temporarily impaired social recognition, which was no longer impaired by injections of $V1_a$ antagonist after recovery, suggesting that the vasopressin-ir projections of the BNST and MeA relay information from the vomeronasal organ to systems involved in social behaviors in intact animals (Bluthe and Dantzer 1992b). This could also occur in voles.

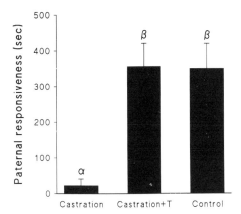

Figure 11.10. Time (±SEM) spent on paternal behavior during a 10-minute testing period by castrated voles, castrated voles treated with testosterone (T), and intact voles. Greek letters indicate significant differences by ANOVA.

Conclusion

Although the changes in the density of the vasopressin-ir projections of the BNST and MeA in paternal prairie vole males suggest that these fibers influence parental responsiveness, aggressive behavior, and pair bonding, it is puzzling why mating does not affect the density of these projections in females, which exhibit the same behavioral changes (Getz et al. 1981). The answer to this question may encourage us to reconsider the significance of sex-related differences in the brain. Sex-related differences in the medial preoptic area and the hypothalamus (e.g., in the size of cell clusters or the distribution of certain transmitters) are thought to be related to differences in the regulation of male and female sexual behavior and/or to the release of gonadotropic hormones (Kelley 1988; Yahr 1988; De Vries 1990; Breedlove 1992). Given the functions identified with the areas where such differences are found, these speculations make sense. They tend, however, to strengthen the notion that sex-related differences in the brain serve mainly to generate sex differences in physiology and behavior. Functional studies of the sexually dimorphic vasopressin-ir projections of the BNST and MeA suggest that the opposite may be equally true – that is, that differences in the brain may enable males and females to display certain hormone-sensitive behaviors in remarkably similar ways even though their hormonal and other physiological conditions differ dramatically.

Acknowledgment

This review was written while the author was funded by NIMH grant ROI MH47538.

References

Adan, R. A. H., and Burbach, J. P. H. (1992). Regulation of vasopressin and oxytocin gene expression by estrogen and thyroid hormone. *Prog. Brain Res.* 92: 127–136.

Albers, H. E., Rowland, C. M., and Ferris, C. F. (1991). Arginine–vasopressin immunoreactivity is not altered by photoperiod or gonadal hormones in the Syrian hamster (*Mesocricetus auratus*). *Brain Res.* 539: 137–142.

Audigier, S., and Barberis, C. (1985). Pharmacological characterization of two specific binding sites for neurohypophyseal hormones in hippocampal synaptic plasma membranes of the rat. *EMBO J.* 4: 1407–1412.

Axelson, J. F., and Van Leeuwen, F. W. (1990). Differential localization of estrogen receptors in various vasopressin synthesizing nuclei of the rat brain. *J. Neuroendocrinol.* 2: 209–216.

Bamshad, M., Novak, M. A., and De Vries, G. J. (1993). Sex and species differences in the vasopressin innervation of sexually naive and parental prairie voles, *Microtus ochrogaster* and meadow voles, *Microtus pennsylvanicus. J Neuroendocrinol.* 5: 247–255.

Bamshad, M., Novak, M. A., and De Vries, G. J. (1994). Cohabitation alters vasopressin innervation and paternal behavior in prairie voles, *Microtus ochrogaster.* *Physiol Behav.* 66: 751–758.

Baum, M. J., and Vreeburg, J. T. M. (1973). Copulation in castrated male rats following combined treatment with estradiol and dihydrotestosterone. *Science* 182: 293–285.

Bayer, S. A. (1980). Quantitative [³H]thymidine radiographic analysis of neurogenesis in the rat amygdala. *J. Comp. Neurol.* 194: 845–875.

Bayer, S. A. (1987). Neurogenetic and morphogenetic heterogeneity in the bed nucleus of the stria terminalis. *J. Comp. Neurol.* 265: 47–64.

Bittman, E. L., Bartness, T. J., Goldman, B. D., and De Vries, G. J. (1991). Suprachiasmatic and paraventricular control of photoperiodism in Siberian hamsters. *Am. J. Physiol.* 260: R90–R101.

Bluthe, R.-M., and Dantzer R. (1990). Social recognition does not involve vasopressinergic neurotransmission in female rats. *Brain Res.* 535: 301–304.

Bluthe, R.-M., and Dantzer, R. (1992a). Chronic intracerebral infusions of vasopressin and vasopressin antagonists modulate behavioral effects of interleukin-1 in rat. *Brain Res. Bull.* 29: 897–900.

Bluthe, R.-M., and Dantzer R. (1992b). Role of the vomeronasal system in the vasopressinergic modulation of social recognition in rats. *Brain Res.* 604: 205–210.

Bluthe, R.-M., Schoenen, J., and Dantzer, R. (1990). Androgen-dependent vasopressinergic neurons are involved in social recognition in rats. *Brain Res.* 519: 150–157.

Bohus, B. (1977). Postcastration masculine behavior in the rat: The role L of hypothalamo-hypophyseal peptides. *Exp. Brain Res.* 28: R8.

Boyd, S. K., Tyler, C. J., and De Vries G. J. (1992). Sexual dimorphism in the vasotocin system of the bullfrog (*Rana catesbeiana*). *J. Comp. Neurol.* 325: 313–325.

Breedlove, S. M. (1992). Sexual dimorphism in the vertebrate nervous system. *J. Neurosci.* 12: 4133–4142.

Brown, T. J., Naftolin, F., and MacLusky, N. J. (1992). Sex differences in estrogen receptor binding in the rat hypothalamus: Effects of subsaturating pulses of estradiol. *Brain Res.* 578: 129–134.

Buijs, R. M., Pevet, P., Masson-Pevet, M., Pool, C. W., De Vries, G. J., Canguilhem, B., and Vivien-Roels, B. (1986). Seasonal variation in vasopressin innervation in the brain of the European hamster (*Cricetus cricetus*). *Brain Res.* 371: 193–196.

Caffé, A. R., and Van Leeuwen, F. W. (1983). Vasopressin-immunoreactive cells in the dorsomedial hypothalamic region, medial amygdaloid nucleus and locus coeruleus of the rat. *Cell Tiss. Res*. 233: 23–33.

Caffé, A. R., Van Leeuwen, F. W., and Luiten, P. G. M. (1987). Vasopressin cells in the medial amygdala of the rat project to the lateral septum and ventral hippocampus. *J. Comp. Neurol*. 261: 237–252.

Caffé, A. R., Van Ryen, P. C., Van der Woude, T. P., and Van Leeuwen, F. W. (1989). Vasopressin and oxytocin systems in the brain and L upper spinal cord of *Macaca fascicularis*. *J. Comp. Neurol*. 287: 302–325.

Caldwell, J. D., Greer, E. R., Johnson, M. F., Prange, A. J., and Pedersen, C. A. (1987). Oxytocin and vasopressin immunoreactivity in hypothalamic and extra-hypothalamic sites in late pregnant and postpartum rats. *Neuroendocrinology* 46: 39–47.

Carlson, N. R., and Thomas, G. J. (1968). Maternal behavior of mice with limbic lesions. *J. Comp. Physiol. Psychol*. 66: 731–737.

Carter, D. A., and Murphy, D. (1993). Regulation of vasopressin (VP) gene expression in the bed nucleus of the stria terminalis: Gonadal steroid-dependent changes in VP mRNA accumulation are associated with alternations in mRNA poly (A) tail length but are independent of the rate of VP gene transcription. *J. Neuroendocrinol*. 5: 509–515.

Catt, K. J., Harwood, J. P., Aguilera, G., and Dufau, M. L. (1979). Hormonal regulation of peptide receptors and target cell responses. *Nature* 280: 109–116.

Chow, S. Y., Sakai, R. R., Witcher, J. A., Adler, N. T., and Epstein, A. N. (1992). Sex and sodium intake in the rat. *Behav. Neurosci*. 106: 172–180.

Commins, D., and Yahr, P. (1984). Adult testosterone levels influence the morphology of a sexually dimorphic area in the mongolian gerbil brain. *J. Comp. Neurol*. 224: 132–140.

Cooper, K. E., Kasting, N. W., Lederis, K., and Veale, W. L. (1979). Evidence supporting a role for endogenous vasopressin in natural suppression of fever in the sheep. *J. Physiol*. 295: 33–45.

Corodimas, K. P., Rosenblatt, J. S, Canfield, M. E., and Morrell, J. I. (1993). Neurons located in the lateral habenula mediate the hormonal onset of maternal behavior in rats. *Soc. Neurosci. Abstr*. 18: 889.

Crenshaw, B. L., and De Vries, G. J. (1991). Vasopressin innervation of the medial preoptic area in gerbils originates in the suprachiasmatic nucleus. *Soc. Neurosci. Abstr*. 17: 1231.

Crenshaw, B. L., De Vries, G. J., and Yahr, P. I. (1992). Vasopressin innervation of sexually dimorphic structures of the gerbil forebrain under various hormonal conditions. *J. Comp. Neurol*. 322: 589–598.

Dantzer, R., Bluthe, R.-M., and Kelley, K. W. (1991). Androgen-dependent vaso-pressinergic neurotransmission attenuates interleukin-1 induced sickness behavior. *Brain Res*. 557: 115–120.

Dantzer, R., Koob, G. F., Bluthe, R. M., and Le Moal, M. (1988). Septal vasopressin modulates social memory in male rats. *Brain Res*. 457: 143–147.

Davidson, J. M. (1966). Characteristics of sex behaviour in male rats following castration. *Anim. Behav*. 14: 266–272.

DeBold, J. E., and Miczek, K. A. (1984). Aggression persists after gonadectomy in female rats. *Horm. Behav*. 18: 177–190.

Del Abril, A. Segovia, S., and Guillamon, A. (1987). The bed nucleus of the stria terminalis in the rat: Regional sex differences controlled by gonadal steroids early after birth. *Dev. Brain Res*. 32: 295–300.

Demotes-Mainard, J., Chauveau, J., Rodriguez, R., Vincent, J. D., and Poulain, D. A., (1986). Septal release of vasopressin in response to osmotic, hypovolemic and electrical stimulation in rats. *Brain Res*. 381: 314.

De Vries, G. J. (1990). Sex differences in neurotransmitter systems. *J. Neuroendocrinol.* 2: 1–13.

De Vries, G. J., and Al-Shamma, H. S. (1990). Sex differences in hormonal responses of vasopressin pathways in the rat brain. *J. Neurobiol.* 21: 686–693.

De Vries, G. J., Best, W., and Sluiter, A. A. (1983). The influence of androgens on the development of a sex difference in the vasopressinergic innervation of the rat lateral septum. *Dev. Brain Res.* 8: 377–380.

De Vries, G. J., and Buijs, R. M. (1983). The origin of vasopressinergic and oxytocinergic innervation of the rat brain with special reference to the lateral septum. *Brain Res.* 273: 307–317.

De Vries, G. J., Buijs, R. M., and Sluiter, A. A. (1984). Gonadal hormone actions on the morphology of the vasopressinergic innervation of the adult rat brain. *Brain Res.* 298: 141–145.

De Vries, G. J., Buijs, R. M., and Swaab, D. F. (1981). Ontogeny of the vasopressinergic neurons of the suprachiasmatic nucleus and their extrahypothalamic projections in the rat brain: Presence of a sex difference in the lateral septum. *Brain Res.* 218: 67–78.

De Vries, G. J., Buijs, R. M., Van Leeuwen, F. W., Caffé, A. R., and Swaab, D. F. (1985). The vasopressinergic innervation of the brain in normal and castrated rats. *J. Comp. Neurol.* 233: 236–254.

De Vries, G. J., Crenshaw, B. D., and Al-Shamma, H. A. (1992). Gonadal steroid modulation of vasopressin pathways. In: *Oxytocin in Maternal, Sexual, and Social Behaviors*, N.Y. Acad. Sci. Vol. 652, ed. C. A. Pedersen, J. D. Caldwell, G. F. Jirikowski, and T. R. Insel, pp. 387–396. New York: New York Academy of Science.

De Vries, G. J., Duetz, W. Buijs, R. M., Van Heerikhuize, J., and Vreeburg, J. T. M. (1986). Effects of androgens and estrogens on the vasopressin and oxytocin innervation of the adult rat brain. *Brain Res.* 399: 296–302.

De Vries, G. J., Wang, Z. X., Bullock, N. A., and Numan, S. (1994). Sex differences in the effects of testosterone and its metabolites on vasopressin messenger RNA levels in the bed nucleus of the stria terminalis of rats. *J. Neurosci.* 14: 1789–1794.

De Wied, D. (1969). Effects of peptide hormones on behavior. In *Frontiers in Neuroendocrinology*, ed. W. F. Ganong and L. Martini, pp. 97–140. Oxford: Oxford University Press.

Disturnal, J. E., Veale, W. L., and Pittman, Q. J. (1985a). Electrophysiological analysis of potential arginine vasopressin projections to the ventral septal area of the rat. *Brain Res.* 342: 162–167.

Disturnal, J. E., Veale, W. L., and Pittman, Q. J. (1985b). Modulation by arginine vasopressin of glutamate excitation in the ventral septal area of the rat brain. *Can. J. Phsyiol. Pharmacol.* 65: 30–35.

Dluzen, D. E., Ramirez, V. D., Carter, C. S., and Getz, L. L. (1981). Male vole urine changes luteinizing hormone-releasing hormone and norepinephrine in female olfactory bulb. *Science* 212: 573–575.

Dorsa, D. M., Brot, M. D., Shewey, L. M., Meyers, K. M., Szot, P., and Miller, M. A. (1988). Interaction of a vasopressin antagonist with vasopressin receptors in the septum of the rat brain. *Synapse* 2: 205–211.

Dubois-Dauphin, M., Tribollet, E., and Dreifuss, J. J. (1989). Distribution of neurohypophysial peptides in the guinea pig brain: An immunocytochemical study of the vasopressin-related glycopeptide. *Brain Res.* 496: 45–65.

Epstein, Y. M., Caster, S. M., Glick, N., Sivan, N., and Ravid, R. (1983). Changes in hypothalamic and extrahypothalamic vasopressin content of water-deprived rats. *Cell Tiss. Res.* 233: 99–111.

Ferris, C. F., Delville, Y., Grzonka, Z., Luber-Narod, J., and Insel, T. R. (1993). An iodinated vasopressin (V_1) antagonist blocks flank marking and selectively labels neural binding sites in golden hamsters. *Physiol. Behav.* 54: 737–747.

Fleischer, S., and Slotnick, B. M. (1978). Disruption of maternal behavior in rats with lesions of the septal area. *Physiol. Behav.* 21: 189–200.

Fliers, E., Guldenaar, S. E. F., van der Wal, N., and Swaab, D. F. (1986). Extra-hypothalamic vasopressin and oxytocin in the human brain: Presence of vasopressin cells in the bed nucleus of the stria terminalis. *Brain Res.* 375: 363–367.

Getz, L. L., Carter, C. S., and Gavish, L. (1981). The mating system of prairie vole, *Microtus ochrogaster*: Field and laboratory evidence for pair-bonding. *Behav. Ecol. Sociobiol.* 8: 189–194.

Gonzalez, A., and Smeets, W. J. A. J. (1992a). Distribution of vasotocin- and meso-tocin-like immunoreactivities in the brain of South African clawed frog *Xenopus-laevis. J. Chem. Neuroanat.* 5: 465–479.

Gonzalez, A., and Smeets, W. J. A. J. (1992b). Comparative analysis of the vaso-tocinergic and mesotocinergic cells and fibers in the brain of two amphibians, the Anuran *Rana ridibunda* and the Urodele *Pleurodeles waltlii. J. Comp. Neurol.* 315: 53–73.

Gorski, R. A. (1984). Critical role for the medial preoptic area in the sexual differentiation of the brain. *Prog. Brain Res.* 61: 129–146.

Guillamon, A., Segovia, S., and Del Abril, A. (1988). Early effects of gonadal steroids on the neuron number in the medial posterior and the lateral divisions of the bed nucleus of the stria terminalis in the rat. *Dev. Brain Res.* 44: 281–290.

Harris, V. H., and Sachs, B. D. (1975). Copulatory behavior in male rats following amygdaloid lesions. *Brain Res.* 86: 514–518.

Hermes, M. L. H. J., Buijs, R. M., Masson-Pevet, M., Van der Woude, T. P., and Pevet, P. (1989). Central vasopressin infusion prevents hibernation in the European hamster (*Cricetus cricetus*). *Proc. Natl. Acad. Sci. (USA)* 86: 6408–6411.

Hines, M., Allen, L. S., and Gorski, R. A. (1992). Sex differences in subregions of the medial nucleus of the amygdala and the bed nucleus of the stria terminalis of the rat. *Brain Res.* 579: 321–326.

Irvin, R. W., Szot, P., Dorsa, D. M., Potegal, M., and Ferris, C. F. (1990). Vasopressin in the septal area of the golden hamster controls scent marking and grooming. *Physiol. Behav.* 48: 693–699.

Jakab, R. L., and Leranth, C. (1991). Convergent vasopressinergic and hippocampal input onto somatospiny neurons of the rat lateral septal area. *Neuroscience* 40: 413–421.

Kasting, N. W. (1989). Criteria for establishing a physiological role for brain peptides a case in point: The role of vasopressin in thermoregulation during fever and anti-pyresis. *Brain Res. Rev.* 14: 143–153.

Kasting, N. W., Veale, W. L., and Cooper, K. E. (1980). Convulsive and hypothermic effects of vasopressin in the brain of the rat. *Can. J. Physiol. Pharmacol.* 58: 316–319.

Kelley, D. B. (1988). Sexually dimorphic behavior. *Annu. Rev Neurosci.* 11: 225–251.

Koolhaas, J. M., Moor, E., Hiemstra, Y., and Bohus, B. (1991). The testosterone-dependent vasopressinergic neurons in the medial amygdala and lateral septum: Involvement in social behaviour of male rats. In *Vasopressin*, Colloque INSERM, Vol. 208, ed. S. Jard and R. Jamison, pp. 213–219. Paris: John Libbey Eurotext Ltd.

Landgraf, R., Neumann, I., and Pittman, Q. J. (1991). Septal and hippocampal release of vasopressin and oxytocin during late pregnancy and parturition in the rat. *Neuroendocrinology* 54: 378–383.

Landgraf, R., Neumann, I., and Schwarzberg, H. (1988). Central and peripheral release of vasopressin and oxytocin in the conscious rat after osmotic stimulation. *Brain Res.* 457: 219–225.

Larsson, K., Södersten, P., and Beyer, C. (1973). Sexual behavior in male rats treated with estrogen in combination with dihydrotestosterone. *Horm. Behav.* 4: 289–299.

Lee, J. H., Jordan, C. L., and Arnold, A. P. (1989). Critical period for androgenic regulation of soma size of sexually dimorphic motoneurons in rat lumbar spinal cord. *Neurosci. Lett.* 98: 79–84.

Lepri, J. J., and Wysocki, C. J. (1987). Removal of the vomeronasal organ disrupts the activation of reproductive function in female voles. *Physiol. Behav.* 40: 349–355.

Lieberburg, I., and McEwen, B. S. (1975). Brain cell nuclear retention of testosterone metabolites 5α-dihydrotestosterone and estradiol-17β in adult rat. *Endocrinology* 100: 588–597.

Malsbury, C. W., and McKay, K. (1987). A sex difference in the pattern of substance P-like immunoreactivity in the bed nucleus of the stria terminalis. *Brain Res.* 420: 365–370.

Malsbury, C. W., and McKay, K. (1989). Sex difference in the substance P-immunoreactive innervation of the medial nucleus of the amygdala. *Brain Res. Bull.* 23: 561–567.

Mayes, C. R., Watts, A. G., McQueen, J. K., Fink, G., and Charlton, H. M. (1988). Gonadal steroids influence neurophysin II distribution in the forebrain of normal and mutant mice. *Neuroscience* 25: 1013–1022.

Merker, G., Blähser, S., and Zeisberger, E. (1980). Reactivity pattern of vasopressin-containing neurons and its relation to the antipyretic reaction in the pregnant guinea pig. *Cell Tiss. Res.* 212: 47–61.

Micevych, P. E., Akesson, T., and Elde, R. (1988). Distribution of cholecystokinin-immunoreactive cell bodies in the male and female rat: II. Bed nucleus of the stria terminalis and amygdala. *J. Comp. Neurol.* 269: 381–391.

Miller, M. A., De Vries, G. J., Al-Shamma, H. A., and Dorsa, D. M. (1992). Rate of decline of vasopressin immunoreactivity and messenger RNA levels in the bed nucleus of the stria terminalis. *J. Neurosci.* 12: 2881–2887.

Miller, M. A., Urban, J. A., and Dorsa, D. M. (1989a). Steroid dependency of vasopressin neurons in the bed nucleus of the stria terminalis by in situ hybridization. *Endocrinology* 125: 2335–2340.

Miller, M. A., Vician, L., Clifton, D. K., and Dorsa, D. M. (1989b). Sex differences in vasopressin neurons in the bed nucleus of the stria terminalis by in situ hybridization. *Peptides* 10: 615–619.

Mizukami, S., Nishizyka, M., and Arai, Y. (1983). Sexual difference in nuclear volume and its ontogeny in the rat amygdala. *Exp. Neurol.* 79: 569–575.

Moore, F. L. (1992). Evolutionary precedents for behavioral actions of oxytocin and vasopressin. In *Oxytocin in Maternal, Sexual, and Social Behaviors*, N.Y. Acad. Sci. Vol. 652, ed. C. A. Pedersen, J. D. Caldwell, G. F. Jirikowski, and T. R. Insel, pp. 156–165. New York: New York Academy of Science.

Naftolin, F., Ryan, K. J., Davies, I. J., Reddy, V. V., Flores, F., Petro, D., Kuhn, M., White, R. J., Takaota, Y., and Wolin, L. (1975). The formation of estrogens by central neuroendocrine tissue. *Rec. Prog. Horm. Res.* 31: 295–316.

Nance, D. M., Shryne, J., and Gorski, R. A. (1974). Septal lesions: Effects on lordosis behavior and pattern of gonadotropin release. *Horm. Behav.* 5: 73–81.

Naylor, A. M., Pittman Q. J., and Veale W. L (1988). Stimulation of vasopressin release in the ventral septum of the rat brain suppresses prostaglandin E_1 fever. *J. Physiol. Lond.* 399: 177–189.

Nishizuka, M., and Arai, Y. (1983). Regional difference in sexually dimorphic synaptic organization of the medial amygdala. *Exp. Brain Res.* 49: 462–465.

Numan, M. (1988). Maternal behavior. In *The Physiology of Reproduction*, ed. E. Knobil and J. Neill, pp. 1569–1645. New York: Raven Press.

Ostrowski, N. L., Lolait, S. J., Bradley, D. J., O'Carroll, A.-M., Brownstein, M. J., and Young, W. S., III. (1992). Distribution of V1a and V2 vasopressin receptor messenger ribonucleic acids in rat liver, kidney, pituitary and brain. *Endocrinology* 131: 533–535.

Pedersen, C. A., Asche, J. A., Monroe, Y. L., and Prange Jr., A. J. (1982). Oxytocin induces maternal behavior in virgin female rats. *Science* 216: 648–649.

Pittman, Q. J., Malkinson, T. J., Kasting, N. W., and Veale, W. L. (1988). Enhanced fever following castration: Possible involvement of brain arginine vasopressin. *Am. J. Physiol.* 254: R513–R517.

Poulin, P., and Pittman, Q. J. (1991). Septal arginine vasopressin (AVP) receptor regulation in rats depleted of septal AVP following long-term castration. *J. Neurosci.* 11: 1531–1539.

Propper, C. R., Jones, R. E., and Lopez, K. H. (1992). Distribution of arginine vasotocin in the brain of the lizard *Anolis carolinensis*. *Cell Tiss. Res.* 267: 391–398.

Raggenbass, M., Tribollet, E., and Dreiffuss, J. J. (1988). Electrophysiological and autoradiographical evidence of V1 vasopressin receptors in the lateral septum of the rat brain. *Proc. Natl. Acad. Sci. (USA)* 84: 7778–7783.

Roselli, C. E. (1991). Sex differences in androgen receptors and aromatase activity in microdissected regions of the rat brain. *Endocrinology* 128: 1310–1316.

Roselli, C. E., and Fasasi, T. A. (1992). Estradiol increases the duration of nuclear androgen receptor occupation in the preoptic area of the male rat treated with dihydrotestosterone. *J. Steroid Biochem. Mol. Biol.* 42: 161–168.

Smeets, W. J. A. J., Sevensma, J. J., and Jonker, A. J. (1990). Comparative analysis of vasotocin-like immunoreactivity in brain of the turtle *Pseudemys scripta elegans* and the snake *Python regius*. *Brain Behav. Evol.* 35: 65–84.

Södersten, P. (1980). A way in which oestradiol might play a role in the sexual behaviour of male rats. *Horm. Behav.* 14: 271–274.

Södersten, P., De Vries, G. J., Buijs, R. M., and Melin, P. (1985). A daily rhythm in behavioural vasopressin sensitivity and brain vasopressin concentrations. *Neurosci. Lett.* 58: 37–41.

Stoll, C. J., and Voorn P. (1985). The distribution of hypothalamic and extrahypothalamic vasotocinergic cells and fibers in the brain of a lizard, *Gekko gecko*: Presence of a sex difference. *J. Comp. Neurol.* 239: 193–204.

Szot, P., and Dorsa, D. M. (1993). Differential timing and sexual dimorphism in the expression of the vasopressin gene in the developing rat brain. *Dev. Brain Res.* 73: 177–183.

Tribollet, E., Audigier, S., Dubois-Dauphin, M., and Dreiffus, J. J. (1988). Gonadal steroids regulate oxytocin receptors but not vasopressin receptors in the brain of male and female rats: An autoradiographic study. *Brain Res.* 511: 129–140

Valcourt, R. J., and Sachs, B. D. (1979). Penile reflexes and copulatory behavior in male rats following lesions in the bed nucleus of the stria terminalis. *Brain Res. Bull.* 4: 131–133.

Van Den Dungen, H. M., Buijs, M., Pool, C. W., and Terlou, M. (1982). The distribution of vastocin and isotocin in the brain of the rainbow trout. *J. Comp. Neurol.* 212: 146–157.

Van Leeuwen, F. W., and Caffé, R (1983). Vasopressin-immunoreactive cell bodies in the bed nucleus of the stria terminalis of the rat. *Cell Tiss. Res.* 228: 525–534.

Van Leeuwen, F. W., Caffé, A. R., and De Vries, G. J. (1985). Vasopressin cells in the bed nucleus of the stria terminalis of the rat: Sex differences and the influence of androgens. *Brain Res.* 325: 391–394.

Versteeg, C. A. M., Cransberg, K., de Jong, W., and Bohus, B. (1983). Reduction of a centrally induced pressor response by neurohypophyseal peptides: The involvement of lower brainstem mechanisms. *Eur. J. Pharmacol.* 94: 133–140.

Viglietti-Panzica, C., Anselmetti, G. C., Balthazart, J., Aste, N., and Panzica, G. C. (1992). Vasotocinergic innervation of the septal region in Japanese quail: Sexual differences and the influence of testosterone. *Cell Tiss. Res.* 267: 261–265.

Voorhuis, T. A. M., Kiss, J. Z., De Kloet, E. R., and De Wied, D. (1988). Testosterone-sensitive vasotocin-immunoreactive cells and fibers in the canary brain. *Brain Res.* 442: 139–146.

Wang, Z. X., Bullock, N. A., and De Vries, G. J. (1993). Sexual differentiation of vasopressin projections of the bed nucleus of the stria terminalis and medial amygdaloid nucleus in rats. *Endocrinology* 132: 2299–2306.

Wang, Z. X., and De Vries, G. J. (1993). Testosterone effects on paternal behavior and vasopressin immunoreactive projections in prairie voles (*Microtus ochrogaster*). *Brain Res.* 631: 156–160.

Wang, Z. X., Ferris, C. F., and De Vries, G. J. (1994). The role of septal vasopressin innervation in paternal behavior in prairie voles (*Microtus ochrogaster*). *Proc. Natl. Acad. Sci. (USA)* 91: 400–404.

Wang, Z. X., Smith, W., Major, D. E., and De Vries, G. J. (1994). Sex and species differences in the effect of cohabitation on vasopressin messenger RNA expression in the bed nucleus of the stria terminalis in prairie voles (*Microtus ochrogaster*) and meadow voles (*Microtus pennsylvanicus*). *Brain Res.* 650: 212–218.

Wilson, S. C. (1982). Parent–young contact in prairie and meadow voles. *J. Mammal.* 63: 300–305.

Winslow, J. T., Hastings, N., Carter, C. S., Harbaugh, C. R., and Insel, T. R. (1993). A role for central vasopressin in pair bonding in monogamous prairie voles. *Nature* 365: 545–547.

Yahr, P. (1988). Sexual differentiation of behavior in the context of developmental psychobiology. In *Handbook of Behavioral Neurobiology*, Vol. 9, ed. E. M. Blass, pp. 197–243. New York: Plenum Press.

Young, W. S., III (1992). Expression of the oxytocin and vasopressin genes. *J. Neuroendocrinol.* 4: 527–540.

Zasorin, N. L., Malsbury, C. W., and Pfaff, D. W. (1975). Suppression of lordosis in the hormone-primed female hamster by electrical stimulation of the septal area. *Physiol. Behav.* 14: 595–599.

Zhou, L., Blaustein, J. D., and De Vries, G. J. (1994). Distribution of androgen receptor immunoreactivity in vasopressin-immunoreactive and oxytocin-immunoreactive neurons in the male rat brain. *Endocrinology* 134: 2622–2627.

Part III

Cellular and molecular mechanisms regulated by sex steroids

12

Neurosteroids and neuroactive steroids

PAUL ROBEL AND ETIENNE-EMILE BAULIEU

Introduction

The relationships between steroid hormones and brain function have been envisioned mostly within the framework of endocrine mechanisms, as responses elicited by secretory products of steroidogenic endocrine glands, borne by the bloodstream, and exerting actions on the brain.

In fact, the brain is a target organ for steroid hormones. Intracellular receptors involved in the regulation of specific gene transcription have been identified in neuroendocrine structures, with each class of receptor having a unique distribution pattern in the complex anatomy of the brain (Fuxe et al. 1981; McEwen 1991a). Mechanisms involving nuclear receptors account for most steroid-induced feedback and many behavioral effects, for the regulation of the synthesis of several neurotransmitters, hormone-metabolizing enzymes, and hormone and neuromediator receptors, and for the organizational effects on neural circuitry that occur during development and persist into adulthood.

Local target tissue metabolism is an important factor in the mechanism of action of sex steroid hormones. Not only might such metabolism be involved in the regulation of intracellular hormone levels, but it might also provide an essential contribution to the cellular response. The brain is a site of extensive steroid metabolism. Aromatization and 5α-reduction represent major routes of androgen metabolism (Naftolin et al. 1975; Celotti et al. 1979; McLusky et al. 1984). The importance of these two pathways lies in the fact that they give rise to metabolites with considerable biological activity and thus are involved in the mechanism by which circulating androgens influence neuroendocrine function and behavior. However, the metabolites thus formed (estradiol, dihydrotestosterone) still operate as ligands of steroid hormone receptors.

Progesterone (PROG) is also converted to several metabolites in the brain, particularly 5α-dihydroprogesterone (5α-DH PROG), which binds with significant affinity to PROG receptors and exerts progesterone-like effects on neuroendocrine functions such as gonadotropin regulation and sexual behavior (Cheng and Karavolas 1975), and 3α-hydroxy-5α-pregnan-20-one (allopregnanolone, 3α,5α-TH PROG), an extremely potent sedative-hypnotic agent (Selye 1941).

The characterization of pregnenolone (PREG) and dehydroepiandrosterone (DHEA) in the rat brain, as nonconjugated steroids and their sulfate (S) and fatty acid (L) esters, which are present at higher concentrations in brain than in blood, has led to a reconsideration of steroid–brain interrelationships (reviewed by Robel et al. 1991).

The accumulation of DHEA, PREG, and their conjugates in brain appears to be independent of adrenal and gonadal production, as shown by the persistence of these steroids in the brain up to 1 month after gland ablation or pharmacological suppression. This contrasts with testosterone and corticosterone, which readily disappear after removal of the corresponding endocrine glands. This observation led to the discovery of a steroid biosynthetic pathway in the central nervous system (CNS).

The term *neurosteroids* was applied in 1981 to those steroids synthesized in the brain either *de novo* from cholesterol or by *in situ* metabolism of bloodborne precursors (Baulieu 1981). These steroids had already been chemically characterized in other tissues, but what was peculiar was their site of synthesis, and hence their eventual involvement in autocrine and paracrine processes that might ultimately regulate brain functioning. This chapter provides a current view of the biosynthesis of neurosteroids in the brain, as well as their possible mechanisms of action and physiological roles. The peripheral nervous system has not been included, although PREG is also synthesized and metabolized in Schwann cells and may produce active metabolites (Morfin et al. 1992; Akwa et al. 1993b). The endocrine effects of peripheral steroid hormones on neuronal activity have been extensively reviewed (McEwen 1991a) and remain outside the scope of this presentation.

A brief history

Several years elapsed between the discovery of DHEA and PREG accumulation in the brain of the rat, as well as in several other mammalian species, including humans, and the conclusive demonstration of *de novo* steroid biosynthesis. The first step in steroid synthesis is the conversion of cholesterol (CHOL) to PREG. Cytochrome P-450scc (scc denotes side-chain cleavage) is found in the mitochondria of steroidogenic endocrine cells as part of a ternary complex with adrenodoxin reductase and adrenodoxin (the cholesterol desmolase complex). Bovine and rat P-450scc enzymes have been purified, and specific antisera have been generated. The corresponding immunoglobulins were employed using an immunohistochemical technique for the detection of cytochrome P-450scc in rat and human brain (Le Goascogne et al. 1987, 1989; Iwahashi et al. 1990). Specific immunohistochemical staining was detected in white matter throughout the brain. The results fulfilled all the criteria of specificity, and they were consistent with the detection of an antigen with the expected molecular size of P-450scc on Western blots (Warner et al. 1989b). It was nevertheless necessary to demonstrate side-chain cleavage activity. Mitochondria from oligodendrocytes were shown to contain this enzymatic activity, converting [^{3}H]CHOL to [^{3}H]PREG. Since oligodendrocytes are the glial cells that synthesize myelin and produce white matter, these results confirmed the immunological localization of P-450scc. The presence of PREG throughout the brain can be explained by the generalized distribution of oligodendrocytes in the CNS (Hu et al. 1987). Mitochondrial PREG formation also occurs in the C6-2B glioma cell line (Papadopoulos et al. 1992). This combination of immunohistochemical and biochemical evidence led to the conclusion that brain cells can perform steroid biosynthesis from CHOL, thus justifying the term "neurosteroids" adopted in numerous publications.

The biosynthesis of PREG from sterol precursors was confirmed by the incubation of newborn rat glial cell cultures in the presence of [^{3}H]mevalonate (MVA). These cells undergo a process of differentiation in culture resembling that *in vivo*, which can be followed by the measurement of 2′,3′-cyclic nucleotide 3′-phosphodiesterase (CNPase) activity. After 15 days in culture, CNPase activity and the biosynthesis of PREG increased in parallel, reaching their highest level on day 21 (Jung-Testas et al. 1989). At that point, the cells labeled by antibodies directed against galactocerebroside, another marker of oligodendrocyte differentiation, were also positive for P-450scc. Cells incubated with [^{3}H]MVA in the presence of aminoglutethimide (AG), a potent inhibitor of P-450scc, accumulated [^{3}H]CHOL in their cytoplasm. Upon release of AG blockade, [^{3}H]CHOL was readily converted to [^{3}H]PREG (Hu et al. 1989), thus confirming the role of P-450scc in PREG biosynthesis.

Neurosteroid metabolism in the brain

Both PREG and DHEA are found in part as their sulfate esters (PREGS and DHEAS, respectively), and the concentration of DHEAS exceeds that of DHEA (Fig. 12.1). The conversion in the brain of Δ5-3β-hydroxy steroids to their S esters is very likely, but is not documented, although preliminary evidence has been obtained for a small amount of sulfotransferase activity (K. Rajkowski, unpublished results). In contrast, the major conjugation forms of PREG and DHEA are their L esters (PREGL and DHEAL, respectively). The acyltransferase responsible for their formation is enriched in the microsomal fraction (Vourc'h et al. 1992). Its activity is highest at the time of myelin formation. It is very likely that several isoforms exist, since CHOL- and corticosterone-esterifying activities,

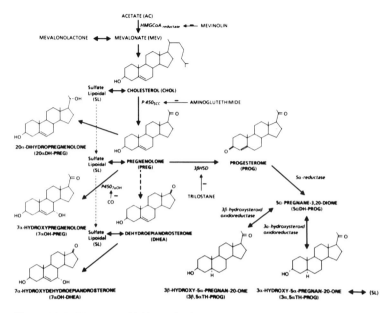

Figure 12.1. Neurosteroid biosynthesis and metabolism in the rat brain. Dashed arrows indicate metabolic conversions not yet formally demonstrated.

distinct from Δ5-3β-hydroxysteroid acyltransferase, have been reported in the rat brain. [³H]PREG can be converted to [³H]PROG, and DHEA to androst-4-ene-3,17-dione by a Δ5-3β-hydroxysteroid dehydrogenase isomerase (3β-HSD) enzyme that is inhibited by specific steroid inhibitors such as trilostane. Three isoforms of 3β-HSD are known in the rat; types I and II are both expressed in the adrenals, gonads, and adipose tissue, whereas type III is specific to the liver. The brain isoform(s) has not yet been isolated (Pelletier et al. 1992).

The corresponding 20α-dihydro derivatives can be formed from either PREG or PROG, although PROG is converted mainly to 5α-reduced metabolites. Two isoforms of 5α-reductase, the enzyme involved in this conversion, have been cloned. mRNA encoding the type 2 isozyme is more abundant than type 1 mRNA in most male reproductive tissues, whereas the type 1 isozyme predominates in peripheral tissues, including the brain (Normington and Russell 1992). The 5α-reduced metabolite of PROG, 5α-DH PROG, is in turn converted to 3α- and 3β-hydroxy-5α-pregnan-20-ones. The corresponding hydroxysteroid oxidoreductases in the brain have not yet been cloned.

Finally, both PREG and DHEA give rise to large amounts of polar metabolites, the respective 7α-hydroxylated compounds (Warner et al. 1989b; Akwa et al. 1992) of unknown biological significance. These pathways of neurosteroid metabolism are illustrated in Figure 12.1.

Table 12.1. *Enzymes of steroid metabolism in rat brain cells*

Enzyme	Substrate → product	Newborn mixed glial[a]	Fetus	
			Astrocytes	Neurons
P-450scc	CHOL → PREG	+	nd	nd
P-450$_{17\alpha}$	PREG → DHEA	nd	nd	nd
	PROG → ADIONE			
20α-HOR	PREG → 20α-DH PREG	+	+	+
	PROG → 20α-DH PROG			
P-450$_{7\alpha}$	PREG → 7α-OH PREG	+	+	nd
	DHEA → 7α-OH DHEA			
3β-HSD	PREG → PROG	+	+	(+)
	DHEA → ADIONE			
5α-Reductase	PROG → DH PROG	+	+	+
3ξ-HOR	DH PROG → TH PROG	3α >> 3β	3β > 3α	3β >> 3α

Abbreviations: P-450, cytochrome P-450; 20α-HOR, 20α-hydroxysteroid oxidoreductase; 3β-HSD, Δ^5-3β-hydroxysteroid dehydrogenase-isomerase; 3ξ-OR, 3α- or 3β-hydroxysteroid oxidoreductase; PREG, pregnenolone; PROG, progesterone; DHEA, dehydroepiandrosterone; ADIONE, androst-4-ene-3,17-dione; DH PROG, 5α-pregnane-3,20-dione; TH PROG, 3α- or 3β-hydroxy-5α-pregnan-20-one; nd, not detected; (+), detected at low cell density.
[a]Consist predominantly of oligodendrocytes.

Localization of steroid-metabolizing enzymes in rat brain cells

The CHOL–desmolase complex appears localized almost exclusively in oligodendrocytes (Table 12.1). The P-450$_{17\alpha}$ enzyme, a 17α-hydroxylase with 17,20-desmolase activity, which is responsible for the conversion of PREG to DHEA, has not yet been localized in the brain, so the origin of DHEA, DHEAS, and DHEAL accumulated there remains unexplained. P-450scc activity is notable in all types of glial cells, but negligible in neurons. Low 3β-HSD activity is present in glial cells and neurons.

The lack of key enzymes indicates that neurons are unable to convert CHOL to PROG, whereas 5α-reductase and 3α-hydroxysteroid oxidoreductase activities are present in both glial cells and neurons (Kabbadj et al. 1993). While glial cells from newborn rats mainly reduce 5α-DH PROG in the 3α-position, neurons and fetal astrocytes produce more of the 3β-hydroxy isomer.

Regulation of neurosteroid metabolism

Little is known about the mechanisms regulating neurosteroid formation. Both cAMP and glucocorticosteroids have been shown to enhance PREG formation from [³H]MVA in mixed glial cell cultures, but these effects might be related to an acceleration of cell differentiation *in vitro* (Jung-Testas et al. 1989). CHOL side-chain cleavage activity is regulated in the C6 glioma cell line by a mitochondrial benzodiazepine receptor, which increases intramitochondrial cholesterol transport, thereby increasing the substrate availability to P-450scc, as previously described in the adrenal glands and gonads. It would seem likely that steroidogenesis in the brain is under some type of control by second messengers like those in classic steroidogenic organs, but the trophic factors, their sources, and the physiological stimuli are not known (Guarneri et al. 1992).

The metabolism of PREG and DHEA by astroglial cells (and probably also by neurons) is regulated by cell density: 3β-HSD activity is strongly inhibited at high cell density (Akwa et al. 1993a, unpublished observations).

Mechanisms of neurosteroid action: genomic effects

Considering that PROG can be synthesized by glial cells, the demonstration of an estrogen-inducible PROG receptor in cultured oligodendrocytes of male and female rats suggests a classical intracellular mechanism of action in an autocrine/paracrine manner (Jung-Testas et al. 1991).

The effects of PROG and estradiol (E_2) were tested on mixed glial cell primary cultures. Cell growth was inhibited by PROG and stimulated by E_2 (Jung-Testas et al. 1992). Both hormones induced dramatic morphological changes of both oligodendrocytes and astrocytes, producing a more differentiated phenotype. This was accompanied by an increased synthesis of myelin basic protein in oligodendrocytes and of glial fibrillary acidic protein in astrocytes.

Nongenomic, membrane receptor–mediated activities of neurosteroids

The anesthetic and sedative properties of steroids like PROG were reported by Hans Selye as early as 1941. It was initially assumed that they acted by partitioning into membrane lipids and altering neuronal function indirectly, as a consequence of a membrane-disordering effect similar to that reported for CHOL, barbiturates, and inhalation anesthetics (reviewed by Paul and Purdy 1992). Indeed, PREG and DHEA are structurally related to CHOL. The hydrophobic fatty esters of PREG and DHEA are also good candidates, as are those of CHOL, for integration into plasma membranes, and are recovered mainly in myelin after subcellular fractionation of brain homogenates (Robel et al. 1991). The S esters of neurosteroids are amphipathic molecules, and they also could become integral compo-

Table 12.2. *Ligand specificity of the synaptosomal binding component*

Steroid sulfates	Isomers	Relative competition ratio
DHEA		100
PREG		43
17-OH PREG		15
ADIOL-3-mono		77
CHOL		0
TH PROG	$5\alpha,3\alpha$-	468
	$5\alpha,3\beta$-	95
TESTO		0
Androsterone	$5\alpha,3\alpha$-	26
	$5\alpha,3\beta$-	125
	$5\beta,3\alpha$-	7
	$5\beta,3\beta$-	45
Estradiol-3-mono-		12
Estradiol-17-mono-		45
Estradiol-3,17-di-		17

Note: ADIOL and estradiol can be esterified at either position 3 or 17 (3-mono- or 17-mono-) or both (3,17-di-). Noncompetitors: DHEA, $5\alpha,3\alpha$-TH PROG up to 25 *M;* GABA, muscimol, flunitrazepam, bicuculline, picrotoxin, mebubarbital up to 1 m*M.*
Abbreviations: ADIOL, androst-5-ene-$3\beta,17$-diol; TESTO, testosterone. For other abbreviations, see Table 12.1.

nents of cell membranes. Indeed, intracranial injection of several steroidal compounds, including PREGS and DHEAS, by iontophoresis and/or pressure produced an excitatory response from some neurons in the septopreoptic area of the guinea pig brain (reviewed by Baulieu et al. 1987). Responses displayed a short latency of onset and fast decay, suggesting an action at the membrane level.

However, several observations contradict the hypothesis of a simple membrane-disordering mechanism of steroid action. Early structure–activity studies showed that the reduced metabolites of PROG [and of deoxycorticosterone (DOC)] were markedly more potent sedative-anesthetic agents than their less polar precursor. Also, unlike the 3α-hydroxy isomers, the 3β-hydroxy isomers of similar polarity were shown to be inactive. Furthermore, a synaptosomal component of proteinaceous nature, displaying a saturable binding affinity for DHEAS and PREGS, has been reported (reviewed by Majewska 1992). [^{3}H]DHEAS also binds to purified synaptosomal membranes, and the results are best fitted by a model involving a single class of saturable binding sites (Sancho et al. 1991). The relative competition ratios of several steroid sulfates have been reported, following the order $3\alpha,5\alpha$-TH PROGS > DHEAS > PREGS (Table 12.2).

Neurosteroids and the GABA_A receptor

The γ-aminobutyric acid type A (GABA$_A$) receptor is an oligomeric protein complex that, when activated by agonists, increases neuronal membrane conductance to Cl$^-$ ions, resulting in membrane hyperpolarization and reduced neuronal excitability. A number of centrally active drugs, including convulsants, anticonvulsants, anesthetics, and anxiolytics, bind to distinct but interacting domains of this receptor complex to modulate Cl$^-$ conductance (Fig. 12.2) (Olsen and Tobin 1990).

Both inhibitory and excitatory steroids excite or inhibit GABA activity, respectively, an action mediated by GABA$_A$-receptor in nerve cell membranes. Inhibitory steroid metabolites, such as 3α,5α-TH PROG, 3α,21-dihydroxy-5α-pregnan-20-one, and tetrahydrodeoxycorticosterone (3α,5α-TH-DOC), both mimic and enhance the effects of GABA. These steroids potentiate both benzodiazepine and muscimol binding, whereas they inhibit the binding of the convulsant *t*-butyl bicyclophosphorothionate (TBPS). Pharmacological evidence indicates that ste-

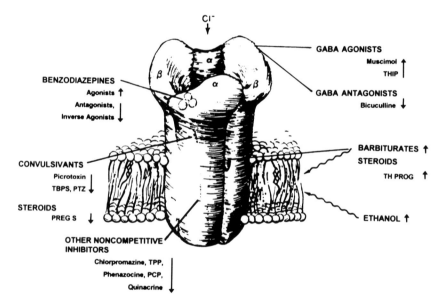

Figure 12.2. A simplified model of the GABA$_A$ receptor. This scheme does not represent either the exact assembly of subunits or the stoichiometry of ligand recognition sites associated with subunits. The continuous lines indicate the categories of sites on the major α- and β-subunits. The γ-subunit (not represented) is necessary for the binding of benzodiazepines. The wavy lines indicate a possible modification of membrane structure. Arrows indicate the stimulation or potentiation (↑) and the inhibition (↓) of GABAergic neurotransmission by the ligands listed. Abbreviations: GABA, γ-aminobutyric acid; THIP, 4,5,6,7-tetrahydroisoxazolo[5,4c]pyridin-3-ol; TBPS, *t*-butyl bicyclophosphorothionate; PTZ, pentylene tetrazole; TPP, tetraphenylphosphonium; PCP, phencyclidine. Adapted from Schwartz (1988).

roid interaction site(s) are distinct from those of both barbiturates and benzodiazepines. The most active inhibitory molecules operate in the low nanomolar range. The 3α-hydroxyl group is an absolute structural requirement. A planar conformation of the A–B ring junction (5α-H) is preferred; however, 5β-H derivatives still display significant activity.

PREGS and DHEAS are prototypical naturally excitatory neurosteroids (Demirgören et al. 1991; Majewska 1990, 1992). At low micromolar concentrations, they antagonize $GABA_A$ receptor–mediated $^{36}Cl^-$ uptake into synaptoneurosomes and Cl^- conductance in cultured neurons. PREGS bimodally modulates [^{3}H]muscimol binding to synaptosomal membranes, slightly potentiates benzodiazepine binding, and inhibits the binding of the convulsant [^{35}S]TBPS to the $GABA_A$ receptor chloride channel. Conversely, PREGS is displaced from its synaptosomal binding site by barbiturates at millimolar concentrations. Moreover, despite mutual competition in these membranes, DHEAS and PREGS behave differently in their capacity to displace [^{35}S]TBPS and [1-^{3}H]phenyl-4-t-butyl 2,6,7-trioxabicyclo[2.2.2]octane ([^{3}H]TBOB) and to enhance [^{3}H]benzodiazepine binding. Furthermore, unlike the sulfate esters of Δ5-3β-hydroxy steroids, those of 3α-hydroxy steroids behave as their unconjugated counterparts, potentiating GABA-induced Cl^- transport in synaptoneurosomes (El Etr et al. 1992). Therefore, distinct sites for neurosteroids, mediating distinct allosteric modes of interaction, seem to exist on the $GABA_A$ receptor or in its membrane vicinity. Nevertheless, one must keep in mind the complexity of $GABA_A$ receptors when interpreting these results, and transfection experiments with defined wild-type or mutated $GABA_A$ receptor subunits should provide more direct evidence for their sites of interaction with neurosteroids.

$GABA_A$ receptors in vivo

The behavioral effects of the GABA-antagonistic neurosteroid PREGS contrast with the hypnotic actions of 3α,5α-TH PROG. When injected intracerebroventricularly, PREGS (8μg/10 μl) shortens the sleep time in rats under pentobarbital hypnosis. Large amounts of PREGS, injected intraperitoneally, have the same effect, exhibiting a dose-dependent relationship (Robel et al. 1991; Majewska 1992).

Several reports have indicated a modulatory role of steroids in models of aggressiveness. DHEA inhibits the aggressive behavior of castrated male mice against lactating female intruders (Baulieu et al. 1987). To eliminate the possibility that the activity of DHEA was related to its conversion to metabolites with documented androgenic or estrogenic potency, behavioral experiments were repeated with the DHEA analog 3α-methyl androst-5-en-17-one (CH_3-DHEA) which, although not demonstrably estrogenic or androgenic in rodents, inhibited the aggressive behavior of castrated mice in a dose-dependent manner, at least as efficiently as DHEA. Both molecules produced a marked and significant decrease of PREGS concentrations in the brain of castrated mice (Young et al. 1991). It is tempting to speculate that DHEA and CH_3-DHEA, by decreasing PREGS levels in brain, might increase the GABAergic tone, which is reportedly involved in the control of aggressiveness.

The memory-enhancing effects of DHEAS and of PREGS in male mice have been documented. When administered intracerebroventricularly after training, they showed memory-enhancing effects in foot-shock avoidance training. PREGS was the most potent, showing effects at 3.5 fmol per mouse (Flood et al. 1992). Infusion of PREGS (12 fmol) into the nucleus basalis magnocellularis of the rat after the acquisition trial enhanced memory performance in a two-trial memory task (Mayo et al. 1993). Conversely, TH PROG (6fmol) disrupted performance when injected before the acquisition trial. A role for neurosteroids in memory processes subserved by this brain region is of interest in view of the implication of this structure in neurodegenerative processes leading to memory loss.

Other membrane effects of neurosteroids

In contrast to GABA-activated Cl^- currents, glycine-activated Cl^- currents are unaffected by $3\alpha,5\alpha$-TH PROG in cultured chick spinal cord neurons (Wu et al. 1990) (Table 12.2). PROG inhibits glycine responses at high concentrations ($EC_{50} = 16$ μM). PROG also modulates a neuronal nicotinic acetylcholine receptor reconstituted in *Xenopus* oocytes ($EC_{50} = 2.9$ μM).

A PROG-3-(O-carboxymethyl)oxime–BSA conjugate also inhibited the currents evoked by acetylcholine, indicating that this steroid interacts with a site located on the extracellular part of the membrane. Radioiodinated BSA conjugated to PROG had previously been shown to bind to nerve cell membranes with an affinity in the nanomolar range.

Neurosteroid physiology

If the neuromodulatory activity of neurosteroids is not merely pharmacological, then their concentrations in brain, at least in defined biological situations, should be in the range of "neuroactive" concentrations. Radioimmunoassays have been developed for measuring PREG, DHEA, their S and L esters, PROG, 5α-DH PROG, and $3\alpha,5\alpha$-TH PROG in plasma and tissues, including brain (Robel et al. 1991; Paul and Purdy 1992). Neither PREG nor DHEA disappears from brain after removal of the adrenals and gonads. Moreover, they undergo prominent circadian variations with an acrophase at the beginning of the dark period, preced-

Table 12.3. *Neurosteroid effects on brain cell surface events*

Event	Receptor	Steroid	Effective concentration (nM)	Reference
Binding to synaptosomal membranes				
		PROG	10^1	Towle and Sze (1983)
		PROG-3-BSA	3×10^1	Ke and Ramirez (1990)
		DHEAS	3×10^3	Demirgören et al. (1991)
		PREGS	2×10^4	Majewska et al. (1990)
		TH PROGS	8×10^2	Sancho et al. (1991)
Allosteric modulation of drug–membrane interactions				
Positive	GABA$_A$	TH PROG	10^1	Paul and Purdy (1992)
		TH PROGS	10^2	El Etr et al. (1992)
Negative		PREGS	10^3–10^5	Majewska (1992)
		DHEAS		
Positive	Glutamate NMDA	PREGS	10^2–10^3	Wu et al. (1991)
Negative	Glutamate QUIS	PROG		Smith (1991)
Negative	Glycine	PROG	10^2	Wu et al. (1990)
		PREGS		
Negative	Nicotinic	PROG	1–10^2	Valera et al. (1992)
		PROG-3-BSA		
Positive	Oxytocin	PROG	0.1	Schumacher et al. (1990)
Negative	Sigma	PROG	10^2	Su et al. (1988)
		PREGS	10^3	

Abbreviations: BSA, bovine serum albumin; NMDA, *N*-methyl D-aspartate; QUIS, Quisqualate. For other abbreviations see Table 12.1.

ing the corticosterone acrophase. PROG also persists, in the low nanomolar range, in the brain of male and female rats after the removal of the adrenals and gonads (Corpéchot et al. 1993), consistent with the operation of a biosynthetic pathway from CHOL, observed in glial cells in culture.

The brain concentrations of PROG and 3α,5α-TH PROG differ markedly between sexes. In particular, whereas the amounts of TH PROG are negligible in intact males, they are in the 10–100 nM range in cyclic and pregnant females, concentrations that are sufficient to exert clear-cut potentiation of GABAergic neurotransmission (Corpéchot et al. 1993). The ovary seems to serve a dual role here: it secretes PROG, which is converted to TH PROG in the brain, concurrently with endogenous PROG, and it secretes estradiol, which may increase 3α-hydroxysteroid oxidoreductase activity in certain brain regions (Cheng and Karavolas 1973).

The concentrations of neurosteroids in brain also depend on stress (Paul and Purdy 1992) and on behavioral situations such as heterosexual exposure (Baulieu et al. 1987).

Conclusion

The effects of steroids on CNS neurotransmitter functions involve both genomic and nongenomic actions (Baulieu 1981; McEwen 1991a,b). The $GABA_A$ receptor complex is a target for naturally occurring steroids, as shown by subcellular, cellular, and *in vivo* experiments. Since several steroids accumulate in brain, independently (at least in part) of the contribution of steroidogenic glands, and their presence can be related to steroid biosynthetic pathways in brain, their designation as "neurosteroids" is justified. Their concentrations in brain appear to be controlled by endogenous brain mechanisms, although a modulatory role of peripheral steroid hormones (e.g., ovarian E_2 secretion) is likely. The level of neurosteroids in brain is consistent with their playing a role in the physiological neuromodulation of situations such as the estrous cycle, pregnancy, and stress and in influencing sexual behavior, mood, memory, developmental, and aging processes.

Obviously, further work is required to establish firmly the physiological significance of neurosteroids and to define their sites of interaction in molecular terms. The data at hand strongly suggest paracrine/autocrine functions of neurosteroids (Fig. 12.3).

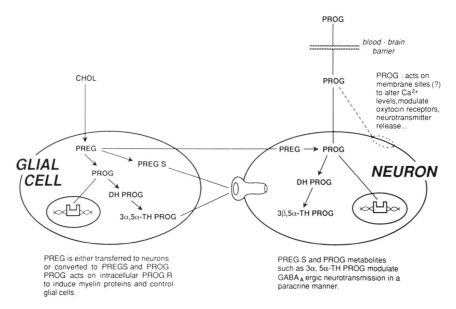

Figure 12.3. A schematic view of the nongenomic and genomic mechanisms of neurosteroid action in neural tissue. The cross-talk between glial cells and neurons implies the provision of steroid precursors and the modulation of $GABA_A$ receptors by steroid metabolites of glial origins.

Acknowledgments

We thank Krzysztof Rajkowski for careful revision of the manuscript, Corinne Legris, Françoise Boussac, and Jean-Claude Lambert for editorial assistance.

References

Akwa, Y., Morfin, R. F., Robel, P., and Baulieu, E. E. (1992). Neurosteroid metabolism: 7α-Hydroxylation of dehydroepiandrosterone and pregnenolone by rat brain microsomes. *Biochem. J.* 288: 959–964.

Akwa, Y., Sananès, N., Gouézou, M., Robel, P., Baulieu, E. E., and Le Goascogne, C. (1993a). Astrocytes and neurosteroids: Metabolism of pregnenolone and dehydroepiandrosterone – Regulation by cell density. *J. Cell. Biol.* 121: 135–143.

Akwa, Y., Schumacher, M., Jung-Testas, I., and Baulieu, E. E. (1993b). Neurosteroids in rat sciatic nerves and Schwann cells. *C.R. Acad. Sci. Paris* 316: 410–414.

Baulieu, E. E. (1981). Steroid hormones in the brain: Several mechanisms? In *Steroid Hormone Regulation of the Brain*, ed. K. Fuxe, J. A. Gustafsson, and L. Wetterberg, pp. 3–14. Oxford: Pergamon Press.

Baulieu, E. E., Robel, P., Vatier, O., Haug, M., Le Goascogne, C., and Bourreau, E. (1987). Neurosteroids: Pregnenolone and dehydroepiandrosterone in the brain. In *Receptor Interactions*, Vol. 48, ed. K. Fuxe and L. F. Agnati, pp. 89–104. Basingstoke: Macmillan.

Celotti, I., Massa, R., and Martini, L. (1979). Metabolism of sex steroids in the central nervous system. In *Endocrinology*, Vol. 1, ed. L. J. De Groot, G. F. Cahill, Jr., E. Steinberger, and A. I. Winegrad, pp. 41–53. New York: Grune & Stratton.

Cheng, Y. G., and Karavolas, H. J. (1973). Conversion of progesterone to 5α-pregnane-3,20-dione and 3α-hydroxy-5α-pregnan-20-one by rat medial basal hypothalami and the effects of estradiol and stage of estrous cycle on the conversion. *Endocrinology* 93: 1157–1162.

Cheng, Y. G., and Karavolas, H. J. (1975). Subcellular distribution and properties of progesterone (Δ4-steroid) 5α-reductase in rat medial basal hypothalamus. *J. Biol. Chem.* 250: 7997–8003.

Corpéchot, C., Young, J., Calvel, M., Wehrey, C., Veltz, J. N., Touyer, G., Mouren, M., Prasad, V. V. K., Banner, C., Sjövall, J., Baulieu, E. E., and Robel, P. (1993). Neurosteroids: 3α-Hydroxy-5α-pregnan-20-one and its precursors in the brain, plasma, and steroidogenic glands of male and female rats. *Endocrinology* 133: 1003–1009.

Demirgören, S., Majewska, M. D., Spivak, C. E., and London, E. D. (1991). Receptor binding and electrophysiological effects of dehydroepiandrosterone sulfate, an antagonist of the GABA$_A$ receptor. *Neuroscience* 45: 127–135.

El Etr, M., Corpéchot, C., Young, J., Akwa, Y., Robel, P., and Baulieu, E. E. (1992). Modulating effect of steroid sulfates on muscimol-stimulated ^{36}Cl uptake. *Eur. J Neurosci.* (suppl. 5), abstr. 2215.

Flood, J. F., Morley, J. E., and Roberts, E. (1992). Memory-enhancing effects in male mice of pregnenolone and steroids metabolically derived from it. *Proc. Natl. Acad. Sci. (USA)* 89: 1567–1571.

Fuxe, K., Gustafsson, J. A., and Wetterberg, L. (1981). *Steroid Hormone Regulation of the Brain*. Oxford: Pergamon Press.

Guarneri, P., Papadopoulos, V., Pan, B., and Costa, E. (1992). Regulation of pregnenolone synthesis in C6-2B glioma cells by 4′-chlorodiazepam. *Proc. Natl. Acad. Sci. (USA)* 89: 5118–5122.

Hu, Z. Y., Jung-Testas, I., Robel, P., and Baulieu, E. E. (1987). Neurosteroids: Oligodendrocyte mitochondria convert cholesterol to pregnenolone. *Proc. Natl. Acad. Sci. (USA)* 84: 8215–8219.

Hu, Z. Y., Jung-Testas, I., Robel, P., and Baulieu, E. E. (1989). Neurosteroids: Steroidogenesis in primary cultures of rat glial cells after release of aminoglutethimide blockade. *Biochem. Biophys. Res. Commun.* 161: 917–922.

Iwahashi, K., Ozaki, H. S., Tsubaki, M., Ohnishi, J. I., Takeuchi, Y., and Ichikawa, Y. (1990). Studies of the immunohistochemical and biochemical localization of the cytochrome P-450scc-linked monooxygenase system in the adult rat brain. *Biochem. Biophys. Acta* 1035: 182–189.

Jung-Testas, I., Hu, Z. Y., Baulieu, E. E., and Robel, P. (1989). Neurosteroids: Biosynthesis of pregnenolone and progesterone in primary cultures of rat glial cells. *Endocrinology* 125: 2083–2091.

Jung-Testas, I., Renoir, J. M., Gasc, J. M., and Baulieu, E. E. (1991). Estrogen-inducible progesterone receptor in primary cultures of rat glial cells. *Exp. Cell. Res.* 193: 12–19.

Jung-Testas, I., Renoir, J. M., Greene, G. L., and Baulieu, E. E. (1992). Demonstration of steroid hormone receptors and steroid action in primary cultures of glial cells. *J. Steroid Biochem. Mol. Biol.* 41: 621–631.

Kabbadj, K., El Etr, M., Baulieu, E. E., and Robel, P. (1993). Pregnenolone metabolism in rodent embryonic neurons and astrocytes. *Glia* 7: 170–175.

Ke, F. C., and Ramirez, V. D. (1990). Binding of progesterone to nerve cell membranes of rat brain using progesterone conjugated to ^{125}I-bovine serum albumin as ligand. *Neurochemistry* 54: 467–472.

Le Goascogne, C., Robel, P., Gouézou, M., Sananès, N., Baulieu, E. E., and Waterman, M. (1987). Neurosteroids: Cytochrome P450scc in rat brain. *Science* 237: 1212–1215.

Le Goascogne, C., Gouézou, M., Robel, P., Defaye, G., Chambaz, E., Waterman, M. R., and Baulieu, E. E. (1989). The cholesterol side-chain cleavage complex in human brain white matter. *J. Neuroendocrinol.* 1: 153–156.

Majewska, M. D. (1992). Neurosteroids: Endogenous bimodal modulators of the GABA$_A$ receptor – Mechanism of action and physiological significance. *Prog. Neurobiol.* 38: 379–395.

Majewska, M. D., Demirgören, S., and London, E. D. (1990). Binding of pregnenolone sulfate to rat brain membranes suggests multiple sites of steroid action at the GABA$_A$ receptor. *Eur. J. Pharmacol.* 189: 307–315.

Mayo, W., Dellu, F., Robel, P., Cherkaoui, J., Le Moal, M., Baulieu, E. E., and Simon, H. (1993). Infusion of neurosteroids into the nucleus basalis magnocellularis affects cognitive processes in the rat. *Brain Res.* 607: 324–328.

McEwen, B. (1991a). Steroid hormones are multifunctional messengers in the brain. *Trends Endocrinol. Metab.* 2: 62–67.

McEwen, B. (1991b). Non-genomic and genomic effects of steroids on neural activity. *Trends Pharmacol. Sci.* 12: 141–147.

McLusky, N. J., Philip, A., Harlburt, C., and Naftolin, F. (1984). Estrogen metabolism in neuroendocrine structures. In *Metabolism of Hormonal Steroids in the Neuroendocrine Structures*, ed. F. Celotti and F. Naftolin, pp. 105–116. New York: Raven Press.

Morfin, R., Young, J., Corpéchot, C., Egestad, B., Sjövall, J., and Baulieu, E. E. (1992). Neurosteroids: Pregnenolone in human sciatic nerves. *Proc. Natl. Acad. Sci. (USA)* 89: 6790–6793.

Naftolin, F., Ryan, K. J., Davies, I. J., Reddy, V. V., Flores, F., Retro, Z., Kahn, M.,

White, R. Y., Takaoka, Y., and Wolin, L. (1975). The formation of estrogens by central neuroendocrine tissues. *Rec. Prog. Horm. Res.* 31: 295–315.

Normington, K., and Russell, D. W. (1992). Tissue distribution and kinetic characteristics of rat steroid 5α-reductase isozymes. *J. Biol. Chem.* 267: 19548–19554.

Olsen, R. W., and Tobin, A. J. (1990). Molecular biology of GABA$_A$ receptors. *FASEB J.* 4: 1469–1480.

Papadopoulos, V., Guarneri, P., Krueger, K. E., Guidotti, A., and Costa, E. (1992). Pregnenolone biosynthesis in C6-2B glioma cell mitochondria: Regulation by a mitochondrial diazepam binding inhibitor receptor. *Proc. Natl. Acad. Sci. (USA)* 89: 5113–5117.

Paul, S. M., and Purdy, R. H. (1992). Neuroactive steroids. *FASEB J.* 6: 2311–2322.

Pelletier, G., Dupont, E., Simard, J., Luu-The, V., Bélanger, A., and Labrie, F. (1992). Ontogeny and subcellular localization of 3β-hydroxysteroid dehydrogenase (3β-HSD) in the human and rat adrenal, ovary and testis. *J. Steroid Biochem. Mol. Biol.* 43: 451–467.

Robel, P., Akwa, Y., Corpéchot, C., Hu, Z. Y., Jung-Testas, I., Kabbadj, K., Le Goascogne, C., Morfin, R., Vourc'h, C., Young, J., and Baulieu, E. E. (1991). Neurosteroids: Biosynthesis and function of pregnenolone and dehydroepiandrosterone in the brain. In *Brain Endocrinology*, ed. M. Motta, pp. 105–131. New York: Raven Press.

Sancho, M. J., Eychenne, B., Young, J., Corpéchot, C., and Robel, P. (1991). Binding of steroid sulfates to synaptosomal membranes. Paper presented at the 73rd Annual Meeting of the Endocrine Society. Washington DC, June 19–22 (abstr 1039).

Schumacher, M., Coirini, H., Pfaff, D. W., and McEwen, B. S. (1990). Behavioral effects of progesterone associated with rapid modulation of oxytocin receptors. *Science* 250: 691–694.

Schwartz, R. D. (1988). The GABA$_A$ receptor-gated ion channel: Biochemical and pharmacological studies of structure and function. *Biochem. Pharmacol.* 17: 3369–3375.

Selye, H. (1941). The anesthetic effect of steroid hormones. *Proc. Soc. Exp. Biol. Med.* 46: 116–121.

Smith, S. S. (1991). Progesterone administration attenuates excitatory amino acid responses of cerebellar Purkinje cells. *Neuroscience* 42: 309–320.

Su, T. P., London, E. D., and Jaffe, J. H. (1988). Steroid binding at Σ receptors suggests a link between endocrine, nervous, and immune systems. *Science* 240: 219–221.

Towle, A. C., and Sze, P. Y. (1983). Steroid binding to synaptic plasma membrane: Differential binding of glucocorticoids and gonadal steroids. *J. Steroid. Biochem.* 18: 135–143.

Valera, S., Ballivet, M., and Bertrand, D. (1992). Progesterone modulates a neuronal nicotinic acetylcholine receptor. *Proc. Natl. Acad. Sci. (USA)* 89: 9949–9953.

Vourc'h, C., Eychenne, B., Jo, D. H., Raulin, J., Lapous, D., Baulieu, E. E., and Robel, P. (1992). Δ5-3β-Hydroxysteroid acyl transferase activity in the rat brain. *Steroids* 57: 210–215.

Warner, M., Strömstedt, M., Möller, L., and Gustafsson, J. A. (1989b). Distribution and regulation of 5α-androstane-3β,17β-diol hydroxylase in the rat nervous system. *Endocrinology* 124: 2699–2706.

Warner M., Tollet, P., Strömstedt, M., Carlström, K., and Gustafsson, J. A. (1989a). Endocrine regulation of cytochrome P-450scc in the rat brain and pituitary gland. *J. Endocrinol.* 122: 341–349.

Wu, F. S., Gibbs, T. T., and Farb, D. H. (1990). Inverse modulation of γ-aminobutyric acid- and glycine-induced currents by progesterone. *Mol. Pharmacol.* 37: 597–602.

Wu, F. S., Gibbs, T. T., and Farb, D. H. (1991). Pregnenolone sulfate: A positive allosteric modulator at the *N*-methyl-D-aspartate receptor. *Mol. Pharmacol.* 40: 333–336.

Young, J., Corpéchot, C., Haug, M., Gobaille, S., Baulieu, E. E., and Robel, P. (1991). Suppressive effects of dehydroepiandrosterone and 3β-methyl-androst-5-en-17-one on attack towards lactating female intruders by castrated male mice: II. Brain neurosteroids. *Biochem. Biophys. Res. Commun.* 174: 892–897.

13

Estrogen synthesis and secretion by the songbird brain

BARNEY A. SCHLINGER AND ARTHUR P. ARNOLD

Steroid hormones act on the brain to influence its organization during development and its activity in adulthood, thereby regulating behavior and physiology. Achieving these effects on the brain is often the culmination of a complex set of events within the endocrine system. During development, the sequence begins with the differentiation of steroid-secreting organs and their expression of steroid-synthetic enzymes. In adulthood, it continues with the regulation of the activities of one or more of these enzymes by pituitary trophic factors. After secretion, but before the steroid reaches targets within specific brain cells, the hormone is subject to a variety of regulatory influences. These can include steps to inactivate the molecule by peripheral catabolism and excretion. The presence of carrier proteins in blood can limit the availability of free steroid to enter tissues. Having reached a target organ, the steroid may encounter additional enzymes that catalyze changes in its structure, rendering the molecule inactive. Alternatively, the steroid may be converted to a molecule with increased biological activity or one that functions along an alternative steroid-activating pathway. When these transformations are complete, the steroid is available to influence cellular function by interacting with intracellular protein receptors. Once bound to ligand, the steroid receptor can bind to specific DNA hormone response elements to influence the transcription of specific genes. The active steroid may also influence cell function without changing gene expression by interacting directly with cell membranes or with other cellular processes.

Although we have learned a great deal about the synthesis, secretion, transport, metabolism, and action of hormones, in some cases these processes are not fully understood and new discoveries have indicated that the boundaries between processes are vague. For example, although the gonads and adrenals have been considered the primary or exclusive steroidogenic organs, the concentrations of some steroids fluctuate in the brain independently of levels in the blood and persist in the brain of adult rats following castration and adrenalectomy (Baulieu 1991; see

Chapter 12 by Robel and Baulieu, this volume). There is now convincing evidence that some of these steroids are synthesized from cholesterol within the brain itself (Jung-Testas et al. 1989). Thus, the concept that the brain is simply a target of steroids secreted by peripheral organs is oversimplified.

Historically, our concepts about the interactions between steroids and brain have been advanced by studies in certain animal models. A research goal in several laboratories, including our own, has been to elucidate the endocrine processes underlying sex steroid effects on the brain in species of songbirds. Research of the past 15–20 years has shown that singing depends on complex interactions between the nervous and endocrine systems. These interactions are necessary to ensure the proper sex-specific development of brain regions that control singing behavior, the learning of species-specific songs, and the expression of these songs within appropriate social contexts. Although we know that steroids influence these behaviors and their neural substrate, it has remained a challenge to elucidate the underlying physiology that provides steroids to the brains of developing and adult birds. Table 13.1 summarizes some of the unusual and/or conflicting experimental results concerning estrogen action and physiology in songbirds.

In this chapter, we describe the results of our studies indicating that estrogen synthesis occurs at sufficiently high levels in the brain of male songbirds to control circulating levels of this steroid. Because the synthesis of estrogen in the brain may be tied to the development and expression of singing behavior, we will first briefly describe features of song, the neural circuitry underlying song and song learning, properties of estrogen action on the songbird brain, and aspects of estrogen physiology in songbirds that do not fit patterns established from studies of other species. With this background, we will then describe what our studies have revealed about the physiology of estrogens in songbirds. Although the results of our research provide a new view of estrogen synthesis and secetion, our studies have generated many new questions, which will be explored in the conclusion.

Singing behavior

Song is a vocal behavior used for communication by many bird species. Of all birds, songbirds (order: Passeriformes; subgroup: oscine) are thought to possess some of the greatest vocal capabilities. These oscine songbirds may be the most recently evolved of all birds (Welty 1975), distinguished from the "suboscine" passeriform species and other bird orders by (among other traits) their highly developed syrinx, or vocal organ, and a unique neural circuit that controls the syringeal musculature (as described later).

Song is used primarily in two social contexts, territorial aggression and mate acquisition. In many cases, breeding males sing to defend their territory against

Table 13.1. *Unusual physiological attributes involving estrogen or estrogen action in oscine birds*

1. Estrogen circulates at relatively high levels in males of several songbird species (Hutchison et al. 1984; Weichel et al. 1986; Marler et al. 1987, 1988; Adkins-Regan et al. 1990; Schlinger et al. 1991; Schlinger and Arnold 1992a).
2. There is conflicting evidence for an unusually large peak of circulating estradiol in male but not female zebra finches at ages when the brain regions controlling song are sensitive to the masculinizing effects of estrogen (Hutchison et al. 1984; Adkins-Regan et al. 1990; Schlinger and Arnold 1992a).
3. Estrogen can both masculinize song and demasculinize other male reproductive behaviors in the zebra finch (Adkins-Regan and Ascenzi 1987).
4. Efforts to block masculine song development by castration have been unsuccessful (Arnold 1975; Kroodsma 1986; Marler et al. 1988; Adkins-Regan and Ascenzi 1990).
5. Castration or exposure to short-day photoperiods (and attendant gonadal regression) do not necessarily eliminate circulating estrogen from male songbirds and might increase circulating estrogen levels (Marler et al. 1988; Adkins-Regan et al. 1990; Brenowitz et al. 1991; Schlinger and Arnold, 1991b; Schlinger et al. 1992).
6. Plasma estrogen level increases after ovariectomy in female zebra finches (Adkins-Regan et al. 1990).
7. Although only very few estrogen-concentrating cells or cells containing estrogen receptors are observed in zebra finch telencephalon using autoradiographic or immunohistochemical techniques (Arnold and Saltiel 1979; Gahr et al. 1987; Nordeen et al. 1987), cytosolic estrogen receptors can be detected in some song control nuclei of castrated males after treatment with an aromatase inhibitor (Walters et al. 1988).
8. Although neural aromatase activity in vertebrates is often most active in limbic brain regions, particularly the hypothalamus–preoptic area, activity is present in zebra finches throughout the telencephalon, including regions that express the highest specific aromatase activity in brain (Vockel et al. 1990).
9. Behaviors known to be at least partially estrogen-dependent persist in some songbirds after castration or during photoperiod-induced gonadal regression (Prove 1974; Arnold 1975; Harding 1983; Moore and Kranz 1983; Logan and Wingfield 1990; Logan 1992; Schwabl 1992; Wingfield and Monk 1992).
10. Compounds commonly used as estrogen receptor antagonists fail to block estrogen action in the developing songbird brain, but function normally in peripheral tissues (Mathews et al. 1988; Mathews and Arnold 1990).
11. Treatment of developing male zebra finches with aromatase inhibitors does not prevent masculine development of the neural circuitry controlling song (Balthazart et al. 1993; Wade and Arnold 1994).

other males (Lack 1943; Thorpe 1961; Catchpole 1982). In addition, the song identifies a suitable male to unpaired females and stimulates their sexual interest (Kroodsma 1976). However, there can be great variability in song expression based on sex, reproductive condition, and the kind of territory a song is used to defend. For example, song can be a behavior of both males and females of species in which both sexes defend the territory or when the sexes require constant auditory contact (Farabaugh 1982). In other cases, males or both sexes sing outside of the breeding season in defense of nonbreeding territories (Schwabl 1992). This complex expression of song, crossing boundaries of sex and social context, makes song unique as a sex steroid–dependent behavior. In contrast, studies directed at understanding the control of behavior by steroids in other animal groups generally focus on those behaviors that are highly sex-specific and tied to cycles of reproduction. These include the male copulatory behaviors (in mammals, mounting, intromission, and ejaculation) or female copulatory behavior (lordosis). It is important to bear in mind that the diverse functional properties of song in species of songbirds may account for some unusual features observed in the nervous and endocrine control of song behavior described in the following sections.

The neural control of song

The neural system controlling song is composed of discrete nuclei found largely in the telencephalon, but also in the midbrain and hindbrain (Fig. 13.1). The song control system consists of two overlapping circuits: a motor circuit and a circuit involved with song learning. Motoneurons, located in the tracheosyringeal hypoglossal nucleus of the medulla, directly innervate the syringeal musculature. These motoneurons receive input from the robust nucleus of the archistriatum (RA), which is innervated by neurons within the higher vocal center (HVC), located in the dorsal neostriatum. Lesions of HVC or RA eliminate or severely disrupt song (Nottebohm and Nottebohm 1976; Simpson and Vicario 1990). Moreover, neurons in these nuclei are electrically active immediately before the onset of song (McCasland and Konishi 1981; McCasland 1987). These results establish these telencephalic nuclei as critical in the motor control of song.

In the song control circuit (the accessory loop), the HVC projects densely to Area X of the lobus paraolfactorius. Area X innervates the medial nucleus of the

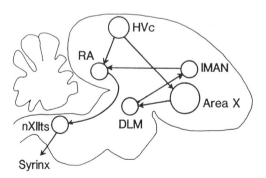

Figure 13.1. A schematic diagram illustrating some of the brain regions that control song and song learning in oscine songbirds. Main descending pathway: HVc, higher vocal center; RA, robust nucleus of the archistriatum; nXIIts, tracheosyringeal portion of the hypoglossal motor nucleus. Accessory loop: lMAN, lateral magnocellular nucleus of the neostriatum; Area X of the lobus parolfactorius; DLM, dorsolateral medial nucleus of the thalamus.

lateral dorsal thalamus (DLM), which projects to the magnocellular nucleus of the neostriatum (MAN). The lateral MAN (lMAN) both receives projections from DLM and also projects to RA. Lesions of Area X or of lMAN in male zebra finches (*Poephila guttata*) before or during song acquisition prevent or delay song learning. Subsequent lesions after song acquisition in adult male zebra finches have little to no effect on song (Bottjer et al. 1984; Sohrabji et al. 1990; Scharff and Nottebohm 1991). Thus, this latter loop is considered essential for song learning, but not for the motor control of song.

One intriguing feature of this song control system is that the sizes of various nuclei in the circuitry vary in proportion to the vocal performance capability of individuals. This result emerges when the circuit is compared between sexes within a species, across species, or across individual males whose song varies in complexity (Arnold et al. 1986). In species such as the zebra finch, where only males sing, the song circuitry is highly sexually dimorphic (Nottebohm and Arnold 1976). In other species, such as the tropical duetting bay wren (*Thryothorus nigricapilla*), a species in which both males and females produce songs of equal complexity, no sex differences have been detected in the song circuit (Canady et al. 1984; Brenowitz et al. 1985; Arnold et al. 1986). These data suggest that factors affecting the organization of the circuit are variable across birds. Because we know that sex steroids influence the development of this system in at least one species (see later), variations in the organization of the song system across and within species are likely mediated by sex steroid hormones.

Hormonal control of song

Organizational effects of sex steroids

The sex differences in the brains of adult zebra finches are quite large. Consequently, this species provides an important animal model by which to understand factors that influence brain development. Phenotypic sexual differentiation, including the differentiation of behavior, is thought to be fully controlled by gonadal secretions (Wilson et al. 1981; MacLusky and Naftolin 1981; Arnold and Gorski 1984; Arnold and Schlinger 1993), while the gonads differentiate in response to genotypic signals (Haqq et al. 1993). Historically, studies of phenotypic sexual differentiation in birds have focused on nonpasseriform species, such as quail and chickens. These studies indicate that the female phenotype develops in response to secretions of the developing ovary, whereas the male phenotype develops in the absence of ovarian secretions (Adkins 1981; Balthazart et al. 1992). Because females are the heterogametic sex in birds (i.e., females – ZW; males – ZZ), some factor, whose gene resides on the W chromosome, must be regulated developmentally to induce ovarian differentiation. This factor might regulate the expression of aromatase, the final requisite enzyme in estrogen biosynthesis, since the synthesis of estrogens is required for differentiation of an ovary in birds (Elbrecht and Smith 1992). After acting locally to determine ovarian structure, estrogens are secreted by the ovary and act throughout the developing embryo to induce the female phenotype.

Estrogens derived from the ovary also act on the developing brain to determine its sex-specific properties. For example, brain regions controlling masculine copulatory behavior in the Japanese quail (*Coturnix c. japonica*) are sensitive to the organizational effects of steroids approximately 12 days after the egg is laid. If a male embryo is exposed to estrogen at this time (by injecting estrogen *in ovo*), the bird appears physically as an adult male, but is unable to exhibit male copulatory behavior when exposed to an adult female (Adkins 1981). Thus, estrogen is thought to demasculinize copulatory behavior in developing female birds. This conclusion appears to extend to songbirds as well, since estrogen can demasculinize some sexual behaviors in the zebra finch (Adkins-Regan and Ascenzi 1987).

Sex steroids are also involved in the development of song control circuitry. If female zebra finches are treated with estradiol (E_2) shortly after hatching, they are able to produce a masculine song as adults (which is improved by adult androgen treatment), and they possess substantially masculine song control circuitry (Gurney and Konishi 1980; Gurney 1981, 1982; Pohl-Apel et al. 1984; Pohl-Apel et al. 1985; Adkins-Regan and Ascenzi 1987; Simpson and Vicario 1991a,b). The effects of testosterone (T) on developing females are similar to those of E_2 (Gurney 1981). However, dihydrotestosterone (DHT) produces only partial masculinization of the female song system (Gurney 1981, 1982; Schlinger and Arnold 1991a), and females treated after hatching with DHT do not sing as adults (Gurney 1981). Since T, but not DHT, is a substrate for aromatization, E_2 is assumed to be the primary hormone responsible for masculinizing song in the zebra finch.

The masculinizing effects of E_2 on the zebra finch song system are paradoxical. Estrogens are fundamental for feminine development in birds, since they are critical to the differentiation of ovarian but not testicular tissue, and they continue to feminize (and demasculinize) other structures including the brain. However, in the case of song, E_2 is a masculinizing hormone. Presumably, estrogens act at the level of the hypothalamus to feminize reproductive behavior and act on the telencephalic song system to masculinize song. Understanding how estrogens can have regionally opposing actions on brain sexual differentiation has been an important force in our search for sites of estrogen synthesis in songbirds.

It is important to add certain caveats here. The ability of E_2 to masculinize the neural circuit controlling song has been demonstrated in only one species, the zebra finch. Because this species exhibits profound sexual dimorphism of behavior, steroid effects observed in this species might not apply to all songbirds. Furthermore, although estrogens can masculinize the female brain, inhibition of estrogen action or estrogen synthesis in posthatching males does not prevent masculine brain development (Mathews et al. 1988; Mathews and Arnold 1990, 1991; Balthazart et al. 1993; Wade and Arnold 1994). It is critical to determine whether estrogen is the endogenous masculinizing hormone in males of other songbird species, as well.

Activational effects of sex steroids

Because song is utilized in the context of reproduction, it is not surprising that it is influenced by sex steroid hormones in adult songbirds. But as we shall see, steroid effects on the expression of adult song may be more complex than effects generally seen on other sex steroid–dependent behaviors.

The expression of masculine reproductive and aggressive behaviors related to reproduction is often tightly linked to the presence of testicular steroids, especially T (Wingfield and Farner 1980; Callard 1983; Wingfield et al. 1987; Schlinger and Callard 1991). In many species, castration can eliminate or reduce sexual behavior, whereas replacement with physiological levels of T can restore behavior (Callard 1983; Harding 1983; Wingfield et al. 1987; Schlinger and Callard 1991). Thus, T is considered the primary activational stimulus for reproductive behavior in males of many vertebrate species. Testosterone also activates singing behavior in songbirds. In many seasonally breeding species (i.e., those species that breed at temperate latitudes and whose hypothalamic–pituitary–gonadal system is regulated by photoperiod), singing behavior is conspicuous during the breeding season, especially at its outset, when males establish territories and compete for the attention of newly arrived females (Lack 1943; Thorpe 1961; Catchpole 1982). At this time the testes are fully enlarged and actively secrete T (Wingfield and Farner 1980; Wingfield et al. 1987). Castration can reduce singing behavior in some songbirds, and singing can be reinstated with exogenous androgens (Arnold 1975; Harding 1983).

Although T is the dominant circulating hormone in adult males of many species, and T replacement can restore sexual and aggressive behaviors in castrated males, T itself might not be the steroid that activates these behaviors. It is readily metabolized in many steroid target tissues, including the brain (Schlinger and Callard 1991). The avian brain expresses two enzymes that convert T into active metabolites, 5α-reductase and aromatase (Schlinger and Callard 1987, 1991). 5α-Reductase converts T to 5α-DHT, a potent androgen in many tissues. Although the extent to which androgen receptors in the brain are occupied by 5α-DHT or by T is unknown, the occupation by 5α-DHT of a significant proportion of these receptors implies a critical, functional role for local 5α-reductase in some tissues (Anderson and Liao 1968; Bruchovsky and Wilson 1968). Aromatase is present in the brain (Hutchison and Steimer 1984; Balthazart 1991; Schlinger and Callard

1991), where it converts T to estradiol-17β, or androstenedione to estrone. In adult males, this transformation is critical in controlling sexual and aggressive behaviors. For example, estrogens can reactivate masculine behavior lost or reduced by castration (Hutchison and Steimer 1984; Balthazart 1991; Schlinger and Callard 1991). Moreover, aromatase is expressed in greatest amounts in diencephalic or limbic structures (e.g., the hypothalamus, preoptic area, septum, and amygdala in mammals or homologous structures in other species), areas that also express estrogen receptors and are critical elements of the neural system controlling these behaviors (Hutchison and Steimer 1984; Blaustein and Olster 1989; Schlinger and Callard 1991). Because estrogens generally circulate at low levels in males, the central synthesis of estrogen from circulating androgen locally in the brain is considered an essential step in the activation of male behavior. Central aromatization is also involved in controlling reproductive behavior (i.e., copulation, aggression, and song) in songbirds. For example, only aromatizable androgens (i.e., T or androstenedione) or a combination of estrogens and non-aromatizable androgens restore reproductive behavior and singing after castration in adult males (Harding 1983; Archawaranon and Wiley 1988; Harding et al. 1988).

Thus, there are many similarities between the hormonal control of reproductive behavior in songbirds and in other vertebrate species. Nevertheless, there are some intriguing differences. For example, many species of songbirds show a second peak of singing behavior in the autumn as they establish nonbreeding territories they will occupy during the winter. Although gonadal activity may increase at this time (Wingfield and Monk 1992), autumnal territorial aggression and song might not be related to increased circulating T (Logan and Wingfield 1990; Logan 1992; Schwabl 1992; Wingfield and Monk 1992). A similar relationship has been observed in some tropical songbirds, which express territorial aggression and song year around (in some cases in both males and females) despite low levels of circulating steroids (Levin and Wingfield 1992). Moreover, in some cases, singing behavior persists after castration (Arnold 1975; Kroodsma 1986). Although the precise mechanisms controlling aggression and song, including the role of estrogen, remain uncertain, it is clear that general assumptions about steroid hormone effects on brain derived from other species may not apply to some species of songbirds.

In summary, estrogens are critical in both the organization of the song system (in zebra finches) and the activation of song in certain behavioral contexts. However, our understanding of the physiology underlying the steroid-dependent regulation of brain development and control of song in adult birds remains controversial. Some of these properties appear to set songbirds apart from other vertebrates. Because estrogen action on or delivery to brain may be unique in songbirds, there might be fundamental differences in the way estrogens are formed, released, or eliminated in songbirds and in other species. To explore the basis of these differences, we have undertaken studies to understand more fully the synthesis and secretion of estrogen in songbirds. The results suggest that the brain is the source of estrogen found in the circulation of adult male and perhaps also ovariectomized female zebra finches, and that the cellular origin of estrogens in the brain might be unique.

The physiology of estrogens in songbirds

We and others have taken several approaches to the study of estrogen synthesis and secretion in adult and developing male songbirds. First, radioimmunoassay has been used to measure plasma estrogen levels (Hutchison et al. 1984; Weichel et al. 1986; Marler et al. 1987, 1988; Adkins-Regan et al. 1990; Schlinger and Arnold 1991b, 1992a; Schlinger et al. 1992a). Second, *in vitro* radioenzymatic assays have been used to measure steroidogenic enzyme activity in tissues, especially the aromatization reaction (Vockel et al. 1990a,b; Schlinger and Arnold 1991b, 1992a; Schlinger et al. 1992a). Third, we have developed methods to measure estrogen synthesis and secretion *in vivo* in adult songbirds (Schlinger and Arnold 1992b, 1993). These methods involve direct injection of [^{3}H]androgen into specific brain regions, body tissues, or the vascular system followed by the purification and definitive identification of radioactive estrogen present in blood draining these tissues. Fourth, the distribution of aromatase has been assessed by immunohistochemistry (Balthazart et al. 1990). Because of our interest in brain aromatase, we have also developed the ability to culture brain tissue from developing zebra finches and measured enzyme activity in these cultures (Arnold 1992; Schlinger et al. 1992b). Finally, we have recently cloned a zebra finch aromatase and are beginning more detailed analyses of this gene and its products in various tissues of both males and females during development and adulthood (Shen et al. in press; Schlinger et al. 1993).

Plasma estrogens

In the early 1980s, it was established that treatment of hatchling female zebra finches with T (an aromatizable androgen) or E$_2$ profoundly masculinized song and the brain regions that control song (Gurney and Konishi 1980; Gurney 1981, 1982). Gurney (1982) observed small masculinizing effects of the nonaromatizable androgen 5α-dihydrotestosterone, which we have confirmed (Schlinger and Arnold 1991a). Because there was evidence for both androgen and estrogen effects on the song system, Gurney postulated that circulating T (presumably from the testes) was being converted to E$_2$ and 5α-DHT in brain (by local aromatization and 5α-reduction) to fully masculinize song in males of this species. This conclusion, however, was not supported by the observations of Hutchison et al. (1984), who measured plasma sex steroid levels of hatchling zebra finches (1–10 days posthatching). These investigators found no sex differences in levels of circulating androgens (testosterone, androstenedione, 5α-DHT), but they did detect much larger quantities of E_2 in the circulation of males than in females of the same age. This result was especially provocative, because the estrogenic masculinization of the mammalian brain is thought to be the result of central aromatization of circulating androgens. Therefore, the data of Hutchison et al. (1984) were the first to suggest that a *circulating* estrogen was responsible for organizing or activating a male reproductive behavior. This conclusion was disputed by Adkins-Regan et al. (1990), who failed to replicate the dramatic plasma E_2 surge in hatchling males found by Hutchison et al. (1984). Therefore, we measured the levels of various sex steroids (E_2, estrone, T, androstenedione, 5α-DHT) in plasma of male and female zebra finches during ages critical for brain sex differentiation (Schlinger and Arnold 1992a). Our conclusions from these data can be summarized as follows: (a) Circulating E_2 level in males during the posthatching period was low, similar to that detected by Adkins-Regan et al. (1990) and lower than that reported by Hutchison et al. (1984); (b) in contrast to E_2, estrone was found to circulate at high levels in both male and female hatchlings; (c) despite the higher level of estrone, we found no significant differences between males and females for estrone or E_2 during the posthatching period (Schlinger and Arnold 1992a). Our results generally agree with those of Adkins-Regan et al. (1990) and suggest that differentially high estrogen delivery to the male brain via the circulation is most likely *not* responsible for the masculinization of song during this period.

Despite the discrepancies in the reports of estrogen content of hatchling blood, each of these studies revealed that total aromatizable androgen level is high in the blood of both males and females (Hutchison et al. 1984; Adkins-Regan et al. 1990; Schlinger and Arnold 1992a). However, because the levels were similar in both sexes, it is unlikely that differentially high glandular secretion of aromatizable substrate in males is responsible for the masculinization of song. Nevertheless, because aromatase is present at high levels in the telencephalon of developing zebra finches (see later), any aromatizable androgen present in blood could serve as substrate for the synthesis of active estrogen by the brain.

Although there remains some uncertainty as to the amount of estrogen present in plasma of male and female zebra finch hatchlings, it is evident that estrogens circulate in large quantities in male songbirds at various ages: during critical periods of brain sexual differentiation, during song learning (Weichel et al. 1986; Marler et al. 1987), and in adulthood (Hutchison et al. 1984; Weichel et al. 1986; Schlinger and Arnold 1991a). For example, plasma E_2 levels in breeding or isolated adult male zebra finches were equal to or exceeded those of adult females under the same housing conditions (Hutchison et al. 1984; Schlinger and Arnold 1991a). Not only are relatively high levels of estrogen present in plasma of normal males, but there is evidence that circulating estrogen is not eliminated by castration (Marler et al. 1988; Adkins-Regan et al. 1990; Schlinger and Arnold 1991b). Marler et al. (1988) showed that E_2 levels in plasma of developing male song sparrows (*Zonotrichia melodia*) (up to 6 months of age) were unaffected by castration. Moreover, despite plasma T levels below 200 pg/ml, E_2 remained as high as 600 pg/ml in plasma of castrated song sparrows as well as in developing male swamp sparrows (*Melospiza georgiana*). Adkins-Regan et al. (1990) examined the effect of castration on adult zebra finches. Once again, androgen levels were reduced up to 4 weeks after castration of males. Nevertheless, in both males and females, E_2 levels in plasma were actually increased more than fivefold after gonadectomy.

We also examined the effect of castration on circulating E_2 levels in adult male zebra finches. However, we included several modifications of the radioimmunoassay procedures previously used. First, we used either separate antibodies to measure E_2 and estrone, or an antibody that cross-reacts with both estrogens. Second, we measured these estrogens either in unchromatographed ether-extracted plasma or after separating steroids using Celite-column chromatography. In addition, we measured estrogens at 3, 28, 60, 90, or 140 days after castration. Our results support previous studies on songbirds showing that castration does not eliminate estrogen from the circulation. While E_2 or estrone occasionally increased (as much as 10-fold) or decreased (as much as 50%), in most cases their levels were comparable to those seen before castration. Finally, it is relevant that estrogen may also persist in the plasma of seasonally breeding male songbirds whose gonads regress under short-day photoperiods and whose plasma androgen levels are correspondingly low (Brenowitz et al. 1991; Schlinger et al. 1992a).

Measures of tissue aromatization *in vitro*

Normally, the presence of sex steroids in the circulation following castration would imply an adrenal source of these hormones (Marler et al. 1988; Adkins-Regan et al. 1990). Mammalian adrenals are known to secrete sex steroids into the circulation during development and in adulthood (Pepe and Albrecht 1990). Castration or exposure to short-day photoperiods can produce adrenal hypertrophy in several species (Kar 1947; Kitay 1968).

To determine whether the adrenals are the source of circulating estrogens in male zebra finches, we quantified aromatase activity using an *in vitro* radioenzymatic procedure ([^{3}H]androstenedione conversion to [^{3}H]estrone and [^{3}H]estradiol) in adrenals of intact and castrated males (Schlinger and Arnold 1991b). In addition, we searched for aromatase in gonadal and nonglandular tissues of adult

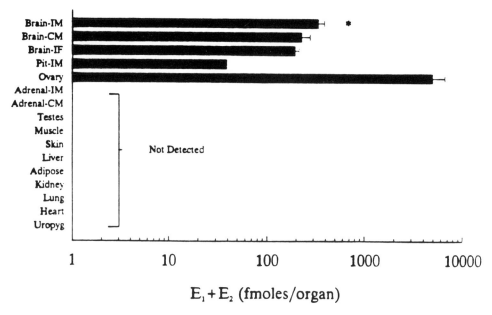

Figure 13.2. Aromatase activity (mean ± SEM; 3–10 determinations per tissue) in whole homogenates of tissues from adult intact (I) or castrated (C) male (M) or female (F) zebra finches. Tissues from castrates were removed 30–140 days after castration. The asterisk denotes significantly different from intact female but not castrated male ($p \leq .05$; E_1, estrone; one-way ANOVA and Duncan's Multiple Range Test). From Schlinger and Arnold (1991b).

and developing males and females (Fig. 13.2; Schlinger and Arnold 1991b, 1992a). Our results can be summarized as follows: (a) Despite using numerous procedural modifications to enhance the capacity to detect estrogen formation (e.g., changing incubation times, pooling large quantities of tissue from several animals, increasing substrate specific activity, searching for estrogen metabolites or conjugates), aromatase activity was either undetected or very poorly expressed in adrenals of intact or castrated adult males or males during the posthatching period (1–10 days of age), and (b) aromatase was undetected in testes, suggesting that the glandular steroidogenic tissues of male songbirds are not important sites of estrogen secretion. These results imply that estrogen present in plasma is synthesized in nonglandular tissues from circulating aromatizable substrate. Therefore, we also searched for aromatase activity in a variety of other non-neural tissues of adult or developing males, including skin, muscle, liver, lung, kidney, adipose, and uropygial gland, but again found no evidence for aromatase activity. As expected, aromatase was present in abundance in ovary of adult females, as well as in females as young as 3 days of age.

In addition to detecting aromatase in ovary from females at all ages, we detected small amounts of aromatase in the male hypothalamus–preoptic area (HPOA) and pituitary, as in other avian species (Fig. 13.2; Schlinger and Callard

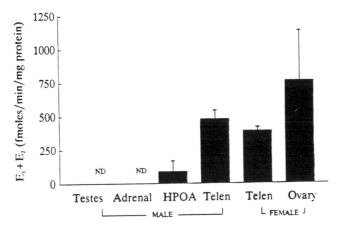

Figure 13.3. Aromatase activity (mean ± SEM) in microsomes from pooled tissues of adult male and female zebra finches. Abbreviations: HPOA, hypothalamus–preoptic area; Telen, telencephalon. From Schlinger and Arnold (1991b).

1987; Schlinger and Callard 1991). However, we were influenced by the studies of Vockel et al. (1990), who found aromatase in nonlimbic regions of the zebra finch telencephalon. However, in those studies, the rate of aromatization in whole homogenates of telencephalon was linear only over very brief incubation times (Vockel et al. 1990). Therefore, we chose to measure activity in microsomal fractions prepared from telencephalon and other tissues, including adrenals and testes (Fig. 13.3). Microsomes from the avian brain are rich in aromatase but depleted of competing enzymes and thus may provide a more accurate measure of tissue aromatization (Schlinger and Callard 1989). Using microsomal preparations, we again failed to detect aromatase activity in adult male adrenals or testes, confirming that these organs are not an important source of estrogen in this species. To our surprise, specific aromatase activity in male telencephalic microsomes was similar to that found in female ovary. This observation, together with the absence of aromatase in testes or adrenals, suggests that the telencephalon itself could be a major source of circulating estrogens in male zebra finches. The presence of relatively large amounts of aromatase in the white-crowned sparrow (*Zonotrichia leucophrys*) telencephalon (Schlinger et al. 1992a) indicates that some results of our studies in the zebra finch may extend to other species of songbirds as well.

Data on aromatase in hatchlings suggest that the brain might also contribute to circulating estrogen during development. Although active aromatase is undetected in testes, adrenals, or other non-neural tissues of hatchlings, aromatase is present at relatively high levels in the telencephalon of hatchling males and females (Fig. 13.4) (Schlinger and Arnold 1992a). Although telencephalic levels in hatchlings are lower than in adults (45 fmol estrogen per minute/milligram micro-

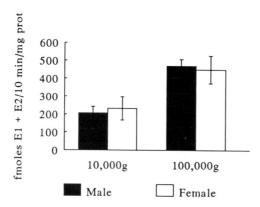

Figure 13.4. Aromatase activity (mean ± SEM) in cellular subfractions of telencephalon of hatchling (4–6 days posthatching) zebra finches ($n = 3$ pools per sex; $n = 6$ telencephalons per pool). 10,000g = synaptosomes and mitochondria; 100,000g = microsomes. From Schlinger and Arnold (1992a).

somal protein), it remains likely that locally formed estrogen has important effects on brain development. For example, similar levels of aromatase have been found in microsomes prepared from the HPOA of reproductively active male quail (Schlinger and Callard 1989), a species in which HPOA aromatase provides sufficient estrogen to activate masculine reproductive and aggressive behaviors when circulating androgens are high (Schlinger and Callard 1991). We assume that sufficient androgenic substrate could also be converted to biologically active estrogen in the brain of hatchling zebra finches, since hatchlings have plasma levels of aromatizable androgen similar to those of adult male zebra finches (Hutchison et al. 1984; Adkins-Regan et al. 1990; Schlinger and Arnold 1992a). In addition, aromatase activity in hatchling zebra finch brain homogenates is more than 100-fold greater than that found in regions of the cerebral cortex and hippocampus of neonatal rhesus monkey, where locally formed estrogen is thought to influence the sexual differentiation of nonreproductive functions (MacLusky et al. 1987).

The brain is generally considered a target of sex steroid hormones. Although central aromatization is recognized to be necessary to regulate neural actions of androgen from peripheral sources (McEwen et al. 1983), the concept of the brain as the *source* of a circulating steroid is new and profoundly influences our thinking about sex steroid hormone physiology in songbirds. This also provides a partial explanation for the absence of castration effects seen in zebra finches and other songbirds and dramatically influences our ideas about the mechanism of sexual differentiation of this species.

Measures of tissue aromatization *in vivo*

To examine further whether circulating estrogens in male zebra finches originate in brain, we developed methods to measure estrogen synthesis *in vivo* in adult zebra finches. The results of these studies strongly support the view that brain synthesis directly controls the circulating level of this estrogen (Schlinger and Arnold 1992b, 1993). These experiments involved the injection of [^{3}H]androgen intravenously or into discrete tissues of anesthetized birds, and then the measurement of [^{3}H]estrogen in blood entering or leaving the brain (in carotid or jugular plasma, respectively). In addition, we quantified the [^{3}H]estrogen content of various tissues after these injections. Presumably, tissues containing a high [^{3}H]estrogen level shortly after vascular or tissue injection of [^{3}H]androgen represent sites of estrogen synthesis, and tissues containing more estrogen in venous blood than in arterial blood must secrete estrogen to enrich the vascular estrogen content.

We observed that [^{3}H]estrogens were present in jugular blood after direct brain injections of as little as 165 pg [^{3}H]androgen injected into the dorsal telencephalon, a brain region rich in aromatase (Schlinger and Arnold 1992b). By contrast, [^{3}H]estrogens were undetected in the jugular after injections of as much as 1 ng [^{3}H]androgen into the aromatase-poor cerebellum. This suggests that nonspecific (non-neural) aromatization probably does not contribute [^{3}H]estrogens to the vasculature after brain androgen injection. Thus, we conclude that [^{3}H]estrogens formed in the dorsal telencephalon can be secreted into the vasculature. In these experiments, brain tissue was extracted to determine the [^{3}H]estrogen content. Consistent with our measures of jugular estrogens, there was considerably more estrogen in whole-brain homogenates after telencephalic than after cerebellar [^{3}H]androgen injection. This confirms the view that sites of estrogen synthesis and secretion can be localized by measuring both tissue and vascular [^{3}H]estrogen content.

We also made direct injections of [^{3}H]androgen (1–36 ng) into peripheral glandular tissues (testes, ovaries) and into the adrenals of intact or castrated males. We then determined the quantity of [^{3}H]estrogen in carotid blood or that remaining in the injected organ (Schlinger and Arnold 1992b, 1993). [^{3}H]Estrogen was routinely found in carotid blood after injection of androgen into the ovary, but never after injection into the testes (Schlinger and Arnold 1992b). Some [^{3}H]estrogen was detected in the carotid of both intact and castrated males after injection into the adrenal, suggesting that aromatase might be present in this organ (Schlinger and Arnold 1992b, 1993). Nevertheless, essentially no [^{3}H]estrogen could be extracted from the adrenals, although some [^{3}H]androgen was still present. This suggests that the estrogen observed in carotid blood after adrenal androgen injection had been synthesized elsewhere, probably in brain, since [^{3}H]estrogen was extracted in large quantities from the telencephalon of adrenal-injected birds. These results confirm that neither the testes nor adrenals of males are important estrogen-synthetic organs.

Despite our observation that estrogen reached the bloodstream after direct injection of androgen into the brain, we assume that the androgenic substrate of brain aromatase is normally derived from the circulation. Therefore, we made direct injections of [³H]androgen into the jugular of intact males and females or castrated males and then examined blood from either the carotid artery or the jugular vein.

We detected significant quantities of [³H]estrogens in the carotid of females, but not males (Fig. 13.5) (Schlinger and Arnold 1992b, 1993), which corroborates the presence of a significant site of aromatase in the female ovary and the lack of peripheral aromatase in males. The amounts of [³H]estrogens found in jugular blood were much higher than those in carotid blood (Schlinger and Arnold 1992b, 1993), indicating that the estrogen content of blood is enriched upon passage through the male's brain (Fig. 13.5). This suggests that androgen enters the brain from the vasculature, where it is converted to estrogen, which is released back into the bloodstream. Similar treatment of castrated males with physiological levels of [³H]androgen produced similar results (Schlinger and Arnold 1993), from which we conclude that estrogen synthesis and secretion by the brain is a normal process in intact adult male zebra finches. Similar results in the male brown-headed cowbird (*Molothrus ater*) indicate that brain estrogen synthesis and secretion might be a property of other songbird species as well (B. Schlinger and A. Arnold, unpublished observations). Our data indicate that estrogen derived from brain might also reach the vasculature of adult female zebra finches (Schlinger and Arnold 1993); however, estrogen physiology is more complex in females, since they possess two sites of aromatization.

The quantity of [³H]estrogen extracted from the telencephalon of both males and females after vascular or tissue injections of [³H]androgen is quite impres-

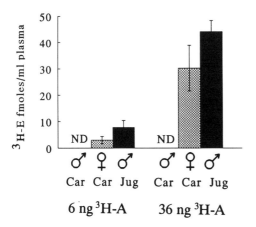

Figure 13.5. Carotid (Car) or jugular (Jug) plasma levels of [³H]estrogen (estrone plus estradiol) recovered 5 minutes after 6 or 36 ng [³H]androstenedione or [³H]testosterone (³H-A) was injected into the jugular of adult male or female zebra finches. ND denotes not detected; mean ± SEM of three separate experiments. From Schlinger and Arnold (1992b).

sive. High levels of [^{3}H]estrogen were extracted from the telencephalon of castrated males whose endogenous androgens were replaced by [^{3}H]androgen. These results suggest that high levels of estrogens are probably always present in the telencephalon of adult zebra finches.

Our measurements of aromatase activity *in vivo* provide particularly strong evidence for the physiological importance of estrogen synthesis by the brain of songbirds. These experiments were conducted under physiological conditions that did not involve the optimization procedures normally required to demonstrate aromatization *in vitro* (e.g., the addition of cofactors, the purification of the enzyme in microsomal preparations, or the varying of incubation parameters such as time and substrate concentrations). In addition, they demonstrate that the brain contributes estrogen to plasma despite the presence of various enzymes that compete for androgenic substrate or catabolize estrogen (see later). We conclude that estrogen synthesis and secretion by brain are fundamental properties of sex steroid physiology in these bird species.

Cell types in brain that express aromatase

So far, we have reviewed several properties of brain aromatase in the zebra finch that differ from the pattern seen in many other vertebrates. For example, aromatase activity is present in the telencephalon, outside of the diencephalic structures traditionally thought to contain aromatase. Moreover, aromatase activity can be higher in some telencephalic structures than in the preoptic area, the brain region often considered to contain the greatest amount of aromatase (Schlinger and Callard 1991). Thus, it was particularly interesting that Balthazart et al. (1990) failed to observe aromatase outside of the preoptic area of the adult zebra finch by immunohistochemical analysis using an antibody prepared against human placental aromatase. It was difficult to account for this absence of labeling, since aromatase activity is elevated in many regions of the telencephalon (Vockel et al. 1990).

In an effort to identify which cells in the telencephalon express aromatase and to try to account for this absence of immunostaining in the telencephalon, we utilized two new approaches to study aromatase in the zebra finch brain. First, we cultured brain tissue from developing zebra finches (1–6 days of age) and measured enzyme activity in these cells *in vitro* (Schlinger et al. 1992b). Second, we isolated an aromatase cDNA probe from a zebra finch ovarian library and conducted *in situ* hybridization histochemical and Northern blot analyses of mRNA from cultured cells (Schlinger et al. 1993, in press a). Details of the characterization of this probe have been published (Shen et al. 1994). Aromatase is expressed at very high levels in these cell cultures. In addition, at least two cell types possess aromatase activity and aromatase mRNA *in vitro*. As expected from biochemical and immunohistochemical studies indicating the presence of aromatase in neurons in other species (Canick et al. 1986; Schlinger and Callard 1989; Shinoda et al. 1989a,b; Balthazart et al. 1990, 1991; Naftolin et al. 1990; Sanghera et al. 1991), neurons in culture expressed high activity and possessed large quantities of aromatase mRNA. Moreover, aromatase mRNA was localized in both neuronal somata and processes (Schlinger et al. 1993, in press a). However, treatments that reduced or eliminated neurons in these cultures had little effect on aromatase activity and did not eliminate aromatase mRNA. For example, neurons were removed from the cultures by vigorous shaking, leaving behind pure cultures of non-neuronal cells presumed to be astrocytes, since they were large and flat. These cells attached readily to the substrate and continued to divide *in vitro*. Despite the absence of neurons, these cultures continued to demonstrate aromatase activity. In addition, aromatase mRNA can be detected on Northern blots containing RNA extracted from non-neuronal cultures. Finally, these cells were labeled by the aromatase cDNA probe using *in situ* hybridization histochemical procedures. These data indicate that both neurons and non-neuronal cells express aromatase in the zebra finch telencephalon.

High levels of 5β-reductase, which is thought to inactivate T in the avian brain (Hutchison and Steimer 1981), and low levels of 5α-reductase can also be detected in these mixed cell cultures. Moreover, like aromatase, the activities of both reductases can be measured in glia-enriched cultures (Schlinger et al., in press b), suggesting that all three enzymes may be co-localized in glia. It is also possible that they might be co-expressed by neurons as well, but we can definitively identify only aromatase in neurons at present.

As discussed previously, these enzymes could interact in brain by competing for the same substrate, and perhaps by synthesizing metabolites that directly interfere with the activities of other enzymes (Schumacher et al. 1991). Our ability to evaluate the nature of such interactions has been limited, because the activity of these enzymes in brain is generally measured in homogenates or cellular subfractions in which the cells and cellular organelles are artificially mixed. The advantage of the cell culture system is that we can study the interactions of these enzymes in living cells. Our results indicate that, despite the presence of 5β-reductase, estrogen is formed in appreciable quantity by cells in these preparations, even when the substrate concentration is below physiological levels (Schlinger et al., in press b). This suggests that 5β-reductase interferes little with the synthesis of estrogen *in vivo*. By contrast, 5α-reductase is difficult to measure in cell cultures. Thus, the abundant expression of aromatase and 5β-reductase reduces the quantity of active androgen, such as T, but results in little formation of active androgenic metabolites such as 5α-DHT. This raises the interesting possibility that the expression of both aromatase and 5β-reductase, together with the limited expression of 5α-reductase, limits the availability of active androgen to androgenic targets in the zebra finch telencephalon.

Further questions and conclusions

There is no question that the brain of songbirds is a rich target for sex steroids. Both androgen and estrogen receptors are present in a variety of brain structures during development and adulthood. During development, sex steroids help regulate the sex-specific organization of brain structures that control several behaviors in adult birds, including song and song learning, as well as copulatory behaviors. Neural structures controlling other sex-specific adult behaviors such as the performance of elaborate courtship displays, the construction of complex nests, and participation in parental care might also be influenced developmentally by steroids in some songbird species. In adults, steroids are also important in the seasonal regulation of reproduction, the expression of sexual behavior and territorial aggression, and the performance of song.

The zebra finch brain has evolved some unique features by which to process steroids. The expression of large amounts of aromatase in the telencephalon, and its presence in glia, are properties of brain aromatase that may be unique to songbirds. Our results indicate that the brain of adult male and perhaps female zebra finches can enrich the vascular supply of estrogen. We must conclude, then, that the estrogen acting on the zebra finch brain is synthesized in brain. However, the contribution of estrogen-synthesizing neurons or glia to specific estrogen-sensitive neural circuits remains unknown.

Numerous other questions remain unanswered. For example, the source of the androgen substrate necessary for brain aromatization is unknown. We assume that testicular or adrenal androgen provides the substrate for brain aromatization; however, if this were the case, the level of estrogen in the blood of male songbirds would parallel that of aromatizable androgen. Because this is often not the case (Hutchison et al. 1984; Weichel et al. 1986; Marler et al. 1987, 1988; Adkins-Regan et al. 1990; Schlinger and Arnold 1992a), other factors must also regulate blood estrogen levels. To address this question, we have begun to measure other steroidogenic enzymes in zebra finch tissues during development and in adulthood (Schlinger and Arnold 1992a). We have also identified 3β-hydroxysteroid dehydrogenase/isomerase activity in cultures of the developing bird brain (Schlinger et al. 1992b). Whether the zebra finch brain expresses additional steroidogenic enzymes, and thus synthesizes androgens *de novo*, remains an intriguing possibility.

It is also possible that small amounts of aromatase, undetectable by our assays, are co-expressed with androgen-synthetic enzymes in testes or male adrenals. By virtue of their proximity to a source of androgen, even small quantities of aromatase might contribute a significant amount of estrogen to blood. *In situ* hybridization histochemical and Northern blot analyses using the zebra finch aromatase cDNA probe in developing zebra finch brain tissue might detect expression of the aromatase gene. However, even if peripheral aromatase is present in males, we still must determine why the brain synthesizes large amounts of estrogen.

The presence of estrogen receptors in the zebra finch song system is also controversial (Gahr et al. 1987; Nordeen et al. 1987; Walters et al. 1988); autoradiographic and immunohistochemical studies indicate few estrogen receptors in the telencephalic song system (Gahr et al. 1987; Nordeen et al. 1987), whereas biochemical measures suggest greater abundance (Walters et al. 1988). If these autoradiographic and immunohistochemical studies are correct, then estrogen receptors and aromatase are not co-localized throughout the telencephalon. This would lead us to consider alternative mechanisms of action for estrogens synthesized in brain. For example, we have previously shown that aromatase is present within synaptosomal preparations of quail HPOA (Schlinger and Callard 1989), suggesting that locally formed estrogens might influence synaptic function or directly modify postsynaptic membrane responses. We would like to know whether aromatase is also localized in terminals within the zebra finch telencephalon where newly synthesized estrogen could act through a nongenomic mechanism. The presence of aromatase mRNA in processes of cultured telencephalic neurons provides some evidence that the aromatase protein might be present in processes and terminals.

We also wonder where estrogens act to control song behavior. As described earlier, song is used in both aggressive and sexual contexts. Nevertheless, no neuroanatomical connections have been identified between nuclei of the song system and several brain regions (e.g., the preoptic area, nucleus taeniae) known to be involved in the control of sexual and aggressive behavior in birds (Hutchison and Steimer 1984; Schlinger and Callard 1991). Perhaps such connections will eventually be identified, although it is also possible that the song control nuclei are themselves centers for the motivational control of singing behavior. Since aromatase is present at high levels in the preoptic area, nucleus taeniae, and many other regions of the telencephalon (Vockel et al. 1990), integration of estrogen action on several brain structures is probably required for the expression of singing behavior. Whether some of these behaviors might also be regulated by estrogens formed in brain (Schlinger et al. 1992a) or by nonsteroidal factors during the nonbreeding season remains a fascinating unresolved problem.

What is the functional role of preoptic aromatase in the male zebra finch? Since low levels of estrogen circulate in males of most species, brain aromatase presumably supplies estrogens locally to estrogen-sensitive neural circuits (Schlinger and Callard 1991). We wonder if estrogen in the blood of male zebra finches, presumably synthesized predominantly in the telencephalon, can then act on non-telencephalic estrogen targets in brain. Alternatively, estrogen might diffuse from areas of high concentration in the telencephalon to more distant estrogen targets in brain without entering the circulation. Estrogen formed in the brain might also act on non-neural estrogen targets. Thus, a great deal has yet to be learned about the mechanisms of estrogen delivery to estrogen targets in species of songbirds.

The high level of telencephalic aromatase (Fig. 13.4) and circulating aromatizable androgen present during development in both male and female zebra finches suggests that estrogen is available in the brains of both sexes. Presumably, hypothalamic brain regions controlling copulatory behaviors are demasculinized in females by ovarian estrogens during development. However, why this same estrogen does not also masculinize the female song circuitry is unknown. Furthermore, we still do not understand the process by which estrogen is available to the telencephalon of males, and not females, during the critical period of sexual differentiation of the song control circuitry. Additional telencephalic mechanisms might be present to activate estrogen in males or inactivate estrogen in females to cause sex-specific masculinization of singing behavior. One possibility is that estrogen-catabolic enzymes are expressed in brain to control local estrogen levels (Schlinger and Arnold 1992a). It will be important to determine whether sex differences in estrogen degradation are present in the zebra finch brain. This could lead to a situation in which females would synthesize estrogen, only to degrade it immediately to protect themselves from its effects.

Finally, why is aromatase expressed at such high levels in the telencephalon of both male and female songbirds? Perhaps the answer lies in the presumed role of estrogen both in the development and activation of song behavior and in the complex expression of song observed across songbird species. Telencephalic estrogen synthesis might be involved in the organization of song circuits in other species, in addition to the zebra finch. There is considerable diversity in the expression of song behavior by both males and females across species. Perhaps the expression of aromatase in the telencephalon provides the estrogen required to organize the song system in males and/or females of those species in which song is favored by selection. Similarly, estrogen might function to activate song behavior in many adult songbirds. Since the seasonal and behavioral contexts in which song can be expressed are diverse across species, the availability of estrogen from telencephalic synthesis might provide the hormonal substrate necessary to activate song under these various conditions. Additional studies on brain aromatization in ecologically diverse species of songbirds will very likely reveal a great deal about the hormonal control of brain development and behavior.

Acknowledgments

We thank Dr. William Raum of the Population Research Center for assistance with radioimmunoassay procedures. The work described here was supported by NIH Grants DC00217, NS08649, and HD19445 and NSF Grants BNS-9020953 and IBN-9120776.

References

Adkins, E. (1981). Early organizational effects of hormones. In *Neuroendocrinology of Reproduction: Physiology and Behavior*, ed. N. T. Adler, pp. 159–228. New York: Plenum Press.

Adkins-Regan, E., Abdelnabi, M., Mobarak, M., and Ottinger, M. A. (1990). Sex steroid levels in developing and adult male and female zebra finches. *Gen. Comp. Endocrinol.* 78: 93–109.

Adkins-Regan, E., and Ascenzi, M. (1987). Social and sexual behaviour of male and female zebra finches treated with oestradiol during the nestling period. *Anim. Behav.* 35: 1100–1112.

Adkins-Regan, E., and Ascenzi, M. (1990). Sexual differentiation of behavior in the zebra finch: Effect of early gonadectomy or androgen treatment. *Horm. Behav.* 24: 114–127.

Anderson, K. M., and Liao, S. (1968). Selective retention of dihydrotestosterone by prostatic nuclei. *Nature* 219: 277–279.

Archawaranon, M., and Wiley, R. H. (1988). Control of aggression and dominance in white-throated sparrows by testosterone and its metabolites. *Horm. Behav.* 22: 497–517.

Arnold, A. P. (1975). The effects of castration and androgen replacement on song, courtship, and aggression in zebra finches (*Poephila guttata*). *J. Exp. Zool.* 191: 309–326.

Arnold, A. P. (1992). Developmental plasticity in neural circuits controlling bird song: Sexual differentiation and the neural basis of learning. *J. Neurobiol.* 23: 1506–1528.

Arnold, A. P., Bottjer, S. W., Brenowitz, E. A., Nordeen, E. J., and Nordeen, K. W. (1986). Sexual dimorphisms in the neural vocal control system in songbirds: Ontogeny and phylogeny. *Brain Behav. Evol.* 28: 22–31.

Arnold, A. P., and Gorski, R. A. (1984). Gonadal steroid induction of structural sex differences in the central nervous system. *Annu. Rev. Neurosci.* 7: 413–442.

Arnold, A. P., and Schlinger, B. A. (1993). Sexual differentiation of brain and behavior: The zebra finch is not just a flying rat. *Brain Behav. Evol.* 42: 231–241.

Arnold, A. P., and Saltiel, A. (1979). Sexual differences in pattern of hormone accumulation in the brain of a songbird. *Science* 205: 702–705.

Balthazart J. (1991). Testosterone metabolism in the avian hypothalamus. *J Steroid Biochem. Mol. Biol.* 40: 557–570.

Balthazart, J., Absil, P., Fiasse, V., and Ball, G. F. (1993). Effects of estrogens on the sexual differentiation of brain and behavior in zebra finches: Studies with an aromatase inhibitor. *Soc. Neurosci. Abstr.* 19: 172.

Balthazart, J., De Clerck, A., and Foidart, A. (1992). Behavioral demasculinization of female quail is induced by estrogens: Studies with the new aromatase inhibitor, R76713. *Horm. Behav.* 26: 179–203.

Balthazart, J., Foidart, A., Surlemont, C., and Harada, N. (1991). Distribution of aromatase-immunoreactive cells in the mouse forebrain. *Cell Tiss. Res.* 263: 71–79.

Balthazart, J., Foidart, A., Surlemont, C., Vockel, A., and Harada, N. (1990). Distribution of aromatase in the brain of the Japanese quail, Ring dove, and Zebra Finch: An immunocytochemical study. *J. Comp. Neurol.* 301: 276–288.

Baulieu, E.-E. (1991). Neurosteroids: A function of the brain. In *Neurosteroids and Brain Function*, Fidia Research Foundation Symposium Series, Vol. 8, ed. E. Costa and S. M. Paul, pp. 63–73. New York: Thieme Medical Publishers.

Blaustein, J. D., and Olster, D. H. (1989). Gonadal steroid hormone receptors and social behaviors. In *Advances in Comparative and Environmental Physiology*, Vol. 3, ed. J. Balthazart, pp. 31–103. Berlin: Springer.

Bottjer, S. W., Miesner, E. A., and Arnold, A. P. (1984). Forebrain lesions disrupt development but not maintenance of song in passerine birds. *Science* 224: 901–903.

Brenowitz, E. A., Arnold, A. P., and Levin, R. N. (1985). Neural correlates of female song in tropical duetting birds. *Brain Res.* 343: 104–112.

Brenowitz, E. A., Nalls, B., Wingfield, J. C., and Kroodsma, D. E. (1991). Seasonal changes in avian song nuclei without seasonal changes in song repertoire. *J. Neurosci.* 11: 1367–1374.

Bruchovsky, N., and Wilson, J. D. (1968). The conversion of testosterone to 5-androstan-17β-ol-3-one by rat prostate *in vivo* and *in vitro*. *J. Biol. Chem.* 243: 2012–2021.

Callard, G. V. (1983). Androgen and estrogen action in the vertebrate brain. *Am. Zool.* 23: 607–620.

Canady, R., Kroodsma, D., and Notebohm, F. (1984). Population differences in complexity of a learned skill are correlates with the brain space involved. *Proc. Natl. Acad. Sci. (USA)* 81: 6232–6234.

Canick, J. A., Vaccaro, D. E., Livingston, E. M., Leeman, S. E., Ryan, K. J., and Fox, T. O. (1986). Localization of aromatase and 5α-reductase to neuronal and non-neuronal cells in the fetal rat hypothalamus. *Brain Res.* 372: 277–282.

Catchpole, C. K. (1982). The evolution of bird sounds in relation to mating and spacing behavior. In *Acoustic Communication in Birds*, ed. D. E. Kroodsma, E. H. Miller, and H. Ouellet, pp. 297–319. New York: Academic Press.

Elbrecht, A., and Smith, R. G. (1992). Aromatase enzyme activity and sex determination in chickens. *Science* 255: 467–470.

Farabaugh, S. M. (1982). The ecological and social significance of duetting. In *Acoustic Communication in Birds*, ed. D. E. Kroodsma, E. H. Miller, and H. Ouellet, pp. 85–124. New York: Academic Press.

Gahr, M., Flugge, G., and Guttinger, H.-R. (1987). Immunocytochemical localization of estrogen binding neurons in the songbird brain. *Brain Res.* 402: 172–177.

Gurney, M. (1981). Hormonal control of cell form and number in the zebra finch song system. *J. Neurosci.* 1: 658–673.

Gurney, M. (1982). Behavioral correlates of sexual differentiation in the zebra finch song system. *Brain Res.* 231: 153–172.

Gurney, M., and Konishi, M. (1980). Hormone induced sexual differentiation of brain and behavior in zebra finches. *Science* 208: 1380–1382.

Haqq, C. M., King, C.-Y., Donahoe, P. K., and Weiss, M. A. (1993). SRY recognizes converted DNA sites in sex-specific promoters. *Proc. Natl. Acad. Sci. (USA)* 90: 1097–1101.

Harding, C. F. (1983). Hormonal influences on avian aggressive behavior. In *Hormones and Aggressive Behavior*, ed. B. Svare, pp. 435–468. New York: Plenum Press.

Harding, C. F., Walters, M. J., Collado, D., and Sheridan, K. (1988). Hormonal specificity and activation of social behavior in male red-winged blackbirds. *Horm. Behav.* 22: 402–418.

Hutchison, J. B., and Steimer, T. (1981). Brain 5β-reductase: A correlate of behavioral sensitivity to androgen. *Science* 213: 244–246.

Hutchison, J. B., and Steimer, T. (1984). Androgen metabolism in the brain: Behavioral correlates. *Prog. Brain. Res.* 61: 23–51.

Hutchison, J. B., Wingfield, J. C., and Hutchison, R. E. (1984). Sex differences in plasma concentrations of steroids during the sensitive period for brain differentiation in the zebra finch. *J. Endocrinol.* 103: 363–369.

Jung-Testas, I., Hu, Z. Y., Bailieu, E.-E., and Robel, P. (1989). Neurosteroids: Biosynthesis of pregnenolone and progesterone in primary cultures of rat glial cells. *Endocrinology* 125: 2083–2091.

Kar, A. B. (1947). The adrenal cortex, testicular relations in the fowl: The effect of castration and replacement therapy on the adrenal cortex. *Anat. Rec.* 99: 177–197.

Kitay, J. I. (1968). The effects of estrogen and androgen on the adrenal cortex of the rat. In *Functions of the Adrenal Cortex*, Vol. 2, ed. K. W. McKerns, pp. 775–811. Amsterdam: North-Holland.

Kroodsma, D. E. (1976). Reproductive development in a female songbird: Differential stimulation by quality of male song. *Science* 192: 574–575.

Kroodsma, D. E. (1986). Song development by castrated march wrens. *Anim. Behav.* 34: 1572–1575.

Lack, D. (1943). *The Life of the Robin.* London: Witherby.

Levin, R. N., and Wingfield, J. C. (1992). The hormonal control of territorial aggression in tropical birds. *Ornis Scand.* 23: 284–291.

Logan, C. A. (1992). Testosterone and reproductive adaptations in the autumnal territoriality of northern mockingbirds (*Mimus polglottos*). *Ornis Scand.* 23: 277–283.

Logan, C. A., and Wingfield, J. C. (1990). Autumnal territorial aggression is independent of plasma testosterone in mockingbirds. *Horm. Behav.* 24: 568–581.

MacLusky, N. J., Clark, A. S., Naftolin, F., and Goldman-Rakic, P. S. (1987). Estrogen formation in the mammalian brain: Possible role of aromatase in sexual differentiation of the hippocampus and neocortex. *Steroids* 50: 461–474.

MacLusky, N. J., and Naftolin, F. (1981). Sexual differentiation of the central nervous system. *Science* 211: 1294–1303.

Marler, P., Peters, S., Ball, G. F., Dufty, A. M., Jr., and Wingfield, J. C. (1988). The role of sex steroids in the acquisition and production of birdsong. *Nature* 336: 770–772.

Marler, P., Peters, S., and Wingfield, J. (1987). Correlations between song acquisition, song production, and plasma levels of testosterone and estradiol in sparrows. *J. Neurobiol.* 18: 531–548.

Mathews, G. A., and Arnold, A. P. (1990). Antiestrogens fail to prevent the masculine ontogeny of the zebra finch song system. *Gen. Comp. Endocrinol.* 80: 48–58.

Mathews, G. A., and Arnold, A. P. (1991). Tamoxifen's effects on the zebra finch song system are estrogenic, not antiestrogenic. *J. Neurobiol.* 22: 957–969.

Mathews, G. A., Brenowitz, E. A., and Arnold, A. P. (1988). Paradoxical hypermasculinization of the zebra finch song system by an antiestrogen. *Horm. Behav.* 22: 540–551.

McCasland, J. S. (1987). Neuronal control of bird song production. *J. Neurosci.* 7: 23–39.

McCasland, J. S., and Konishi, M. (1981). Interaction between auditory and motor activities in an avian song control nucleus. *Proc. Natl. Acad. Sci. (USA)* 78: 7815–7819.

McEwen, B. S., Bigeon, A., Davis, P. G., Krey, L. C., Luine, V. N., McGinnis, M. Y., Paden, C. M., Parsons, B., and Rainbow, T. C. (1983). Steroid hormones: Humoral signals which alter brain cell properties and functions. *Rec. Prog. Horm. Res.* 30: 41–92.

Moore, M. C., and Kranz, R. (1983). Evidence for androgen independence of male mounting behavior in white-crowned sparrows (*Zonotrichia leucophrys gambelii*). *Horm. Behav.* 17: 414–423.

Naftolin, F., Leranth, C., and Balthazart, J. (1990). Ultrastructural localization of aromatase immunoreactivity in hypothalamic neurons. *Endocrine Soc. Abstr.* 192.

Nordeen, K. W., Nordeen, E. J., and Arnold, A. P. (1987). Estrogen accumulation in zebra finch song control nuclei: Implications for sexual differentiation and adult activation of song behavior. *J. Neurobiol.* 18: 569–582.

Nottebohm, F., and Arnold, A. P. (1976). Sexual dimorphism in vocal control areas of the songbird brain. *Science* 194: 211–213.

Nottebohm, F., and Nottebohm, M. E. (1976). Left hypoglossal dominance in the control of canary and white-crowned sparrow song. *J. Comp. Physiol.* 108: 171–192.

Pepe, G. J., and Albrecht, E. D. (1990). Regulation of the primate fetal adrenal cortex. *Endocrine Rev.* 11: 151–176

Pohl-Apel, G. (1985). The correlation between the degree of brain masculinization and song quality in estradiol-treated female zebra finches. *Brain Res.* 336: 381–383.

Pohl-Apel, G., Sossinka, G., and Sossinka, R. (1984). Hormonal determination of song capacity in females of the zebra finch: Critical phase of treatment. *Z. Tierpsychol.* 64: 330–336.

Prove, E. (1974). Der Einfluss von Kastration und Testosteronsubstitution auf das sexualverhalten mannlicher Zebrafinken (*Taeniophygia guttata castanotis* Gould). *J. Ornithol.* 115: 338–347.

Sanghera, M. K., Simpson, E. R., McPhaul, M. J., Kozlowski, G., Conley, A. J., and Lephart, E. D. (1991). Immunocytochemical distribution of aromatase cytochrome P450 in the rat brain using peptide-generated polyclonal antibodies. *Endocrinology* 129: 2834–2844.

Scharff, C., and Nottebohm, F. (1991). A comparative study of the behavioral deficits following lesions of various parts of the zebra finch song system: Implications for vocal learning. *J. Neurosci.* 11: 2896–2913.

Schlinger, B. A., Amur-Umarjee, S., Campagnoni, A. T., and Arnold, A. P. (in press a). Neuronal and non-neuronal cells express high levels of aromatase in primary cultures of developing zebra finch telencephalon. *J. Neurosci.*

Schlinger, B. A., Amur-Umarjee, S., Campagnoni, A. T., and Arnold, A. P. (in press b). 5β-Reductase and other androgen-metabolizing enzymes in mixed primary cultures of developing zebra finch telencephalon. *J. Neuroendocrinol.*

Schlinger, B. A., and Arnold, A. P. (1991a). Androgen effects on development of the zebra finch song system. *Brain Res.* 561: 99–105.

Schlinger, B. A., and Arnold, A. P. (1991b). Brain is the major site of estrogen synthesis in a male songbird. *Proc. Natl. Acad. Sci. (USA)* 88: 4191–4194.

Schlinger, B. A., and Arnold, A. P. (1992a). Plasma sex steroids and tissue aromatization in hatchling zebra finches: Implications for the sexual differentiation of singing behavior. *Endocrinology* 130: 289–299.

Schlinger, B. A., and Arnold, A. P. (1992b). Circulating estrogens in a male songbird originate in the brain. *Proc. Natl. Acad. Sci. (USA)* 89: 7650–7653.

Schlinger, B. A., and Arnold, A. P. (1993). Estrogen synthesis *in vivo* in the adult zebra finch: Additional evidence that circulating estrogens can originate in brain. *Endocrinology* 133: 2610–2616.

Schlinger, B. A., and Callard, G. V. (1987). A comparison of aromatase, 5α- and 5β-reductase activities in the brain and pituitary of male and female quail (*C. c. japonica*). *J. Exp. Zool.* 242: 171–180.

Schlinger, B. A., and Callard G. V. (1989). Localization of aromatase in synaptosomal and microsomal subfractions of quail (*Coturnix japonica*) brain. *Neuroendocrinology* 49: 434–441.

Schlinger, B. A., and Callard, G. V. (1991). Brain–steroid interactions and the control of aggressiveness in birds. In *Neuroendocrine Perspectives*, Vol. 9, ed. E. E. Muller and R. M. MacLeod, pp. 1–43. New York: Springer.

Schlinger, B. A., Popper, P., Shen, P. J., Micevych, P. E., Campagnoni, A., and Arnold, A. P. (1993). Quantitative analysis of aromatase mRNA in the zebra finch brain. *Soc. Neurosci. Abstr.* 19: 171.

Schlinger, B. A., Slotow, R. H., and Arnold, A. P. (1992a). Plasma estrogens and brain aromatase in winter white-crowned sparrows. *Ornis Scand.* 23: 292–297.

Schlinger, B. A., Vanson, A. M., Amur-Umarjee, S., Campagnoni, A., and Arnold, A. P. (1992b). 3β-Hydroxysteroid dehydrogenase/isomerase activity in primary cultures from developing zebra finch telencephalon. *Soc. Neurosci. Abstr.* 18: 231.

Schumacher, M., Hutchison, R. E., and Hutchison, J. B. (1991). Inhibition of hypothalamic aromatase activity by 5β-dihydrotestosterone. *J. Neuroendocrinol.* 3: 221–226.

Schwabl, H. (1992). Winter and breeding territorial behavior and levels of reproductive hormones of migratory European robins. *Ornis Scand.* 23: 271–276.

Shen, P., Campagnoni, C. W., Kampf, K., Schlinger, B. A., Arnold, A. P., and Campagnoni, A. T. (in press). Isolation and characterization of a zebra finch aromatase cDNA: *In situ* hybridization reveals high aromatase expression in brain. *Mol. Brain Res.*

Shinoda, K., Sakamoto, N., Osawa, Y., and Pearson, J. (1989a). Aromatase neurons in the monkey forebrain demonstrated by antibody against human placental antigen X-P$_2$ (hPAX-P$_2$). *Soc. Neurosci. Abstr.* 15: 232.

Shinoda, K., Yagi, H., Fujita, H., Osawa, Y., and Shiotani, Y. (1989b). Screening of aromatase-containing neurons in rat forebrain: An immunohistochemical study with antibody against human placental antigen X-P$_2$ (hPAX-P$_2$). *J. Comp. Neurol.* 290: 502–515.

Simpson, H. B., and Vicario, D. S. (1990). Brain pathways for learned and unlearned vocalizations differ in zebra finches. *J. Neurosci.* 10: 1541–1556.

Simpson, H. B., and Vicario, D. S. (1991a). Early estrogen treatment alone causes female zebra finches to produce male-like vocalizations. *J. Neurobiol.* 22: 755–776.

Simpson, H. B., and Vicario, D. S. (1991b). Early estrogen treatment of female zebra finches masculinizes the brain pathway for learned vocalization. *J. Neurobiol.* 22: 777–793.

Sohrabji, F., Nordeen, K. W., and Nordeen, E. J. (1990). Selective impairment of song learning following lesions of a forebrain nucleus in juvenile zebra finches. *Neural Behav. Biol.* 53: 51–63.

Thorpe, W. H. (1961). *Bird-Song: The Biology of Vocal Communication and Expression in Birds.* Cambridge University Press.

Vockel, A., Prove, E., and Balthazart, J. (1990a). Sex- and age-related differences in the activity of testosterone metabolizing enzymes in microdissected nuclei of the zebra finch brain. *Brain Res.* 511: 291–302.

Vockel, A., Prove, E., and Balthazart, J. (1990b). Effects of castration on testosterone metabolizing enzymes in the brain of male and female zebra finches. *J. Neurobiol.* 21: 808–825.

Wade, J., and Arnold, A. P. (1994). Post-hatching inhibition of aromatase activity does not alter sexual differentiation of the zebra finch song system. *Brain Res.* 639: 347–350.

Walters, M. J., McEwen, B. S., and Harding, C. F. (1988). Estrogen receptor levels in hypothalamic and vocal control nuclei in the male zebra finch. *Brain Res.* 459: 37–43.

Weichel, K., Schwager, G., Heid, P., Guttinger, H. R., and Pesch, A. (1986). Sex differences in plasma steroid concentrations and singing behavior during ontogeny in canaries (*Serinus canaria*). *Ethology* 73: 281–294.

Welty, J. C. (1975). *The Life of Birds.* Philadelphia: Saunders.

Wilson, J. D., George, F. W., and Griffin, J. E. (1981). The hormonal control of sexual development. *Science* 211: 1278–1284.

Wingfield, J. C., Ball, G. F., Dufty, Jr. A. M., Hegner, R. E., and Ramenofsky, M. (1987). Testosterone and aggression in birds. *Am. Sci.* 75: 602–608.

Wingfield, J. C., and Farner, D. S. (1980). Control of seasonal reproduction in temperate-zone birds. *Prog. Reprod. Biol.* 5: 62–101.

Wingfield, J. C., and Monk, D. (1992). Control and context of year-round territorial aggression in the non-migratory Song Sparrow (*Zonotrichia melodia morphna*). *Ornis Scand.* 23: 298–303.

14

Neurobiological regulation of hormonal response by progestin and estrogen receptors

JEFFREY D. BLAUSTEIN, MARC J. TETEL, AND
JOHN M. MEREDITH

Introduction

Ovarian hormones have many cellular actions in the central nervous system that result in changes of behaviors and reproductive physiology (Blaustein and Olster 1989). One approach to unraveling the cellular processes by which ovarian hormones act on the brain has been the study of hormonal regulation of female sexual behavior. While the induction of such behavior usually requires stimulation by ovarian hormones, the specific hormonal conditions required for the stimulation and inhibition of sexual behavior vary in accord with the antecedent hormonal conditions in each species. These hormonal conditions include patterns as different as the sequential presence of estradiol and progesterone in rats and guinea pigs (Dempsey et al. 1936; Boling and Blandau 1939), the sequential presence of progesterone and estradiol in sheep (Robinson 1954), the presence of estradiol alone in prairie voles (Dluzen and Carter 1979), and the presence of testosterone metabolized neuronally to estradiol in musk shrews (Rissman 1991). We have studied the hormonal regulation of sexual behavior in rats and guinea pigs by the sequential presence of estradiol and progesterone. In ovariectomized guinea pigs injected with estradiol and progesterone, as during the estrous cycle, the period of sexual receptivity lasts for approximately 8 hours. While the specific cellular endpoints may vary in each of these species, the fundamental cellular processes by which hormones act are likely to be similar in all species.

Estradiol and progesterone have numerous specific effects on neuronal physiology (Pfaff and Schwartz-Giblin 1988). For example, estradiol and/or progesterone regulate second-messenger systems (Etgen et al. 1992; also see Chapter 16 by Hoffman et al., this volume), neuropeptides (Simerly et al. 1989; also see Chapters 5 by Flanagan and McEwen, 6 by Hammer and Cheung, 7 by Popper et al., 9 by Akesson and Micevych, and 11 by De Vries, this volume), neurotransmitter/neuropeptide receptors (Johnson 1992; also see Chapters 5 by Flanagan and McEwen, 6 by Hammer and Cheung, and 7 by Popper et al., this volume), neuritic outgrowth *in vitro* (Toran-Allerand 1991 and Chapter 17, this volume), and synapse formation (Matsumoto 1991). While the specific relationship of each of these cellular responses to sexual behavior is not known, each suggests possible mechanisms by which sex steroid hormones may influence brain function and behavior.

In early models of steroid hormone action based on studies of peripheral reproductive tissues and tumor cell lines, it was suggested that unoccupied receptors were located in the cytoplasm, reaching the cell nucleus only after occupation by steroid hormone (Gorski et al. 1968; Jensen et al. 1968). However, it is now thought that both unoccupied and occupied estrogen receptors (ER) and progestin receptors (PR) are present almost exclusively in cell nuclei (King and Greene 1984; Welshons et al. 1984). After activation, usually by binding to their cognate ligand, receptors are transformed, and the hormone–receptor complex may then cause changes in gene expression, leading to alterations in protein synthesis and consequently to altered cellular functions (Gorski et al. 1986). This model is consistent with the idea that sex steroid hormone receptors belong to a large superfamily of receptors that function as transcription factors.

Like peripheral reproductive tissues, some neurons have sex steroid hormone receptors (Stumpf 1968; Eisenfeld 1969). The neuroanatomical distribution of ER and PR, and the heterogeneous distribution of neurons containing these receptors, have been described in rats, guinea pigs, and many other species using *in vitro* binding assay of microdissected tissues (e.g., Rainbow et al. 1982), autoradiography (e.g., Stumpf 1970), immunocytochemistry (Blaustein et al. 1988; Blaustein and Turcotte 1989a; DonCarlos et al. 1991; Fox et al. 1991; Dellovade et al. 1992; Tobet et al. 1993; Li et al. 1994), and *in situ* hybridization (Lauber et al. 1990; Simerly et al. 1990). High concentrations of ER are present in discrete regions of forebrain (including the anterior hypothalamus, preoptic area, mediobasal hypothalamus, and amygdala) and midbrain, all sites at which estradiol has effects on behavior and physiology. PR are induced by estradiol in many of these areas.

In this review of the cellular processes involved in hormonal regulation of sexual behavior, we will discuss the regulation of steroid hormone receptors by ovarian hormones in the context of hormonal regulation of sexual behavior, as well as a series of preliminary experiments in which we have begun to delineate the neuronal populations that respond directly and indirectly to ovarian steroid hormones. We will then present the results of studies on the regulation of steroid hormone receptors by afferent input. Finally, we will discuss data demonstrating the presence of steroid hormone receptors in unexpected subcellular neuronal locations.

Necessity of ER and PR for the induction of sexual behavior

As in non-neural tissues, the neural effects of estradiol and progesterone on sexual behavior in rodents are believed to require interaction with intracellular, hormone-specific receptors (Blaustein and Olster 1989). The necessity of neural ER for many of the neuronal actions of estradiol has been shown by a variety of correlational experiments and through the use of steroid hormone antagonists. Estradiol stimulation results in cell nuclear binding of ER in areas containing intracellular ER. Treatments with antiestrogens that block the binding of estradiol to hypothalamic ER also block the neural actions of estradiol on sexual behavior (Blaustein and Olster 1989). In many areas, the concentration of PR increases after estradiol priming (MacLusky and McEwen 1978; Blaustein and Feder 1979a; Moguilewsky and Raynaud 1979). Treatment of estradiol-primed animals with progesterone causes a transient accumulation of cell nuclear PR (Blaustein and Feder 1980), which remains elevated with a time course that correlates quite well with the duration of sexual receptivity. Progesterone facilitation of sexual behavior can be blocked by administration of a progestin antagonist (Brown and Blaustein 1984; Etgen and Barfield 1986) as well as by treatment with PR antisense oligonucleotides that block the synthesis of PR (Ogawa et al. 1992). Therefore, the binding of estradiol and progesterone to their respective intracellular receptors is a critical step in the mechanism of hormonal regulation of sexual behavior.

Site of action of estradiol for sexual behavior in guinea pigs

The results of earlier studies in which estradiol or progesterone was implanted into the brain suggested that the mediobasal hypothalamus was a site for progesterone facilitation of sexual receptivity in estradiol-primed guinea pigs (Morin and Feder 1974). We assessed progesterone-facilitated sexual behavior in guinea pigs implanted bilaterally with small (33-gauge) cannulae containing dilute estradiol (Delville and Blaustein 1991). Only cannulae located at the rostral and ventral aspect of the ventrolateral hypothalamus (VLH) induced a behavioral response to progesterone. The VLH, as we use the term in this chapter, includes the ER- and PR-rich area adjacent to the ventromedial hypothalamus in guinea pigs. This area contains the ventrolateral nucleus of the hypothalamus and its surround, and at rostral levels the steroid hormone receptor–immunoreactive cells form a crescent between the arcuate nucleus and the fornix. Because PR immunoreactivity is estradiol-dependent, animals were perfused and sections were immunostained for progestin receptors. Only the presence of PR immunoreactivity in the rostral–ventral VLH was correlated with response to progesterone (Fig. 14.1). Thus, stimulation of a population of ovarian steroid hormone–sensitive neurons

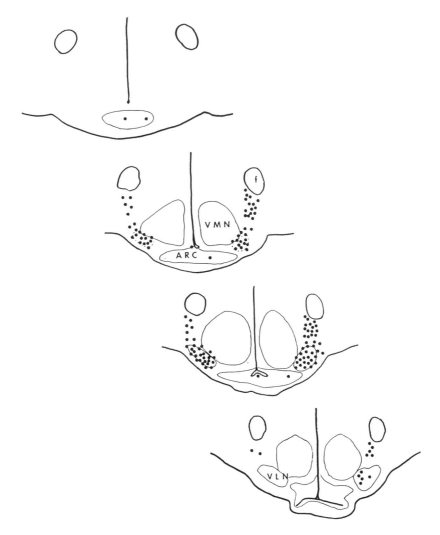

Figure 14.1. Reconstruction of the PR immunoreactivity observed at different levels in estradiol-implanted guinea pigs that became sexually receptive after progesterone injection. The scores of PR-IR cell counts within the rostral–ventral aspects of the VLH correlate with sexual receptivity. Abbreviations: VMN, ventromedial nucleus of the hypothalamus; VLN, ventrolateral nucleus of the hypothalamus; ARC, arcuate nucleus. Single dot represents one PR-IR neuron. After Delville and Blaustein (1991).

within the VLH by very low levels of estradiol may be sufficient to induce sexual behavior in response to progesterone in female guinea pigs. The fact that virtually all cells in which estradiol induces PR immunoreactivity in this area contain ER immunoreactivity (Blaustein and Turcotte 1989b; Warembourg et al. 1989) suggests that estradiol and progesterone act in the same cells.

It should be emphasized that the steroid receptor–rich VLH just described may be functionally homologous with the ventromedial hypothalamic area in rats.

However, in guinea pigs, there is a wider distribution of ER- and PR-immunoreactive (IR) neurons. Tract-tracing studies in guinea pigs suggest that the sex steroid receptor–IR region of the VLH has projections similar to those of the ventromedial nucleus of the hypothalamus (VMN) in rats (Ricciardi and Blaustein 1994). Furthermore, the neuropeptide substance P (SP) is found in PR-IR neurons in the VLH of guinea pigs (Nielsen and Blaustein 1990), as well as in ER-containing neurons in the VMN in rats (Akesson and Micevych 1988).

Heterogeneous regulation of steroid receptors

The distribution of ER and PR in the brain is quite heterogeneous. Some neuroanatomical regions have many cells containing ER, while others have none. Cells containing high concentrations of ER-IR or PR-IR are often located in the vicinity of cells containing low concentrations or undetectable levels of these receptors. Furthermore, many factors, including hormonal and environmental stimuli, can markedly affect the concentration of receptors in subpopulations of neurons. One goal of research on steroid receptor–containing neurons has been to define subpopulations of steroid receptor–IR cells based on their neurotransmitter phenotype or afferent or efferent projections. In some cases, it has been determined that a particular cell phenotype responds in a specified way, while other phenotypes do not. This idea is best exemplified in a non-neuronal tissue, the rabbit oviduct (Hyde et al. 1989). Differences in cell phenotype are more obvious in this tissue than in the brain, and estradiol has opposite effects on the different cell types. While estradiol increases PR immunoreactivity in some cell types, it has no effect or actually decreases PR immunoreactivity in other cell types. Unfortunately, neuronal phenotype is not as straightforward as cellular phenotype in the oviduct, which hinders efforts to categorize the cellular response to a particular hormone treatment.

Progesterone-induced decrease of PR concentration

Following termination of sexual receptivity, downregulation of hypothalamic PR occurs (Blaustein and Feder 1979b; Brown and Blaustein 1985; Blaustein and Olster 1989), and guinea pigs become refractory to progesterone for facilitation of sexual receptivity. We have suggested that downregulation of PR is a critical cellular event that leads to the termination of behavioral estrus and to the ensuing behavioral refractoriness to progesterone. To determine if behaviorally relevant progesterone treatments selectively downregulate PR in specific neuroanatomical areas, ovariectomized guinea pigs were injected with behaviorally effective doses of estradiol benzoate followed by oil or progesterone 42 hours later (Blaustein and Turcotte 1990), and animals were perfused 4, 12, or 24 hours after progesterone.

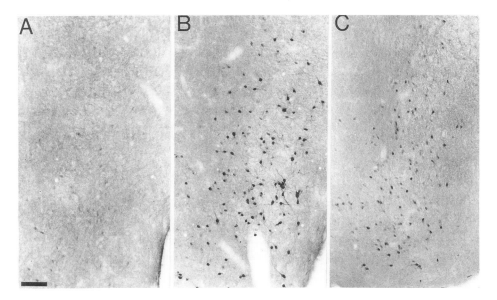

Figure 14.2. PR immunostaining in the rostral aspect of the VLH of ovariectomized guinea pigs injected with (A) oil vehicle (0 hours) and oil vehicle (42 hours), perfused 24 hours later; (B) estradiol benzoate (0 hours) and oil vehicle (42 hours), perfused 24 hours later; or (C) estradiol benzoate (0 hours) and progesterone (42 hours), perfused 24 hours later. Scale bar, 100 μm. After Blaustein and Turcotte (1990).

As shown previously (Blaustein et al. 1988; DonCarlos et al. 1989), estradiol dramatically increased the number of PR-IR cells in defined regions of the preoptic area and hypothalamus. By 24 hours after progesterone injection, a decrease in PR immunoreactivity had occurred in all neuroanatomical areas studied. However, the number of PR-IR cells in the VLH decreased substantially by just 12 hours after progesterone injection (Fig. 14.2). This suggests that the VLH may be particularly responsive to the effects of progesterone on downregulation of PR.

Induction of PR immunoreactivity and sex behavior by pulses of estradiol

Another experiment that demonstrated the site-specific effects of hormones on steroid receptors utilized an estradiol pulse paradigm. The administration of small, discrete pulses of estradiol-17β is as effective as that of a much larger single dose of the esterified, longer-acting estradiol benzoate in priming ovariectomized rats (Sodersten et al. 1981) and guinea pigs (Olster and Blaustein 1990) to display progesterone-facilitated lordosis. Immunocytochemistry was used to determine whether PR-containing cells within particular subareas of the hypothalamus and preoptic area are differentially responsive to these two estradiol treatments (Olster and Blaustein 1990). Estradiol pulse treatment induced fewer

PR-IR cells in the arcuate nucleus and medial preoptic–anterior hypothalamic nuclei than did treatment with estradiol benzoate. In contrast, PR immunoreactivity in the medial preoptic area, the periventricular preoptic area, and the VLH did not differ between groups of animals receiving the two different estradiol treatments. This suggests that PR-IR cells in the VLH (and medial and periventricular preoptic area) are particularly responsive to the behaviorally sufficient estradiol pulse treatment.

Recent work has suggested that substance P (sP) may play a role in the hormonal induction of lordosis in rats (Dornan et al. 1987). When female guinea pigs were treated with subcutaneous Silastic capsules delivering high chronic levels of estradiol, 36% of the PR-IR cells in the VLH also contained sP immunoreactivity (Olster and Blaustein 1992), but when female guinea pigs were injected with two small, behaviorally effective pulses of estradiol-17β, 53% of the PR-IR cells contained sP immunoreactivity. This is consistent with the idea that subpopulations of PR-IR cells may be differentially responsive to estradiol depending on the mode of administration.

Estradiol pulses and ER immunoreactivity

We have hypothesized that the increased effectiveness of estradiol-17β pulses in inducing sexual behavior could be due to differential regulation of a subpopulation of ER-containing cells by the two estradiol treatments. Because the concentration of receptors often predicts sensitivity to hormones (Blaustein and Olster 1989), we hypothesized that the greater effectiveness of pulsatile estradiol, as opposed to chronic high levels of estradiol, could be due to the upregulation of ER by the first estradiol pulse, thereby increasing sensitivity to estradiol in some neurons.

We examined the effects of a single low dose of estradiol-17β (2 μg; similar to the first pulse of a multiple-pulse regimen) on ER. Because binding of the H 222 antibody to ER in *para*-formaldehyde-fixed tissue sections seems to be decreased by the occupation of the ER (Blaustein 1993), we used the ER-21 antiserum (Blaustein 1992a; provided by G. Greene), which apparently binds to both occupied and unoccupied ER.

In contrast to our prediction, ER immunoreactivity using the ER-21 antiserum decreased within 4 hours of a single injection of estradiol-17β in some areas (e.g., the VLH; Fig. 14.3). In other areas (e.g., the medial preoptic area), there was no consistent change in the number of ER-IR cells. This apparent downregulation is consistent with reports that the level of ER mRNA in the rat hypothalamus decreases following treatment with estradiol (Lauber et al. 1991; Simerly and Young 1991). While the results do not support the idea that pulses of estradiol increase ER-IR in some neurons, they do not rule out the possibility that upregula-

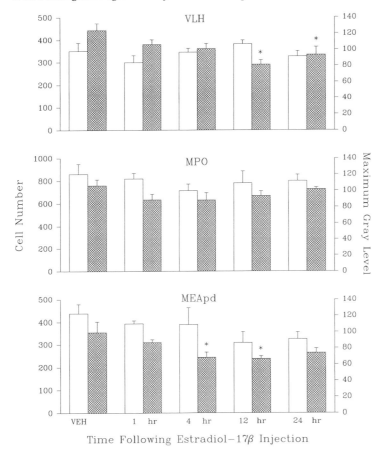

Figure 14.3. Estradiol downregulation of ER immunoreactivity. Cell number (open bars) and maximum pixel density per cell (hatched bars) in the VLH, the medial preoptic area (MPOA), and the posterodorsal portion of the medial amygdala (MEApd) of female guinea pigs treated with 2 μg of estradiol-17β sc at various times before perfusion. Asterisk denotes $p < .05$ contrasted with vehicle controls.

tion of ER might occur in a brain region not examined or in a specific subpopulation of ER-IR cells.

Effects of estradiol on Fos immunoreactivity

It is apparent that not all neurons in any given region contain the same concentration of ER or PR. Therefore, heterogeneous sensitivity and response to estradiol is expected between neuroanatomical areas and among neurons within a particular area. While a great deal of attention has been given to finding neurotransmitters in steroid receptor–containing neurons, neurons without cell nuclear ER can also respond either directly or indirectly to estradiol. To begin to identify populations of neurons that respond to behaviorally effective hormonal treatments, we have utilized Fos immunocytochemistry. Fos, the protein product of the proto-oncogene c-*fos*, is minimally expressed in cells exhibiting basal levels of activity, but it is induced genomically within minutes upon extracellular stimulation (Morgan and Curran 1991). Thus, immunocytochemical detection of Fos can be used as a marker for neurons that respond to a particular hormonal or environmental stimulus (Sagar et al. 1988).

Previous studies that attempted to identify neurons which respond to particular hormonal treatments using this technique in female rats obtained somewhat conflicting results. The number of Fos-IR cells was shown to increase in the medial preoptic area, medial amygdala nucleus, and VMN from 12 to 48 hours after injection of a large dose of estradiol (100 μg/kg; Insel 1990). After injection of a lower dose of estradiol benzoate (10 μg/kg), the Fos-IR cell number increased only in the anterior medial preoptic area. In contrast, another study in which rats were injected with 5 μg estradiol and 5 μg estradiol benzoate revealed no increase in the number of Fos-IR cells in the preoptic area or hypothalamus within 24 hours (Gibbs et al. 1990). However, when rats were injected with a behaviorally effective dose of estradiol benzoate (5 μg 44 hours before perfusion), the number of Fos-IR cells increased by more than 600% in the medial preoptic area and by more than 200% in the dorsomedial hypothalamus (A. Auger and J. Blaustein, unpublished observations). The fact that Fos immunoreactivity increased in cells in the dorsomedial hypothalamus following estradiol benzoate treatment, despite the relative lack of estrogen receptors in this area, suggests that the action of estradiol may be indirect in at least some of these cells. Suprisingly, there was no effect of hormone priming on Fos expression observed 44 hours after estradiol benzoate treatment in the VMN or the adjacent medial tuberal area, both areas containing high concentrations of ER-IR cells. However, because it is likely that particular neurons and particular brain regions respond with different latencies (Jennes et al. 1992) than others, we cannot rule out the possibility that Fos immunoreactivity increases at an earlier or later time in these regions.

Noradrenergic inputs onto ER-containing neurons

Of the many neurotransmitters that influence female sexual behavior in rodents, norepinephrine has a particularly well-established role (Nock and Fedar 1981; Crowley et al. 1989). For example, a decrease in the synthesis of norepinephrine by inhibition of dopamine β-hydroxylase or by noradrenergic antagonists blocks the hormonal induction of lordosis, and stimulation of noradrenergic receptors increases sexual behavior. Infusion of noradrenergic agonists into the mediobasal hypothalamus facilitates lordosis, while noradrenergic antagonists implanted in this same area inhibit lordosis in guinea pigs and rats (Fernandez-Guasti et al. 1985; Etgen 1990; Malik et al. 1993).

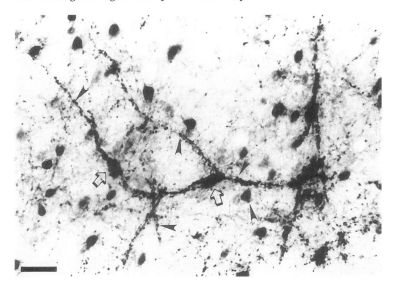

Figure 14.4. A small population of ER-IR cell bodies (open arrows) and processes in the dorsomedial hypothalamus that appear to be completely enveloped by dopamine β-hydroxylase varicosities (arrowheads), suggestive of noradrenergic terminals. Scale bar, 25 μm. After Tetel and Blaustein (1991).

Steroid receptor–containing neurons are innervated by afferent fibers containing a wide range of neurotransmitters (Brown et al. 1990). Noradrenergic varicosities, which are putative sites of norepinephrine release, are closely associated with many ER-IR neurons throughout the preoptic area and hypothalamus (Fig. 14.4; Tetel and Blaustein 1991). In fact, almost 80% of the ER-IR neurons in the VLH have closely associated noradrenergic varicosities. These findings extend the earlier observation of catecholaminergic terminals surrounding [3H]estradiol-labeled cells in the rat hypothalamus (Heritage et al. 1980).

Noradrenergic regulation of ER and PR

There is evidence that this putative noradrenergic input onto ER-IR neurons might regulate the concentration of steroid hormone receptors in subpopulations of these neurons, providing a mechanism for altering the sensitivity of these neurons to steroid hormones. *In vitro* binding studies have revealed that noradrenergic antagonists decrease ER and/or estradiol-induced PR in the mediobasal hypothalamus of guinea pigs and rats (Blaustein and Olster 1989; Blaustein 1992b). We reported that ER-IR neurons with closely associated noradrenergic varicosities in the rostral aspect of the VLH had more intense immunostaining, suggestive of higher ER concentration, than did ER-IR neurons lacking this association (Tetel and Blaustein 1991). This correlation was not observed in the caudal VLH, suggesting a regional specificity of this apparent regulation in the rostral VLH.

Activation of neurons by vaginal-cervical stimulation of neurons

We used Fos immunocytochemistry to study the afferent regulation of forebrain and midbrain neurons by tactile stimuli associated with female reproduction in rats. Vaginal-cervical stimulation (VCS), normally provided by the male during copulation, elicits many behavioral and endocrine changes associated with female reproduction, such as increasing the intensity of lordosis (Diakow 1975), decreasing the duration of sexual receptivity (Blandau et al. 1941), and inducing pseudopregnancy (Butcher et al. 1972). We identified neurons that respond to VCS in hormone-primed (5 μg estradiol benzoate followed 44 hours later by 500 μg progesterone), freely mating females. The sensitivity of the Fos immunocytochemical technique was titrated so that few Fos-IR cells were present in control animals, making it possible to identify individual neurons responding to VCS (Tetel et al. 1993a). Dramatic increases in Fos expression were observed in discrete forebrain and midbrain areas (Fig. 14.5) following VCS induced either by mating or manually with a probe (Tetel et al. 1993a). Similar results have been observed by other groups using similar experimental paradigms (Pfaus et al. 1993; Rowe and Erskine 1993).

Effects of steroid hormones on VCS-induced neuronal response

In another study, we examined the role of steroid hormones in VCS-induced Fos expression. The effects of manual VCS on Fos immunoreactivity were compared in animals that were hormone-primed, as in the previous studies, or injected with vehicle. Suprisingly, while no effect of hormones on VCS-induced Fos immunoreactivity was detected in any other VCS-responsive brain region, hormonal priming actually *decreased* by half the number of VCS-induced Fos-IR cells in the VMN (M. Tetel et al., unpublished observations). This decrease may represent an inhibitory response of these neurons to VCS following hormonal priming, which was previously found in studies involving electrophysiological techniques (Chan et al. 1984). Thus, the suppression of Fos induction by VCS in VMN neurons, which contain dense steroid hormone receptors, suggests that hormones can alter the neuronal responsiveness to environmental stimuli such as VCS. While the results of this experiment, taken together with results of electrophysiological experiments (Haskins and Moss 1983; Chan et al. 1984), suggest that hormones act centrally to alter the responsiveness of hypothalamic neurons to VCS, we cannot rule out the possibility that estradiol and progesterone act peripherally to elicit their effects on VCS responsiveness. An effect of estradiol acting peripherally on sexual receptivity has been suggested by recent studies using the steroidal antiestrogen ICI 182,780, which decreased sexual receptivity without having any apparent effect on binding of estradiol in the brain (Wade et al. 1993). Neverthe-

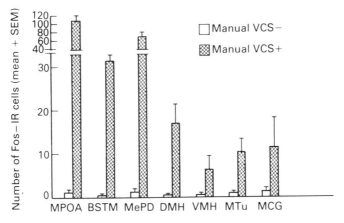

Figure 14.5. Mean number of Fos-IR cells ($\pm$ SEM) per section in the presence (manual VCS+) or absence (manual VCS−) of VCS. Abbreviations: MPOA, medial preoptic area; BSTM, medial bed nucleus of the stria terminalis; MePD, posterodorsal portion of the medial amygdala; DMH, dorsomedial hypothalamus; VMH, ventromedial hypothalamus; MTu, medial tuberal nucleus and surrounding area; MCG, midbrain central gray. After Tetel et al. (1993b).

less, whether hormones act centrally or peripherally, the results demonstrate that hormonal priming decreases the number of neurons in the VMN that respond to VCS.

The presence of ER within VCS-responsive neurons

All of the brain regions that respond to VCS by increasing Fos immunoreactivity also contain steroid hormone receptors (Stumpf 1970; Pfaff and Keiner 1973). We used a double-label immunofluorescent technique to label Fos and ER in order to determine whether any of the VCS-responsive neurons contained ER immunoreactivity. In fact, we found that the majority of the manual VCS-induced Fos-IR cells in the medial division of the bed nucleus of the stria terminalis, medial preoptic area, and posterodorsal portion of the medial amygdala also contained ER immunoreactivity (Figs. 14.6 and 14.7). Furthermore, many of the VCS-responsive cells in the VMN and midbrain central gray contained ER immunoreactivity as well. These findings suggest that hormonal and sensory information associated with female reproduction converge within subpopulations of neurons. This integration of sensory and hormonal information may occur by "cross-talk" (Tzukerman et al. 1991) between the two transcription factors, Fos and ER. Perhaps VCS alters the sensitivity of these cells to estradiol through regulation of ER by Fos and other immediate early gene products. In support of this putative mechanism, immediate early gene products can regulate the activity of ER (Doucas et al. 1991), as well as other steroid hormone receptors (Shemshedini et al. 1991).

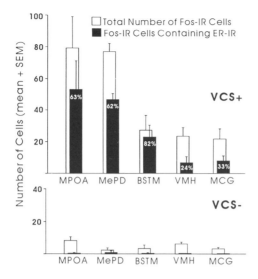

Figure 14.6. Mean number of Fos-IR cells (± SEM) in the presence (VCS +) or absence (VCS −) of VCS. The percentage shown within the filled bars represents the proportion of VCS-induced Fos-IR cells that also contained ER immunoreactivity. For abbreviations, see legend to Figure 14.5.

Alternatively, estradiol may influence the responsiveness of neurons to VCS by acting through an estrogen response element in the c-*fos* gene (Weisz and Rosales 1990). The finding that hormonal priming alters the number of VMN neurons responding to VCS provides support for this mechanism of integration.

It has been suggested that noradrenergic neurons mediate tactile stimuli provided by the male, such as VCS (Hansen et al. 1980). In support of this hypothesis, lesions of the ventral noradrenergic bundle, which innervates the hypothalamus (Moore and Bloom 1979), block VCS-induced pseudopregnancy in rats (Hansen et al. 1980). As already discussed, noradrenergic neurons interact with steroid hormone–sensitive neurons in the hypothalamus and preoptic area. These findings, taken together with the fact that stimulation of noradrenergic receptors increases Fos immunoreactivity in the brain (Bing et al. 1992), suggest that noradrenergic neurons could be involved in the VCS-induced increase in Fos immunoreactivity observed in ER-IR neurons.

Cytoplasmic steroid hormone receptors

There is abundant evidence that one of the neuronal mechanisms of action by which estradiol and progesterone influence sexual behavior involves cell nuclear ER and PR acting genomically in cell nuclei as transcription factors. Most studies in peripheral reproductive tissues (e.g., Stumpf et al. 1983; King and Greene

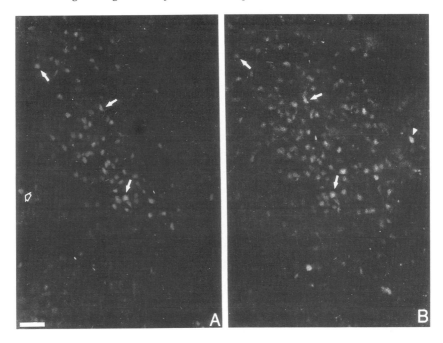

Figure 14.7. (A) VCS-Induced Fos-IR cells and (B) ER-IR cells from the medial bed nucleus of the stria terminalis. White arrows in A and B point to cells containing Fos and ER immunoreactivity. Open arrow in A points to cell containing Fos immunoreactivity only, and white arrowhead in B points to cell containing ER immunoreactivity only. Scale bar, 50 μm.

1984; Perrot-Applanat et al. 1985; Ennis et al. 1986; Warembourg et al. 1986; Gasc et al. 1989), as well as in the brain (e.g., Cintra et al. 1986; Sar and Parikh 1986; Gahr et al. 1987; Guennoun et al. 1987; DonCarlos et al. 1989; Axelson and Van Leeuwen 1990; Brown et al. 1990; Liposits et al. 1990; Herbison and Theodosis 1991; Kallo et al. 1992), have found nearly exclusive cell nuclear localization of ER and PR.

In contrast, recent immunocytochemical studies provide further potential sites of action for steroid hormones other than cell nuclei within neurons. In our initial studies of neuronal PR (Blaustein et al. 1988) and ER immunoreactivity (Blaustein and Turcotte 1989a), as well as in recent studies in many other species, including rats (ER: Blaustein 1992a; androgen receptors: Wood and Newman 1993), opossums (Fox et al. 1991), musk shrews (Dellovade et al. 1992), hamsters (Li et al. 1994), ferrets (Tobet et al. 1993), quail (Balthazart et al. 1989), and sheep (Lehman et al. 1993), receptors have also been observed in the perikaryal cytoplasm and in cytoplasmic processes with the appearance of both axons and dendrites (Fig. 14.8). There is tremendous heterogeneity in the appearance of cytoplasmic immunostaining. While in some cases it is present more than 500 μm

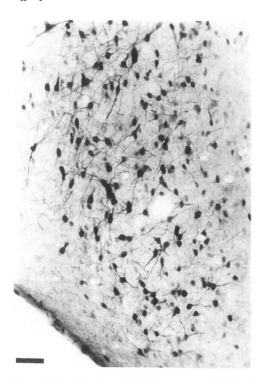

Figure 14.8. Cytoplasmic ER immunoreactivity in the VLH of an ovariectomized guinea pig brain, visualized by the diaminobenzidine–peroxidase technique in freezing microtome-cut sections. Note the heterogeneity in the presence of cytoplasmic reaction product among cells. Scale bar, 50 μm.

from the cell nucleus, the cytoplasmic immunostaining is *often* considerably lighter than the cell nuclear staining. Furthermore, in some neurons the cytoplasmic reaction product is restricted to the perikaryal cytoplasm, while in other neurons it is abundant in cytoplasmic processes. Finally, it is quite common for cells containing abundant cytoplasmic immunoreactivity to be located adjacent to cells completely lacking cytoplasmic immunoreactivity.

This finding is not idiosyncratic to a particular antibody. While in earlier studies in which ER immunoreactivity was reported in neuronal cytoplasm, the same antibody – the H 222 monoclonal antibody (Greene et al. 1984) with an epitope in the ligand binding region of the estrogen receptor – was used, we extended the results using other antisera (Blaustein 1992a). We observed a similar pattern of ER immunoreactivity in rat brain using either a polyclonal antiserum against the "hinge" region between the steroid binding domain and the DNA binding domain of the rat ER (ER 715; Furlow et al. 1990) or a polyclonal antiserum raised against the 21–amino acid N-terminus of the rat ER (Blaustein 1992a; ER-21, gift of G. Greene).

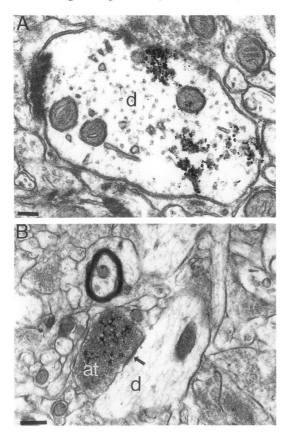

Figure 14.9. Electron microscope photomicrograph of ER immunoreactivity in a dendrite and axon terminal of neurons in guinea pig brain. (A) ER-IR reaction product is scattered within a distal dendrite (d) within the VLH of a female guinea pig. Scale bar, 0.2 μm. (B) ER-IR reaction product is associated with clear round vesicles in an axon terminal (at) within the VLH of a female guinea pig. ER-IR was visualized by the silver-intensified diaminobenzidine–peroxidase technique. Note the absence of reaction product in surrounding structures. Scale bar, 0.5 μm.

Ultrastructural localization of ER

Having established that ER immunoreactivity is present in the perikaryal cytoplasm and cytoplasmic processes of some neurons, we used an ultrastructural technique to determine whether these cytoplasmic processes were axons or dendrites (Blaustein et al. 1992), a distinction that could not be made at the light microscopic level. We observed abundant reaction product in VLH and midbrain central gray in both proximal and small distal dendrites (Fig. 14.9), and in axon terminals, containing predominantly round, and in some cases flattened, clear vesicles. Within the somata, reaction product was observed in association with rough endoplasmic reticulum. Similar patterns are seen when either the H 222 antibody or the antiserum directed against the N-terminus of the receptor (ER 21) is used, suggesting that the ER immunoreactivity observed in the dendrites and axon terminals represents the entire receptor, not just a cleavage product.

In another study that adds support to the idea that some steroid receptors are present in axon terminals, ovariectomized guinea pigs were first injected intra-cerebroventricularly with the axonal transport inhibitor colchicine in a typical procedure that enhances neuropeptide immunostaining in neuronal perikarya. This treatment resulted in the appearance of ER and PR immunoreactivity in cell nuclei and cytoplasm in areas typically lacking these receptors in guinea pigs (ER and PR in the dorsomedial thalamic region; PR in the medial amygdala; Blaustein and Olster 1993). These results suggest that ER and PR may be transported down axons in some neurons at a rapid enough rate that they do not ordinarily accumulate within the soma of the neurons. Although we cannot rule out the possibility that the appearance of ER and PR immunoreactivity is due to an effect of colchicine that does not involve direct inhibition of axonal transport of steroid receptors, the results are consistent with the finding that ER and PR immunoreactivity is transported to axon terminals.

While we were initially surprised to observe ER in cytoplasm, cytoplasmic localization of transcription factors is quite common (Whiteside and Goodbourn 1993). Because the ER is synthesized in the cytoplasm and activation of nuclear localization signals may then direct the receptors into cell nuclei (Whiteside and Goodbourn 1993), it is actually expected that some portion of the total cellular receptors will be extranuclear at any time. It has been shown that second messengers can influence the phosphorylation and activation of these nuclear localization sites. For example, Power et al. (1991) suggested that stimulation of dopamine receptors may phosphorylate these nuclear localization signals, resulting in movement of the receptors from cytoplasm to cell nuclei. Thus, the proportion of receptors present in cell nuclei versus cytoplasm could be influenced by a number of factors, including activation by second messengers, perhaps via phosphorylation.

Functions of receptors in cytoplasmic processes

In addition to the general issue of why ER should be present outside the cell nucleus, the ultrastructural findings raise the question of why ER are found in specific, unpredicted subcellular locations, such as the dendritic processes. If ER are found to be associated with polyribosomes, they could be involved in mediating the actions of estradiol on synapse formation (Matsumoto 1991).

The presence of ER in axon terminals suggests possibilities for direct estradiol action at the presynaptic terminal. Direct effects of estradiol on neurotransmitter release have been suggested by experiments that have examined the effects of estradiol on the efflux of catecholamines from synaptosomes (Nixon et al. 1974), release of catecholamines from the hypothalamus in organ culture (Paul et al. 1979), and amphetamine-induced dopamine release in the striatum (Becker 1990). Alternatively, as with dendritic ER, axon terminal ER could be involved in the action of estradiol on synapse formation in the hypothalamus.

Finally, as already discussed, models for membrane receptor mechanisms of action of steroid hormones on the brain have received attention and support. There is a good deal of evidence for rapid effects of steroid hormones on neuronal physiology, membrane receptors for steroid hormones, and non-nuclear effects of steroid hormones on physiology and behavior (Kelly et al. 1977; Towle and Sze 1983; Ramirez et al. 1985; Nabekura et al. 1986; Becker 1990; Ke and Ramirez 1990; McEwen 1991; Frye et al. 1992). Since the techniques that we used to study cytoplasmic receptors were not designed to determine whether any of the receptors are associated with postsynaptic membranes, it is possible that some of the ER or PR in dendrites and axon terminals represent receptors being transported to function as membrane receptors.

Summary

Steroid hormone receptors and steroid-sensitive neurons are heterogeneously distributed throughout the nervous system. Superimposed on this heterogeneous distribution is the regulation of receptor concentrations in subpopulations of these neurons. A population of neurons can have high concentrations of ER or PR in one physiological state and low concentrations in another. For example, in the absence of estradiol there are few PR-IR neurons in the brain, but exposure to estradiol produces a dramatic increase in PR immunoreactivity in many of these ER-IR neurons. Thus, this population of neurons is likely to be responsive only to estradiol under one physiological condition, but to estradiol and progesterone under another. It has been argued that this induction of PR mediates the responsiveness of the neurons to progesterone for facilitation of sexual behavior. While progesterone generally downregulates PR in the brain, this occurs more rapidly in the VLH than in other areas. Similarly, we have shown that a single pulse of estradiol decreases the concentration of its own receptor, but only in some neuroanatomical areas. A challenge facing researchers in this field is to determine the phenotypes of cells that respond differentially to particular hormonal treatments.

In this chapter, we have also examined the potential role of neurotransmitters in the regulation of hormone-dependent sexual behavior and the associated cellular events. We discussed experiments suggesting that norepinephrine regulates ER and PR in the hypothalamus. The results of a neuroanatomical study suggest that direct noradrenergic innervation could regulate the concentration of ER in subpopulations of these neurons.

We have reviewed the results of experiments in which Fos immunocytochemistry was used to determine neuronal pathways sensitive to steroid hormones and peripheral stimulation. We presented studies that revealed an increase in the expression of Fos in areas associated with reproduction following steroid hormone or VCS. In one neuroanatomical area, the presence of steroid hormones decreases the neuronal responsiveness to tactile stimuli such as VCS. Consistent with the idea that this regulation is due to a direct effect of steroid hormones, we have observed that the majority of the cells that express Fos following VCS also contain ER. These data suggest that hormonal and sensory information associated with female reproduction may converge in subpopulations of ER-containing neurons.

Finally, while it is generally agreed that estradiol and progesterone act, in part, through genomic processes, it appears that there are also receptors for each hormone at other neuronal sites. While the majority of ER have been observed in the cell nucleus (suggestive of genomic actions) there is a smaller, but consistent population of ER located in the perikaryal cytoplasm and cytoplasmic processes, including dendrites and axon terminals. Therefore, the tremendous heterogeneity in the cell-by-cell regulation of steroid hormone receptors in the brain may be complicated by heterogeneous regulation even among subcellular locations within individual neurons. Thus, neuronal sensitivity to a particular hormone could fluctuate in response to changing physiological demands, not only among neurons but also in subcellular structures within each neuron.

References

Akesson, T. R., and Micevych, P. E. (1988). Estrogen concentration by substance P-immunoreactive neurons in the medial basal hypothalamus of the female rat. *J. Neurosci. Res.* 19: 412–419.

Axelson, J. F., and VanLeeuwen, F. W. (1990). Differential localization of estrogen receptors in various vasopressin synthesizing nuclei of the rat brain. *J. Neuroendocrinol.* 2: 209–216.

Balthazart, J., Gahr, M., and Surlemont, C. (1989). Distribution of estrogen receptors in the brain of the Japanese Quail: An immunocytochemical study. *Brain Res.* 501: 205–214.

Becker, J. B. (1990). Direct effect of 17-beta-estradiol on striatum: Sex differences in dopamine release. *Synapse* 5: 157–164.

Bing, G. Y., Stone, E. A., Zhang, Y., and Filer, D. (1992). Immunohistochemical studies of noradrenergic-induced expression of c-*fos* in the rat CNS. *Brain Res.* 592: 57–62.

Blandau, R. J., Boling, J. L., and Young, W. C. (1941). The length of heat in the albino rat as determined by the copulatory response. *Anat. Rec.* 79: 453–463.

Blaustein, J. D. (1992a). Cytoplasmic estrogen receptors in rat brain: Immunocytochemical evidence using three antibodies with distinct epitopes. *Endocrinology* 131: 1336–1342.

Blaustein, J. D. (1992b). Modulation of sex steroid receptors by neurotransmitters: Relevant techniques. *Neuroprotocols* 1: 42–51.

Blaustein, J. D. (1993). Estrogen receptor immunoreactivity in rat brain: Rapid effects of estradiol injection. *Endocrinology* 132: 1218–1224.

Blaustein, J. D., and Feder, H. H. (1979a). Cytoplasmic progestin receptors in guinea pig brain: Characteristics and relationship to the induction of sexual behavior. *Brain Res.* 169: 481–497.

Blaustein, J. D., and Feder, H. H. (1979b). Cytoplasmic progestin receptors in female guinea pig brain and their relationship to refractoriness in expression of female sexual behavior. *Brain Res.* 177: 489–498.

Blaustein, J. D., and Feder, H. H. (1980). Nuclear progestin receptors in guinea pig brain measured by an in vitro exchange assay after hormonal treatments that affect lordosis. *Endocrinology* 106: 1061–1069.

Blaustein, J. D., King, J. C., Toft, D. O., and Turcotte, J. (1988). Immunocytochemical localization of estrogen-induced progestin receptors in guinea pig brain. *Brain Res.* 474: 1–15.

Blaustein, J. D., Lehman, M. N., Turcotte, J. C., and Greene, G. (1992). Estrogen receptors in dendrites and axon terminals in the guinea pig hypothalamus. *Endocrinology* 131: 281–290.

Blaustein, J. D., and Olster, D. H. (1989). Gonadal steroid hormone receptors and social behaviors. In *Advances in Comparative and Environmental Physiology*, Vol. 3, ed. J. Balthazart, pp. 31–104. Berlin: Springer.

Blaustein, J. D., and Olster, D. H. (1993). Colchicine-induced accumulation of estrogen receptor and progestin receptor immunoreactivity in atypical areas in guinea-pig brain. *J. Neuroendocrinol.* 5: 63–70.

Blaustein, J. D., and Turcotte, J. C. (1989a). Estrogen receptor-immunostaining of neuronal cytoplasmic processes as well as cell nuclei in guinea pig brain. *Brain Res.* 495: 75–82.

Blaustein, J. D., and Turcotte, J. C. (1989b). Estradiol-induced progestin receptor immunoreactivity is found only in estrogen receptor-immunoreactive cells in guinea pig brain. *Neuroendocrinology* 49: 454–461.

Blaustein, J. D., and Turcotte, J. C. (1990). Down-regulation of progestin receptors in guinea pig brain: New findings using an immunocytochemical technique. *J. Neurobiol.* 21: 675–685.

Boling, J. L., and Blandau, R. J. (1939). The estrogen–progesterone induction of mating responses in the spayed female rat. *Endocrinology* 25: 359–364.

Brown, T. J., and Blaustein, J. D. (1984). Inhibition of sexual behavior in female guinea pigs by a progestin receptor antagonist. *Brain Res.* 301: 343–349.

Brown, T. J., and Blaustein, J. D. (1985). Loss of hypothalamic nuclear-bound progestin receptors: Factors involved and the relationship to heat termination in female guinea pigs. *Brain Res.* 358: 180–190.

Brown, T. J., Maclusky, N. J., Leranth, C., Shanabrough, M., and Naftolin, F. (1990). Progestin receptor-containing cells in guinea pig hypothalamus: Afferent connections, morphological characteristics, and neurotransmitter content. *Mol. Cell. Neurosci.* 1: 58–77.

Butcher, R. L., Fugo, N. W., and Collins, W. E. (1972). Semicircadian rhythm in plasma levels of prolactin during early gestation in the rat. *Endocrinology* 90: 1125–1127.

Chan, A., Dudley, C. A., and Moss, R. L. (1984). Hormonal and chemical modulation of ventromedial hypothalamic neurons responsive to vaginocervical stimulation. *Neuroendocrinology* 38: 328–336.

Cintra, A., Fuxe, K., Harfstrand, A., Agnati, L. F., Miller, L. S., Greene, J. L., and Gustafsson, J. A. (1986). On the cellular localization and distribution of estrogen receptors in the rat tel- and diencephalon using monoclonal antibodies to human estrogen receptor. *Neurochem. Intern.* 8: 587–595.

Crowley, W. R., O'Connor, L. H., and Feder, H. H. (1989). Neurotransmitter systems and social behavior. In *Advances in Comparative and Environmental Physiology*, Vol. 3, ed. J. Balthazart, pp. 162–208. Berlin: Springer.

Dellovade, T. L., Blaustein, J. D., and Rissman, E. F. (1992). Neural distribution of estrogen receptor immunoreactive cells in the female musk shrew. *Brain Res.* 595: 189–194.

Delville, Y., and Blaustein, J. D. (1991). A site for estradiol priming of progesterone-facilitated sexual receptivity in the ventrolateral hypothalamus of female guinea pigs. *Brain Res.* 559: 191–199.

Dempsey, E. W., Hertz, R., and Young, W. C. (1936). The experimental induction of oestrus (sexual receptivity) in the normal and ovariectomized guinea pig. *Am. J. Physiol.* 116: 201–209.

Diakow, C. (1975). Motion picture analysis of rat mating behavior. *J. Comp. Physiol. Psychol.* 88: 704–712.

Dluzen, D. E., and Carter, C. S. (1979). Ovarian hormones regulating sexual and social behaviors in female prairie voles, *Microtus ochrogaster*. *Physiol. Behav.* 23: 597–600.

DonCarlos, L. L., Greene, G. L., and Morrell, J. I. (1989). Estrogen plus progesterone increases progestin receptor immunoreactivity in the brain of ovariectomized guinea pigs. *Neuroendocrinology* 50: 613–623.

DonCarlos, L. L., Monroy, E., and Morrell, J. I. (1991). Distribution of estrogen receptor-immunoreactive cells in the forebrain of the female Guinea Pig. *J. Comp. Neurol.* 305: 591–612.

Dornan, W. A., Malsbury, C. W., and Penney, R. B. (1987). Facilitation of lordosis by injection of substance P into the midbrain central gray. *Neuroendocrinology* 45: 498–506.

Doucas, V., Spyrou, G., and Yaniv, M. (1991). Unregulated expression of c-Jun or c-Fos proteins but not Jun-D inhibits oestrogen receptor activity in human breast cancer derived cells. *EMBO J.* 10: 2237–2245.

Eisenfeld, A. J. (1969). Hypothalamic oestradiol-binding macromolecules. *Nature* 224: 1202–1203.

Ennis, B. W., Stumpf, W. E., Gasc, J.-M., and Baulieu, E.-E. (1986). Nuclear localization of progesterone receptor before and after exposure to progestin at low and high temperatures: Autoradiographic and immunohistochemical studies of chick oviduct. *Endocrinology* 119: 2066–2075.

Etgen, A. M. (1990). Intrahypothalamic implants of noradrenergic antagonists disrupt lordosis behavior in female rats. *Physiol. Behav.* 48: 31–36.

Etgen, A. M., and Barfield, R. J. (1986). Antagonism of female sexual behavior with intracerebral implants of antiprogestin RU 38486: Correlation with binding to neural progestin receptors. *Endocrinology* 119: 1610–1617.

Etgen, A. M., Ungar, S., and Petitti, N. (1992). Estradiol and progesterone modulation of norepinephrine neurotransmission: Implications for the regulation of female reproductive behavior. *J. Neuroendocrinol.* 4: 255–271.

Fernandez-Guasti, A., Larsson, K., and Beyer, C. (1985). Potentiative action of α- and β-adrenergic receptor stimulation in inducing lordosis behavior. *Pharmacol. Biochem. Behav.* 22: 613–617.

Fox, C. A., Ross, L. R., Handa, R. J., and Jacobson, C. D. (1991). Localization of cells containing estrogen receptor-like immunoreactivity in the Brazilian opossum brain. *Brain Res.* 546: 96–105.

Frye, C. A., Mermelstein, P. G., and Debold, J. F. (1992). Evidence for a non-genomic action of progestins on sexual receptivity in hamster ventral tegmental area but not hypothalamus. *Brain Res.* 578: 87–93.

Furlow, J. D., Ahrens, H., Mueller, G. C., and Gorski, J. (1990). Antisera to a synthetic peptide recognize native and denatured rat estrogen receptors. *Endocrinology* 127: 1028–1032.

Gahr, M., Flugge, G., and Guttinger, H.-R. (1987). Immunocytochemical localization of estrogen-binding neurons in the songbird brain. *Brain Res.* 402: 173–177.

Gasc, J. M., Delahaye, F., and Baulieu, E. E. (1989). Compared intracellular localization of the glucocorticosteroid and progesterone receptors: An immunocytochemical study. *Exper. Cell Res.* 181: 492–504.

Gibbs, R. B., Mobbs, C. V., and Pfaff, D. W. (1990). Sex steroids and Fos expression in rat brain and uterus. *Mol. Cell. Neurosci.* 1: 29–40.

Gorski, J., Toft, D., Shyamala, G., Smith, D., and Notides, A. (1968). Hormone receptors: Studies on the interaction of estrogen with the uterus. *Rec. Prog. Horm. Res.* 24: 45–80.

Gorski, J., Welshons, W. V., Sakai, D., Hansen, J., Walent, J., Kassis, J., Shull, J., Stack, G., and Campen, C. (1986). Evolution of a model of estrogen action. *Rec. Prog. Horm. Res.* 42: 297–329.

Greene, G. L., Sobel, N. B., King, W. J., and Jensen, E. V. (1984). Immunochemical studies of estrogen receptors. *J. Steroid Biochem.* 20: 51–56.

Guennoun, R., Reyss-Brion, M., and Gasc, J.-M. (1987). Progesterone receptors in hypothalamus and pituitary during the embryonic development of the chick: Regulation by sex steroid hormones. *Dev. Brain Res.* 37: 1–9.

Hansen, S., Stanfield, E. J., and Everitt, B. J. (1980). The role of ventral bundle noradrenergic neurones in sensory components of sexual behavior and coitus-induced pseudopregnancy. *Nature* 286: 152–154.

Haskins, J. T., and Moss, R. L. (1983). Action of estrogen and mechanical vaginocervical stimulation on the membrane excitability of hypothalamic and midbrain neurons. *Brain Res. Bull.* 10: 489–496.

Herbison, A. E., and Theodosis, D. T. (1991). Neurotensin-immunoreactive neurons in the rat medial preoptic area are oestrogen-receptive. *J. Neuroendocrinol.* 3: 587–589.

Heritage, A. S., Stumpf, W. E., Sar, M., and Grant, L. D. (1980). Brainstem catecholamine neurons are target sites for sex steroid hormones. *Science* 207: 1377–1379.

Hyde, B. A., Blaustein, J. D., and Black, D. L. (1989). Differential regulation of progestin receptor immunoreactivity in the rabbit oviduct. *Endocrinology* 125: 1479–1483.

Insel, T. R. (1990). Regional induction of c-fos-like protein in rat brain after estradiol administration. *Endocrinology* 126: 1849–1853.

Jennes, L., Jennes, M. E., Purvis, C., and Nees, M. (1992). c-fos expression in noradrenergic A2-neurons of the rat during the estrous cycle and after steroid hormone treatments. *Brain Res.* 586: 171–175.

Jensen, E. V., Suzuki, T., Kawasima, T., Stumpf, W. E., Jungblut, P. W., and de Sombre, E. R. (1968). A two-step mechanism for the interaction of estradiol with rat uterus. *Proc. Natl. Acad. Sci. (USA)* 59: 632–638.

Johnson, A. E. (1992). The regulation of oxytocin receptor binding in the ventromedial hypothalamic nucleus by gonadal steroids. In *Oxytocin in Maternal, Sexual, and Social Behaviors*, N.Y. Acad. Sci. Vol. 652, ed. C. A. Pedersen, J. D. Caldwell, G. F. Jirikowski, and T. R. Insel, pp. 357–373. New York: New York Academy of Science.

Kallo, I., Liposits, Z., Flerko, B., and Coen, C. W. (1992). Immunocytochemical characterization of afferents to estrogen receptor-containing neurons in the medial preoptic area of the rat. *Neuroscience* 50: 299–308.

Ke, F. C., and Ramirez, V. D. (1990). Binding of progesterone to nerve cell membranes of rat brain using progesterone conjugated to I-125-bovine serum albumin as a ligand. *J. Neurochem.* 54: 467–472.

Kelly, M. J., Moss, R. L., Dudley, C. A., and Fawcett, C. P. (1977). The specificity of the response of preoptic-septal area neurons to estrogen: 17α-estradiol versus 17β-estradiol and the response of extrahypothalamic neurons. *Exp. Brain Res.* 30: 43–52.

King, W. J., and Greene, G. L. (1984). Monoclonal antibodies localize oestrogen receptor in the nuclei of target cells. *Nature* 307: 745–747.

Lauber, A. H., Mobbs, C. V., Muramatsu, M., and Pfaff, D. W. (1991). Estrogen receptor messenger RNA expression in rat hypothalamus as a function of genetic sex and estrogen dose. *Endocrinology* 129: 3180–3186.

Lauber, A. H., Romano, G. J., Mobbs, C. V., and Pfaff, D. W. (1990). Estradiol regulation of estrogen receptor messenger ribonucleic acid in rat mediobasal hypothalamus: An in situ hybridization study. *J. Neuroendocrinol.* 2: 605–611.

Lehman, M. N., Ebling, F. J. P., Moenter, S. M., and Karsch, F. J. (1993). Distribution of estrogen receptor-immunoreactive cells in the sheep brain. *Endocrinology* 133: 876–886.

Li, H. Y., Blaustein, J. D., DeVries, G. J., and Wade, G. N. (1994). Estrogen receptor-immunoreactivity in hamster brain: Preoptic area, hypothalamus and amygdala. *Brain Res.* 61: 304–312.

Liposits, Z., Kallo, I., Coen, C. W., Paull, W. K., and Flerko, B. (1990). Ultrastructural analysis of estrogen receptor immunoreactive neurons in the medial preoptic area of the female rat brain. *Histochemistry* 93: 233–239.

MacLusky, N. J., and McEwen, B. S. (1978). Oestrogen modulates progestin receptor concentrations in some rat brain regions but not in others. *Nature* 274: 276–278.

Malik, K. F., Morrell, J. I., and Feder, H. H. (1993). Effects of clonidine and phentolamine infused into the medial preoptic area and medial basal hypothalamus of the guinea pig. *Neuroendocrinology* 57: 177–188.

Matsumoto, A. (1991). Synaptogenic action of sex steroids in developing and adult neuroendocrine brain. *Psychoneuroendocrinology* 16: 25–40.

McEwen, B. S. (1991). Non-genomic and genomic effects of steroids on neural activity. *Trends. Pharmacol. Sci.* 12: 141–147.

Moguilewsky, M., and Raynaud, J. P. (1979). The relevance of hypothalamic and hypophyseal progestin receptor regulation in the induction and inhibition of sexual behavior in the female rat. *Endocrinology* 105: 516–522.

Moore, R. Y., and Bloom, F. E. (1979). Central catecholamine neuron systems: Anatomy and physiology of the norepinephrine and epinephrine systems. *Ann. Rev. Neurosci.* 2: 113–168.

Morgan, J. I., and Curran, T. (1991). Stimulus-transcription coupling in the nervous system: Involvement of the inducible proto-oncogenes fos and jun. *Annu. Rev. Neurosci.* 14: 421–451.

Morin, L. P., and Feder, H. H. (1974). Hypothalamic progesterone implants and facilitation of lordosis behavior in estrogen-primed ovariectomized guinea pigs. *Brain Res.* 70: 81–93.

Nabekura, J., Oomura, Y., Minami, T., Mizuno, Y., and Fukuda, A. (1986). Mechanism of the rapid effect of 17β-estradiol on medial amygdala neurons. *Science* 233: 226–228.

Nielsen, K. H., and Blaustein, J. D. (1990). Many progestin receptor-containing neurons in the guinea pig ventrolateral hypothalamus contain substance-P: Immunocytochemical evidence. *Brain Res.* 517: 175–181.

Nixon, R. L., Janowsky, D. S., and Davis, J. M. (1974). Effects of progesterone, β-estradiol, and testosterone on the uptake and metabolism of ^{3}H-norepinephrine,

^{3}H-dopamine and ^{3}H-serotonin in rat brain synaptosomes. *Res. Comm. Chem. Pathol. Pharmacol.* 7: 233–236.

Nock, B., and Feder, H. H. (1981). Neurotransmitter modulation of steroid action in target cells that mediate reproduction and reproductive behavior. *Neurosci. Biobehav. Rev.* 5: 437–447.

Ogawa, S., Olazabal, U. E., and Pfaff, D. W. (1992). Behavioral change after local administration of antisense sequence for progesterone receptor mRNA in female rat hypothalamus. In *Antisense Strategies*, N.Y. Acad. Sci. Vol. 660, ed. R. Baserga and D. T. Denhardt, pp. 298–299. New York: New York Academy of Science.

Olster, D. H., and Blaustein, J. D. (1990). Biochemical and immunocytochemical assessment of neural progestin receptors following estradiol treatments that eliminate the sex difference in progesterone-facilitated lordosis in guinea-pigs. *J. Neuroendocrinol.* 2: 79–86.

Olster, D. H., and Blaustein, J. D. (1992). Estradiol pulses induce progestin receptors selectively in substance P-immunoreactive neurons in the ventrolateral hypothalamus of female guinea pigs. *J. Neurobiol.* 23: 293–301.

Paul, S. M., Axelrod, J., Saavedra, J. M., and Skolnick, P. (1979). Estrogen-induced efflux of endogenous catecholamines from the hypothalamus *in vitro*. *Brain Res.* 178: 499–505.

Perrot-Applanat, M., Logeat, F., Groyer-Picard, M. T., and Milgrom, E. (1985). Immunocytochemical study of mammalian progesterone receptor using monoclonal antibodies. *Endocrinology* 116: 1473–1484.

Pfaff, D., and Keiner, M. (1973). Atlas of estradiol-concentrating cells in the central nervous system of the female rat. *J. Comp. Neurol.* 151: 121–158.

Pfaff, D. W., and Schwartz-Giblin, S. (1988). Cellular mechanisms of female reproductive behaviors. In *The Physiology of Reproduction*, ed. E. Knobil, and J. Neill, pp. 1487–1568. New York: Raven Press.

Pfaus, J. G., Kleopoulos, S. P., Mobbs, C. V., Gibbs, R. B., and Pfaff, D. W. (1993). Sexual stimulation activates c-*fos* within estrogen concentrating regions of the female rat forebrain. *Brain Res.* 624: 253–267.

Power, R. F., Mani, S. K., Codina, J., Conneely, O. M., and O'Malley, B. W. (1991). Dopaminergic and ligand-independent activation of steroid hormone receptors. *Science* 254: 1636–1639.

Rainbow, T., Parsons, B., MacLusky, N., and McEwen, B. (1982). Estradiol receptor levels in rat hypothalamic and limbic nuclei. *J. Neurosci.* 2: 1439–1445.

Ramirez, V. D., Kim, K., and Dluzen, D. (1985). Progesterone action on the LHRH and the nigrostriatal dopamine neuronal systems: In vitro and in vivo studies. *Rec. Prog. Horm. Res.* 41: 421–472.

Ricciardi, K. H. N., and Blaustein, J. D. (1994). Projections from the ventrolateral hypothalamic neurons containing progestin receptor- and substance P-immunoreactivity to specific forebrain and midbrain areas in female guinea pigs. *J. Neuroendocrinol.* 6: 135–144.

Rissman, E. (1991). Evidence that neural aromatization of androgen regulates the expression of sexual behaviour in female musk shrews. *J. Neuroendocrinol.* 3: 441–448.

Robinson, J. T. (1954). Necessity for progesterone with estrogen for the induction of recurrent estrus in the ovariectomized ewe. *Endocrinology* 55: 403–408.

Rowe, D. W., and Erskine, M. S. (1993). c-*fos* proto-oncogene activity induced by mating in the preoptic area, hypothalamus, and amygdala in the female rat: Role of afferent input via the pelvic nerve. *Brain Res.* 621: 25-34.

Sagar, S. M., Sharp, F. R., and Curran, T. (1988). Expression of c-*fos* protein in brain: Metabolic mapping at the cellular level. *Science* 240: 1328–1330.

Sar, M., and Parikh, I. (1986). Immunohistochemical localization of estrogen receptor in rat brain, pituitary and uterus with monoclonal antibodies. *J. Steroid Biochem.* 24: 497–503.

Shemshedini, L., Knauthe, R., Sassonecorsi, P., Pornon, A., and Gronemeyer, H. (1991). Cell-specific inhibitory and stimulatory effects of Fos and Jun on transcription activation by nuclear receptors. *EMBO J.* 10: 3839–3849.

Simerly, R. B., Chang, C., Muramatsu, M., and Swanson, L. W. (1990). Distribution of androgen and estrogen receptor messenger RNA-containing cells in the rat brain: An *in situ* hybridization study. *J. Comp. Neurol.* 294: 76–95.

Simerly, R. B., and Young, B. J. (1991). Regulation of estrogen receptor messenger ribonucleic acid in rat hypothalamus by sex steroid hormones. *Mol. Endocrinol.* 5: 424–432.

Simerly, R. B., Young, B. J., Capozza, M. A., and Swanson, L. W. (1989). Estrogen differentially regulates neuropeptide gene expression in a sexually dimorphic olfactory pathway. *Proc. Natl. Acad. Sci. (USA)* 86: 4766–4770.

Sodersten, P., Eneroth, P., and Hansen, S. (1981). Induction of sexual receptivity in ovariectomized rats by pulse administration of oestradiol-17β. *J. Endocrinol.* 89: 55–62.

Stumpf, W. E. (1968). Estradiol-concentrating neurons: Topography in the hypothalamus by dry-mount autoradiography. *Science* 162: 1001–1003.

Stumpf, W. E. (1970). Estrogen–neurons and estrogen–neuron systems in the periventricular brain. *Am. J. Anat.* 129: 207–218.

Stumpf, W. E., Gasc, J. M., and Baulieu, E. E. (1983). Progesterone receptors in pituitary and brain: Combined autoradiography-immunohistochemistry with tritium-labeled ligand and receptor antibodies. *Mikroskopie* 40: 359–363.

Tetel, M. J., and Blaustein, J. D. (1991). Immunocytochemical evidence for noradrenergic regulation of estrogen receptor concentrations in the guinea pig hypothalamus. *Brain Res.* 565: 321–329.

Tetel, M. J., Getzinger, M. J., and Blaustein, J. D. (1993a). Fos expression in the rat brain following vaginal-cervical stimulation by mating and manual probing. *Neuroendocrinology* 5: 397–404.

Tetel, M. J., Getzinger, M. J., and Blaustein, J. D. (1993b). Fos expression in the rat brain following vaginal-cervical stimulation by mating and manual probing. *J. Neuroendocrinol.* 5: 397–404.

Tobet, S. A., Chickering, T. W., Fox, T. O., and Baum, M. J. (1993). Sex and regional differences in intracellular localization of estrogen receptor immunoreactivity in adult ferret forebrain. *Neuroendocrinology* 58: 316–324.

Toran-Allerand, C. D. (1991). Organotypic culture of the developing cerebral cortex and hypothalamus: Relevance to sexual differentiation. *Psychoneuroendocrinology* 16: 7–24.

Towle, A. C., and Sze, P. Y. (1983). Steroid binding to synaptic plasma membrane: Differential binding of glucocorticoids and gonadal steroids. *J. Steroid Biochem.* 18: 135–143.

Tzukerman, M., Zhang, X., and Pfahl, M. (1991). Inhibition of estrogen receptor activity by the tumor promoter 12-*O*-tetradeconylphorbol-13-acetate: A molecular analysis. *Mol. Endocrinol.* 5: 1983–1992.

Wade, G. N., Blaustein, J. D., Gray, G. M., and Meredith, J. M. (1993). ICI 182,780: A "pure" antiestrogen that affects behaviors and energy balance in rats without affecting the brain. *Am. J. Physiol.* 265: R1392–R1398.

Warembourg, M., Jolivet, A., and Milgrom, E. (1989). Immunohistochemical evidence of the presence of estrogen and progesterone receptors in the same neurons of the guinea pig hypothalamus and preoptic area. *Brain Res.* 480: 1–15.

Warembourg, M., Logeat, F., and Milgrom, E. (1986). Immunocytochemical localization of progesterone receptor in the guinea pig central nervous system. *Brain Res.* 384: 121–131.

Weisz, A., and Rosales, R. (1990). Identification of an estrogen response element upstream of the human c-Fos gene that binds the estrogen receptor and the AP-1 transcription factor. *Nucl. Acids Res.* 18: 5097–5106.

Welshons, W. V., Lieberman, M. E., and Gorski, J. (1984). Nuclear localization of unoccupied oestrogen receptors. *Nature* 307: 747–749.

Whiteside, S. T., and Goodbourn, S. (1993). Signal transduction and nuclear targeting: Regulation of transcription factor activity by subcellular localization. *J. Cell Sci.* 104: 949–955.

Wood, R. I., and Newman, S. W. (1993). Intracellular partitioning of androgen receptor immunoreactivity in the brain of the male syrian hamster: Effects of castration and steroid replacement. *J. Neurobiol.* 24: 925–938.

15

Molecular actions of steroid hormones and their possible relations to reproductive behaviors

MONA FREIDIN AND DONALD PFAFF

A new generation of neurobiological work beyond the classical demonstrations of hormone–behavior relations by Beach, Young, and their colleagues has provided the launching platform for serious molecular analyses of some of the neuroendocrine mechanisms involved. Following the determination of cellular targets for estradiol in the brain (Pfaff 1968), a circuit for estrogen-dependent lordosis behavior was elucidated (Pfaff 1980). Principles of steroid hormone/central nervous system/behavior mechanisms include the apparent universality among vertebrates of certain nuclear binding features; the conservation of many endocrine and biochemical reactions from simpler non-neural tissues; the modular construction of the neural circuit for reproductive behavior; and the economic use of neural information within the circuit on both the sensory and motor sides (Pfaff et al. 1994). The explosion of molecular techniques and knowledge surrounding nuclear hormone receptors enables us to analyze brain mechanisms further for steroid-influenced behaviors using mRNA hybridization, genomic structure, protein–DNA binding, *in vitro* transcription, antisense DNA, and neurotrophic viral vector technologies.

Receptor structure

Steroid hormones influence mammalian tissues by promoting cell development and differentiation as well as acute cellular functions. In target tissues, steroids act by entering the cell and binding to specific receptor proteins. These receptors, which belong to a class of ligand-inducible transcription factors, regulate transcription by interacting with short *cis*-acting DNA elements – termed hormone response elements (HRE) – in the promoter regions of specific genes (Carson-Jurica et al. 1990; Yamamoto 1985).

Steroid receptors share physical as well as functional properties. However, selective differences in the mechanism of action of each receptor molecule must exist to allow a unique response to specific hormones. This chapter focuses on the role of estrogen and progesterone in gene transcription, with an emphasis on progesterone and its receptor and on proenkephalin gene expression in the female rat brain.

Cloning of steroid receptors allowed for a rapid advancement in our understanding of the structure–function relationships of these molecules. Different receptors from various species have been cloned by standard techniques, including the glucocorticoid, estrogen, androgen, progesterone, and vitamin D receptors (Danielsen et al. 1986; Evans 1988; Gronemeyer et al. 1987; Koike et al. 1987; Yamamoto 1985). Examination of the cDNAs for these proteins revealed highly conserved sequences and permitted the identification and cloning of other related receptor genes, such as the mineralocorticoid and retinoic acid receptors (Carson-Jurica et al. 1990; Petkovich et al. 1987; Yamamoto 1985). Molecular analyses of the receptor genes revealed the existence of a superfamily of receptors including steroids, vitamin D, retinoic acid, thyroid hormone, and other transcription factors. Despite differences in their ligands, these receptors seem to share some aspects of mechanistic pathways as transcriptional factors in regulating cellular functions.

Sequence comparison among different receptors and mutational studies have identified functional domains correlated with positive and negative modulation of transcription, nuclear translocation, DNA binding, and hormone binding (Carson-Jurica et al. 1990; Lucas and Granner 1992). The DNA binding domain is the most highly conserved region. It consists of 66–68 amino acids, of which 20 are conserved throughout the family. In particular, nine cysteine residues are conserved and form two "zinc finger" domains (Carson-Jurica et al. 1990; Freedman et al. 1988; Lucas and Granner 1992). The presence of zinc in the finger structures formed by the cysteines has been confirmed by X-ray absorption and visible light spectroscopy for the glucocorticoid receptor (GR; Freedman et al. 1988). Transfection studies using receptor chimeras from different steroid receptors have shown that gene specificity can be conferred by the DNA binding module. Target gene response to GR can be converted to that of an estrogen receptor (ER) by altering 3 amino acids in the first zinc finger region, while a single amino acid substitution permits recognition of both estrogen and glucocorticoid DNA response elements (Umesono and Evans 1989).

In addition to the DNA binding domain, two other regions of functional similarity have been identified. These two relatively conserved sequences, approximately 42 and 22 amino acids each, correspond to hormone binding domains and are located in the C-terminal portion of the steroid receptor protein. The hormone binding domain is not as highly conserved as the DNA binding region, with 30–60% homology across different classes of receptors. It does, however, display a high degree of homology across species for a particular receptor (Carson-Jurica et al. 1990; Lucas and Granner 1992). For example, the steroid binding domain of the mouse progesterone receptor (PR) shares 85 and 96% homology with human and rabbit PR in nucleotide and amino acid sequences, respectively (Scholt et al. 1991). Insertion, deletion, and point mutations in the hormone binding region of glucocorticoid and ER resulted in the loss of hormone binding activity, indicating the functional sensitivity of this region to even single amino acid changes (Kumar et al. 1986; Rusconi and Yamamoto 1987). Because of this response to manipulation, precise mapping of the hormone binding domain by mutational analysis has not been completed.

Under physiological conditions, steroid receptors are activated only by binding of their appropriate hormone ligand. Binding of a steroid hormone to its cognate receptor has been shown to increase DNA binding of the receptor to its HRE, to regulate transcription, and to stimulate the translocation of GR receptors from the cytoplasm to the nucleus (Curtis and Korach 1990; Kumar and Chambon 1988; Madan and DeFranco 1993; Picard et al. 1990; Read et al. 1988). As demonstrated by gel retardation assay, treatment of human PR with progesterone or R5020 (a PR agonist) increased binding between the receptor and an oligonucleotide containing a consensus progesterone response element (PRE) (Bagchi et al. 1988). *In vitro* transfection assays have shown that dexamethasone regulates transcription by binding to the GR, leading to complex formation of the receptor with a glucocorticoid response element (GRE) in promoter sequences linked to reporter genes (Chalepakis et al. 1988; Picard et al. 1990). Unlike other steroid receptors, which are concentrated in the cell nucleus, the GR is located primarily in the cytoplasm until it is activated by a glucocorticoid. Hormone binding causes the translocation of ligand-bound receptor from the cytoplasm to the nucleus (Chalepakis et al. 1988; Picard et al. 1990; Picard and Yamamoto 1987; Picard et al. 1988). This steroid-dependent nuclear translocation signal appears to be at least partly controlled by the hormone binding domain of the GR (Picard et al. 1990; Picard et al. 1988).

It should be noted that although hormones activate steroid receptors *in vivo*, ligand binding is not a necessary condition for receptor function. The ER, in particular, has been shown to bind its HRE with almost equal affinity whether in the estrogen-bound or ligand-free state (Curtis and Korach 1990; Murdoch et al. 1990). Cytosolic and nuclear extracts obtained from mouse and rat uterus treated either *in vivo* or *in vitro* with estradiol bind vitellogenin estrogen response element (ERE) oligonucleotides with similar association and dissociation constants (Curtis and Korach 1990). Heat-activating estrogen-free ER also failed to increase its binding to the ERE. Thus, by itself, the hormone binding domain of the ER might not regulate ER activity merely through sequence-specific DNA binding. Other factors must also be involved.

The chicken PR does not necessarily require hormone to undergo the allosteric changes necessary to bind DNA. Treatment of partially purified chicken PR with progesterone resulted in only a 1.5-fold increase of binding affinity to a 23-base-pair PRE, and only a 2-fold decrease in affinity for a nonspecific DNA sequence (Rodriguez et al. 1989). Gel shift analyses of these PR–PRE complexes showed no differences, further supporting the conclusion that progesterone is not necessary for sequence-specific binding of the chicken PR. Finally, heat or salt activation of highly purified human PR and rat liver GR also promotes binding of ligand-free receptors to their target HRE (Beato 1986; Klein-Hitpass et al. 1990).

Transcriptional regulation

Current concepts of hormone regulation of transcription are based on a model of *cis*-elements interacting with *trans*-acting factors. The DNA sequences of HREs complexed with steroid receptors represent the *cis*-elements and *trans*-acting factors of this paradigm, respectively. Studies of the mechanisms of the hormonal control of transcription have focused on identifying the DNA response elements in the 5′ promoter regions of the genes controlling a particular receptor.

HREs for steroid receptors from diverse genes have been identified and characterized. Based on comparisons of these sequences, consensus HREs for each receptor type have been derived. Many HREs are palindromic and consist of an inverted repeat sequence separated by an optimal number of nucleotides (Table 15.1). Surprisingly, it has been shown that different receptors can regulate gene function through identical consensus response elements. As shown in Table 15.1, the consensus GRE sequence is the same as the HRE for the mineralocorticoid, androgen, and progesterone receptors. In addition, the thyroid receptor and retinoic acid receptor response elements are nearly identical, except for the spacing between separated sequences.

Functional HREs with nonconsensus sequences have been described. The human PR contains two promoters that were found to be estrogen-inducible in transient transfection assays; yet these promoters did not contain any perfect, classical ERE sequences (Kastner et al. 1990). In addition, the phosphoenolpyruvate carboxykinase gene contains a functional GRE that does not contain the critical sequence TGTTCT (Peterson et al. 1988). These GREs act as positive promoters in the absence of glucocorticoid, suggesting that other cellular proteins might be involved. Rat and human proopiomelanocortin (POMC) genes also have been reported to contain atypical GREs that negatively and positively regulate transcription (Fremeau et al. 1986; Gagner and Drouin 1985; Israel and Cohen 1985). Finally, the presence of an HRE is not sufficient to confer steroidal regulation of activity. Sabol and co-workers found three consensus GRE sequences in a putative promoter region of the rat proenkephalin gene, yet were unable to induce transcription of a reporter gene from this promoter in transient transfection experiments (Joshi and Sabol 1991).

As illustrated in Table 15.1, HREs are often palindromic. This dyad structure

Table 15.1. *Typical high-affinity DNA HRE sequences*

Receptor group	HRE sequence
Glucocorticoid Progesterone Mineralocorticoid Androgen	$\frac{TG}{GT}$AACAnnnTGTTCT
Estrogen	AGGTCAnnnTGACCT (inverted repeat)
Thyroid hormone Retinoic acid	GGTCA........GGTCA (direct repeats)

Note: Derived from the promoters of strongly hormone-responsive genes in various cell types, these are typical HRE DNA sequences. Consensus steroid hormone receptor response sequences are summarized from Carson-Jurica et al. (1990) and Lucas and Granner (1992). The DNA sequence for the GRE (and the PRE) has alternative bases at the 5' end.

led many workers to propose that hormone receptors bind as dimers to the DNA. Gel retardation studies showed that two GR molecules bind to a synthetic GRE as a dimer, with one receptor bound per half-site (Tsai et al. 1988). Methylation interference studies showed that the contact points of the ER with the ERE displayed a perfect twofold rotational symmetry (Klein-Hitpass et al. 1989). A point mutation in one-half of the ERE palindrome decreased competition efficiency fivefold – far more than would be predicted if each monomer bound independently. This suggests that members of a dimer can assemble in a cooperative fashion.

The heat shock proteins are a class of cytosolic molecules known to associate with steroid receptors before their activation by hormone. It is currently held that one particular heat shock protein, hsp90, functions to repress the DNA binding activity of receptors in the absence of hormone. The hsp90 forms a large complex with both GR and PR by interacting with their hormone binding domains. The complexing of the appropriate ligand with GR or PR causes release of the activated receptor and allows DNA binding to proceed (Carson-Jurica et al. 1990; Denis et al. 1988). Surprisingly, the binding of progesterone and dissociation of PR from the PR–hsp90 complex promotes the formation of PR dimers in the absence of DNA. A positive correlation was observed between the capacity of PR isoforms to oligomerize in solution and their capacity to bind DNA (DeMarzo et al. 1991).

Regulation of gene activity by estrogen and progesterone

The mouse mammary tumor virus (MMTV) has been used as a prototypical steroid-inducible gene. The MMTV promoter contains a glucocorticoid- and progestin-inducible region from base pair -201 to -69, in which consensus GRE and PRE sequences are found; androgen and mineralocorticoid responsive sequences are present, as well. However, glucocorticoid and progesterone exert only positive controls on MMTV promoter activity, and estrogen has no effect on MMTV-linked reporter genes. Thus, this gene is of limited value in examining the interactions of estrogen and progesterone on transcriptional activity.

The PR gene

The human PR gene has been studied intensely by several groups. The PR is unusual among the various steroid receptors in that it is expressed in two forms: a lower molecular weight form of 80,000–90,000 daltons (PR_A) and a larger molecular weight form of approximately 108,000–120,000 daltons (PR_B). PR_A is an N-terminally truncated variant of PR_B. The two receptors show some functional differences depending on the species or tissue in which they are located and the promoter they are controlling (Gronemeyer et al. 1991; Kastner et al. 1990). Cloning of the cDNA and genes to the human, chicken, and mouse PR revealed a single-copy gene that produces multiple transcripts (Gronemeyer et al. 1991; Scholt et al. 1991). Two in-frame ATG codons were found in the cDNA sequence that generated transcripts corresponding to the A and B forms of the receptor (Carson-Jurica et al. 1990; Kastner et al. 1990). More important, transcription from chimeric genes revealed the existence of two promoter regions upstream and downstream of the second ATG site, from which different classes of PR mRNAs could be derived.

Estrogen and progesterone, respectively, exert positive and negative regulatory controls on PR gene transcription. Treatment of human breast cancer cell lines with estradiol increased both PR mRNA and protein levels, while treatment with R5020 decreased mRNA and protein content (Alexander et al. 1989; Read et al. 1988). Both promoter regions of the human PR gene are functionally estrogen-inducible, although neither contains a perfect consensus ERE sequence (Kastner et al. 1990). By contrast, the rabbit PR contains multiple ER and PR binding sites as determined by DNase I protection assays. However, only a single ERE conferred physiological regulation in transient transfection assays (Savouret et al. 1991). Of particular note, this ERE also mediated the downregulation of transcription by progesterone agonists even though it lacked detectable PR binding activity. The mechanism by which progesterone might negatively control PR expression has yet to be determined.

In the brain, PR mRNA is induced by estrogen in the ventromedial hypothalamus (VMH), but not in the amygdala, of genetic females (Romano et al. 1989a), but not males (Lauber et al. 1991b). Thus, the mechanisms controlling basal transcription in the brain must be determined, and the differential effect of estrogen on various brain regions and in both sexes must be described. Significant differences have been observed between a putative ERE on the PR gene and the consensus ERE using gel shift assays to analyze ER–DNA binding (A. Lauber and D. Pfaff, unpublished results). In particular, the AP-1 sequence overlaps the PR–ERE, providing opportunities for transmitter–hormone interaction. In addition, we have cloned and sequenced the rat PR promoter (X. S. Wu-Peng et al., unpublished observations) to facilitate future investigations of protein–DNA binding and transcriptional assays. The behavioral importance of new PR mRNA synthesis is shown by the application of antisense DNA directed against PR mRNA into the VMH, which markedly reduced lordosis and courtship behaviors (Ogawa et al. 1994).

Steroid hormone receptors are phosphoproteins, with phosphorylation occurring mainly on serine residues (Moudgil 1990). Phosphorylation may play a functional role in receptor activation, since hormone binding rapidly increases the degree of phosphorylation of several types of steroid receptors *in vivo*. Although the PR is highly phosphorylated in the absence of hormone, stimulation of cultured cells with progesterone increased PR phosphorylation and target gene transcription (Beck et al. 1992). This activation of transcription by progesterone-mediated phosphorylation of PR appears to be augmented by cAMP-dependent protein kinases (Beck et al. 1992; Denner et al. 1990). Levels of the receptors themselves, however, appear to be unaffected or decreased by various protein kinase activators. This suggests that the transcriptional activity of PR might be modulated through phosphorylation of PR and/or the phosphorylation of accessory proteins involved in PR-regulated transcription.

The preproenkephalin gene

The preproenkephalin (PPE) gene also appears to be regulated by steroid hormones (see Chapter 6 by Hammer and Cheung, this volume). Treatment of ovariectomized rats with estradiol rapidly increased PPE mRNA in the VMH in a dose-dependent manner, without affecting the amygdala or the caudatoputamen (Lauber et al. 1990; Romano et al. 1988, 1989b). However, Sprague–Dawley male rats given the same estradiol treatments exhibited no effect (Romano et al. 1990), while treatment of males with a high dose of estrogen begun immediately upon castration increased VMH PPE mRNA in Fisher 344 rats, a strain particularly sensitive to estrogen (Hammer et al.1993).

Recent results suggest that an alternative PPE transcript might be involved in some neurons. Certain neurons of the basal forebrain and reticular thalamus, as well as a small number of VMH neurons, synthesize a large amount of nuclear RNA representing part of the first PPE intron (Brooks et al. 1993) and utilize a novel transcriptional PPE start site (Kilpatrick et al. 1990). The first PPE intron is a candidate region for transcriptional controls, being largely unmethylated and containing sexually differentiated DNase hypersensitivity sites (Funabashi et al. 1993). This transcript is estrogen-sensitive in VMH neurons, and gel retardation assays reveal the presence of robust protein–DNA binding (P. Brooks et al., unpublished data). The relative importance of intronic sequences, compared with the conventional somatic promoter, can best be approached by *in vivo* promoter analyses in actual brain tissue. For example, *in vivo* promoter analyses using a viral vector (Kaplitt et al. 1991, 1993) with the PPE promoter driving a reporter gene reveal that a 2,700-base fragment supports regionally specific transcription in the rat brain (Kaplitt et al., in press). A similar estrogen effect on transcription is observed in the medial basal hypothalamus (J. Yin et al., unpublished results).

Whether the ER and the PR interact directly with the PPE gene remains to be determined. Estrogen is concentrated in enkephalin-immunoreactive hypothalamic neurons (Akesson and Micevych 1991), and the 5′ flanking promoter region of the PPE gene contains at least one consensus ERE and an additional half-palindromic ERE linked to an AP-1 site. Using gel shift methodology (Zhu and Pfaff 1994), we detected ER-like binding to selected ERE DNA sequences in the PPE promoter region; however, the results were not identical to those expected from the consensus ERE (Y.-S. Zhu and D. Pfaff, unpublished results). In addition, the 5′ flanking region contains glucocorticoid-responsive sequences, as determined by transient transfection assays, though consensus GREs were not evident (Joshi and Sabol 1991). These functional GRE domains may, in fact, be regulated by progesterone when the PPE gene is placed in the appropriate environment, since the steroid response elements for GR and PR are identical and their activities can be altered by neighboring promoter sequences (Carson-Jurica et al. 1990).

Steroid hormones might also affect PPE transcription through indirect routes. For example, the ER and PR might regulate PPE expression through activation of other transcription factor genes (e.g., c-*fos* and c-*jun* proto-oncogenes), through interactions with other factors or in conjunction with weaker enhancer elements that have yet to be detected. Both the ER and GR interact positively with c-*fos* and c-*jun* regulation of cAMP-inducible gene function. Co-transfection of c-*jun*, *jun*B, or *jun*D expression vectors augmented GR-dependent transcription from either MMTV long terminal repeat or synthetic GRE in cultured T-cells (Maroder et al. 1993). This effect was also cell-specific, because c-*fos* and c-*jun* blocked GR-dependent transcription in other cell types. The ovalbumin gene promoter has an ER-inducible element containing a half-palindromic ERE, which could also constitute an AP-1 site. This sequence was observed to bind a Fos–Jun complex, yet co-transfection experiments revealed that c-*fos*, c-*jun*, and ER co-activated the albumin promoter, suggesting that the ER acts as an accessory factor in regulating transcription at this site (Gaub et al. 1990).

Cyclic AMP regulation of PPE transcription has been well described. Two cAMP response elements, ENCRE I and ENCRE II, have been mapped in the 5' promoter region of the PPE gene, and these elements have been shown to act synergistically in the induction of PPE transcription by cAMP and phorbol ester (Comb et al. 1988). The transcription factors AP-1 and AP-2 bind to sequences flanking and overlapping the ENCRE II site on the gene (Comb et al. 1988). In addition, co-transfection of cDNA for the catalytic subunits of cAMP-dependent protein kinase with a vector containing the PPE promoter dramatically increased reporter gene activity (Huggenvik et al. 1991). This response was dependent on the cAMP response elements present in the PPE sequence. These results suggest that cAMP actions converge on the PPE gene at discrete loci to regulate transcription.

PPE gene expression is tightly correlated with reproductive behavior in the female rat (Lauber et al. 1990). The product of PPE gene expression, enkephalin, could act through δ-receptors to regulate courtship or reproduction, since intraventricular injection of a δ-agonist increased lordosis (Pfaus and Pfaff 1992). In the future, it will be necessary to determine how hormone-triggered genomic changes produce the estrogenic induction and tissue- and sex-specific expression of PPE mRNA.

References

Akesson, T. R., and Micevych, P. E. (1991). Endogenous opioid-immunoreactive neurons of the ventromedial hypothalamic nucleus concentrate estrogen in male and female rats. *J. Neurosci. Res*. 28: 359–366.

Alexander, I. E., Clarke, C. L., Shine, J., and Sutherland, R. L. (1989). Progestin inhibition of progesterone receptor gene expression in human breast cancer cells. *Mol. Endocrinol*. 3: 1377–86.

Bagchi, M. K., Elliston, J. F., Tsai, S. Y., Edwards, D. P., Tsai, M-J., and O'Malley, B. W. (1988). Steroid hormone-dependent interaction of human progesterone receptor with its target enhancer element. *Mol. Endocrinol*. 2: 12221–12229.

Beato, W. T. (1986). Steroid-free glucocorticoid receptor binds specifically to mouse mammary tumour virus DNA. *Nature* 324: 688–691.

Beck, C. A., Weigel, W. L., and Edwards, D. P. (1992). Hormone and cellular modulators of protein phosphorylation of transcriptional activity, DNA binding and phosphorylation of human progesterone receptor. *Mol. Endocrinol*. 6: 607–20.

Brooks, P. J., Funabashi, T., Kleopoulos, S. P., Mobbs, C. V., and Pfaff, D. W. (1993). Cell-specific expression of preproenkephalin intronic heteronuclear RNA in the rat forebrain. *Mol. Brain Res*. 19: 22–30.

Carson-Jurica, M. A., Schrader, W. T., and O'Malley, B. W. (1990). Steroid receptor family: Structure and functions. *Endocrine Rev*. 11(2): 201–220.

Chalepakis, G., Arnemann, J., Slalter, E., Bruller, H. J., Gross, B., and Beato, M. (1988). Differential gene activation by glucocorticoids and progestins through the hormone regulatory element of mouse mammary tumor virus. *Cell* 53: 371–382.

Comb, M., Mermond, N., Hyman, S. E., Pearlberg, J., Ross, M. E., and Goodman,

H. E. (1988). Proteins bound at adjacent DNA elements act synergistically to regulate human proenkephalin cAMP inducible transcription. *EMBO J.* 7(12): 3793–3805.

Curtis, S. W., and Korach, K. S. (1990). Uterine estrogen receptor interaction with estrogen-responsive DNA sequences in vitro: Effects of ligand binding on receptor DNA complexes. *Mol. Endocrinol.* 4: 276–286.

Danielsen, M., Northrop, J. P., and Ringold, G. M. (1986). The mouse glucocorticoid receptor: Mapping of functional domains by cloning sequencing and expression of wild-type and mutant receptor proteins. *EMBO J.* 5: 2513–2522

DeMarzo, A. M., Beck, C. A., Onate, S. A., and Edwards, D. P. (1991). Dimerization of mammalian progesterone receptor occurs in the absence of DNA and is related to the release of the 90KD heat shock protein. *Proc. Natl. Acad. Sci. USA* 88: 72–76.

Denis, M., Poellinger, L., Wikstom, A.-C. and Gustafsson, J.-A. (1988). Requirement of hormone for thermal conversion of glucocorticoid receptor to a DNA binding state. *Nature* 333: 686–88.

Denner, L. A., Weigel, N. L., Maxwell, B. L., Schrader, W. T., and O'Malley, B. W. (1990). Regulation of progesterone receptor-mediated transcription by phosphorylation. *Science* 250: 1740–1743.

Evans, R. M. (1988). The steroid and thyroid hormone receptor superfamily. *Science* 240: 889–93.

Freedman, L. P., Luisi, B. F., Korszun, Z. R., Busavappa, R., Sigler, P. B., and Yamamoto, K. R. (1988). The function and structure of the metal coordination sites with the glucocorticoid receptor DNA binding domain. *Nature* 334: 543–545.

Fremeau, R. T., Lundblad, J. R., Pritchelt, D. B., Wilcox, J. N., and Roberts, J. L. (1986). Regulation of pro-opiomelanocortin gene transcription in individual cell nuclei. *Science* 234: 1265–1267.

Funabashi, T., Brooks, P. J., Mobbs, C. V., and Pfaff, D. W. (1993). Tissue-specific DNA methylation and DNase hypersensitive sites in the promoter and transcribed regions of the rat preproenkephalin gene. *Mol. Cell. Neurosci.* 4: 499–509.

Gagner, J-P. and Drouin, J. (1985). Opposite regulation of proopiomelanocortin gene transcription by glucocorticoid and CRIT. *Mol. Cell. Endocrinol.* 40: 25–31.

Gaub, M-P., Bellard, M., Scheuer, I., Chambon, P., and Sassone-Corsi, P. (1990). Activation of the ovalbumin gene by the estrogen receptor involves the Fos–Jun complex. *Cell* 63: 1267–1276.

Gronemeyer, H., Meyer, M-E., Bocquel, M.-T, Kastner, P., Turcotte, B., and Chambon, P. (1991). Progestin receptor: Isoforms and antihormone action. *J. Steroid Biochem. Mol. Biol.* 40(1–3): 271–278.

Gronemeyer, H., Turcotte, B., Quirin-Stricker, C., Bocquel, M. T., Meyer, M. E., Krozowski, Z., Jeltsch, J. M., Lerouge, T., Garnier, J. M., and Chambon, P. (1987). The chicken progesterone receptor: Sequence, expression, and functional analysis. *EMBO J.* 6: 3985–3989.

Hammer, R. P., Bogic, L., and Handa, R. J. (1993). Estrogenic regulation of proenkephalin mRNA expression in the ventromedial hypothalamus of the adult male rat. *Mol. Brain Res.* 19: 129–134.

Huggenvik, J., Collard, M., Stofko, R., Seasholtz, A., and Uhler, M. (1991). Regulation of the human enkephalin promoter by two isoforms of the catalytic subunit of cyclic adenosine 3′,5′-monophosphate-dependent protein kinase. *Mol. Endocrinol.* 5: 921–930.

Israel, A., and Cohen, S. N. (1985). Hormonally mediated negative regulation of human pro-opiomelanocortin gene expression after transfection into mouse L cells. *Mol. Cell Biol.* 5: 2443.

Joshi, J., and Sabol, S. L. (1991). Proenkephalin gene expression C6 rat glioma cells:

Potentiation of cyclic adenosine 3′,5′-monophosphate-dependent transcription by glucocorticoids. *Mol. Endocrinol.* 5: 1069–1080.

Kaplitt, M. G., Kleopoulos, S. P., Hanlon, B. A., Rabkin, S. D., and Pfaff, D. W. (in press). *In vivo* promoter analysis for rat preproenkephalin studies with defective viral vectors. *Proc. Natl. Acad. Sci. (USA)*.

Kaplitt, M. G., Pfaus, J. G., Kleopoulos, S. P., Hanlon, B. A., Rabkin, S. D., and Pfaff, D. W. (1991). Expression of a functional foreign gene in adult mammalian brain following in vivo transfer via herpes simplex virus type 1 defective viral vector. *Mol. Cell. Neurosci.* 2: 320–330.

Kaplitt, M. G., Rabkin, S., and Pfaff, D. W. (1993). Molecular alterations in nerve cells: Direct manipulation and physiological mediation. In *Current Topics of Neuroendocrinology*, Vol. 11, ed. M. Imura, pp. 169–191. Berlin: Springer.

Kastner, P., Krust, A., Turcotte, B., Stropp, U., Tora, L., Gronemeyer, H., and Chambon, P. (1990). Two distinct estrogen-regulated promoters generate transcripts encoding two functionally different progesterone receptor forms A and B. *EMBO J.* 9: 1603–1614.

Kilpatrick, D., Zinn, S., Fitzgerald, M., Higuchi, H., Sabol, S., and Meyerhardt, J. (1990). Transcription of the rat and mouse proenkephalin genes is initiated at distinct sites in spermatogenic and somatic cells. *Mol. Cell. Biol.* 10: 3717–3726.

Klein-Hitpass, L., Tsai, S. Y., Green, G. L., Clark, J. H., Tsai, M.-J., and O'Malley, B. W. (1989). Specific binding of estrogen to the estrogen response element. *Mol. Cell. Biol.* 9(1): 43–49.

Klein-Hitpass, L., Tsai, S. Y., Weigel, N. L., Allan, G. F., Riley, D., Rodriguez, R., Schrader, W. T., Tsai, M.-J., and O'Malley, B. W. (1990). The progesterone receptor stimulates cell-free transcription by enhancing the formation of a stable preinitiation complex. *Cell* 60: 247–253.

Koike, S., Sakai, M., and Muramatsu, M. (1987). Molecular cloning and characterization of the rat estrogen receptor cDNA. *Nucl. Acids Res.* 15: 2499–2513.

Kumar, V., and Chambon, P. (1988). The estrogen receptor binds tightly to its responsive element as a ligand-induced homodimer. *Cell* 55: 145–156.

Kumar, V., Green, S., Staub, A., and Chambon, P. (1986). Localization of the oestradiol-binding and putative DNA-binding domains of the human estrogen receptor. *EMBO J.* 5: 2231–2235.

Lauber, A. H., Mobbs, C. V., Miramatsu, M., and Pfaff, D. W. (1991a). Estrogen receptor messenger RNA expression in the rat hypothalamus as a function of genetic sex and estrogen dose. *Endocrinology* 129: 3180–3186.

Lauber, A. H., Romano, G. L., Mobbs, C. V., Howell, R. D., and Pfaff, D. W. (1990). Oestradiol induction of proenkephalin messsenger RNA in hypothalamus: Dose-response and relation to reproductive behavior in the female rat. *Mol. Brain Res.* 8: 47–54.

Lauber, A. H., Romano, G. J., and Pfaff, D. W. (1991b). Sex differences in estradiol regulation of progestin receptor mRNA in rat mediobasal hypothalamus as demonstrated by in situ hybridization. *Neuroendocrinology* 53: 608–613.

Lucas, P. C., and Granner, D. K. (1992). Hormone response domains in gene transcription. *Annu. Rev. Biochem.* 61: 1131–1173.

Madan, A. P., and DeFranco, D. B. (1993). Bidirectional transport of glucocorticoid receptors across the nuclear envelope. *Proc. Natl. Acad. Sci. (USA)* 90: 3588–3592.

Maroder, M., Farina, A. R., Vacca, A., Felli, M. P., Meco, D., Screpanti, I., Frati, L., and Gulino, A. (1993). Cell-specific bifunctional role of Jun oncogene family members on glucocorticoid receptor dependent transcription. *Endocrinology* 7: 570–584.

Moudgil, V. K. (1990). Phosphorylation of steroid hormone receptors. *Biochem. Biophys. Acta Mol. Cell. Res.* 1055: 243–258.

Murdoch, F. E., Meier, D. A., Furlow, J. D., Grunwald, K. A., and Gorski, J. (1990). Estrogen receptor binding to a DNA response element in vitro is not dependent on estradiol. *Biochemistry* 29: 8377–8385.

Ogawa, S., Olazabal, U. E., Oarhar, I. S., and Pfaff, D. W. (1994). Effects of intra-hypothalamic administration of antisense DNA for progesterone receptor mRNA on reproductive behavior and progesterone receptor immunoreactivity in female rat. *J. Neurosci.* 14: 1766–1774.

Peterson, D. D., Magnuson, M. A., and Granner, D. K. (1988). Location and characterization of two widely separated glucocorticoid response elements in the phosphoenolpyruvate carboxykinase gene. *Mol. Cell Biol.* 8: 96–104.

Petkovich, M., Brand, N. J., Krust, A., and Chambon, P. (1987). A human retinoic acid receptor which belongs to the family of nuclear receptors. *Nature* 330: 444–446.

Pfaff, D. W. (1968). Uptake of ^{3}H-17β-estradiol in the female rat brain: An autoradiography study. *Endocrinology* 82: 1149–1155.

Pfaff, D. W. (1980). *Estrogens and Brain Function: Neural Analysis of a Hormone-Controlled Mammalian Reproductive Behavior*. New York: Springer.

Pfaff, D. W., Schwartz-Giblin, S., McCarthy, M. M., and Kow, L.-M. (1994). Cellular and molecular mechanisms of female reproductive behaviors. In *The Physiology of Reproduction*, 2d ed., Vol. 2, ed. E. Knobil and J. Neill, pp. 107–220. New York: Raven Press.

Pfaus, J. G., and Pfaff, D. W. (1992). Mu, delta, and kappa opioid receptor agonists selectively modulate sexual behaviors in the female rat: Differential dependence on progesterone. *Horm. Behav.* 26: 457–473.

Picard, D., Kumar, V., Chambon, P., and Yamamoto, K. R. (1990). Signal transduction by steroid hormones: Nuclear localization is differently regulated in estrogen and glucocorticoid receptors. *Cell* 1: 291–299.

Picard, D., Salser, S. J., and Yamamoto, K. R. (1988). A movable and regulable inactivation function within the steroid binding domain of the glucocorticoid receptor. *Cell* 54: 1073–1080.

Picard, D., and Yamamoto, K. R. (1987). Two signals mediate hormone-dependent nuclear localization of the glucocorticoid receptor. *EMBO J.* 6: 3333–3340.

Read, L. D., Snider, C. E., Miller, J. S., Greene, G. L., and Katzenellenbogen, B. S. (1988). Ligand-modulated regulated progesterone receptor messenger ribonucleic acid and protein in human cancer cell lines. *Mol. Endocrinol.* 2: 263–271.

Rodriguez, R., Carson, M. A., Weigel, N. L., O'Malley, B. W., and Schrader, W. T. (1989). Hormone-induced changes in the in vitro DNA-binding activity of the chicken progesterone receptor. *Mol. Endocrinol.* 3: 356–362.

Romano, G. J., Harlan, R. E., Shivers, B. D., Howells, R. D., and Pfaff, D. W. (1988). Estrogen increases proenkephalin messenger ribonucleic acid levels in the ventromedial hypothalamus of the rat. *Mol. Endocrinol.* 2: 1320–1328.

Romano, G. J., Krust, A., and Pfaff, D. W. (1989a). Expression and oestrogen regulation of progesterone receptor mRNA in neurons of the mediobasal hypothalamus: An in situ hybridization study. *Mol. Endocrinol.* 3: 1295–1300.

Romano, G. J., Mobbs, C. V., Howells, R. D., and Pfaff, D. W. (1989b). Estrogen regulation of proenkephalin gene expression in the ventromedial hypothalamus of the rat: Temporal qualities and synergisms with progesterone. *Mol. Brain Res.* 5: 51–58.

Romano, G. J., Mobbs, C. V., Lauber, A. H., Howells, R. D., and Pfaff, D. W. (1990). Differential regulation of proenkephalin gene expression by estrogen in the ventromedial hypothalamus of male and female rats: Implications for the molecular basis of a sexually differentiated behavior. *Brain Res.* 536: 63–68.

Rusconi, S., and Yamamoto, K. R. (1987). Functional dissection of the hormone and DNA binding activities of the glucocorticoid receptor. *EMBO J.* 6: 1309–1315.

Savouret, J. F., Baily, A., Misrahi, M., Rauch, C., Redeuilh, G., Chauchereau A., and Milgrom E. (1991). *EMBO J*. 10: 1875–1883.

Scholt, D. R., Shyamala, G., Schneider, W., and Parry, G. (1991). Molecular cloning, sequence analyses, and expression of complementary DNA encoding murine progesterone receptor. *Biochemistry* 30: 7014–7020.

Tsai, S. Y., Carlstedt-Duke, J. A., Weigek, N. L., Dahlman, K., Gustafsson, J.-A., Tsai, M.-J., and O'Malley, B. W. (1988). Molecular interactions of steroid hormone receptor with its enhancer element: Evidence for receptor dimer formation. *Cell* 55: 361–367.

Umesono, K., and Evans, R. M. (1989). Determinants of target gene specificity for steroid/thyroid hormone receptors. *Cell* 57: 1139–1146.

Yamamoto, K. R. (1985). Steroid receptor regulated transcription of specific genes and gene networks. *Annu. Rev. Genetics* 19: 209–229.

Zhu, V.-S., and Pfaff, D. W. (in press). Protein–DNA binding assay for analysis of steroid-sensitive neurons in the mammalian brain. In *Neurobiology of Steroids*, Vol. 22 of Methods of Neurosciences, ed. E. R. de Kleot and W. Santo. San Diego, CA: Academic Press.

16

Effects of sex steroids on the central nervous system detected by the study of Fos protein expression

GLORIA E. HOFFMAN, RULA ABBUD, WEI-WEI LE,
BARBARA ATTARDI, KATHIE BERGHORN, AND
M. SUSAN SMITH

The presence of specific steroid receptors in neurons and evidence of steroid binding provide a basis for understanding the direct effects of steroids on neuronal function. Yet many key actions of gonadal steroids may be transmitted through multisynaptic pathways, in which only a subset of the participants are directly steroid-sensitive. The exploration of steroid actions on multisynaptic neuronal systems is nonetheless important but has been complicated by the lack of experimental techniques for monitoring activity changes across brain areas. In recent years, scientists have recognized that Fos, the protein product of the c-*fos* immediate early gene, is expressed in neurons that are stimulated. Immunocytochemical staining of Fos following specific stimuli has proved valuable for assessing neuronal activity (Morgan and Curran 1989; Sagar et al. 1988), and the results of studies using this approach will be presented here.

Specifically, the studies summarized in this chapter examined certain effects of gonadal steroids on neuronal activity as assessed by Fos staining. Two principal lines of investigation will be discussed. The first is the role of estrogen and progesterone in the luteinizing hormone–releasing hormone (LHRH) neuron activation that drives the preovulatory LH surge. The second is the role of progesterone in modifying neuronal activation by excitatory amino acids in brain systems other than the LHRH neurons. The latter studies point to the global effects that gonadal steroid hormones can have on neuronal function distinct from their actions on the regulation of reproductive function.

Gonadal steroids and LHRH activity

*Verifying the usefulness of Fos expression as a marker for
LHRH activity*

Gonadal steroids play an important role in regulating the reproductive neuroendocrine axis, and their actions within the brain culminate in the alteration of

LHRH release, which in turn changes pituitary release of LH and follicle-stimulating hormone (FSH). Because LHRH neurons lack estrogen and progesterone receptors (Fox et al. 1990; Herbison et al. 1993; Herbison and Theodosis 1992; Lehman and Karsch 1993; Watson et al. 1992), the effects of gonadal steroids on this system are presumed to be indirect. The LHRH neurons are small (10–12 μm in diameter), few in number ($\sim$1,200 per rat), and widely dispersed throughout the forebrain (Hoffman, 1983; Wray et al. 1989; Wray and Hoffman 1986a,b) so conventional means cannot be used to monitor their function. However, the early reports that Fos staining could provide a measure of activity in neurons (Sagar et al. 1988) prompted us to explore this approach for studying LHRH function. We learned that, like many other neurons, LHRH neurons normally express virtually no Fos, which is induced after electrical stimulation (Hoffman et al. 1993), setting the stage for further analysis of LHRH neuronal activity.

Fos expression in LHRH neurons during the estrous cycle

The preovulatory LH surge provides an excellent model for studying LHRH stimulation. The LH surge on proestrus is accompanied by an increase in LHRH output (Levine and Ramirez 1982; Sarkar et al. 1976), but the mechanics responsible for the change of LHRH release are unclear. According to one theory, the LH surge occurs as the result of synchronization of LHRH cell firing (Ojeda and Urbanski 1987). If this theory were true, some neurons would be active and express Fos at all times during the estrous cycle. However, LHRH neurons express Fos only in concert with the LH surge (Figs. 16.1 and 16.2). The presence of Fos in LHRH neurons only at the time of the LH surge supports the theory that increased afferent neuronal stimulation of LHRH neurons, not synchronization of ongoing activity, initiates the LH surge in the rat. This hypothesis is supported by the observation that infusion of catecholamines directly into the preoptic area induces LH release and stimulates Fos expression in LHRH neurons (Plotsky and Rivest 1991). Moreover, in the male rat, administration of pulses of the excitatory amino acid N-methyl-DL-aspartic acid (NMA) both releases LH and stimulates the expression of Fos and another early gene product, Jun, in LHRH neurons (M.S. Smith et al. 1991).

The presence of Fos or related proteins in a subset of LHRH neurons at times other than during an LH surge, as observed in the ferret and hamster (Berriman et al. 1992; Lambert et al. 1992), might suggest the presence of species differences in the patterns of LHRH release. For example in the ferret (an induced ovulator), 20% of LHRH neurons express Fos or related proteins in the unmated male or female. The increase to 60% of LHRH neurons expressing Fos in the female 1 hour after mating could indicate that synchronization of LHRH firing as well as recruitment occurs in this species. In the hamster, LHRH neurons in the caudal

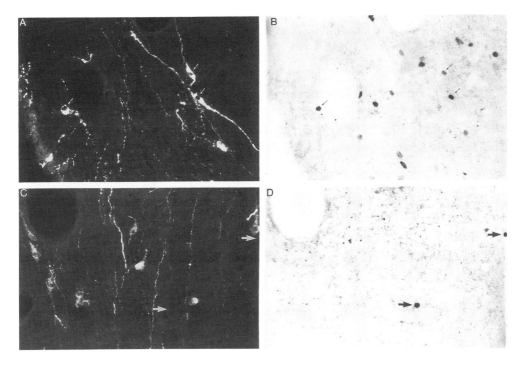

Figure 16.1. Micrographs illustrating the presence of Fos immunoreactivity in LHRH neurons at the time of the LH surge (A, B), but not before the onset of the LH surge (C, D). Panels A and C are fluorescence micrographs showing LHRH neurons; B and D are bright-field images of the same sections showing Fos staining. Thin arrows identify Fos-positive LHRH neurons; thick arrows identify Fos-positive cells that do not contain LHRH neurons.

preoptic area expressed Fos at all times of the estrous cycle, with no marked change in that area at the time of ovulation. This would suggest that a separate, active population of LHRH neurons might contribute to basal activity without participation in the LH surge.

During the course of a normal estrous cycle in the rat, increasing Fos expression in LHRH neurons coincided with the onset of the LH surge (Fig. 16.2) and persisted for a short time after the LH surge had subsided (Lee et al. 1990a, 1992). The confinement of Fos expression in LHRH neurons to the time of the LH surge is observed with other models of the LH surge as well. Induction of precocious puberty in immature animals bearing anterior hypothalamic lesions in the preoptic area is accompanied by Fos expression in LHRH neurons at the time of the LH surge, but not at other times of the cycle (Junier et al. 1992). Furthermore, in either immature rats or ovariectomized adult rats treated with estrogen and progesterone, the concordance of Fos expression in LHRH neurons with the LH surge is maintained (Hoffman et al. 1990, 1992; Lee et al. 1990b, 1992).

During the rising phase of the LH surge on proestrus, examination of Fos ex-

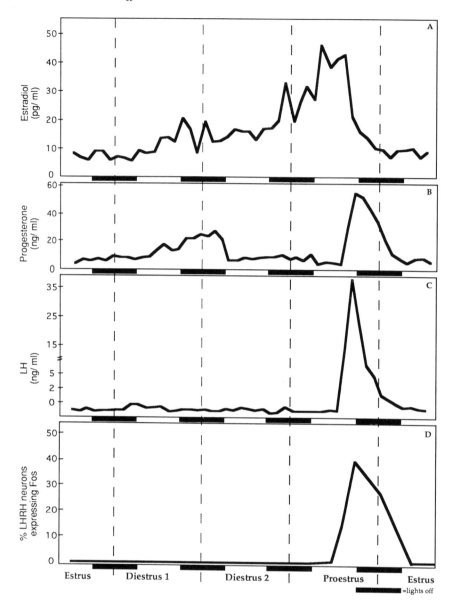

Figure 16.2. Changes in plasma levels of (A) estrogen, (B) progesterone, and (C) LH, as well as changes in (D) Fos expression in LHRH neurons during the rat estrous cycle. The plasma hormone values were obtained from M. S. Smith et al. (1975); the LHRH data were derived from Lee et al. (1990a).

pression in LHRH neurons helped to define the mechanics of the LH surge. As the surge progressed (determined by increasing plasma LH concentrations), a greater number of LHRH neurons possessed Fos immunoreactivity. The magnitude of Fos expression in LHRH neurons (expressed in terms of the proportion of LHRH neurons containing Fos immunoreactivity) accurately predicted the magnitude of

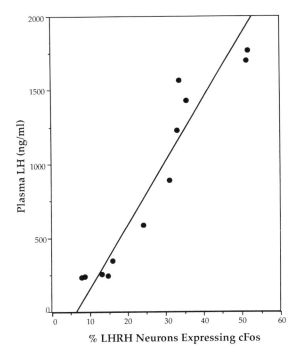

Figure 16.3. Plot showing the relationship between the proportion of LHRH neurons expressing Fos and plasma LH values 30 minutes to 1 hour before rats were killed, which occurred during the ascending limb of the LH surge. Modified from Lee et al. (1992).

LH secretion (Fig. 16.3), suggesting that more LHRH neurons were recruited into the active state as the LH surge progressed. These initial studies provided the basis for the study of steroid activation of LHRH neurons or their afferents.

Sexual dimorphism of steroid-induced LH surges

In rats, the capacity to manifest an LH surge is sexually dimorphic. Phasic secretion of LH in response to gonadal steroids is observed only in females. Since the pituitary of both males and females can respond to LHRH stimulation, the likely site of the dimorphism is the central nervous system. LHRH neurons are present in equal numbers in the male and female rat (Wray et al. 1986a), and few differences in LHRH cell morphology between the sexes have been noted. Moreover, since LHRH neurons are devoid of receptors for estrogen or progesterone (Fox et al. 1990; Herbison and Theodosis 1992; Lehman and Karsch 1993; Shivers et al. 1983; Watson et al. 1992), the lack of responsiveness of males to steroid stimulation must arise from a failure of LHRH neurons to receive appropriate stimulatory signals from an extrinsic steroid-sensitive afferent source. While we do not yet know precisely which neurons are responsible for stimulating the LHRH neurons, examination of the preoptic area of immature males treated with estrogen and

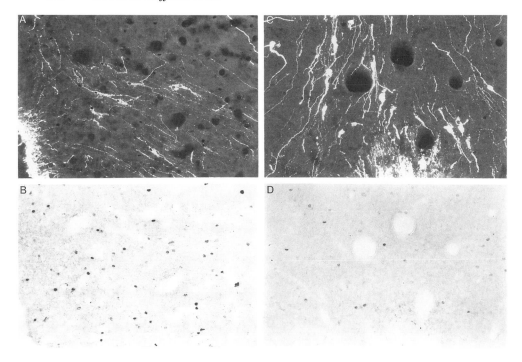

Figure 16.4. Presence of Fos in LHRH neurons of immature female rats (A, B) after estrogen and progesterone treatment but not in steroid-treated immature male rats (C, D). LHRH is depicted as a white fluorescent cytoplasmic stain, and darkly stained Fos-positive nuclei can be seen in all LHRH cells (A) from the steroid-treated female; a bright-field image of Fos staining in the same section verifies the presence of the immediate early gene product (B). Note that in the steroid-treated male (C, D), the LHRH neurons contain no Fos, and there is a marked reduction in the general pattern of Fos labeling. The sections in A and C were photographed using double exposure to reveal both the fluorescent image and Fos staining viewed with bright-field optics.

progesterone shows not only a difference in LHRH activation (immature males failed to show the marked activation of Fos expression following estrogen and progesterone administration that is elicited in similarly treated immature females, Fig. 16.4), but also a marked difference in the presence of Fos in neurons of the preoptic area other than the LHRH neurons (Fig. 16.5). Presumably, these activated neurons represent a potential afferent source for stimulation of LHRH neurons. The intensely stimulated neurons of the periventricular preoptic area are among those that contain estrogen receptors (Herbison and Theodosis 1992), raising the possibility that the steroid relay could be channeled through that neuronal population.

Progesterone's role in altering LHRH activity

The rise of plasma progesterone level that coincides with the onset of the LH surge (Fig. 16.2) (M.S. Smith et al. 1975) is thought to be responsible for syn-

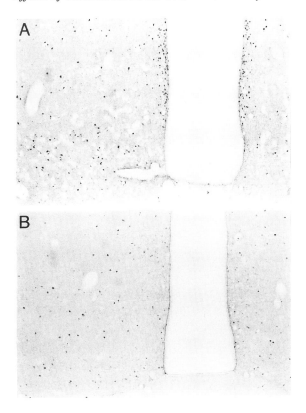

Figure 16.5. (A) Fos immunoreactivity within neurons of the periventricular zone of the preoptic area was detected in immature female rats that exhibited an LH surge after steroid pretreatment. (B) Male rats similarly treated with estrogen and progesterone failed to show an LH surge, and the preoptic area neurons failed to exhibit Fos staining. The sections shown in this figure were taken from the rats shown in Figure 16.4.

chronizing and enhancing the LH surge (Attardi 1984). A pituitary site of action is unequivocal (Attardi 1984). Yet the role that progesterone plays in stimulating LHRH neurons (if any) was unclear until the application of immediate early gene detection strategies revealed the patterns of LHRH neuronal activation.

Two approaches were used to address this issue. The first was to compare the activation of LHRH neurons in proestrous rats that were untreated or pretreated (1230 hours on proestrus) with the progesterone antagonist RU 486. Blockade of progesterone effects not only profoundly reduced the magnitude of the LH surge (as had been previously reported; Brown-Grant and Naftolin 1972; Caligaris et al. 1972; De Paolo and Barraclough 1979) but also reduced the degree to which LHRH neurons expressed Fos (Fig. 16.6). Fewer LHRH neurons were activated at the time of the LH surge, and the intensity of Fos staining was attenuated in the absence of progesterone effects (Fig. 16.7). That progesterone and not other ovarian factors enhanced LHRH activation was examined in ovariectomized or immature female rats administered either estrogen alone or estrogen followed 24 hours

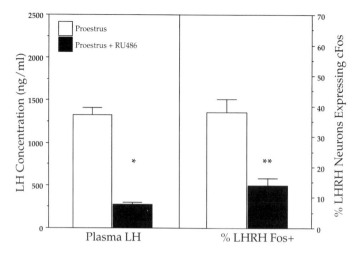

Figure 16.6. Effects of treatment of proestrous rats with the progesterone receptor blocker RU 486 reveals a reduction of peak plasma LH (*, $p \leq .0001$), as well as a concomitant reduction of Fos labeling in LHRH neurons (**, $p \leq .0004$). From Lee et al. (1990b).

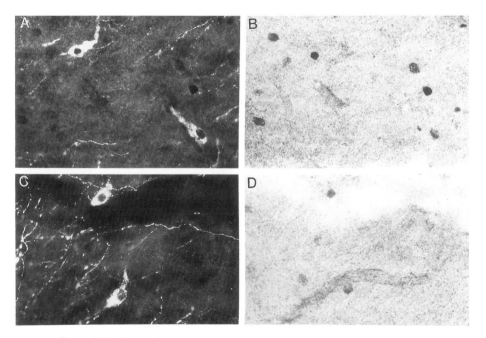

Figure 16.7. Comparison between Fos labeling in LHRH neurons from (A, B) proestrous rats that were otherwise untreated but killed at the time of the LH surge, and (C, D) proestrous rats pretreated with RU 486 at 1230 hours on proestrus showing a reduction in the intensity of Fos labeling at the time of the LH surge when progesterone receptors were blocked. From Hoffman et al. (1993).

later by progesterone. When estrogen-primed animals were treated with progesterone to mimic more closely the natural course of hormonal secretion, the magnitude of the LH surge was greater than with estrogen treatment alone. As in the proestrous model, the magnitude of expression of Fos in terms of both the number of LHRH neurons stimulated and the intensity of Fos staining was greater after treatment with both steroids than after treatment with estrogen alone (Hoffman et al. 1990; Lee et al. 1990b).

The actions of progesterone not only augment the release of LH at the time of the surge, but also promote its decline. Without progesterone, the LH surge declines more gradually and animals display daily LH surges on subsequent days (Legan et al. 1975). An important question to be addressed is whether the presence of progesterone terminates the responsiveness of the pituitary to daily activation by LHRH or whether the LHRH neuronal activation itself is blocked. Our initial analysis of Fos expression in LHRH neurons during the estrous cycle would suggest that since no Fos expression was detected in LHRH neurons on the afternoon of estrus, a blockade of LHRH activation by progesterone had occurred. However, that conclusion is valid only if daily surges of LH are accompanied by daily changes in LHRH neuronal Fos expression. In ovariectomized rats primed with estradiol benzoate (5 μg sc, 1230 hours, day 0), followed by a second dose of estradiol benzoate 24 hours later (50 μg sc, 1230 hours, day 1), LH surges occur on the afternoons of both day 2 and day 3 (Fig. 16.8). Fos was expressed in LHRH neurons on the afternoon of day 2, was low on the morning of day 3, and was again observed on the afternoon of day 3 (Fig. 16.9) (Hoffman et al. 1993). Thus, we can conclude that the lack of Fos labeling in LHRH neurons after exposure to progesterone most likely reflects a termination of the activational signals to the LHRH neurons.

Role of progesterone in modulating cortical activation during lactation

Altered cortical function in lactating rats

In the preceding section, we described the advances in our understanding of the stimulatory actions of gonadal steroids that were derived from changes in immediate early gene expression in the LHRH neuron system. Expression of Fos, as a marker for neuronal activation, has also been extremely useful in identifying changes in extrahypothalamic neuronal function in association with other reproductive states such as lactation.

The effects of the suckling stimulus are varied; it induces maternal behavior (Numan 1988), while at the same time shifting cortical EEG pathways toward synchronization typical of sleep (Lincoln et al. 1980; Wakerley et al. 1988). The

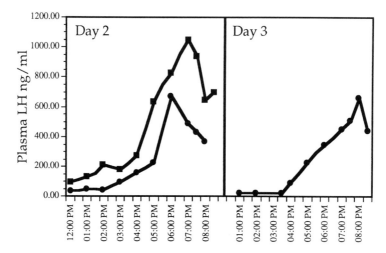

Figure 16.8. Mean plasma LH values in two groups of rats primed with 5 μg estradiol benzoate on day 0 (0900 hours), administered 50 μg estradiol benzoate on day 1 (0900 hours), and bled on either day 2 (group 1, squares) or days 2 and 3 (group 2, circles). Note the presence of LH surges on the afternoons of both days 2 and 3.

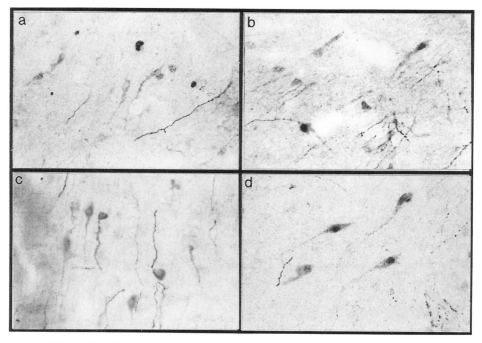

Figure 16.9. Fos is absent in LHRH neurons on the morning (A) and evening (C) of day 2, and activation of LHRH neurons at the time of the LH surges on day 2 (B) or day 3 (D). From Hoffman et al. (1993).

lactating rat shows attenuated responses to a number of stresses as well. The endocrine profile of the lactating rat is one of high plasma progesterone concentrations and low estrogen levels (Lee et al. 1989a; Smith and Neill 1977), and this particular hormonal pattern may be responsible for modulating cortical neuronal activity.

Several clinical and experimental studies have demonstrated that progesterone suppresses epileptiform activity, while estrogen increases its incidence (Backstrom et al. 1984; Mattson and Cramer 1985; Roscizewska et al. 1989). These effects of ovarian steroids may be the result of changes in the electrophysiological properties of *N*-methyl D-aspartate (NMDA) receptors, as suggested by experiments performed on cerebellar Purkinje cells. Estradiol potentiated the responses of these cells to NMDA and quisqualate (S.S. Smith 1989), whereas progesterone attenuated such responses (S.S. Smith 1989; S.S. Smith et al. 1987). These data suggest that estrogen and progesterone might modulate the activity of NMDA receptors. Another type of receptor implicated as a target for the effects of progesterone is the A form of the γ-aminobutyric acid (GABA$_A$) receptor. Progesterone and its metabolites can potentiate the inhibitory effects of GABA by acting at the same site as do barbiturates (Barker et al. 1987; Harrison et al. 1989; Majewska et al. 1986). The rapid onset of the response to ovarian steroids suggests that these effects do not require changes in gene expression but are a result of direct action at the cell membrane. It is also possible that the effects of steroids are mediated by nuclear receptors. Changes in the number of steroid receptors during lactation have been reported; lactating mice have fewer estrogen receptor–immunoreactive cells in the limbic system than do virgin or pregnant mice (Koch and Ehret 1989). No data are available regarding possible changes in progesterone receptors. Irrespective of the mechanism of action of steroids on the nervous system, the high levels of progesterone and low levels of estradiol associated with lactation might be important in altering neuronal responsiveness.

We have used excitatory amino acids (EAAs) as tools for studying the effects of lactation on the activity of the central nervous system. For example, hypothalamic neuronal responsiveness to EAAs is altered by the suckling stimulus, as LHRH neurons are refractory to stimulation by NMA in lactating rats (Abbud and Smith 1993; Pohl et al. 1989). In addition, instead of the typical stimulatory effect of EAAs on prolactin secretion, EAAs inhibit prolactin secretion in lactating animals. Our first indication that lactation also altered cortical responsiveness to EAAs was based on behavioral observations. Whereas NMA induced behavioral responses in cycling rats, characterized principally by hyperactivity, lactating rats were unresponsive to NMA, suggesting a suckling-induced alteration in cortical responsiveness. We explored the effects of lactation on cortical function by comparing responsiveness to EAAs in cycling rats during diestrus and in lactating rats suckling eight pups during day 10 postpartum using Fos labeling as a marker of

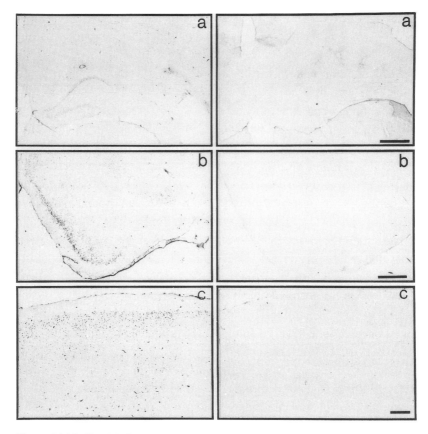

Figure 16.10. Fos labeling in (A) the dentate gyrus (DG), (B) piriform corex, and (C) parietal cortex after systemic injections of NMA (40 mg/kg iv) in cycling (*left*) or lactating (*right*) rats. Cycling rats were studied on diestrus 1 of the estrous cycle, and the lactating rats suckling eight pups were studied on day 10 postpartum. Fos appears as black dots at this low magnification. From Abbud et al. (1993).

neuronal activation (Abbud et al. 1992). EAAs stimulate Fos expression in many brain areas, including some where EAA receptors are abundant (Morgan and Curran 1991; Morgan and Linnoila 1991; Sonnenberg et al. 1989). We also examined the patterns of Fos expression in the cortex of cycling and lactating rats in response to systemic injections of NMA (Abbud et al. 1992; Abbud and Smith 1991). We compared the responses to NMA with those of kainate, an EEA that binds to a different class of glutamate receptors (Barnard and Henley 1990) and induces a behavioral pattern consisting of "wet dog shakes" (Olney 1980). NMA induced significant Fos expression in the hippocampus (dentate gyrus) and paleo-cortex and neocortex in the cycling animal, but little if any Fos was induced in these same areas in lactating animals (Fig. 16.10), consistent with the lack of behavioral responses (hyperactivity) to NMA. A similar suppression of cortical activation was observed when NMA was administered into the third ventricle of lactating animals. The absence of Fos expression in the lactating rats was specific to cortical structures, since Fos was induced in many areas of the hypothalamus and brain stem in response to NMA. In contrast to the cortical refractoriness observed in response to NMA during lactation, kainate induced similar patterns of Fos labeling in the hippocampus and cortex of cycling and lactating rats (Fig. 16.11). These results reflected the similarity in behavioral responses (induction of wet dog shakes) between the two groups. Thus, there are no apparent deficits in kainate receptor–mediated cortical activation during lactation; rather, the inhibitory effects of lactation on cortical activation are specific to effects mediated by NMDA receptors (Abbud et al. 1992).

The pathways and mechanisms by which lactation inhibits NMDA receptor–mediated cortical activation are unknown. However, both neural and hormonal mechanisms may contribute to this inhibition. To understand the effects of lactation on cortical responsiveness to NMA, it is important to consider the mechanisms by which EAAs stimulate Fos expression in cortical neurons. It is difficult to conclude that the effects of EAAs are exerted directly at the cortex rather than

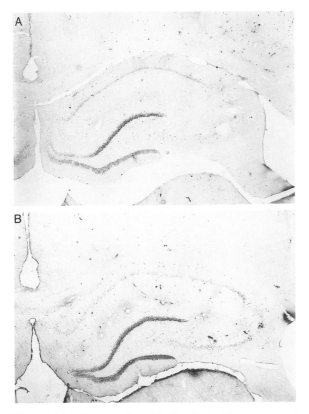

Figure 16.11. Pattern of Fos labeling in the hippocampus after systemic injection of kainate (1.5–2.5 mg/kg iv) to cycling (A) and lactating (B) rats. From Abbud et al. (1993).

arising secondary to the activation of sensory brain stem pathways, particularly since EAAs might not cross the blood–brain barrier (Price et al. 1978). Induction of Fos in brain areas that are inside the blood–brain barrier might be due to EAA activation of afferent pathways. That Fos labeling is not always observed in brain areas where NMDA receptors are known to be present supports the notion that the effects of NMA on the cortex may be indirect. For example, the CA1 layer of the hippocampus, an area that expresses high levels of NMDA receptors, showed no Fos expression in response to NMA. In any case, the suckling-induced deficits were specific to NMDA receptor–mediated events.

Role of progesterone in suppression of cortical activation by EEAs

We investigated the factors involved in the suppression of cortical activation in lactating rats in response to NMA by examining the effects of high levels of progesterone associated with the suckling stimulus (Abbud et al. 1993). It is well established that the high levels of progesterone during lactation result from the effects of sucking-induced prolactin secretion acting on the ovary to maintain corpus luteum function (Lee et al. 1989b; M.S. Smith 1981). Thus, we determined the degree of cortical activation, as measured by Fos expression in response to NMA administration after removal of the suckling stimulus and/or during blockade of progesterone's effects by ovariectomy or treatment with RU 486. To assess the degree of Fos expression, we used a 4-point scale of staining intensity: 0 = no Fos, 1 = light Fos, 2 = medium Fos, and 3 = maximum Fos (shown in Fig. 16.12 for the piriform cortex, which was chosen as representative of most cortical areas). A similar scale was used to assess Fos expression in the hippocampus, which was analyzed separately. Cycling rats expressed a maximum level of Fos labeling, equal to 3, in the piriform cortex after NMA treatment, while lactating rats expressed either no (0) or little (1) Fos labeling (Fig. 16.13). In lactating animals that were either ovariectomized or treated with RU 486 (Fig. 16.14), Fos intensity ranged between 2 and 3 on the 4-point scale after NMA treatment. Therefore, significant but not complete recovery of cortical activation occurs in the presence of the suckling stimulus if progesterone's effects are blocked. We then studied the effects of pup removal on Fos expression in the cortex in response to NMA. Significant recovery of cortical activation was not observed until 24 hours after pup removal. At that time, the majority of animals exhibited a Fos labeling intensity of 2, representing significant but not complete recovery of cortical activation (Fig. 16.15). However, complete recovery in the piriform cortex was uniformly observed only after removal of both the suckling stimulus and blockade of progesterone's effects (Fig. 16.16), since almost all of the animals in these groups exhibited a Fos intensity of 3. The recovery of cortical activation accompanied the recovery of the behavioral responses to NMA. Animals that exhibited a Fos labeling intensity of 3 in the piriform cortex and hippo-

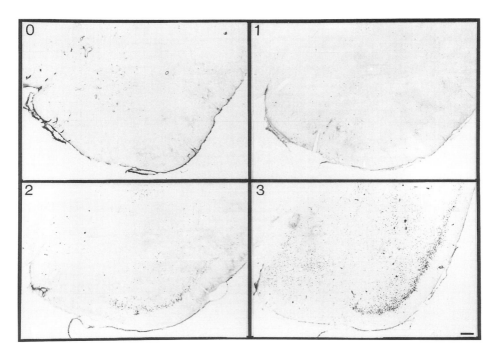

Figure 16.12. Photomicrographs illustrating the Fos intensity scale used for the piriform cortex. 0 = no Fos labeling; 1 = light Fos labeling intensity; 2 = medium Fos labeling intensity; 3 = high Fos labeling intensity. From Abbud et al. (1993).

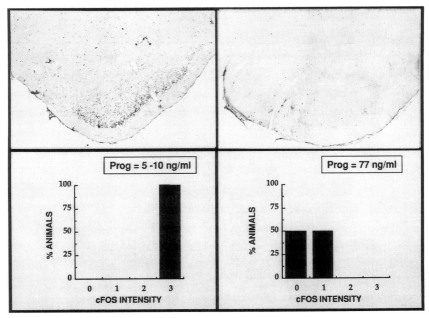

Figure 16.13. Intensity of the Fos labeling in response to NMA in the piriform cortex of cycling (*left*, n = 6) and lactating rats (*right*, n = 4). The top panels show photomicrographs of the piriform cortex of a representative cycling rat (*left*) and lactating rat (*right*) treated with four injections of NMA (40 mg/kg iv) at 10-minute intervals. The bottom panels show the percentage of animals exhibiting different Fos labeling intensities in each group. Included in the figure are typical progesterone values (Prog) for the group. From Abbud et al. (1993).

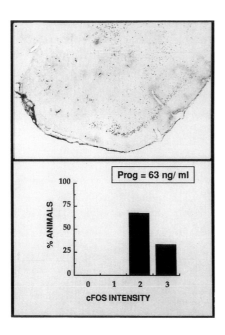

Figure 16.14. Effect of treatment with RU 486 on the pattern of Fos expression in the piriform cortex of lactating rats. The top panel shows a photomicrograph of a representative animal, and the bottom histogram shows the degree of Fos induction expressed as a percentage of animals exhibiting a given Fos labeling intensity ($n = 6$). Similar results were obtained for ovariectomized lactating dams. From Abbud et al. (1993).

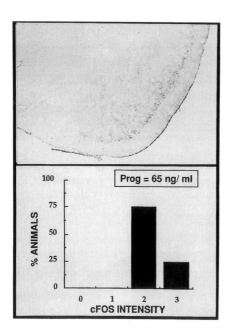

Figure 16.15. *Top*: The effect of pup removal for 24 hours on Fos staining in a representative animal showing a Fos labeling intensity of 2. *Bottom*: A histogram showing the percentage of animals exhibiting a given Fos labeling intensity rating. From Abbud et al. (1993).

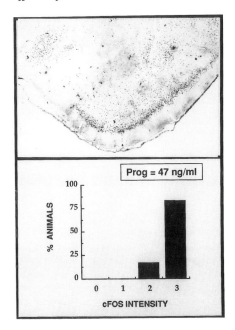

Figure 16.16. Effect of treatment with RU 486 on the pattern of Fos labeling in the piriform cortex of lactating rats with pups removed for 24 hours ($n = 6$ in each group). The photomicrographs from representative animals show a Fos labeling intensity of 3. From Abbud et al. (1993).

campus showed the most pronounced behavioral response. However, animals with a Fos labeling intensity of 2 or less showed only partial recovery of the hyperactive behavior after treatment with NMA.

In general, the recovery of hippocampal responsiveness to NMA paralleled that exhibited by cortex (Abbud et al. 1993). None of the lactating animals showed any Fos expression in the hippocampus in response to NMA (Fig. 16.17b). In the presence of a suckling stimulus but absence of progesterone, the intensity of Fos ranged from 0 to 2 in ovariectomized animals (Fig. 16.17c) and from 1 to 3 in RU 486–treated rats (Fig. 16.17d). There was a smaller degree of recovery in the hippocampus than in the piriform cortex when progesterone's effects were blocked or after removal of the suckling stimulus for 24 hours. Removal of the suckling stimulus for 24 hours and blockade of progesterone's effects resulted in complete recovery of cortical activation in response to NMA. However, hippocampal recovery was still not complete at this time, since only 40–50% of the animals showed a maximum Fos intensity.

The results reveal that both the suckling stimulus and high levels of progesterone contribute to the cortical refractoriness during lactation (Abbud et al. 1993). To determine whether high progesterone levels alone could alter cortical activation in response to NMDA receptor activation, we studied animals on day

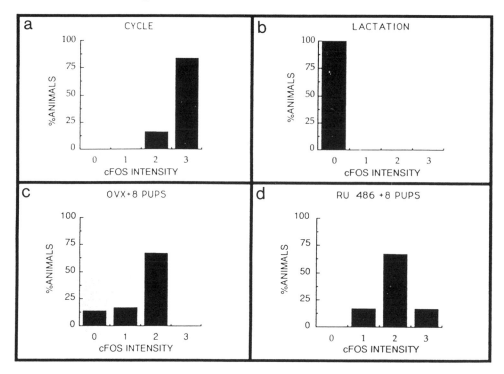

Figure 16.17. Histograms showing the percentage of animals expressing a given level of Fos labeling intensity in the hippocampus of (a) cycling rat, (b) lactating rat, (c) ovariectomized (OVX) dam + 8 pups, and (d) RU 486–treated dam + 8 pups. The same 4-point scale of Fos labeling intensity was used as shown in Figure 16.11. From Abbud et al. (1993).

10 postpartum that were nursing two pups, a minimum suckling stimulus associated with elevated levels of progesterone (Smith and Neill 1977). We also examined animals during days 13–14 of pregnancy, a state characterized by high levels of progesterone (Devorshak-Harvey et al. 1987; Morishige et al. 1973). No deficits were observed in response to NMA in the pregnant rat (Fig. 16.18). Cortical activation, as judged by behavioral responses and Fos expression, was similar to that observed in cycling rats. Thus, high levels of progesterone alone cannot induce cortical refractoriness to NMDA receptor activation. These data suggest that the neuronal changes brought about by the suckling stimulus are necessary for the inhibitory effects of progesterone to be observed. Whether the suckling stimulus and progesterone act on similar or different neural pathways to affect cortical activity is presently unknown. It is possible that the suckling stimulus induces changes in neuronal input to sites that are responsive to progesterone, which then permits the inhibitory effects of progesterone to be manifested. There are numerous examples of the inhibitory effects of progesterone or its metabolites on neuronal activity, one being its anesthetic effects at very high doses (Arai et al.

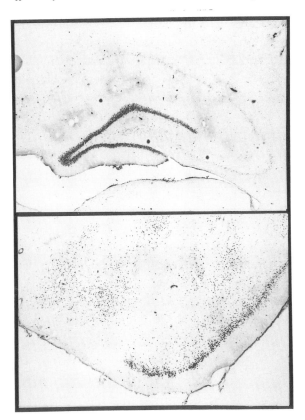

Figure 16.18. Patterns of Fos labeling in the hippocampus and piriform cortex of a pregnant rat ($n = 5$) treated with NMA on day 14 of pregnancy show a Fos labeling intensity of 3. From Abbud et al. (1993a).

1967). In fact, several studies have suggested that progesterone and its metabolites can act at the cell membrane to modulate the activity of excitatory (S.S. Smith 1989, 1991; S.S. Smith et al. 1987) and inhibitory receptors (Harrison et al. 1989; Maggi and Perez 1984; Majewska et al. 1986). While these results support the idea that cortical activity can be modulated by the steroid hormonal environment, they do not cast any light on the molecular mechanisms underlying progesterone action.

In our studies, we were able to restore cortical responsiveness to NMA only partially by treatment with RU 486. It is not known whether RU 486 can also bind to the progesterone modulatory sites of other receptors at the level of the cell membrane. These data could indicate that both receptor- and non-receptor-mediated actions of progesterone, or its metabolites, are responsible for NMA refractoriness. If progesterone exerts some of its effects at the level of the cell membrane, it may act either by decreasing the activity of excitatory receptors or by increasing the activity of inhibitory receptors. Direct modulation of EEA receptors by progesterone is suggested by data from experiments performed in Purkinje cells (S.S. Smith 1991; Smith et al. 1987), which reveal that progesterone attenuated the response to EEAs by a mechanism that was independent of GABA receptor activation. However, it is still possible that progesterone might indirectly modulate NMDA receptor activity by acting on GABA receptors. Several reports have suggested that NMDA receptors and GABA receptors might be inversely regulated. For example, activation of NMDA receptors results in the inhibition of GABA receptors in the hippocampus (Kamphuis et al. 1991; Stelzer et al. 1987). On the other hand, there is ample literature supporting the idea that progesterone and its metabolites can potentiate the activity of an inhibitory type of receptor, the $GABA_A$ receptor. In several brain regions (Barker et al. 1987; Harrison et al. 1987, 1989; Maggi and Perez 1984; S.S. Smith 1989), as well as in *Xenopus* oocytes (Woodward et al. 1992), these steroids act at the barbiturate site (Majewska et al. 1986) or at a novel site of the $GABA_A$ receptor, and the result could explain their anesthetic effects. Such an interaction of progesterone, GABA receptors, and NMDA receptors could account for the deficit in NMDA, but not kainate, receptor activation during lactation. In addition, activation of GABA receptors results in the hyperpolarization of the cell membrane, which might lead to blockade of the NMDA receptor channel by magnesium. Hyperpolarization of the cell membrane does not block kainate receptors. In this regard, there are data showing that lactation results in an increase in GABA levels in the cerebrospinal fluid. Removal of pups for 6 hours significantly decreases GABA concentration, which recovers upon pup replacement (Qureshi et al. 1987). However, these data provide no evidence as to the source of the increase in GABA concentrations.

It is possible that the deficits in cortical responsiveness to NMA during lactation are due to changes in NMDA receptor function in the hippocampus and cortex. Estrogen has been shown to increase NMDA receptor agonist binding sites in the hippocampal CA1 region (Weiland 1992). Thus, during lactation, low estrogen levels could be accompanied by a decrease in NMDA agonist sites. Another possibility is that changes in afferent input to the cortex and hippocampus might mediate this inhibition. The pathways activated by the suckling stimulus travel from the nipple up the spinal cord to the brain stem and then project to cortical and hypothalamic areas (Tindall et al. 1967). Suckling might activate inputs that would either inhibit an excitatory drive or excite an inhibitory drive to the cortex. The inhibitory effects of progesterone could be manifested at the site of the activated input, such as the brain stem. Another possibility is that the suckling stimulus itself might alter neuronal input to the cortex such that the inhibitory effects of progesterone on cortical neurons can be observed. In either case, NMDA receptor activation could be rendered ineffective and result in the apparent lack of neuronal activation in the target areas in response to NMA. We have examined whether areas of the brain stem known to modulate cortical excitability were differentially stimulated to express Fos in lactating and cycling rats after NMA or kainate treatment (Abbud et al. 1994). We focused on the catecholaminergic and serotonergic pathways, both of which project broadly to the cortex and hippocampus. Double-label immunocytochemistry for Fos and dopamine β-hydroxylase (DβH) was used to study the pattern of activation in catecholaminergic cell bodies. DβH-Positive cell bodies in the A1/C1 (ventrolateral medulla), A2/C2 (nucleus tractus solitarius), and A5 regions expressed Fos in response to both NMA and kainate in diestrous and lactating rats. In contrast, although NMA induced Fos expression in many cell bodies in the locus coeruleus of diestrous rats (Fig. 16.19A), little Fos was observed in cells in this area of lactating rats (Fig. 16.19B). However, kainate was equally effective in inducing Fos expression in the locus coeruleus of diestrous (Fig. 16.19C) and lactating rats (Fig. 16.19D). Similarly, the induction of Fos expression in the dorsal raphe in response to NMA was attenuated during lactation (Fig. 16.20B) when compared with diestrus (Fig. 16.20A), whereas Fos expression in this area in response to kainate was similar during diestrus (Fig. 16.20C) and lactation (Fig. 16.20D).

These results demonstrate that induction of Fos expression by NMA, but not

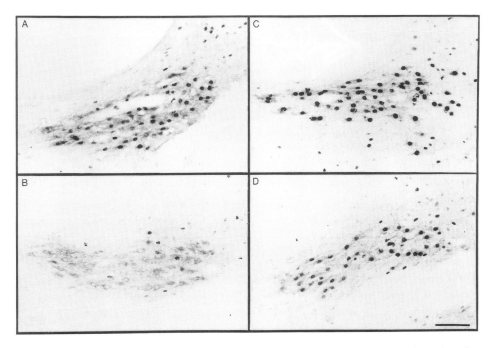

Figure 16.19. Pattern of Fos expression in the locus coeruleus shown in horizontal section in response to NMA (A, B) and kainate (C, D) in diestrous (A, C) and lactating (B, D) rats. Tissue sections were processed for double-label immunocytochemistry to detect the presence of Fos and dopamine β-hydroxylase (DβH). Fos is shown as a black nuclear stain, while the presence of DβH is illustrated by the lighter cytoplasmic stain. From Abbud et al. (1994).

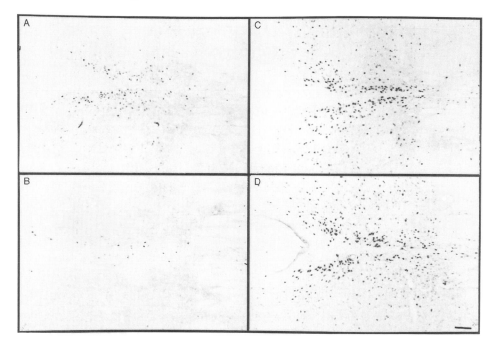

Figure 16.20. Fos expression in the dorsal raphe shown in horizontal section after treatment with NMA (A, B) or kainate (C, D) in diestrous (A, C) and lactating (B, D) rats. From Abbud et al. (1994).

kainate, is greatly attenuated in the locus coeruleus and dorsal raphe of lactating rats. However, other catecholaminergic cell bodies in the brain stem were activated to a similar degree in diestrous and lactating rats in response to NMA. Both the locus coeruleus and dorsal raphe serve modulatory roles in the cortex and hippocampus and have been implicated in the regulation of cortical EEG, and therefore the state of arousal of the animal (Berridge and Foote 1991; Kocsis and Vertes 1992; McGinty and Harper 1976; Steriade 1992; Trulson and Jacobs 1979; Vanderwolf 1988). The attenuation of locus coeruleus and dorsal raphe responsiveness to NMA could also contribute to the deficits in cortical and hippocampal activation during lactation. Neurons of the locus coeruleus also possess steroid receptors, particularly those for estrogen (Sar and Stumpf 1981). Whether the low levels of estrogen found in the lactating rat evoke changes in locus coeruleus neurons that alter their responsiveness to stimulation remains to be determined. Reports of the co-localization of progesterone receptors with serotonin in dorsal raphe neurons (Bethea 1993) could explain how high levels of progesterone present during lactation might modulate cortical responsiveness to NMDA receptor activation at the site of afferent projections to the cortex, as well as at the site of cortical neurons.

Conclusions

It is thought that the presence of Fos represents neuronal activation and correlates with the changing neurotransmitter release in response to a stimuli. It is important to recognize that Fos does not cause neurotransmitter release per se. This is most easily understood from a temporal perspective. Whereas changes in neuro-transmitter release occur within seconds to minutes after the onset of a stimulus (e.g., behavioral changes in response to NMA treatment occur within a few minutes of administration and neuroendocrine hormone release in response to stimulation has a similar time course), synthesis of immediate early gene products takes considerably longer, with peak detection of the product 60–90 minutes after stimulation (Hoffman et al. 1993). Thus, release of the neurotransmitter occurs (and may cease) long before Fos protein has been synthesized. However, for some neuronal systems, expression of immediate early gene products might play a role in regulating the dynamics of transmitter synthesis that follows a burst of activity. This regulatory role would depend on whether the DNA binding site recognizing Fos complexes is present on the transmitter gene. For example, the promoter region of the tyrosine hydroxylase, dynorphin, and enkephalin genes contains AP-1 binding sites, and Fos expression can increase the transcription rate of their mRNAs (Draisci and Iadarola 1989; Icard-Liepkalns et al. 1992). In addition to having a potential effect on transmitter gene expression, it is equally feasible that the stimulation of the immediate early gene products evokes changes in the synthesis of more general cell constituents (synaptic vesicles, microtubules, metabolic enzymes, etc.). While scientists are actively pursuing this avenue of research, the paucity of knowledge about the targets of Fos need not curtail the use of Fos to study neuronal activation. Irrespective of its target, the latent image of cellular activity that Fos staining engenders provides a powerful tool for exploring steroid actions within the brain.

Acknowledgments

The authors greatly appreciate the technical expertise of Maria Freilino, Jamee Bonnett, and Caroline Phalin, as well as the photographic skills of Thomas C. Waters. This work was supported by USPHS Awards NS 28730 and HD 14643.

References

Abbud, R., Hoffman, G. and Smith, M. (1993). Cortical refractoriness to NMA stimulation in the lactating rat: Recovery after pup removal and blockade of progesterone receptors. *Brain Res.* 604: 16–23.
Abbud, R., Hoffman, G. E., and Smith, M. S. (1994). Lactation-induced deficits in

NMDA receptor-mediated cortical and hippocampal activation: Changes in NMDA receptor gene expression and brainstem activation. *Mol. Brain Res.* 25: 323–332.

Abbud, R., Lee, W.-S., Hoffman, G., and Smith, M. (1992). Lactation inhibits hippocampal and cortical expression of c-*fos* in response to NMDA but not kainate receptor agonists. *Mol. Cell. Neurosci.* 3: 244–250.

Abbud, R., and Smith, M. (1991). Differences in the LH and prolactin responses to multiple injections of kainate, as compared to NMA, in cycling rats. *Endocrinology* 129: 3254–3258.

Abbud, R., and Smith, M. (1993). Altered LH and prolactin responses to excitatory amino acids during lactation. *Neuroendocrinology* 58: 454–464.

Arai, Y., Hiroi, M., Gorski, M. J., and Gorski, R. (1967). Influence of intravenous progesterone administration on the cortical electroencephalogram of the female rat. *Neuroendocrinology.* 2: 275–282.

Attardi, B. (1984). Progesterone modulation of the luteinizing hormone surge: Regulation of hypothalamic and pituitary progestin receptors. *Endocrinology* 115: 2113–2122.

Backstrom, T., Zetterlund, B., Bloom, S., and Romano, M. (1984). Effect of IV progesterone infusions on the epileptic discharge frequency in women with partial epilepsy. *Acta Neurol. Scand.* 69: 240–248.

Barker, J. L., Harrison, N. L., Lange, G. D., Owen, D. G. (1987). Potentiation of γ-aminobutyric-acid-activated chloride conductance by a steroid anaesthetic in cultured rat spinal neurones. *J. Physiol.* 386: 485–501.

Barnard, E. A., and Henley, J. M. (1990). The non-NMDA receptors: Types, protein structure and molecular biology. *TIPS* 11: 500–507.

Berridge, C. W., and Foote, S. L. (1991). Effects of locus coeruleus activation on electroencephalographic activity in neocortex and hippocampus. *J. Neurosci.* 11: 3135–3145.

Berriman, S. J., Wade, G. N., and Blaustein, J. D. (1992). Expression of fos-like proteins in gonadotropin releasing hormone neurons of Syrian hamsters: Effects of estrous cycles and metabolic fuels. *Endocrinology* 131: 2222–2228.

Bethea, C. L. (1993). Colocalization of progestin receptors with serotonin in raphe neurons in macaque. *Neuroendocrinology* 57: 1–6.

Brown-Grant, K., and Naftolin, F. (1972). Facilitation of luteinizing hormone secretion in the female rat by progesterone. *J. Endocrinol.* 53: 37–45.

Caligaris, L., Astrada, J. J., and Taleisnik, S. (1972). Influence of age on the release of luteinizing hormone induced by oestrogen and progesterone in immature rats. *J. Endocrinol.* 55: 97.

De Paolo, L. V., and Barraclough, C. A. (1979). Dose dependent effects of progesterone on the facilitation and inhibition of spontaneous gonadotropin surges in estrogen treated ovariectomized rats. *Biol. Reprod.* 21: 1015–1023.

Devorshak-Harvey, E., Bona-Gallo, A., and Gallo, R. V. (1987). The relationship between declining plasma progesterone levels and increasing luteinizing hormone pulse frequency in late gestation in the rat. *Endocrinology* 120: 1597–1601.

Draisci, G., and Iadarola, M. (1989). Temporal analysis of increases in c-*fos*, preprodynorphin and pre-proenkephalin mRNAs in rat spinal cord. *Mol. Brain Res.* 6: 31–37.

Fox, S., Harlan, R., Shivers, B., and Pfaff, D. (1990). Chemical characterization of neuroendocrine targets for progesterone in the female rat brain and pituitary. *Neuroendocrinology* 51: 276–283.

Harrison, N. L., Majewska, M. D., Harrington, J. W., and Barker, J. L. (1987). Structure–activity relationships for steroid interaction with the γ-aminobutyric acid–A receptor complex. *J. Pharmacol. Exp. Ther.* 241: 346–353.

Harrison, N. L., Meyers, D. E. R., Owen, D. G., and Barker, J. L. (1989). Mechanistic studies of anesthetic steroids and barbiturates. *Anesthesiology* 71: A594.

Herbison, A. E., Robinson, J. E., and Skinner, D. C. (1993). Distribution of estrogen receptor-immunoreactive cells in the preoptic area of the ewe: Co-localization with glutamic acid decarboxylase but not luteinizing hormone-releasing hormone. *Neuroendocrinology* 57: 751–759.

Herbison, A. E., and Theodosis, D. T. (1992). Localization of oestrogen receptors in preoptic neurons containing neurotensin but not tyrosine hydroxylase, cholecystokinin or luteinizing hormone-releasing hormone in male and female rat. *Neuroscience* 50: 283–298.

Hoffman, G. E. (1983). LHRH neurons and their projections. In *Structure and Function of Peptidergic and Aminergic Neurons*, ed. Y. Sano, Y. Ibata, and E. A. Zimmerman, pp. 183–202. Tokyo: Japan Scientific Societies Press.

Hoffman, G. E., Lee, W. S., Attardi, B., Yann, V., and Fitzsimmons, M. D. (1990). LHRH neurons express c-fos after steroid activation. *Endocrinology* 126: 1736–1741.

Hoffman, G. E., Smith, M. S., and Fitzsimmons, M. D. (1992). Detecting steroidal effects on immediate early gene expression in the hypothalamus. *Neuroprotocols* 1: 52–66.

Hoffman, G. E., Smith, M. S., and Verbalis, J. G. (1993). c-Fos and related immediate early gene products as markers for neuronal activity in neuroendocrine systems. *Front. Neuroendocrinol.* 14: 173–213.

Icard-Liepkalns, C., Biguet, N. F., Vyas, S., Robert, J. J., Sassone, C., Mallet, P., and Mallet, J. (1992). AP-1 complex and c-*fos* transcription are involved in TPA provoked and trans-synaptic inductions of the tyrosine hydroxylase gene: Insights into long-term regulatory mechanisms. *J. Neurosci. Res.* 32: 290–298.

Junier, M. P., Wolf, A., Hoffman, G. E., Ma, Y. G., and Ojeda, S. R. (1992). Effect of hypothalamic lesions that induce precocious puberty on the morphological and functional maturation of the luteinizing hormone releasing hormone neuronal system. *Endocrinology* 131: 787–798.

Kamphuis, W., Gorter, J. A., and Lopes da Silva, F. (1991). A long-lasting decrease in the inhibitory effect of GABA on glutamate responses of hippocampal pyramidal neurons induced by kindling epileptogenesis. *Neuroscience* 41: 425–431.

Koch, M., and Ehret, G. (1989). Immunocytochemical localization and quantitation of estrogen-binding cells in male and female (virgin, pregnant, lactating) mouse brain. *Brain Res.* 489: 101–112.

Kocsis, B., and Vertes, R. P. (1992). Dorsal raphe neurons: Synchronous discharge with the theta rhythm of the hippocampus in the freely behaving rat. *J. Neurophysiol.* 68: 1463–1467.

Lambert, G. M., Rubin, B. S., and Baum, M. J. (1992). Sex difference in the effect of mating on c-fos expression in luteinizing hormone-releasing hormone neurons of the ferret forebrain. *Endocrinology* 131: 1473–1480.

Lee, L. R., Haisenleder, D. J., Marshall, J. C., and Smith, M. S. (1989a). Effects of progesterone on pulsatile luteinizing hormone (LH) secretion and LH subunit messenger ribonucleic acid during lactation in the rat. *Endocrinology* 124: 2128–2134.

Lee, L. R., Haisenleder, D. J., Marshall, J. C., and Smith, M. S. (1989b). Expression of alpha-subunit and luteinizing hormone (LH) beta messenger ribonucleic acid in the rat during lactation and after pup removal: Relationship to pituitary gonadotropin-releasing hormone receptors and pulsatile LH secretion. *Endocrinology* 124: 776–782.

Lee, W. S., Smith, M. S., and Hoffman, G. E. (1990a). Luteinizing hormone releasing hormone (LHRH) neurons express c-Fos during the proestrous LH surge. *Proc. Natl. Acad. Sci. (USA)* 87: 5163–5167.

Lee, W. S., Smith, M. S., and Hoffman, G. E. (1990b). Progesterone enhances the surge of luteinizing hormone by increasing the activation of luteinizing hormone-releasing hormone neurons. *Endocrinology* 127: 2604–2606.

Lee, W. S., Smith, M. S., and Hoffman, G. E. (1992). c-*fos* activity identifies recruitment of LHRH neurons during the ascending phase of the proestrous luteinizing hormone surge. *J. Neuroendocrinol.* 4: 161–166.

Legan, S. J., Coon, G. A., and Karsch, F. J. (1975). Role of estrogen as initiator of daily LH surges in the ovariectomized rat. *Endocrinology* 96: 50–56.

Lehman, M. N., and Karsch, F. J. (1993). Do gonadotropin-releasing hormone, tyrosine hydroxylase-, and b-endorphin-immunoreactive neurons contain estrogen receptors? A double label immunocytochemical study in the Suffolk ewe. *Endocrinology* 133: 887–895.

Levine, J. E., and Ramirez, V. D. (1982). Luteinizing hormone-releasing hormone release during the rat estrous cycle and after ovariectomy, as estimated with push–pull cannulae. *Endocrinology* 111: 1439–1448.

Lincoln, D. W., Hentzen, K., Hin, T., van der Schoot, P., Clarke, G., and Summerlee, A. J. (1980). Sleep: A prerequisite for reflex milk ejection in the rat. *Exp. Brain Res.* 38: 151–162.

Maggi, A., and Perez, J. (1984). Progesterone and estrogens in rat brain: Modulation of GABA (γ-aminobutyric acid) receptor activity. *Eur. J. Pharmacol.* 103: 165–168.

Majewska, M. D., Harrison, N. L., Schwartz, R. D., Barker, J. L., and Paul, S. M. (1986). Steroid hormone metabolites are barbiturate-like modulators of the GABA receptor. *Science* 232: 1004–1007.

Mattson, R. H., and Cramer, J. A. (1985). Epilepsy, sex hormones, and antiepileptic drugs. *Epilepsia* 26 (Suppl. 1): S40–S51.

McGinty, D. J,. and Harper, R. M. (1976). Dorsal raphe neurons: Depression of firing during sleep in cats. *Brain Res.* 101: 569–575.

Morgan, J. I., and Curran, T. (1989). Stimulus-transcription coupling in neurons: Role of cellular immediate-early genes. *TINS* 12: 459–462.

Morgan, J. I., and Curran, T. (1991). Proto-oncogene transcription factors and epilepsy. *TIPS* 12: 343–349.

Morgan, P. F., and Linnoila, M. (1991). Regional induction of c-Fos mRNA by NMDA: A quantitative in-situ hybridization study. *NeuroReport* 2: 251–254.

Morishige, W. K., Pepe, G. P., and Rothchild, I. (1973). Serum luteinizing hormone, prolactin and progesterone levels during pregnancy in the rat. *Endocrinology* 92: 1527–1530.

Numan, M. (1988). Maternal behavior. In *The Physiology of Reproduction*, ed. E. Knobil and J. D. Neill, pp. 1569–1649. New York: Raven Press.

Ojeda, S., and Urbanski, H. (1987). Puberty in the rat. In *The Physiology of Reproduction*, ed. E. Knobil and J. Neill, pp. 1697–1738. New York: Raven Press.

Olney, J. W. (1980). Kainic acid and other excitotoxins: A comparative analysis. In *Glutamate as a Neurotransmitter*, ed. G. DiChiara and G. L. Gessa, pp. 375–384. New York: Raven Press.

Plotsky, P., and Rivest, S. (1991). Estrogen-dependent activation of c-Fos in LHRH immunopositive neurons by norepinephrine. *Endocrine Soc. Abstr.* 74: 99, no.190.

Pohl, C., Lee, L., and Smith, M. (1989). Qualitative changes in luteinizing hormone and prolactin responses to *N*-methyl-aspartic acid during lactation in the rat. *Endocrinology* 124: 1905–1911.

Price, M. T., Olney, J. W., Mitchell, M. V., Fuller, T., and Cicero, T. J. (1978). LH releasing action of *N*-methyl aspartate is blocked by GABA or taurine but not by dopamine antagonists. *Brain Res.* 158: 461–465.

Qureshi, G. A., Hansen, S., and Sodersten, P. (1987). Offspring control of cerebrospinal fluid GABA concentrations in lactating rats. *Neurosci. Lett.* 75: 85–88.

Roscizewska, D., Buntner, B., Guz, I., and Zawisza, L. (1989). Ovarian hormones, anticonvulsant drugs, and seizures during the menstrual cycle in women with epilepsy. *J. Neurol. Neurosurg. Psychiatr.* 49: 47–51.

Sagar, S. M., Sharp, F. R., and Curran, T. (1988). Expression of c-fos protein in brain: Metabolic mapping at the cellular level. *Science* 240: 1328–1331.

Sar, M., and Stumpf, W. E. (1981). Central noradrenergic neurones concentrate 3H-oestradiol. *Nature (London)* 289: 500–502.

Sarkar, D. M., Chiappa, S. A., and Fink, G. (1976). Gonadotropin-releasing hormone surge in pro-oestrous rats. *Nature (London)* 264: 461–466.

Shivers, B. D., Harlan, R. E., Morrell, J. I., and Pfaff, D. W. (1983). Absence of oestradiol concentration in cell nuclei of LHRH-immunoreactive neurones. *Nature (London)* 304: 345–347.

Smith, M. S. (1981). The effects of high levels of progesterone secretion during lactation on the control of gonadotropin secretion in the rat. *Endocrinology* 109: 1509–1517.

Smith, M. S., Freeman, M. E., and Neill, J. D. (1975). The control of progesterone secretion during the estrous cycle and early pseudopregnancy in the rat: Prolactin and gonadotropin levels associated with rescue of the corpus luteum of pseudopregnancy. *Endocrinology* 96: 219–227.

Smith, M. S., Lee, W. S., Abbud, R., and Hoffman, G. E. (1991). Use of c-*fos* as a marker of neuronal activation reveals sex differences in responsiveness to NMA. *Soc. Neurosci. Abstr.* 17: 907, no.361.10.

Smith, M. S., and Neill, J. D. (1977). Inhibition of gonadotropin secretion during lactation in the rat: Relative contribution of suckling and ovarian steroids. *Biol. Reprod.* 17: 255–261.

Smith, S. S. (1989). Progesterone enhances inhibitory responses of cerebellar Purkinje cells mediated by the GABA-A receptor subtype. *Brain Res. Bull.* 23: 317–322.

Smith, S. S. (1991). Progesterone administration attenuates excitatory amino acid responses of cerebellar Purkinje cells. *Neuroscience* 42: 309–320.

Smith, S. S., Waterhouse, B. D., Chapin, J. K., and Woodward, D. J. (1987). Progesterone alters GABA and glutamate responsiveness: A possible mechanism for its anxiolytic action. *Brain Res.* 400: 353–359.

Sonnenberg, J. L., Mitchelmore, C., Macgregor-Leon, P. F., Hempstead, J., Morgan, J. I., and Curran, T. (1989). Glutamate receptor agonists increase the expression of Fos, Fra, and AP-1 DNA binding activity in the mammalian brain. *J. Neurosci. Res.* 24: 72–80.

Stelzer, A., Slater, N. T., and Bruggencate, G. T. (1987). Activation of NMDA receptors blocks GABAergic inhibition in an in vitro model of epilepsy. *Nature* 326: 698–701.

Steriade, M. (1992). Basic mechanisms of sleep generation. *Neurology* 4: 9–17.

Tindall, J. S., Knaggs, G. S., and Turvey, A. (1967). The afferent path of the milk ejection reflex in the brain of the guinea pig. *J. Endocrinol.* 38: 337-349.

Trulson, M. E., and Jacobs, B. L. (1979). Raphe unit activity in freely moving cats: Correlation with level of behavioral arousal. *Brain Res.* 163: 135–150.

Vanderwolf, C. H. (1988). Cerebral activity and behavior: Control by central cholinergic and serotonergic systems. *Intern. Rev. Neurobiol.* 30: 225–340.

Wakerley, J. B., Clarke, G., and Summerlee, A. J. S. (1988). Milk ejection and its control. In *The Physiology of Reproduction*, ed. E. Knobil, J. Neill, L. L. Ewing, G. S. Greenwald, C. L. Markert, and D. W. Pfaff, pp. 2283–2321. New York: Raven Press.

Watson, R. E., Langub, M. C., and Landis, J. W. (1992). Further evidence that most luteinizing hormone releasing hormone neurons are not directly estrogen-responsive: Simultaneous localization of luteinizing hormone-releasing hormone and estrogen receptor immunoreactivity in the guinea pig brain. *J. Neuroendocrinol.* 4: 311–317.

Weiland, N. G. (1992). Estradiol selectively regulates agonist binding sites on the *N*-methyl-D-aspartate receptor complex in the CA1 region of the hippocampus. *Endocrinol.* 131: 662–668.

Woodward, R. M., Polenzani, L., and Miledi, R. (1992). Effects of steroids on γ-aminobutyric acid receptors expressed in xenopus oocytes by poly(A)[+] RNA from mammalian brain and retina. *Mol. Pharmacol.* 41: 89–103.

Wray, S., Grant, P., and Gainer, H. (1989). Evidence that cells expressing luteinizing hormone-releasing hormone mRNA in the mouse are derived from progenitor cells in the olfactory placode. *Proc. Natl. Acad. Sci. (USA)* 86: 8132–8136.

Wray, S., and Hoffman, G. E. (1986a). A developmental study of the quantitative distribution of LHRH neurons in postnatal male and female rats. *J. Comp. Neurol.* 252: 522–531.

Wray, S., and Hoffman, G. E. (1986b). Postnatal morphological changes in rat LHRH neurons correlated with sexual maturation. *Neuroendocrinology* 43: 93–97.

17

Developmental interactions of estrogens with neurotrophins and their receptors

C. DOMINIQUE TORAN-ALLERAND

Introduction

Appropriate development and maintenance of neurons and their interconnections are fundamental to central nervous system (CNS) function and plasticity in mammals. The molecules that regulate these processes during development as well as in mature neurons are largely unknown and minimally characterized. Two classes of substances implicated in the regulation of development, survival, and neurotransmitter production of neurons in telencephalic regions that subserve learning, memory, and general cognition are estrogen and the family of endogenous growth- and survival-promoting proteins, the neurotrophins, nerve growth factor (NGF), brain-derived neurotrophin factor (BDNF), neurotrophin-3 (NT-3), and neurotrophin-4/5 (NT-4/5).

The organization of neural circuits controlling a broad spectrum of behavioral functions is permanently influenced by exposure of the developing CNS to gonadal steroid hormones: estrogens and androgens. Exposure to these steroids during pre- and postnatal "critical periods" profoundly and permanently influences the organization of the nervous system. The consequences of this hormonal exposure are evident in striking sexual dimorphisms of neuronal structure and function, which underlie the observed sex differences in behaviors, including those related to cognition. The cellular and molecular mechanisms by which gonadal steroids permanently influence the structural and functional organization of the developing brain remain ill-defined, although it seems clear that effects on cellular aspects of neuronal differentiation are likely to be involved (reviewed by Juraska 1991; Toran-Allerand 1991a). Estrogen enhances the growth and differentiation of neurons and their processes (neurites) within the developing hypothalamus, preoptic area, and cerebral cortex *in vivo* and *in vitro* (reviewed by Toran-Allerand 1984, 1991b), as well as in steroid target regions of the adult CNS following injury through steroid deprivation, axotomy, or deafferentation (Matsumoto and Arai 1981; Frankfurt et al. 1990).

The estrogen receptor

Estradiol binds with high affinity to intranuclear receptors in neurons that are widely distributed throughout developing forebrain regions, such as the hypothalamus, preoptic area, cerebral cortex, striatum, septum/nuclei of the diagonal band of Broca, and hippocampus and amygdala, among others. The estrogen receptor is a ligand-modulated, transcription-regulating factor and member of the superfamily of genes encoding receptors for steroid hormones, retinoic acid, thyroid hormone, and vitamin D_3 (Evans 1988). The estrogen receptor is a highly conserved, single-copy gene whose transcription product represents a relatively rare mRNA population (Koike et al. 1987; White et al. 1987). The promoter regions of several estrogen-inducible genes, including vitellogenin (Klein-Hitpass et al. 1986), c-*fos* (Weisz and Rosales 1990), prolactin (Waterman et al. 1988), and β-luteinizing hormone (Shupnik and Rosenzweig 1991), contain specific DNA sequences of nucleotides, termed an estrogen-specific regulatory or responsive element (ERE), that mediates cell-specific trans-activation by the estrogen receptor–ligand complex. The ERE is usually found as a 13-base-pair inverted repeat (palindrome), separated by 3 intervening base pairs, with the consensus sequence, 5′-GGTCANNNTGACC-3′, but can sometimes be found as direct repeats or even as half-palindromes (Gaub et al. 1990). Activation of estrogen-responsive genes alters the transcriptional efficiency of estrogen-inducible genes, regulating mRNA and protein synthesis in order to effect a cascade of protein–DNA interactions that ultimately enhance or suppress expression of those genes or gene networks (reviewed by Landers and Spelsberg 1992). The specific genes involved, as well as the cascade of molecular and cellular events that follow estrogen binding to the receptor, are poorly understood. The factors influencing transcriptional and translational control of the estrogen receptor in the CNS remain essentially unknown.

Possible mechanisms underlying estrogen actions

Despite the profound effects of estrogen in early neural development, the molecular mechanisms underlying these actions are unknown. A critical and, as yet, unanswered question, for example, is whether the actions of estrogen on neurite growth are mediated directly on growth-related target genes or involve intermediate steps (Toran-Allerand 1984). Since technical aspects have precluded tracing estrogen-responsive neurites to their cell bodies of origin, we do not even know whether the estrogen receptor–containing neurons themselves respond to estrogen by neurite enhancement. Estrogen might also regulate neurite growth and differentiation, as well as neuronal survival, through intermediate steps that involve interactions with growth factors and/or their receptors (Toran-Allerand 1984, 1991b; Toran-Allerand et al. 1988). To address this issue, we have chosen to focus on the developing forebrain, a known target of both estrogen and the growth factor family of neurotrophins, as well as a site of estrogen and neurotrophin synthesis.

Our investigation, however, has been rendered more difficult by the fact that both estrogen and neurotrophins can influence many of the same genes associated with neurite growth, such as tau microtubule-associated protein (Drubin et al 1985; Aletta et al. 1990; Ferreira and Caceres 1991; Hanemaaijer and Ginsburg 1991) and GAP-43 (Federoff et al. 1988; Lustig et al. 1991). Their site of action, however, need not be the same and might range from transcription through translation and post-translational protein processing. Interactions between estrogen and growth factors in extraneural targets such as the female reproductive tract [epidermal growth factor (EGF); Mukku and Stancel 1985; Nelson et al. 1991; Ignar-Trowbridge et al. 1992] and mammary tumor cell lines [EGF and insulin-like growth factor-1 (IGF-1); Dickson and Lippman 1987; Read et al. 1989] have been implicated in the mediation of an increasing number of estrogen-induced differentiative processes.

Neurotrophins and their receptors

Like estrogen target neurons, the distribution of neurons expressing neurotrophins and their cognate receptors is widespread throughout the developing and adult CNS.

Neurotrophins

The neurotrophins NGF, BDNF, and NT-3 are members of a homologous family of proteins that have important growth and trophic actions on the development, survival, and maintenance of the central and peripheral (PNS) nervous systems (reviewed by Barde 1989, 1990). Despite their strong similarity, these peptides appear to exhibit restricted spatial and temporal patterns of expression, different sequential actions, and different stages of central and peripheral target responsiveness. While NGF, BDNF, and NT-3 mRNAs are reportedly expressed in distinct neuronal populations of the developing and adult brain, we have recently documented co-expression of the mRNA encoding at least two neurotrophins in the developing forebrain (Miranda et al. 1993b). Although neurotrophin functions in the CNS are ill-defined, particularly during development, forebrain regions such as the hippocampus and neocortex are particularly rich in their mRNA (reviewed by Barde 1990). NGF, BDNF, and NT-3 have been implicated in the development, survival, and plasticity of cholinergic and noncholinergic basal forebrain neurons and of dopaminergic neurons of the substantia nigra (Gnahn et al. 1983; Hefti 1986; Williams et al. 1986; Kromer 1987; Cavicchioli et al. 1991; Knusel et al. 1991, 1992; Hyman et al. 1991). Impairments of spatial memory retention associated with an age-dependent atrophy of cholinergic neurons in the basal forebrain of aged rats have been ameliorated by NGF infusions (Fischer et al. 1991).

Neurotrophin receptors

Neurotrophin-responsive neurons express high- and low-affinity receptors. The "low-affinity" NGF, or *pan*-neurotrophin receptor (p75), binds NGF, BDNF, and NT-3 in common with equally low affinity (reviewed by Chao 1992). Although the precise functions of p75 remain unclear and controversial, some studies suggest that, in certain situations, the presence of p75 might be a requisite for NGF action (Hempstead et al. 1991) and might confer ligand discrimination and specificity (Rodriguez-Tebar et al. 1992). The biological activities of the neurotrophins, however, are mediated by a family of ligand-specific, high-affinity trans-membrane receptors with an intracellular tyrosine kinase domain, which transduce the effects of their ligands to the nucleus. These receptors are members of the proto-oncogene *trk* subfamily of glycoproteins, *trkA, trkB*, and *trkC*, which are preferentially expressed in neural tissues and are required for signal transduction of NGF, BDNF, and NT-3, respectively (reviewed by Chao 1992; Miranda et al. 1993a). Neurotrophin binding results in autophosphorylation of tyrosine residues and induces tyrosine kinase activity (reviewed by Chao 1992; Schlessinger and Ullrich 1992). Depending on the cells or tissue (e.g., neural or extraneural, normal or transformed) the *trk* family might or might not require complexing with p75 for signal transduction through pathways, which are poorly understood. The cellular and regional distribution of the *trk* family in the developing brain has not yet been fully characterized.

Cells expressing *trkA* bind NGF, NT-3, and NT-4/5 but not BDNF (Hempstead et al. 1991; Kaplan et al. 1991a,b; Klein et al. 1991a,b). Although normal *trkA* gene expression *in vivo* is reportedly restricted to subsets of spinal and cranial sensory neurons and to neurons of the basal forebrain (Martin-Zanca et al. 1990; Kaplan et al. 1991a,b; Vasquez and Ebendal 1991; Holtzman et al. 1992), we have also observed expression of its mRNA throughout the developing forebrain, particularly in the postnatal cerebral cortex and hippocampus (Miranda et al. 1993a,b). In contrast, *trkB* transcripts are multiple (full-length and truncated), differentially regulated during development, and widely distributed in neurons and non-neuronal cells throughout the CNS and PNS (reviewed by Chao 1992). One *trkB* protein product constitutes a tyrosine kinase receptor glycoprotein and, since it does not bind NGF, has been proposed as the high-affinity BDNF receptor (Soppet et al. 1991; Squinto et al. 1991). NT-3 has the highest affinity for another member of this family, *trkC* (Lamballe et al. 1991) and, like *trkB*, comprises several full-length and truncated isoforms (Tsoulfas et al. 1993; Valenzuela et al. 1993). Although signal transduction of the biological effects of NGF in PC12 cells appears to require both p75 and *trkA* (Hempstead et al. 1991; Kaplan et al. 1991a,b; Loeb et al. 1991), it is unclear whether p75 is also required for high-affinity binding to *trkB* and *trkC* and signal transduction of BDNF and NT-3, respectively. The factors regulating neurotrophin and neurotrophin receptor gene transcription and translation have been little studied.

Co-localization of estrogen receptors with neurotrophins and their receptors

We proposed initially that estrogen could interact with neurotrophins and their receptors because of their regional co-distribution within the developing and adult forebrain (Toran-Allerand and MacLusky 1989). This hypothesis received support from the demonstration of co-localization of estrogen receptors (^{125}I-labeled estrogen binding sites) with p75 mRNA and protein (Fig. 17.1) (Toran-Allerand et al. 1992) in cholinergic neurons of the developing and adult rodent basal forebrain (Toran-Allerand et al. 1992). Co-expression of these two receptor systems was also seen in neurons of the developing and adult rodent cerebral cortex, striatum (caudatoputamen), and hippocampus, among others (Miranda et al. 1993).

Since all currently identified neurotrophins bind p75, neurons co-localizing estrogen binding sites and p75 would be likely targets not only of NGF, but of BDNF and NT-3 as well. By analogy, moreover, at least some of these neurons would most likely also co-localize the signal-transducing members of the *trk* family. This, in fact, appears to be the case. The significance of co-expression of estrogen receptors with p75 has been further emphasized by our subsequent findings [using double-label (isotopic/nonisotopic) *in situ* hybridization histochemistry] of co-localization of estrogen receptor mRNA with the mRNAs for all the neurotrophin receptors (p75, *trkA*, and *trkB*) in the developing basal forebrain and two of its projection targets, the hippocampus and cerebral cortex (Fig. 17.2A; Miranda et al. 1993a). Not all estrogen receptor–containing neurons appeared to co-localize p75 in these regions, implying that some neurons might co-express estrogen receptor mRNA and *trkA* or *trkB* mRNA alone. The functions of p75 must be reassessed, since not all neurons expressing *trkA* or *trkB* co-localized p75 (Miranda et al., 1993a). One possibility is that these differential patterns of expression reflect the anatomical basis for observations that some (Cordon-Cardo et

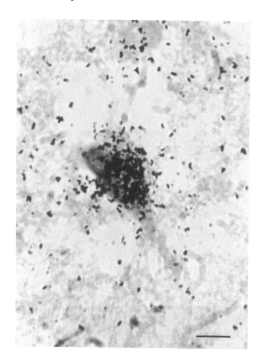

Figure 17.1. Co-localization of estrogen and p75 receptors in neurons of the P12 female mouse septum–diagonal band. The results were obtained by combining ^{125}I-labeled estrogen autoradiography (11β-methoxy[16α-^{125}I]iodoestradiol; discrete concentration of silver grains in the emulsion underlying neuronal nuclei) and immunohistochemistry for p75 (monoclonal antibody IgG 192; dark cytoplasmic reaction product). Scale bar, 1 μm. From Toran-Allerand et al. (1992).

al. 1991; Birren et al. 1992; Nebreda et al. 1991) but not all (Hempstead et al. 1991; Wright et al. 1992) of the biological actions of neurotrophins in peripheral neurons and non-neural cell lines might be transduced independently of p75. Whether a similar situation exists for the responses of developing CNS neurons is, as yet, unknown.

Estrogen receptor mRNA is co-localized not only with neurotrophin receptor mRNA but also with the mRNA for the neurotrophin ligands (Fig. 17.2A) themselves (Miranda et al. 1993a). In addition, groups of forebrain neurons co-expressed the mRNA for both the neurotrophin ligands (NGF, BDNF, NT-3) and their receptors (p75, *trkA*, and *trkB*; Fig. 17.2B) (Miranda et al. 1993b), as well as for two neurotrophin ligand mRNAs (e.g., NT-3 and NGF or NT-3 and BDNF) (Miranda et al. 1993b). Not every neuron expressing neurotrophin ligand mRNA co-localized the mRNA for its cognate receptor. Such interneuronal variability indicates that any autocrine mechanism may be cell-specific, region-specific, and temporally specific. Moreover, since these forebrain regions are all

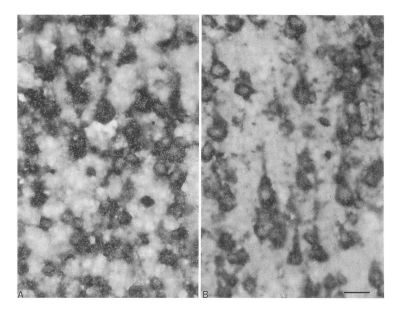

Figure 17.2. Co-localization of estrogen receptor mRNA with BDNF and *trkB* mRNA in the P9 mouse cerebral cortex by combined (isotopic–nonisotopic) *in situ* hybridization histochemistry. (A) Combined bright-field and epipolarization photomicrography, showing co-localization of estrogen receptor (digoxigenin; dark reaction product) and BDNF [^{35}S; silver grains] mRNAs within the same cortical neurons. (B) Co-localization of BDNF (^{35}S) and *trkB* (digoxigenin) mRNAs in the same cortical region as in (A). Hybridization with the two probes leads to significant accumulation of silver grains and cytoplasmic alkaline phosphatase reaction product over most cortical neurons. Scale bar, 20 μm.

targets of estrogen as well, there is a very high probability that neurons co-localizing the mRNA for a neurotrophin ligand and receptor or for two neurotrophins are also responsive to estrogen.

Interactions of estrogen with neurotrophins and their receptors

Like NGF, BDNF, and NT-3, estrogen influences cholinergic enzymes such as choline acetyltransferase (ChAT). ChAT concentration (Luine et al. 1980) and activity (Luine 1985) in the developing septum–diagonal band complex are also responsive to estrogen (Luine 1985) and exhibit a sexual dimorphism in the location and direction of the response (Libertun et al. 1973; Luine and McEwen 1983; Loy and Sheldon 1986; Luine et al. 1986). Like neurotrophins, estrogen has been reported to act as a "growth factor" for basal forebrain cholinergic neurons following transplantation into the anterior chamber of the eye (Honjo et al. 1992).

Both estrogen and the neurotrophins elicit the well-characterized induction of many primary (early) response genes, such as the proto-oncogenes c-*fos* and c-*jun*, which rapidly and transiently alter protein synthesis–independent gene transcription to modulate gene expression in a variety of cell types, including neurons (Wu et al. 1989; Sheng and Greenberg 1990; Weisz et al. 1990; Hershman 1991; Hoffman et al. 1992). Homo- and heterodimerization of Fos and Jun with other cellular oncogene products, particularly members of the *jun* family (Fos/Jun and Jun/Jun), and association of these complexes with DNA at a nucleotide sequence motif known as the AP-1 binding site are important for some of the transcriptional actions of both estrogen and NGF (D'Mello and Heinrich 1991; Hoffman et al. 1992). In the case of estrogen, the AP-1 site might also regulate estrogen receptor interaction with DNA and estrogen regulation of gene expression (Landers and Spelsberg 1992). These interactions enhance or repress transcription of target genes, depending on the response element and the cell phenotype.

Estrogen enhancement of neuronal survival and neurite growth via interactions with endogenous growth factors was initially dismissed as highly improbable in the preneurotrophin era of the late 1970s (Toran-Allerand et al. 1980). At the same time, we suggested that estrogen might stimulate secretion of an "NGF-like substance," which might "self-induce" (by an autocrine effect) neurite growth of the responsive neuron itself or act on the immediately surrounding neurons (by a local paracrine effect) (Toran-Allerand et al. 1980). This hypothesis became more plausible following our observations that synergistic interactions between estradiol and a class of endogenous growth factors, the insulin-related peptides (e.g., insulin and IGF-1), may be important in the mediation or modulation of estrogen-responsive neurite growth *in vitro* (Toran-Allerand et al. 1988).

A critical question raised by the co-localization of estrogen receptors with the mRNA for neurotrophins and their receptors is whether these patterns of co-expression are functionally significant. Computer-assisted sequence analysis of the identified promoter region (Sehgal et al. 1988) of the human p75 (Toran-Allerand et al. 1992), the 5' region flanking the *trk* oncogene breakpoint in the *trkA* gene (Sohrabji et al. 1994), and the promoter and 5' flanking regions of the cloned rodent (Selby et al. 1987; Zheng and Heinrich 1988) and human (Ullrich et al. 1983) NGF genes (Miranda et al. 1993) indicates the presence of sequences with a high degree of homology to the putative vitellogenin and c-*fos* EREs (Fig. 17.3). These findings suggest a molecular basis for the proposed interactions of estrogen with both a neurotrophin (NGF) and its receptors (p75 and *trkA*) and are consistent with the hypothesis that, in some instances, estrogen might regulate steroid-inducible genes in the CNS via interactions with neurotrophins and their receptors.

In order to address the potential for interplay between estrogen and neurotrophins, we have used both *in vivo* and *in vitro* model systems. Preliminary (unpublished) *in situ* hybridization histochemical studies in explants of the new-

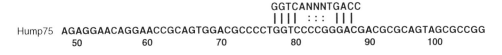

Figure 17.3. Putative ERE sequences in NGF, p75, and *trkA* genes. Computer-assisted comparison of the 13-base palindromic ERE sequence and the human NGF and NGF receptor genes revealed close homology to sequences in the promoter region of the p75 gene, the 5′ region flanking the *trk* oncogene breakpoint in the *trkA* gene, and the promoter region of the NGF gene. Homologous nucleotides are indicated by vertical bars. Humngfr, human p75 sequence; HumtrkA, human *trkA* sequence; HumNGF, human NGF sequence.

born mouse septum–diagonal band and cerebral cortex have suggested that concurrent exposure to very high estrogen levels (100-fold higher than during proestrus) and NGF for 21 days elicits a striking increase in the number of cells expressing estrogen receptor mRNA and in the intensity of the hybridization signal compared with either ligand added separately or not at all. Proestrous levels of estrogen normally downregulate the estrogen receptor in adult neural and extraneural targets (Shughrue et al. 1992). Whether the observed increase in estrogen receptor mRNA resulted from an enhancement of neuronal survival, an increase in gene expression, or both is unclear. Enhancement of neurogenesis is unlikely, since neurons in these regions are postmitotic at this developmental stage. Moreover, exposure of hemicoronal cultures of postnatal day 3 septum–diagonal band and cerebral cortex–striatum to estradiol with or without concurrent exposure to NGF and BDNF for 14 days *in vitro* (D. Toran-Allerand and W. Bentham, unpublished results) resulted in significantly less estrogen receptor mRNA compared with the matched, homologous explant halves receiving estradiol and both neurotrophins concurrently. The apparent influence of these neurotrophins on estrogen receptor mRNA expression in the cortical cultures suggests that extensive co-expression of estrogen receptor mRNA with the mRNA for *trkA* and *trkB* in the CNS is biologically relevant.

Interactions of estrogen with neurotrophins might be critical for the differentiation and survival of neurons co-expressing each receptor system. These ligands might also reciprocally influence each other's actions, whether by regulating receptor and/or ligand availability or by reciprocal regulation at the level of signal transduction or gene transcription. Support for this is derived from studies which suggest that testosterone- and estrogen-induced behavioral defeminization of neonatal female rats can be blocked by intraventricular injection of antibodies to NGF but not by antibodies to insulin or EGF (Yanase et al. 1988; Hasegawa et al. 1991).

Functional consequences of receptor co-expression

In order to test further the hypothesis that estrogen sensitivity is a universal phenotypic feature of all the CNS and PNS targets of neurotrophins, and that estrogen and neurotrophins can reciprocally regulate each other's receptors, we analyzed two prototypical peripheral targets of NGF not previously reported to respond to estrogen: sensory [dorsal root ganglion (DRG)] neurons of the intact female rat (Sohrabji et al. 1994) and PC12 cell cultures (Sohrabji et al. 1994).

DRG

DRG, obtained from adult female rats at proestrus and following ovariectomy with or without estrogen replacement, were studied by non-isotopic *in situ* hybridization and Northern blot analysis for estrogen receptor, p75, and *trkA* mRNA expression (Sohrabji et al. 1994). Estrogen receptor mRNA was unexpectedly expressed in all DRG neurons, along with p75 and *trkA* mRNA. There were striking differences in the patterns of hybridization of each probe under these experimental conditions. While the intensity of the hybridization signal varied greatly during proestrus, ovariectomy resulted in uniform upregulation of estrogen receptor mRNA. This upregulation has been observed in most known adult targets of estrogen, including the uterus and brain (Shughrue et al. 1992). Nuclear receptor binding assays using [^{3}H]moxestrol (RU 2848) revealed that DRG from ovariectomized females expressed fully compatible estrogen binding sites, clearly documenting that estrogen receptor mRNA in DRG is translated into functional binding sites.

In contrast, the expression of p75 and *trkA* mRNA transcripts was co-regulated: upregulated threefold during proestrus and downregulated following ovariectomy. One should emphasize, however, that not only are high levels of estrogen present during the afternoon of proestrus, but high levels of testosterone, prolactin, and follicle-stimulating hormone (FSH) are present as well (Brown-Grant et al. 1970; Neill 1972; Butcher et al. 1974), all of which might also influence transcription of the NGF receptor genes. Any changes in *trkA* and p75 mRNA expression during proestrus must represent a sum total of several hormones exerting diverse control or the same gene. Thus, the responses seen during proestrus and after ovariectomy cannot be compared with those seen after ovariectomy alone and with estrogen replacement. The similarity in the direction and extent of *trkA* mRNA regulation, however, both at proestrus and following estrogen replacement, suggests that *trkA* mRNA upregulation during proestrus might be due specifically to the high titers of estrogen present at this time.

Furthermore, in DRG obtained following ovariectomy and estrogen replacement for 4 or 52 hours, there was a time- and receptor-specific regulation of p75 and *trkA* mRNA. P75 was rapidly downregulated by 4 hours, with a return to basal levels after 52 hours of estrogen replacement. In contrast, estrogen elicited a significant, time-dependent upregulation of *trkA* mRNA that was comparable to the magnitude of the *trkA* mRNA response seen during proestrus.

PC12 cells

Studies have also been carried out on the interactions of estrogen and NGF in differentiated (NGF +) and naive (NGF −) PC12 cells (Sohrabji et al. 1994). PC12 cells, a neural line of rat adrenal pheochromocytoma origin (Greene and Tischler 1982), resemble chromaffin cells and express both the high- (*trkA*) and low-affinity (p75) components of the NGF receptor system. Upon exposure to NGF, PC12 cells cease proliferating and extend long neurites, assuming phenotypic features of sympathetic neurons. Like DRG cells, NGF-deprived (naive) PC12 cells were found to express low levels of estrogen receptor mRNA by both *in situ* hybridization and reverse transcriptase–polymerase chain reaction (Sohrabji et al. 1994). Even more surprising, after exposure to NGF, virtually all PC12 cells intensely expressed estrogen receptor mRNA in a dose-dependent fashion. Translation of the mRNA into receptor protein was confirmed in naive PC12 cells by [³H]moxestrol nuclear binding assay (Sohrabji et al. 1994). NGF treatment increased estrogen receptor levels sixfold. The nonestrogenic (nonaromatizable) gonadal steroid 5α-dihydrotestosterone failed to block specific nuclear moxestrol binding sites. In contrast, in the absence of NGF exposure, estrogen appeared to downregulate PC12 estrogen receptor mRNA and protein, as in the adult brain and uterus.

Studies of estrogen effects on NGF-treated PC12 cells by Northern blot analysis for p75 and *trkA* mRNA expression revealed, as in the DRG, differential regulation of each neurotrophin receptor mRNA as a function of the duration of estrogen exposure. NGF-Primed cells responded to estrogen treatment by a twofold increase in *trkA* mRNA between 4 and 7 days of estrogen exposure. In contrast, there was a transient 50% decrease in p75 mRNA, after 3 days of estrogen treatment, with a return to control levels by 7 days. These patterns resembled those observed following estrogen replacement in DRG neurons. Why these estrogenic actions on PC12 *trkA* mRNA expression required 7 days of exposure is unclear. It should be pointed out, however, that estrogen receptors in naive PC12 cells are present at very low levels, and there simply might not be a sufficient number of receptors to permit responses to estrogen during the first few days of exposure.

While co-localization of the estrogen and neurotrophin receptor systems suggests a possible interactive role for their ligands, the present data from peripheral neurotrophin targets also suggest that estrogen and neurotrophins might influence reciprocally neuronal responsiveness to each other's ligand by regulating the expression of receptor for that ligand. The functional consequences of differential regulation of p75 and *trkA* mRNA by estrogen remain purely speculative. How, or even if, estrogen regulates the encoded proteins is currently unknown, and the individual contributions of each form of the receptor to the functional NGF receptor complex remain unclear and controversial. If *trkA* is the signal-transducing arm of the NGF receptor, then estrogen, by increasing *trkA* mRNA, might alter neuronal responsivity to NGF by altering a critical ratio of p75 and *trkA* molecules, which has been suggested to be important for the efficiency of NGF binding (Battleman et al. 1993). Overexpression of p75 in Sf9 cells expressing *trkA* inhibits autophosphorylation of *trkA* and decreases binding of NGF, emphasizing the apparent importance of the p75/*trkA* ratio (D. Kaplan, personal communication). Thus, by increasing *trkA* mRNA while decreasing p75 mRNA, estrogen might modulate the transduction by receptor tyrosine kinases. Our findings may be viewed as providing an indication that p75 and *trkA* are each responsive to the hormonal state. These observations are of considerable significance, since, other than depolarization and exposure to NGF *in vitro*, no other substance or treatment has been shown to regulate *trkA* mRNA expression, particularly *in vivo*. The possibility of reciprocal regulation of *trkB* and *trkC* mRNA in the CNS by estrogen and other neurotrophins is currently unknown.

Although estrogen appears to have differential effects on p75 and *trkA* in adult DRG neurons and PC12 cells, its actions on p75 appear to be consistent with those reported in the brain of the ovariectomized adult rat and those of another gonadal steroid hormone, testosterone. For example, long-term (16- to 30-day) estrogen replacement with a Silastic implant has been reported to decrease p75 mRNA expression in the hippocampus and septum of the ovariectomized rat (Gibbs and Pfaff 1992). Moreover, testosterone replacement has been shown to downregulate p75 mRNA in Sertoli cells of the adult testis, an androgen target (Persson et al. 1990). The developing rat basal forebrain and cerebellum exhibit sexual dimorphism of p75 mRNA expression (greater in females than in males during the first two postnatal weeks *only*) (Kornack et al. 1991). Such dimorphism might be the consequence of testosterone downregulation during the significant pre- (E18) and postnatal (P1) surges of androgen in the developing male.

Conclusions

In considering the developmental role of estrogen, the general tendency has been to focus only on its importance for sexual differentiation of CNS structure and function. At most, however, sexual differentiation of the CNS should be viewed as a marker of gonadal steroid action during development. In a more general sense, it might perhaps be more appropriate to consider the actions of estrogen as important epigenetic influences involved in neuronal development, survival, plasticity, regeneration, and perhaps even aging, thus expanding the view of gonadal steroid action in the CNS beyond the strict (and restrictive) confines of sexual differentiation.

These observations suggest a potential substrate in the CNS for developmentally critical autocrine and/or local paracrine regulatory mechanisms, as well as for synergistic and reciprocal interactions of estrogen and neurotropins (Fig. 17.4). Estrogen influences on neurotrophin ligand and receptor molecules could alter the responsiveness of both the estrogen and neurotrophin targets and promote neuronal survival, growth, and differentiation of neuromodulatory or neurotransmitter systems. These effects could shift the developmental patterns of the resulting neural networks. The complex interactions of estrogen with autocrine-, paracrine-, or target-derived neurotrophins would thus ensure differentiation and survival of their respective target neurons during early developmental periods,

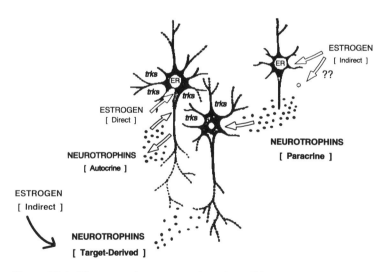

Figure 17.4. Diagrammatic representation of possible mechanisms underlying estrogen-responsive neurite growth and neuronal survival. The actions of estrogen may be mediated directly on growth-related target genes via its receptor. In addition, estrogen may also act through intermediate (indirect) steps that involve interactions with neurotrophins and/or their receptors in estrogen target neurons with autocrine or local paracrine consequences. Redrawn and modified from Toran-Allerand (1984).

when functional synaptic contacts with targets have yet to be made, and during the later stages of differentiation and adulthood (Fig. 17.4). Hence, estrogen could have a critical role in directly mediating or modulating neurotrophin-dependent regulatory actions and in indirectly modulating a wide variety of gene networks. On the other hand, co-localization of more than one neurotrophin mRNA might well reflect the basis for the observed different, often sequential, actions of neurotrophins such as BDNF and NT-3 (survival vs. maturation) and the different developmental stages of CNS neuron responsiveness to neurotrophins (mitosis vs. differentiation) (Collazo et al. 1992; Segal et al. 1992). Moreover, more is known about estrogen than neurotrophin actions in the developing CNS, but the potential significance of the observed patterns of estrogen receptor co-expression with neurotrophins and their receptors for neurotrophin regulation of the estrogen receptor should be examined (Fig. 17.5).

The direction of the responses to estrogen as a consequence of interactions with neurotrophins, however, is unpredictable, because estrogen is known either to enhance or to suppress responsive genes depending on the age, the tissue, or the developmental stage of the animal. Estrogen downregulates estrogen receptor mRNA in the adult CNS, adult DRG, and PC12 cells, but in striking contrast estrogen appears to upregulate estrogen receptor mRNA expression in the developing CNS (unpublished results).

These disparate responses might reflect an important function of the interactions of estrogen and the neurotrophins – namely, neurotrophins as regulatory switches. Thu·, estrogenic upregulation of estrogen receptors in the brain during development (a stage during which estrogen exerts profound organizational influences on the CNS) might result from the high neurotrophin ligand and receptor levels present at that time. In the adult nervous system, the actions of estrogen

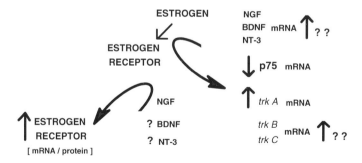

Figure 17.5. Schematic diagram of the possible consequences of co-localization of estrogen receptors with neurotrophin ligands and their receptors in the CNS. While we have evidence only for reciprocal regulation of the estrogen and NGF receptor systems by their ligands in DRG and PC12 cells (Sohrabji et al. 1994), the same responses are likely to occur in the CNS as well and involve all neurotrophin receptors. Co-expression of neurotrophin ligand mRNAs in estrogen targets strongly suggests a role for estrogen in neurotrophin synthesis in the CNS.

might be "freed" from the regulatory influences of neurotrophins, perhaps through (a) a reduction of their level and/or changes in their pattern of expression; (b) changes in the ratio of receptor species isoforms (full vs. truncated; p75 vs. *trks*); (c) changes in the modes or sources (autocrine-, paracrine-, and/or target-derived) of neurotrophin action associated with maturation; or (d) changes in the physiological role(s) of neurotrophins as seen, for example, in nociceptive sensory neuron responses to NGF (Lewin and Mendell 1993). Any or all of these changes might result in the characteristic adult pattern of estrogen downregulation of its own receptor. The upregulation of neurotrophin receptors such as p75 (Higgins et al. 1989) and the return of sensitivity to NGF in basal forebrain regions and in former developmental targets such as the striatum (Gage et al. 1989), which accompanies CNS injury, might explain the apparent "switch" in the direction of estrogen regulation of its receptor back to the developmental pattern seen in ovariectomized, axotomized, or deafferented estrogen target regions (Matsumoto and Arai 1981; Frankfurt et al. 1990). These changes in the direction of estrogen regulation of its receptor following CNS damage result in re-expression of the growth-promoting properties of estrogen on neurites and synapses (synaptogenesis, neuropil enhancement, and dendrite differentiation), which is normally seen only during development and never in the intact adult.

Conversely, it is possible that co-expression of estrogen receptors with neurotrophins and their receptors merely represents an epiphenomenon or another example of the multiple redundancies that appear to abound both during neural development and throughout life. While estrogen and neurotrophins might each regulate the same biological endpoints during differentiation such as neurite growth, the underlying mechanisms might be quite independent. Both estrogen and NGF have been shown to regulate members of the primary response family of genes such as c-*fos* and c-*jun*, which are trans-activating factors that regulate transcription in the nervous system. Documentation of putative ERE sequences in the promoter region (Toran-Allerand et al. 1992) of the cloned human p75 gene (Sehgal et al. 1988), the 5' region flanking the *trk* oncogene breakpoint in the *trkA* gene (Sohrabji et al. 1994), and the promoter and 5' flanking regions (Miranda et al. 1993) of the cloned rodent (Selby et al. 1987; Zheng and Heinrich 1988) and human NGF genes (Ullrich et al. 1983) (Fig. 17.3) suggests more than mere convergence and independent regulation of the same biological endpoints.

Mediation of estrogen action by interactions with neurotrophins and their receptors (or with other locally synthesized growth factor/receptor systems) might represent a universal mechanism by which the effects of hormones such as estrogen, with multiple and varied actions in both neural and extraneural tissues, could act locally and exhibit both regional and temporal specificities. Estrogen might have to interact with other DNA regulatory elements such as neurotrophins, and these interactions could be relevant to understanding not only the mechanisms underlying some of the actions of estrogen in the CNS, but how estrogen elicits such dramatically different responses at different stages of life.

Acknowledgments

The significant contributions of Dr. Rajesh C. G. Miranda and Dr. Farida Sohrabji to various aspects of these studies are gratefully acknowledged. Mr. Wayne D. L. Bentham provided skilled technical assistance throughout the course of many of these studies. The work described in this chapter was supported in part by awards from the USPHS (NIA, NICHD, and NIMH), the NSF, the American Health Assistance Foundation, an ADAMHA Research Scientist Award, a gift from the Maffei family, Milano, Italy, to C.D.T.-A., and an NINDS award to Lloyd A. Greene.

References

Aletta, J., Tsao, H., and Greene, L. A. (1990). How do neurites grow? Clues from NGF-regulated cytoskeletal phosphoproteins. In *Trophic Factors and the Nervous System*, ed. L. A. Horrocks, pp. 203–218, New York: Raven Press.

Barde, Y.-A. (1989). Trophic factors and neuronal survival. *Neuron* 2: 1525–1534.

Barde, Y.-A. (1990). The nerve growth factor family. *Prog. Growth Factor Res.* 2: 237–248.

Battleman, D. S., Geller, A. I. and, Chao, M. V. (1993). HSV-1 vector-mediated gene transfer of the human nerve growth factor receptor p75^{hNGFR} defines high-affinity NGF binding. *J. Neurosci.* 13: 941–951.

Birren, S. J., Verdi, J. M., and Anderson, D. J. (1992). Membrane depolarization induces p140trk and NGF-responsiveness but not p75LNGFR in MAH cells. *Science* 257: 395–397.

Brown-Grant, K., Exley, D., and Naftolin, F. (1970). Peripheral plasma oestradiol and luteinizing hormone concentration during the oestrus cycle of the rat. *J. Endocrinol.* 48: 295–261.

Butcher, R. L., Collins, W. E., and Fugo, N. W. (1974). Plasma concentrations of LH, FSH, prolactin, progesterone and estradiol-17β throughout the 4 day estrous cycle of the rat. *Endocrinology* 94: 1704–1708.

Cavicchioli, L., Flanigan, T. P., Dickson, J. G., Vantini, G., Dal Toso, R., Fuco, M., Walsh, F. S., and Leon, A. (1991). Choline acetyltransferase messenger RNA expression in developing and adult rat brain: Regulation by nerve growth factor. *Mol. Brain Res.* 9: 319–325.

Chao, M. V. (1992). Neurotrophin receptors: A window into neuronal differentiation. *Neuron* 9: 583–593.

Collazo, D., Takahashi, H., and McKay, R. D. G. (1992). Cellular targets and trophic functions of neurotrophin-3 in the developing rat hippocampus. *Neuron* 9: 643–656.

Cordon-Cardo, C., Tapley, P., Jing, S., Nanduri, V., O'Rourke, E., Lamballe, F., Kovary, K., Klein, R., Jones. K. R., Reichardt, L. F., and Barbacid. M. (1991). The *trk* tyrosine protein kinase mediates the mitogenic properties of nerve growth factor and neurotrophin 3. *Cell* 66: 173–183.

D'Mello, S. R., and Heinrich, G. (1991). Nerve growth factor gene expression: Involvement of a downstream AP-1 element in basal and modulated transcription. *Mol. Cell. Neurosci.* 2: 157–167.

Dickson, R. B., and Lippman, M. E. (1987). Estrogenic regulation of growth and polypeptide growth factor secretion in human breast carcinoma. *Endocrin. Rev.* 8: 29–43.

Drubin, D. G., Feinstein, S. C., Shooter, E. M., and Kirschner, M. W. (1985). Nerve growth factor-induced neurite outgrowth in PC12 cells involves the coordinate induction of microtubule assembly and alpha assembly promoting factors. *J. Cell Biol.* 101: 1799–1807.

Evans, R. M. (1988). The steroid and thyroid hormone receptor superfamily. *Science* 240: 889–896.

Federoff, H. J., Grabczyk, E., and Fishman, M. C. (1988). Dual regulation of GAP-43 gene expression by NGF and glucocorticoids. *J. Biol. Chem.* 263: 19290–19295.

Ferreira, A., and Caceres, A. (1991). Estrogen-enhanced neurite growth: Evidence for a selective induction of *tau* and stable microtubules. *J. Neurosci.* 11: 293–400.

Fischer, W., Bjorklund, A., Chen, K. I., and Gage, F. H. (1991). NGF improves spatial memory in aged rats as a function of age. *J. Neurosci.* 11: 1889–1906.

Frankfurt, M., Gould, E., Woolley, C., and, McEwen, B. S. (1990). Gonadal steroids modify spine density in the ventromedial hypothalamus neurons: A Golgi study. *Neuroendocrinology* 51: 530–535.

Gage, F. H., Batchelor, P., Chen, K. S., Chin, D., Higgins, G. A., Koh, S., Deputy, S., Rosenberg, M. B., Fischer, W., and Bjorklund, A. (1989). NGF receptor re-expression and NGF-mediated cholinergic neuronal hypertrophy in the damaged adult neostriatum. *Neuron* 2: 1177–1184.

Gaub, M.-P., Bellard, M., Scheuer, I., Chambon, P., and Sassone-Corsi, P. (1990). Activation of the ovalbumin gene by the estrogen receptor involves the *fos–jun* complex. *Cell* 63: 1267–1276.

Gibbs, R., and Pfaff, D. W. (1992). Effects of estrogen and fimbria/fornix transection on p75NGFR and ChAT expression in the medial septum and diagonal band of Broca. *Exp. Neurol.* 116: 23–39.

Gnahn, H., Hefti, F., Heumann, R., Schwab, M., and Thoenen, H. (1983). NGF-mediated increase in choline acetyltransferase (ChAT) in the neonatal forebrain: Evidence for a physiological role of NGF in the brain? *Dev. Brain Res.* 9: 45–52.

Greene, L. A., and Tischler, A. S. (1982). PC12 pheochromocytoma cultures in neurobiological research. *Adv. Cell. Neurobiol.* 3: 373–414.

Hanemaaijer, R., and Ginzburg, I. (1991). Involvement of mature tau isoforms in the stabilization of neurites in PC12 cells. *J. Neurosci. Res.* 30: 163–170.

Hasegawa, N., Takeo, T., and Sakuma, Y. (1991). Differential regulation of estrogen-dependent sexual development of rat brain by growth factors. *Neurosci. Lett.* 123: 183–186.

Hefti, F. (1986). Nerve growth factor promotes survival of septal cholinergic neurons after fimbrial transections. *J. Neurosci.* 6: 2158–2162.

Hempstead, B. L., Martin-Zanca, D., Kaplan, D. R., Parada, L. F., and Chao, M. V. (1991). High-affinity NGF binding requires co-expression of the *trk* proto-oncogene and the low-affinity NGF receptor. *Nature* 350: 678–683.

Hershman, H. R. (1991). Primary response genes induced by growth factors and tumor promoters. *Annu. Rev. Biochem.* 60: 281–319.

Higgins, G. A., Koh, S., Chen, K. S., and Gage, F. H. (1989). NGF induction of NGF receptor gene expression and cholinergic neuronal hypertrophy within the basal forebrain of the adult rat. *Neuron* 3: 247–256.

Hoffman, G. E., Smith, M. S., and Fitzsimmons, M. D. (1992). Detecting steroid effects on immediate early gene expression in the hypothalamus. *Neuroprotocols* 1: 52–56.

Holtzman, D. M., Li, Y., Parada, L. F., Kinsman, S., Chen, C.-K., Valetta, J., Zhou, J., Long, J. B., and Mobley, W. C. (1992). p140*trk* mRNA marks NGF-responsive forebrain neurons: Evidence that *trk* gene expression is induced by NGF. *Neuron* 9: 465–478.

Honjo, H., Tamura, T., Matsumoto, Y., Kawata, M., Ogino, Y., Tanaka, K., Yamamoto, T., Ueda, S., and Okada, H. (1992). Estrogen as a growth factor to central nervous cells. *J. Steroid Biochem. Mol. Biol.* 41: 633–635.

Hyman, C., Hofer, M., Barde, Y.-A., Juhasz, M., Yancopoulos, G. D., Squinto, S. P., and Lindsay, R. M. (1991). BDNF is a neurotrophic factor for dopaminergic neurons of the substantia nigra. *Nature* 350: 2230–2232.

Ignar-Trowbridge, D. M., Nelson, K. G., Biwell, M. C., Curtis, S. W., Washburn, T. F., McLachlan, J. A., and Korach, K. S. (1992). Coupling of dual signaling pathways: Epidermal growth factor actions involves the estrogen receptor. *Proc. Natl. Acad. Sci. (USA)* 89: 4658–4662.

Juraska, J. M. (1991). Sex differences in "cognitive" regions of the rat brain. *Psychoneuroendocrinology* 16: 105–119.

Kaplan, D. R., Hempstead, B. L., Martin-Zanca, D., Chao, M. V., and Parada, L. F. (1991a). The *trk* proto-oncogene product: Signal transducing receptor for nerve growth factor. *Science* 252: 554–557.

Kaplan, D. R., Martin-Zanca, D., and Parada, L. P. (1991b). Tyrosine phosphorylation and tyrosine kinase activity of the *trk* proto-oncogene product induced by NGF. *Nature* 350: 158–160.

Klein, R., Jing, S., Nanduri, V., O'Rourke, E., and Barbacid, M. (1991a). The *trk* proto-oncogene encodes a receptor for nerve growth factor receptor. *Cell* 65: 189–197.

Klein, R., Nanduri, V., Jing, S., Lamballe, F., Tapley, P., Bryant, S., Cordon-Cardo, C., Jones, K. R., Reichardt, L. F., and Barbacid, M. (1991b). *trkB*, tyrosine protein kinase is a receptor for brain-derived neurotrophic factor and neurotrophin-3. *Cell* 66: 395–403.

Klein-Hitpass, L., Schorpp, M., Wagner, U., and Ryffel, U. (1986). An estrogen-responsive element derived from the 5′ flanking region of *Xenopus* vitellogenin A2 gene functions in transfected human cells. *Cell* 46: 1053–1061.

Knüsel, B., Rabin, S., Widmer, H. R., Hefti, F., and Kaplan, D. R. (1992). Neurotrophin-induced *trk* receptor phosphorylation and cholinergic response in primary cultures of embryonic rat brain neurons. *NeuroReport* 3: 885–888.

Knüsel, B., Winslow, J. W., Rosenthal. A., Burton, L. F., Seid, D. P., Nikolics, K., and Hefti, F. (1991). Promotion of central cholinergic and dopaminergic neuron differentiation by brain-derived neurotrophic factor but not neurotrophin-3. *Proc. Natl. Acad. Sci. (USA)* 88: 961–965.

Koike, S., Sakai, M., and Maramatsu, M. (1987). Molecular cloning and characterization of the rat estrogen receptor cDNA. *Nucl. Acids Res.* 15: 2499–2513.

Kornack, D. R., Lu, B., and Black, I. B. (1991). Sexually dimorphic expression of the NGF receptor gene in the developing rat brain. *Brain Res.* 542: 171–174.

Kromer, L., F. (1987). Nerve growth factor treatment after brain injury prevents neuronal death. *Science* 235: 213–216.

Lamballe, F., Klein, R., and Barbacid, M. (1991). *trkC*, a new member of the *trk* family of tyrosine protein kinases is a receptor for neurotrophin-3. *Cell* 66: 967–979.

Landers, J. P., and Spelsberg, T. C. (1992). New concepts in steroid hormone action: Transcription factors, proto-oncogenes and the cascade model for steroid regulation of gene expression. *Crit. Rev. Eukaryotic Gene Expression* 2: 19–63.

Lewin, G. R., and Mendell, L. M. (1993). Nerve growth factor and nociception. *TINS* 16: 353–359.

Libertun, C., Timiras, P. S., and Kragt, C. L. (1973). Sexual differences in the hypothalamic cholinergic system before and after puberty: Inductory effect of testosterone. *Neuroendocrinology* 12: 73–85.

Loeb, D. M., Maragos, J., Martin-Zanca, D., Chao, M. V., Parada, L. F., and Greene, L. A. (1991). The *trk* proto-oncogene rescues NGF responsiveness in mutant NGF non-responsive PC12 cell lines. *Cell* 66: 961–966.

Loy, R., and Sheldon, R. A. (1986). Sexually dimorphic development of cholinergic enzymes in the rat septo-hippocampal system. *Dev. Brain Res.* 34: 156–160.

Luine, V. N. (1985). Estradiol increases choline acetyltransferase activity in specific basal forebrain nuclei and projection areas of female rats. *Exp. Neurol.* 89: 484–489.

Luine, V. N., and McEwen, B. S. (1983). Sex differences in cholinergic enzymes of diagonal band nuclei in the rat preoptic area. *Neuroendocrinology* 36: 475–482.

Luine, V. N., Park, D., Joh, T., Reis, D., and McEwen, B. S. (1980). Immunochemical demonstration of increased choline acetyltransferase concentration in rat preoptic area after estradiol administration. *Brain. Res.* 191: 273–277.

Luine, V. N., Renner, K. J., and McEwen, B. S. (1986). Sex-dependent differences in estrogen regulation of choline acetyltransferase are altered by neonatal treatments. *Endocrinology* 115: 874–878.

Lustig, R. H., Sudol, M., Pfaff, D. W., and Federoff, H. J. (1991). Estrogen regulation of sex dimorphism in growth-associated protein 43Kda (GAP-43) mRNA in the rat. *Mol. Brain. Res.* 11: 125–132.

Martin-Zanca, D., Barbacid, M., and Parada, L. F. (1990). Expression of the *trk* proto-oncogene is restricted to sensory cranial and spinal ganglia of neural crest origin in mouse development. *Genes Dev.* 4: 683–694.

Matsumoto, A., and Arai, Y. (1981). Neuronal plasticity in the deafferented hypothalamic arcuate nucleus of adult female rats and its enhancement by treatment with estrogen. *J. Comp. Neurol.* 197: 197–206.

Matsumoto, A., Murakami, S., and Arai, Y. (1988). Neurotropic effects of estrogen in the neonatal preoptic area grafted into the adult brain. *Cell. Tissue Res.* 252: 33–37.

Miranda, R. C., Sohrabji, F., and Toran-Allerand, C. D. (1993a). Estrogen target neurons co-localize the mRNAs for the neurotrophins and their receptors during development: A basis for the interactions of estrogen and the neurotrophins, *Mol. Cell. Neurosci.* 4: 510–525.

Miranda, R. C., Sohrabji, F., and Toran-Allerand, C. D. (1993b). Neuronal co-localization of the mRNAs for the neurotrophins and their receptors in the developing CNS suggests the potential for autocrine interactions. *Proc. Natl. Acad. Sci. (USA)*, 90: 6439–6443.

Mukku, V. R., and Stancel, G. M. (1985). Regulation of epidermal growth factor receptor by estrogen. *J. Biol. Chem.* 260: 9820–9824.

Nebreda, A. R., Martin-Zanca, D., Kaplan, D. R., Parada, L. F., and Santos E. (1991). Induction by NGF of meiotic maturation of *Xenopus* oocytes expressing the *trk* proto-oncogene product. *Science* 252: 558–563.

Neill, J. D. (1972). Comparison of plasma prolactin levels in cannulated and decapitated rats. *Endocrinology* 90: 568–572.

Nelson, K. G., Takahashi, T., Bossert, N. L., Walmer, D. K., and McLachlan, J. A. (1991). Epidermal growth factor replaces estrogen in the stimulation of female genital tract growth and differentiation. *Proc. Natl. Acad. Sci. (USA)* 88: 21–25.

Persson, H., Ayer-LeLievre, C., Soder, O., Villar, M, J., Metsis, M., Olson, L., Ritzen, M., and Hokfelt, T. (1990). Expression of β-nerve growth factor receptor mRNA in Sertoli cells down-regulated by testosterone. *Science* 247: 704–707.

Read, L. D., Greene, G. L., and Katzenellenbogen, B. S. (1989). Regulation of estrogen receptor messenger ribonucleic acid and protein levels in human breast cancer cell lines by sex steroid hormones, their antagonists and growth factors. *Mol. Endocrinol.* 3: 295–304.

Rodriguez-Tebar, A., Dechant, G., Gotz, R., and Barde, Y.-A. (1992). Binding of neu-rotrophin-3 to its neuronal receptors and interactions with nerve growth factor and brain-derived neurotrophic factor. *EMBO J.* 11: 917–922.

Schlessinger, J., and Ullrich, A. (1992). Growth factor signaling by receptor tyrosine kinases. *Neuron* 9: 383–391.

Segal, R. A., Takahashi, H., and McKay, R. D. G. (1992). Changes in neurotrophin responsiveness during the development of cerebellar granule cells. *Neuron* 9: 1041–1052.

Sehgal, A., Patil, N., and Chao, M. V. (1988). A constitutive promoter S directs expres-sion of the nerve growth factor receptor gene. *Mol. Cell. Biol.* 8: 3160–3167.

Selby, M. J., Edwards, R., Sharp, F., and Rutter, W. J. (1987). Mouse nerve growth factor gene: Structure and expression. *Mol. Cell. Biol.* 7: 3057–3064.

Sheng, M., and Greenberg, M. E. (1990). The regulation and function of Fos and other immediate early genes in the nervous system. *Neuron* 4: 477–485.

Shughrue, P. J., Refsdal, C. D., and Dorsa, D. M. (1992). Estrogen receptor messenger ribonucleic acid in female rat brain during the estrus cycle: A comparison with ovari-ectomized females and intact males. *Endocrinology* 131: 381–388.

Shupnik, M. A., and Rosenzweig, B. A. (1991). Identification of an estrogen response element in the rat luteinizing hormone beta gene: DNA–estrogen receptor interac-tions and functional analysis. *J. Biol. Chem.* 266: 17084–17091.

Sohrabji, F. Greene, L. A., Miranda, R. C., and Toran-Allerand, C. D. (1994a). Recip-rocal regulation of estrogen and nerve growth factor receptors by their ligands in PC12 cells. *J. Neurobiol.* 22: 974–988.

Sohrabji, F., Miranda, R. C., and Toran-Allerand, C. D. (1994b). Ovarian hormones differentially regulate estrogen and nerve growth factor mRNAs in adult sensory neurons. *J. Neurosci.* 14: 459–471.

Soppet, D., Escandon, E., Maragos, J., Middlemas, D. S., Reid, S. W., Blair, J., Bur-ton, L. E., Stanton, B. R., Kaplan, D. R., Hunter, T., Nikolics, K., and Parada, L. F. (1991). The neurotrophic factor brain-derived neurotrophic factor and neuro-trophin-3 are ligands for the *trkB* tyrosine kinase receptor. *Cell* 65: 895–903.

Squinto, S. P., Stitt, T. N., Aldrich, T. H., Davis, S., Bianco, S. M., Radziejewski, C., Glass, D. J., Masiakowski, P., Furth, M. E., Valenzuela, D. M., DiStefano, P. S., and Yancopoulos, G. D. (1991). *trkB* encodes a functional receptor for brain-derived neurotrophic factor and neurotrophin-3 but not nerve growth factor. *Cell* 65: 885–893.

Toran-Allerand, C. D. (1984). On the genesis of sexual differentiation of the central nervous system: Morphogenetic consequences of steroidal exposure and possible role of α-fetoprotein. *Prog. Brain Res.* 61: 63–98.

Toran-Allerand, C. D. (1991a). Organotypic cultures of the developing cerebral cortex and hypothalamus: Relevance to sexual differentiation. *Psychoneuroendocrinology* 16: 7–24.

Toran-Allerand, C. D. (1991b). Interaction of estrogen with growth factors in the devel-oping central nervous system. In *The New Biology of Steroid Hormones*, ed. R. B. Hochberg and F. Naftolin, pp. 311–319. New York: Raven Press.

Toran-Allerand, C. D., and MacLusky, N. J. (1989). Co-localization of estrogen and NGF receptors: Implications for the basal forebrain. *Soc. Neurosci. Abstr.* 15: 954.

Toran-Allerand, C. D., Gerlach, J. L., and McEwen, B. S. (1980). Autoradiographic localization of ^{3}H-estradiol related to steroid responsiveness in cultures of the hypo-thalamus and preoptic area. *Brain Res.* 184: 517–522.

Toran-Allerand, C. D., Miranda, R. C., Bentham, W., Sohrabji, F., Brown, T. J., Hochberg, R. B., and MacLusky, N. J. (1992). Estrogen receptors co-localize with low-affinity nerve growth factor receptors in cholinergic neurons of the basal fore-brain. *Proc. Natl. Acad. Sci. (USA)* 89: 4668–4972.

Toran-Allerand, C. D., Pfenninger, K., and Ellis, L. (1988). Estrogen and insulin synergism in neurite growth enhancement in vitro: Mediation of steroid effects by interactions with growth factors? *Dev. Brain Res.* 41: 87–91.

Tsoulfas, R., Soppet, D., Escandon, E., Tessarollo, L,, Mendoza-Ramirez, J.-L., Rosenthal, A., Nikolics, K., and Parada, L. F. (1993). The rat *trkC* locus encodes multiple neurogenic receptors that exhibit differential response to neurotrophin-3 in PC12 cells. *Neuron* 10:975–990.

Ullrich, A., Gray, A., Berman, C., Dull, T. J. (1983). Human β-nerve growth factor gene sequence highly homologous to that of mouse. *Nature* 303:821–825.

Valenzuela, D. M., Maisonpierre, P. C., Glass, D. J., Rojas, E., Nunez, L., King, Y., Gies, D. R., Stitt, T. N, Ip, N. Y., and Yancopoulos, G. D. (1993). Alternative forms of rat *TrkC* with different functional capabilities. *Neuron* 10: 963–974.

Vasquez, M. E., and Ebendal, T. (1991). Messenger RNAs for *trk* and the low-affinity NGF receptor in rat basal forebrain. *NeuroReport* 2: 593–596.

Waterman, M. L., Adler, S., Nelson, C., Greene, G. L., Evans, R. M., and Rosenfeld, M. G. (1988). A single domain of the estrogen receptor confers deoxyribonucleic acid binding and transcriptional activation of the rat prolactin gene. *Mol. Endocrinol.* 2: 14–21.

Weisz, A., Cicatiello, L., Persico, E., Scalona, M., and Bresciani, F. (1990). Estrogen stimulates transcription of the *c-jun* protooncogene. *Mol. Endocrinol.* 4: 1041–1050.

Weisz, A., and Rosales, R. (1990). Identification of an estrogen response element upstream of the human c-fos gene that binds the estrogen receptor and the AP-1 transcription factor. *Nucl. Acids Res.* 18: 5097–5106.

White, R., Lees, J. A., Needham, M., Ham, J., and Parker, M. (1987). Structural organization of the mouse estrogen receptor. *Mol. Endocrinol.* 1: 735–744.

Williams, L. R., Varon, S., Peterson, G. M., Wictorin, K., Fischer, W., Bjorklund, A., and Gage, F. H. (1986). Continuous infusion of nerve growth factor prevents basal forebrain neuronal death after fimbria-fornix transection. *Proc. Natl. Acad Sci. (USA)* 83: 923–925.

Wright, E. M., Vogel, K. S., and Davies, A. (1992). Neurotrophic factors promote the maturation of developing sensory neurons before they become dependent on these factors for survival. *Neuron* 9: 139–150.

Wu, B. Y., Fodor, E. J., Edwards, R. H., and Rutter, W. J. (1989). Nerve growth factor induces the proto-oncogene *c-jun* in PC12 cells. *J. Biol. Chem.* 264: 9000–9003.

Yanase, M., Honmura, A., Akaishi T., and Sakuma, Y. (1988). Nerve growth factor-mediated sexual differentiation of the rat hypothalamus. *Neurosci. Res.* 6: 181–185.

Zheng, M., and Heinrich, G. (1988). Structural and functional analysis of the promoter region of nerve growth factor gene. *Mol. Brain Res.* 3: 133–140.

18

Sex steroid influences on cell–cell interactions in the magnocellular hypothalamoneurohypophyseal system

GLENN I. HATTON

Introduction

It should not have been surprising that gonadal steroids exert powerful influences on cell–cell interactions in the magnocellular hypothalamoneurohypophyseal system (HNS) of the rat, but somehow it was. Since this system is responsible for the manufacture and release of oxytocin during parturition and lactation, times when the levels of circulating gonadal steroids show dramatic variations, some steroid involvement in the functioning of HNS could have been anticipated. Perhaps such anticipation was dulled by the observation that the dynamic interactions taking place among the cells of the HNS also occurred in response to manipulations of the animal's hydrational state, which have not been associated traditionally with variations in gonadal steroid output. However, gonadal steroids appear to exert some control over the cellular mechanisms that release both oxytocin and vasopressin in response to dehydration. Estrogens and androgens, under comparable conditions, often have opposite effects on the HNS. This chapter reviews the main structure–function relationships of the HNS, the dynamics of these relationships under physiological conditions of altered peptide hormone demand, and some of the roles possibly played in these functions by gonadal steroids in both males and females.

The magnocellular HNS

The magnocellular HNS is constituted chiefly by the supraoptic (SON) and paraventricular (PVN) nuclei, accessory nuclei in the anterior hypothalamus, and the neurohypophysis or neural lobe (NL) of the posterior pituitary to which the neurons of those hypothalamic nuclei send axonal projections (Fig. 18.1).

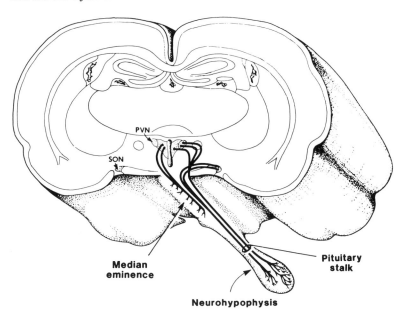

Figure 18.1. Semidiagrammatic representation of the rat HNS. Projections of SON and PVN to the neurohypophysis are shown. Also shown are PVN projections to the external zone of the median eminence. Not shown are the many intra- and extrahypothalamic projections of the magnocellular neurons.

The anterior hypothalamus

The magnocellular neuroendocrine cells of the anterior hypothalamus synthesize and release oxytocin (OXT) or arginine vasopressin (AVP) in response to a variety of physiological demands (reviewed by Hatton 1990; Renaud and Bourque 1991; Crowley and Armstrong 1992). In recent years, these magnocellular neuroendocrine cells have been found to contain other peptides as well – dynorphin, galanin, corticotropin-releasing hormone, and cholecystokinin. Often these are co-localized with AVP and/or OXT in the magnocellular neuroendocrine cells (Vanderhaeghen et al. 1981; Watson et al. 1982; Whitnall et al. 1983; Sawchenko et al. 1984; Rökaeus et al. 1988; Meister et al. 1990).

Wherever these magnocellular neuroendocrine cells are located, be it in the main or the accessory nuclei or clustered along blood vessels, these cell groups tend to be densely packed (Fig. 18.2) and highly vascularized. These features are critical for the functioning and plasticity of this system. The PVN contains a larger variety of cell types than the SON, and as a result studies of the magnocellular neuroendocrine cells per se have tended to focus on the SON rather than on the PVN. However, when similar studies have been carried out on both nuclei, equivalent results have usually been obtained, suggesting that observations regarding the magnocellular neuroendocrine cells of the SON may be extrapolated

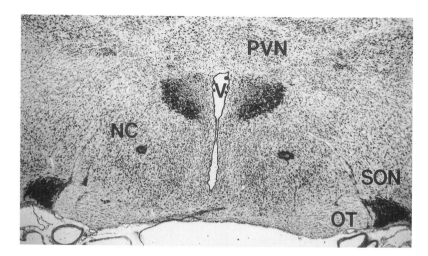

Figure 18.2. Photomicrograph of a Nissl-stained coronal section at the level of the SON and PVN. Also shown is one of the paired internuclear groups of magnocellular neurosecretory neurons, the nucleus circularis (NC). OT, Optic tract; V, third ventricle. Scale bar, 1 mm.

to include those in the PVN. Of course, these two cell groups are positioned quite differently with respect to the third ventricle and the pial surface and do not necessarily have similar afferent or efferent projections. The remainder of this chapter focuses primarily on the SON with its two major neuronal cell types (OXT and AVP) and on the NL, which contains axonal terminals from both the SON and PVN.

Cells of the SON tend to be arranged with somata projecting dendrites ventrally and axons dorsally (as diagrammed in Fig. 18.3). OXT-Containing cells are roughly grouped together in the more anterior and dorsal parts of the SON, while the more ventral and posterior portions of the SON are occupied by groupings of AVP-containing neurons. There is a considerable degree of mixing of these two cell types, however, especially in the middle one-third or so of the nucleus (Hou-Yu et al. 1986). Dendrites of these densely packed cell bodies tend to run in parallel subjacent to the pial surface, either rostrocaudally beneath the SON or ventrolaterally along the external periamygdaloid cortex. A single, dorsally projecting dendrite often gives rise to the cell's axon, which may collateralize before reaching the NL. Under physiological conditions that require little or no hormone release, these magnocellular neuroendocrine cells are separated from their neighbors by fine astrocytic processes. Similarly, glial processes are also interposed between adjacent dendrites under such "basal" conditions. Most of the astrocytic glia whose processes are insinuated between adjacent SON somata or dendrites have cell bodies that are located in a ventral glial lamina just dorsal to the pia

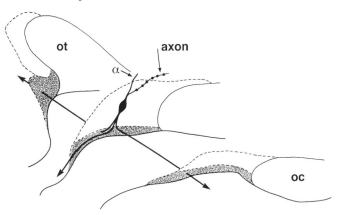

Figure 18.3. Semidiagrammatic representation of the SON (dashed lines) at three rostrocaudal levels. A neuron in the central portion is shown with its beaded axon arising from a primary dendrite (α) and exiting the nucleus dorsomedially. Other dendrites of this cell, as is typical, project first ventrally to the zone adjacent to the pial surface (stippled), where they turn and project rostrocaudally and/or ventrolaterally along the cortical surface. ot, Optic tract; oc, optic chiasm.

mater and ventral to the SON dendrites that project rostrocaudally (see Fig. 18.3). Individual astrocytes, the nuclei of which are found at the pial–glial interface, may extend processes all the way to the dorsalmost parts of the SON (Salm and Hatton 1980; Bonfanti et al. 1993).

The neurohypophysis

The NL of the posterior pituitary consists of neurosecretory axons and their terminals, some non-neurosecretory axons, modified astrocytes called pituicytes, basal lamina, fenestrated capillaries, a few perivascular cells, and some microglia (for a review see Wittkowski 1986). Relationships among the main elements of the NL are depicted in Figure 18.4. Pituicytes comprise approximately 30% of the volume of the NL and, together with the neurosecretory processes, completely dominate the parenchymal side of the basal lamina. Substances released from elements on the parenchymal side must, of course, traverse the basal lamina in order to enter the fenestrated capillaries leading to the general circulation. Under conditions of low demand for hormone release, the neurosecretory endings occupy a relatively small percentage of the basal lamina, the remainder being occupied by pituicytes (Tweedle and Hatton 1987; Beagley and Hatton 1992, 1994). Also under these basal conditions, the pituicytes surround and engulf neurosecretory axons and axonal endings (Tweedle and Hatton 1980a,b, 1982; Beagley and Hatton 1994).

To summarize, under basal conditions, glial processes are interposed between

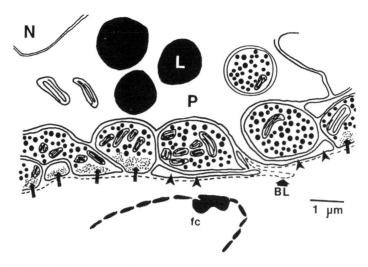

Figure 18.4. Diagrammatic representation of the neurovascular contact zone in rat neuro-hypophysis. The basal lamina (BL), represented by a dashed line, is occupied either by pituicyte processes (arrowheads) or by neurosecretory terminals (arrows) containing microvesicles (stippling). Other neurosecretory granule–containing endings (indicated by large filled circles in the endings) are separated from the BL by pituicyte processes and do not contain microvesicles. Lipid inclusions (L) are shown in the pituicyte (P) cytoplasm; fc, fenestrated capillary.

adjacent neurosecretory cell bodies in the SON proper and between neighboring magnocellular neuroendocrine cell dendrites in the SON dendritic zone. Similarly, in the NL, pituicytes tend to separate the neurosecretory axons from one another, by enclosing them, and from the perivascular contact zone, by occupying a relatively large extent of the basal lamina. This picture of coordinated morphology also extends to the physiology of the HNS. In addition, under basal conditions, OXT and AVP neurons of the SON exhibit slow, irregular firing (0.05–2 Hz) with correspondingly little release of hormone into the blood and relatively low levels of biosynthetic activity.

Activation of the HNS

Physiological activation of the HNS may occur through dehydration, hemorrhage, parturition, lactation and suckling of the young, as well as various types of stress. Of these stimuli, the most well studied are dehydration, either by water deprivation or by administration of hypertonic saline, and lactation with active nursing of offspring (for more detailed reviews see Poulain and Wakerley 1982; Wakerley 1987; Hatton 1990). Any of these stimulus complexes is capable of producing dramatic changes in the activity and structural organization of the HNS (Fig. 18.5).

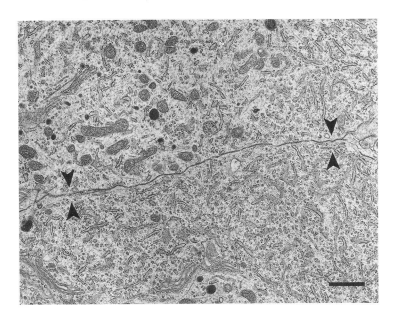

Figure 18.5. Ultrastructural appearance of adjacent SON magnocellular neurons and dendrites under conditions of increased hormone demand. Arrows delineate the extent of membrane apposition between two SON cell bodies. Bar, 1 μm.

Electrophysiological correlates of activation

Dehydration, for example, results in coordinated enhancement of electrical activity in both OXT and AVP neurons, with a majority of these cells responding to the increased demand for peptide release. Activation of this system induces different patterns of electrical activity in the two cell types. Both OXT and AVP cells show slow, irregular activity under basal conditions. OXT neurons respond to dehydration by increasing their firing from the basal rate of <2 Hz to 6–15 Hz and are then described as firing in a "fast, continuous" manner, while AVP neurons in the stimulated state display "phasic bursting" patterns of activity, in which the cells fire a prolonged burst of action potentials and are then silent for a protracted period of time. Individual episodes of either activity or silence may vary from 2 seconds to 2 minutes or more. Regardless of sex or reproductive state of the animal, dehydration stimuli produce similar changes in the electrical activity of OXT and AVP neurons in rats.

Lactation is another activated state for the HNS. Accordingly, OXT neurons are primarily fast, continuous firing cells. These neurons fire brief, intermittent, high-frequency bursts (>50 Hz) in response to the continuous suckling of the pups. Relative synchrony of these bursts provides a bolus of OXT in the blood, which causes contraction of myoepithelial cells of the mammary glands, resulting in milk ejection. The mechanisms involved in synchronizing the OXT cell bursts are not understood entirely; since such extraordinary coordination is absent in males and virgin females, however, it is likely that sex steroids may be involved.

Release of peptides

In addition to these changes in the electrical activity of SON neurons, activation of the HNS is thought to result in the release of OXT and AVP within the nucleus itself. Evidence for this was first reported by Moos et al. (1984), based on observations from isolated SONs maintained *in vitro*. It has also been demonstrated that OXT (Moos et al. 1984) and AVP (Ramirez et al. 1990) could selectively facilitate their own release. Electrophysiological evidence for excitatory actions of OXT and AVP on PVN and SON neurons supports these biochemical results (Inenaga and Yamashita 1986; Yamashita et al. 1987). Intranuclear release of peptide in response to physiological stimuli has been shown *in vivo* using microdialysis (Neumann et al. 1993). At present, two mechanisms appear possible for the release of intranuclear OXT or AVP – one synaptic, the other dendritic. The synaptic mechanism, although more widely accepted, is less likely in this case. OXT-Immunoreactive terminals have been reported in the SON (Theodosis 1985), but they are extremely rare and would probably not be sufficient in number to account for the quantities of OXT measured during HNS activation. Furthermore, AVP-immunoreactive terminals have not been found. However, both OXT and AVP are abundant in dendrites of the SON and can be released from brain slices or from perfused brain by neuronal depolarization (Pow and Morris 1989) and osmotic stimulation (Tweedle et al. 1988b).

Structural plasticity in the HNS

A rather remarkable set of morphological alterations accompanies physiological activation of the HNS. These changes affect the basic intrinsic circuitry of the SON, as well as the neurosecretory terminal access to the neurovascular contact zone in the NL. Many of the features of this structural reorganization of the HNS are common to both dehydration and lactation. At the level of the SON somata, HNS activation induces withdrawal of the glial processes that are normally present between adjacent neurons, resulting in direct soma–somatic appositions (Fig. 18.5). In the lactating rat, this glial withdrawal appears to be coincident with a decreased expression of the astrocyte marker glial fibrillary acidic protein (Salm et al. 1985). Synaptogenesis occurs if the stimulus conditions persist, as in chronic dehydration induced by drinking saline or in lactation/nursing, or if the dehydration is acute and very potent, as with intraperitoneal injections of hypertonic saline (Theodosis et al. 1981; Hatton and Tweedle 1982; Tweedle and Hatton 1984; Modney and Hatton 1989; Beagley and Hatton 1992). The new synapses that form are of a type only rarely seen under basal conditions; these "multiple synapses" consist of one synaptic bouton contacting two or more post-synaptic elements (either somata or dendrites) (Fig. 18.6).

A similar set of neuronal–glial dynamics occurs in the SON dendritic zone when the HNS is activated. Withdrawal of astrocytic processes from between the dendrites allows several neighboring dendrites, from different SON neurons, to come into apposition and form dendritic bundles (Fig. 18.6). Multiple synapses also increase in number in the dendritic zone under certain stimulated conditions, although not all stimuli that induce bundling are effective. For instance, both dendritic bundling and multiple synapses are associated with the events of parturition and subsequent lactation (Perlmutter et al. 1984). However, induction of maternal behavior in virgin rats by the continuous presence of young pups results in dendritic bundling reminiscent of that which occurs in the lactating animal but does not induce increases in axodendritic multiple synapses (Salm et al. 1988). Neither acute water deprivation nor 10 days of drinking hypertonic saline was effective in inducing multiple axodendritic synapse formation, although both conditions produced increased dendritic bundling (Perlmutter et al. 1985). Transcardial perfusion, with hypertonic balanced salt solutions for less than 30 minutes,

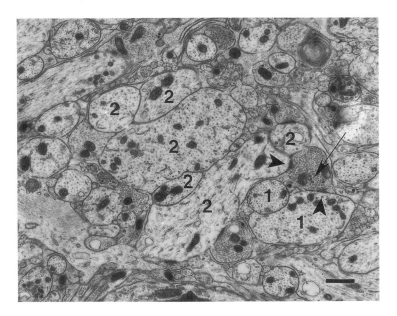

Figure 18.6. Bundles in the dendritic zone of the SON. A small bundle of two dendrites (dendrites numbered with 1's) and a large bundle of six dendrites (dendrites numbered with 2's) are shown. An axon terminal (A) makes synaptic contacts (arrows) with two dendrites, forming a multiple synapse. Scale bar, 1 μm.

however, increased both dendritic bundling and the number of new multiple axo-dendritic synapses (Tweedle et al. 1993).

Retraction of astrocytic processes from between SON dendrites appears to be preparatory to, and permissive of, the formation of dendrodendritic gap junctional connections between SON neurons, at least in lactating mothers and maternally behaving virgins. Dye coupling (indicated by the transfer of a low molecular weight, intracellularly injected dye, such as Lucifer Yellow, to one or more nearby neurons via gap junction) was observed first in the SON and PVN more than a decade ago (Andrew et al. 1981). The incidence of such coupling has been observed repeatedly to be two- to fourfold higher in lactating rats than in untreated virgin females (Hatton et al. 1987; Yang and Hatton 1987; Hatton and Yang 1990, 1993) and appears to be exclusively dendrodendritic. Electrical coupling between SON neurons has also been demonstrated and corresponds to dye coupling between the same neurons from which recordings were made (Yang and Hatton 1988). The incidence of dye coupling can be enhanced by activation of an excitatory amino acid–mediated olfactory system input to the SON in lactating or maternally behaving virgin rats. In both cases dendrites are bundled (Hatton and Yang 1990; Modney et al. 1990). Finally, the quantity of mRNA for the neuronal (liver) gap junction protein connexin-32 in the SON was low in virgins, increased dramatically on the day of parturition, and remained elevated after 14 days of lactation (P. Micevych, P. Popper, and G. Hatton, unpublished observations). While abundant, levels of mRNA for the astrocytic (heart) gap junction protein connexin-43 did not vary across these times (Hatton and Micevych 1992). It is likely that the hormonal events of late pregnancy, or parturition itself, play a role in enhancing the expression of connexin-32 in SON neurons.

At the level of the NL, dehydration and parturition/lactation evoke similar changes in organization – for example, during the process of parturition, at which time about 60% of the OXT and AVP in the NL is released into the blood (Fuchs and Saito 1971). This is accompanied by release of the engulfed neurosecretory axons by the pituicytes, which retract from their positions along much of the basal lamina (Tweedle and Hatton 1982, 1987), allowing axonal endings increased access to the basal lamina–perivascular space, thus facilitating hormone entry into the circulation. Similar effects on pituicytes have been observed with water deprivation or saline drinking (Tweedle and Hatton 1980a, 1987), with intraperitoneal injection of hypertonic saline (Beagley and Hatton 1992), and by *in vitro* stimulation of the acutely isolated NL with either hypertonic artificial cerebrospinal fluid (Perlmutter et al. 1984) or the β-adrenergic agonist isoproterenol (Luckman and Bicknell 1990; Smithson et al. 1990). Thus, activation of the HNS evokes qualitatively similar changes in the NL, irrespective of the particular activating stimulus conditions.

All of the observed changes in HNS structure and function that result from the physiological conditions are entirely reversible. That is, upon weaning of the pups (Theodosis and Poulain 1984) or upon rehydration after water deprivation or saline drinking (Tweedle and Hatton, 1977, 1980a, 1984, 1987; Perlmutter et al. 1985), structural alterations and the electrical activities of HNS neurons return to basal conditions. The only permanent, and apparently irreversible, alteration is the persistence (for nearly a year) of postsynaptic membrane specializations, in the form of annular synapses in freeze-fractured PVN neurons, in rats that had been mothers (Hatton and Ellisman 1982). Such specializations were not seen in males or in virgin females. The functional importance of these membrane specializations, or of their persistence, has not been determined.

Functional significance of the observed plastic changes

The functional roles of physiologically induced alterations in the HNS are not clear. However, one functional consequence of the glial retraction that allows direct apposition of large amounts of neuronal membrane would be an elevated extracellular K^+ concentration, due to reduced spatial buffering by glia. This could contribute to the coordinated increase in SON neuronal excitability. Another possible consequence of the restricted extracellular space and dendritic bundling is the generation of electrical fields around groups of SON neurons, similar to what has been observed in hippocampus and suprachiasmatic nucleus (Taylor and Dudek, 1984; Bouskila and Dudek 1993). Such field effects would synchronize the electrical activity of cells in the group. Gap junctions and multiple synapses also have potential for synchronizing the activity of SON cells. In the case of OXT neurons in lactating rats, these mechanisms may be responsible for the synchrony developed for the milk ejection reflex (reviewed by Wakerley and Ingram 1993). Both the incidence of gap junctions and their strengths (conductances) may be modulated by direct sensory inputs (Hatton and Yang 1990; Modney et al. 1990). Multiple GABAergic synapses in the SON somatic zone (Theodosis et al. 1986b) may also modulate the function of gap junctions. Superficially, the changes that occur in the NL appear to be simply and directly related to facilitating the entry of neurosecretory products into the circulation via removal from the basal lamina of a physical barrier formed by the pituicytes. However, pituicytes seem also to be involved in modulating the number of *en passant* contacts that neurosecretory axons make with the basal lamina (Tweedle et al. 1989).

Direct effects of gonadal steroids on the HNS

Experimental evidence for gonadal steroid actions on the HNS has existed for many years; however, recent research has complicated attempts to understand the mechanisms by which these steroids achieve their effects. For example, nongenomic actions of steroids on membranes of the nervous system are being reported with increasing frequency (e.g., Akaishi and Sakuma 1989; Minami et al. 1990). Since some neurons and glia may have more than one of these types of steroid receptors, simply demonstrating the presence of a nuclear receptor does not preclude the presence of other types.

Effects on the magnocellular nuclei

In the case of the HNS, the presence of nuclear estrogen receptors is reasonably well established in the PVN (Sar and Stumpf 1980). More recent work (Warembourg and Poulain 1991) has demonstrated the co-localization of nuclear estrogen receptors with OXT in virtually all SON neurons in the guinea pig, although progesterone receptors were found rarely in these cells. It seems likely, from the evidence for estrogen receptors in SON neurons of both the mouse (Stumpf and Sar 1975) and the guinea pig (Warembourg and Poulain 1991), that estrogen receptors will eventually be localized within the rat SON.

Effects on somasomatic and axosomatic changes

The SON undergoes structural reorganization when the demand for peptide hormone synthesis and release is powerful and/or protracted. Some of the changes observed in response to physiological stimulation can be induced by continuous intracerebroventricular infusions over several days of OXT or OXT agonists, but not by AVP (Theodosis et al. 1986a). Both increased amounts of direct membrane appositions and numbers of multiple synapses were observed with this treatment. Such changes were subsequently reported to be dependent on sex steroids; OXT infusions were effective in inducing the plastic changes in animals undergoing prolonged diestrus or in ovariectomized rats treated with progesterone (4 days) followed by estradiol (2 days) but did not induce structural changes in normally cycling animals (Montagnese et al. 1990). Studies manipulating sex steroids in males would be of interest, however, since similar plastic changes occur in the SONs of males or females during dehydration (Tweedle and Hatton 1984; Chapman et al. 1986).

It is interesting that there is variation in the number of axosomatic synapses in the arcuate nucleus during the estrous cycle in rats (Garcia-Segura et al. 1988). Declines in the number of such synapses occurred during proestrus and estrus, when circulating estradiol levels rise, and the number of these synapses was reinstated when estradiol was low again. These workers have established that estradiol was the crucial factor in this synaptic remodeling; in ovariectomized rats, estradiol administration resulted in a decreased number of axosomatic synapses.

Progesterone plus estradiol, or progesterone by itself, did not alter the number of these synapses (Perez et al. 1993). GABAergic synapses appeared to account for the increased number of synapses in the arcuate nucleus when circulating estradiol was low (Párducz et al. 1993). The increases in the number of multiple synapses (many of which are GABAergic) and in the incidence of interneuronal coupling in the lactating rat occur when levels of circulating estradiol are low.

Effects on interneuronal coupling

Castration of male rats results in a significant decline in plasma AVP, whereas ovariectomy tends to have the opposite effect on plasma AVP in females, and replacement of testosterone or estradiol reverses the effects of gonadectomy (Crofton et al. 1985). Since an increase in the incidence of dye coupling had been observed after activation of the HNS, either during lactation (Hatton et al. 1987; Yang and Hatton 1987) or dehydration (Cobbett and Hatton 1984), we tested the hypothesis that circulating sex steroids influence this particular type of cell–cell interaction. In the lateral magnocellular portion of the PVN, there is a high concentration of AVP neurons that are visualized easily in coronally prepared hypothalamic slices; it was these cells from which our samples were drawn. Furthermore, we biased our sampling toward AVP neurons by selecting for phasically firing cells. This bias was not absolute, however, since we did include some slow-firing and silent neurons. Castration reduced the incidence of dye coupling to about one-third of its value in the sham-castrated group (Table 18.1). Testosterone replacement with a Silastic capsule implanted at the time of castration prevented this decrease in castrated animals. Indeed, exogenous testosterone appeared to enhance the incidence of coupling. These results are consistent with those of Crofton et al. (1985), indicating that testosterone or one of its metabolites has a profound influence on the cell–cell interactions involved in increasing AVP release. Since estradiol tended to have the opposite effect on AVP release in females (Crofton et al. 1985), testosterone or dihydrotestosterone, rather than aromatized testosterone, probably mediates these effects in males.

In similar studies on the influence of estrogens on coupling, there was a low incidence of Lucifer Yellow dye coupling among neurons of either the SON or the PVN of normally cycling, virgin, female rats (Table 18.1). Ovariectomy has a powerful effect on cell–cell coupling in the SON, resulting in an almost threefold increase in the dye coupling index (see Table 18.1) over sham-ovariectomized animals (Fig. 18.7). Administration of estradiol-17β, but not estradiol-17α, was sufficient to prevent the effect of ovariectomy, although possible involvement of other ovarian steroids cannot be ruled out (Hatton et al. 1992). From these results and from those of decreased coupling in castrated males, we concluded that es-

Table 18.1. *Effect of castration with or without testosterone replacement on dye coupling among PVN magnocellular neurons*

Treatment[a]	No. of injected cells	No. of dye-filled cells	No. of coupled cells	Dye coupling index[b]
Sham	48	58	19	0.33
Castrate	51	54	6	0.11[c]
Castrate (+T)	27	36	16	0.44
Castrate (no T)	35	38	6	0.16[c]

[a]T, Testosterone.
[b]Number of dye-coupled cells per total number of dye-filled cells.
[c]Significantly different from appropriate control, $p < .01$ by χ^2.

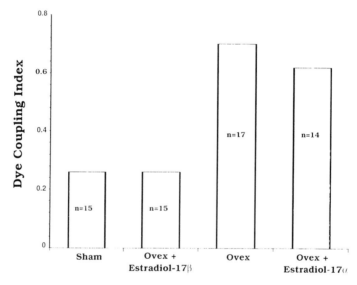

Figure 18.7. Incidence of dye coupling among SON in four groups of rats: sham-ovariectomized (Sham); ovariectomized and treated with estradiol-17β (Ovex + Estradiol-17β); ovariectomized and untreated (Ovex); and ovariectomized and treated with estradiol-17α (Ovex + Estradiol-17α). Dye coupling index is the number of dye-coupled neurons per total number of dye-filled neurons. Groups on the left differ from groups on the right at $p \leq .0001$.

tradiol and testosterone have opposite effects on coupling in the HNS. Furthermore, the inhibitory action of estradiol on coupling in females suggests that the facilitatory effect of testosterone in males is not due to its aromatization to estrogen. However, this conclusion would be wrong if estradiol and testosterone have opposite effects on SON neurons of males and females.

Mechanisms of steroid action

The mechanism(s) by which sex steroids may regulate intercellular coupling is still uncertain. Gap junctional conductances in many systems, including parts of the central nervous system, are enhanced by stimulation that results in increased cAMP (Flagg-Newton et al. 1981; Giaume et al. 1991; Sáez et al. 1991). Notable exceptions to this are the gap junctions in the retina (Miyachi and Murakami 1989; Hampson et al. 1992) and in parturient rat myometrium (Sakai et al. 1992) where cAMP accumulations decrease or block junctional permeability and uncouple the cells. Since estradiol pretreatment downregulates cAMP in striatal neurons in culture (Maus et al. 1990), similar effects may be involved in the regulation of coupling, since removal of estradiol results in an upregulation of cAMP and increased coupling in the SON. Testosterone has been shown to upregulate gap junction mRNA expression in the sexually dimorphic spinal nucleus of the bulbocavernosus (Matsumoto et al. 1991; Fisher and Micevych 1993). This may also occur in the HNS. Furthermore, in the male HNS, treatments that result in greatly reduced seminal vesicle weights and reduced circulating testosterone levels such as severe dehydration (Cobbett and Hatton 1985) also decrease the incidence of dye coupling among magnocellular neurons of the PVN (Cobbett and Hatton 1984).

Effects on the neurohypophysis

Testosterone also affects the functional morphology of the NL (Tweedle et al. 1988a). Castration induced an 8–20% reduction in neural occupation of the basal lamina, as compared with sham controls or testosterone treated castrates. The number of enclosed axons per pituicyte in castrated animals receiving no testosterone replacement was 2.5 times greater. On both measures, the testosterone replacement and sham controls were not significantly different from each other. These changes accompany the removal of testosterone from the circulation and are consistent with decreased activation of the HNS and decreased peptide hormone output. Both the observations of decreased dye coupling (Cobbett et al. 1987) and decreased plasma AVP (Crofton et al. 1985) following castration are consistent with the NL changes. As noted previously, activation of the HNS by physiological stimuli that increase demand for peptide hormone output results in more neural, and therefore less glial, occupation of the basal lamina, and induces pituicytes to enclose fewer neurosecretory axons (Tweedle and Hatton 1980a, 1982, 1987).

It seems likely that pituicyes respond to gonadectomy, since they possess nuclear androgen receptors (Sar and Stumpf 1979) and are mobilized by β-adrenergic stimulation in the isolated NL (Luckman and Bicknell 1990; Smithson et al. 1990). However, the mechanism(s) by which activation of pituicyte androgen receptors might effect release of enclosed axons and a retraction of pituicyte processes from the basal lamina remains to be determined. Cyclic nucleotides may be involved, as they are in the β-adrenergic mobilization of pituicytes (Bicknell et al. 1989), but it is not known whether androgenic stimulation would modulate these nucleotides in pituicytes.

Summary

The HNS participates in such vital processes as water balance, cardiovascular control, parturition, and milk delivery to suckling offspring. Reversible plastic changes in all levels of the HNS – somatic, dendritic, and axon terminal levels of the neurons and glia of the system – are associated with many of these physiological processes. Prominent among this reorganization of the HNS are increased soma–somatic direct appositions, synaptogenesis resulting in the production of new multiple synapses (presynaptic boutons contacting two or more postsynaptic elements), dendritic bundling, interneuronal coupling, and increased occupation of the basal lamina in the neurohypophysis by neurosecretory axon terminals. These alterations are associated with active movement of the astrocytic glia that are an integral part of the structure and function of the HNS. Sex steroid actions exert a strong influence on HNS structure–function relationships; some of the OXT neurons of the HNS have nuclear estrogen receptors, and the neurohypophyseal astrocytes have androgen receptors. Removal of circulating steroids by gonadectomy influences circulating HNS hormone levels, the incidence of interneuronal coupling, the amount of basal lamina occupied by neural as compared with glial membrane, and the capacity of exogenous OXT to induce soma–somatic appositions and multiple synapse formation. When the actions of estrogen and androgen on the HNS have been studied in the same experimental paradigm, they have been found to produce opposite effects. The mechanisms by which gonadal steroids alter HNS functioning are currently under study and are not entirely clear. It is likely, however, that these mechanisms involve the induction of protein synthesis directed at synaptic proliferation or the production of gap junction proteins and activation of enzymes linked to cyclic nucleotides. Other actions may be to induce or inhibit the expression of glial fibrillary acidic protein, as well as membrane mobility proteins such as actin.

Acknowledgments

The author's research is supported by NIH grants RO1 09140 and RO1 16942. I thank P. Rusch for technical assistance, W. Stella for word processing, and Dr.

C. Decavel and Dr. B. G. Stanley for helpful suggestions on an earlier draft of the manuscript.

References

Akaishi, T., and Sakuma, Y. (1989). Effects of estrogen on hypothalamic paraventricular neurons in tissue slices from ovariectomized rats. *Biomed. Res.* 10: 149–155.

Andrew, R. D., MacVicar, B. A., Dudek, F. E., and Hatton, G. I. (1981). Dye transfer through gap junctions between neuroendocrine cells of rat hypothalamus. *Science* 211: 1187–1189.

Beagley, G. H., and Hatton, G. I. (1992). Rapid morphological changes in supraoptic nucleus and posterior pituitary induced by a single hypertonic saline injection. *Brain Res. Bull.* 28: 613–618.

Beagley, G. H., and Hatton, G. I. (1994). Systemic signals contribute to induced morphological changes in the hypothalamo-neurohypophysial system. *Brain Res. Bull.* 33: 211–218.

Bicknell, R. J., Luckman, S. M., Inenaga, K., Mason, W. T., and Hatton, G. I. (1989). β-Adrenergic and opioid receptors on pituicytes cultured from adult rat neurohypophysis: Regulation of cell morphology. *Brain Res. Bull.* 22: 379–388.

Bonfanti, L., Poulain, D. A., and Theodosis, D. T. (1993). Radial glia-like cells in the supraoptic nucleus of the adult rat. *J. Neuroendocrinol.* 5: 1–5.

Bouskila, Y., and Dudek, F. E. (1993). Neuronal synchronization without calcium-dependent synaptic transmission in the hypothalamus. *Proc. Nat'l. Acad. Sci. (USA)* 90: 3207–3210.

Chapman, D. B., Theodosis, D. T., Montagnese, C., Poulain, D. A., and Morris, J. F. (1986). Osmotic stimulation causes structural plasticity of neurone-glia relationships of the oxytocin but not vasopressin secreting neurones in the hypothalamic supraoptic nucleus. *Neuroscience* 17: 679–686.

Cobbett, P., and Hatton, G. I. (1984). Dye coupling in hypothalamic slices: Dependence on *in vivo* hydration state and osmolality of incubation medium. *J. Neurosci.* 4: 3034–3038.

Cobbett, P., and Hatton, G. I. (1985). Hypertonic saline drinking affects seminal vesicle weight in rats. *J. Physiol.* 364: 65P.

Cobbett, P., Yang, Q. Z., and Hatton, G. I. (1987). Incidence of dye coupling among magnocellular paraventricular nucleus neurons in male rats is testosterone dependent. *Brain Res. Bull.* 18: 365–370.

Crofton, J. T., Baer, P. G., Share, L., and Brooks, D. P. (1985). Vasopressin release in male and female rats: Effects of gonadectomy and treatment with gonadal steroid hormones. *Endocrinology* 117: 1195–1200.

Crowley, W. R., and Armstrong, W. E. (1992). Neurochemical regulation of oxytocin secretion in lactation. *Endocrine Rev.* 13: 33–65.

Fisher, R., and Micevych, P. E. (1993). Regulation of connexin 32 in motor networks of mammalian neurons. In *Progress in Cell Research*, Vol. 3, ed. J. E. Hall, G. A. Zampighi, and R. M. Davis, pp. 141–148. Amsterdam: Elsevier.

Flagg-Newton, J. L., Dahl, G., and Loewenstein, W. R. (1981). Cell junction and cyclic AMP: I. Upregulation of junctional membrane permeability and junctional membrane particles by administration of cyclic nucleotides or phosphodiesterase inhibitor. *J. Mem. Biol.* 63: 105–121.

Fuchs, A.-R., and Saito, S. (1971). Pituitary oxytocin and vasopressin content of pregnant rats before, during and after parturition. *Endocrinology* 88: 574–578.

Garcia-Segura, L. M., Hernandez, P., Olmos, G., Tranque, P. A., and Naftolin, F. (1988). Neuronal membrane remodelling during the oestrus cycle: A freeze–fracture study in the arcuate nucleus of the rat hypothalamus. *J. Neurocytol.* 17: 377–383.

Giaume, C., Marin, P., Cordier, J., Glowinski, J., and Premont, J. (1991). Adrenergic regulation of intercellular communications between cultured striatal astrocytes from the mouse. *Proc. Nat'l Acad. Sci. (USA)* 88: 5577–5581.

Hampson, E. C. G. M., Vaney, D. I., and Weiler, R. (1992). Dopaminergic modulation of gap junction permeability between amacrine cells in mammalian retina. *J. Neurosci.* 12: 4911–4922.

Hatton, G. I. (1990). Emerging concepts of structure–function dynamics in adult brain: The hypothalamo-neurohypophysial system. *Prog. Neurobiol.* 34: 437–504.

Hatton, G. I., and Micevych, P. E. (1992). Connexin 32 mRNA levels in the supraoptic nucleus prior to and during lactation. *Soc. Neurosci. Abstr.* 18: 1417.

Hatton, G. I., and Tweedle, C. D. (1982). Magnocellular neuropeptidergic neurons in hypothalamus: Increases in membrane apposition and number of specialized synapses from pregnancy to lactation. *Brain Res. Bull.* 8: 197–204.

Hatton, G. I., and Yang, Q. Z. (1990). Activation of excitatory amino acid inputs to supraoptic neurons: I. Induced increases in dye-coupling in lactating, but not virgin or male rats. *Brain Res.* 513: 264–269.

Hatton, G. I., and Yang, Q. Z. (1994). Incidence of neuronal coupling in supraoptic nucleus: Estimation by neurobiotin and Lucifer Yellow. *Brain Res.* 650: 63–69.

Hatton, G. I., Yang, Q. Z., and Cobbett, P. (1987). Dye coupling among immunocytochemically identified neurons in the supraoptic nucleus: Increased incidence in lactating rat. *Neuroscience* 21: 923–930.

Hatton, G. I., Yang, Q. Z., and Koran, L. E. (1992). Effects of ovariectomy and estrogen replacement on dye coupling among rat supraoptic nucleus neurons. *Brain Res.* 572: 291–295.

Hatton, J. D., and Ellisman, M. H. (1982). A restructuring of hypothalamic synapses is associated with motherhood. *J. Neurosci.* 2: 704–707.

Hou-Yu, A., Lamme, A. T., Zimmerman, E. A., and Silverman, A.-J. (1986). Comparative distribution of vasopressin and oxytocin neurons in the rat brain using a double-label procedure. *Neuroendocrinology* 44: 235–246.

Inenaga, K., and Yamashita, H. (1986). Excitation of neurones in the rat paraventricular nucleus *in vitro* by vasopressin and oxytocin. *J. Physiol.* 370: 165–180.

Luckman, S. M., and Bicknell, R. J. (1990). Morphological plasticity that occurs in the neurohypophysis following activation of the magnocellular neurosecretory system can be mimicked in vitro by β-adrenergic stimulation. *Neuroscience* 39: 701–709.

Matsumoto, A., Arai, Y., Urano, A., and Hyodo, S. (1991). Androgen regulates gap junction mRNA expression in androgen-sensitive motoneurons in the rat spinal cord. *Neurosci. Lett.* 131: 159–162.

Maus, M., Prémont, J., and Glowinski, J. (1990). *In vitro* effects of 17β-oestradiol on the sensitivity of receptors coupled to adenylate cyclase on striatal neurons in primary culture. In *Steroids and Neuronal Activity*, ed. D. Chadwick, pp. 145–155. New York: Ciba Foundation/Wiley.

Meister, B., Villar, M. J., Ceccatelli, S., and Hökfelt, T. (1990). Localization of chemical messengers in magnocellular neurons of the hypothalamic supraoptic and paraventricular nuclei: An immunohistochemical study using experimental manipulations. *Neuroscience* 37: 603–633.

Minami, T., Oomura, T., Nabekura, J., and Fukuda, A. (1990). 17β-Estradiol depolarization of hypothalamic neurons is mediated by cyclic AMP. *Brain Res.* 519: 301–307.

Miyachi, E.-I., and Murakami, M. (1989). Decoupling of horizontal cells in carp and turtle retinae by intracellular injection of cyclic AMP. *J. Physiol.* 419: 213–224.

Modney, B. K., and Hatton, G. I. (1989). Multiple synapse formation: A possible compensatory mechanism for increased cell size in rat supraoptic nucleus. *J. Neuroendocrinol.* 1: 21–27.

Modney, B. K., Yang, Q. Z., and Hatton, G. I. (1990). Activation of excitatory amino acid inputs to supraoptic neurons: II. Increased dye-coupling in maternally behaving virgin rats. *Brain Res.* 513: 270–273.

Montagnese, C., Poulain, D. A., and Theodosis, D. T. (1990). Influence of ovarian steroids on the ultrastructural plasticity of the adult rat supraoptic nucleus induced by central administration of oxytocin. *J. Neuroendocrinol.* 2: 225–231.

Moos, F., Freund-Mercier, M. J., Guerne, Y., Guerne, J. M., Stoeckel, M. E., and Richard, P. (1984). Release of oxytocin and vasopressin by magnocellular nuclei *in vitro*: Specific facilitatory effect of oxytocin on its own release. *J. Endocrinol.* 102: 63–72.

Neumann, I., Russell, J. A., and Landgraf, R. (1993). Oxytocin and vasopressin release within the supraoptic and paraventricular nuclei of pregnant, parturient and lactating rats: A microdialysis study. *Neuroscience* 53: 65–75.

Párducz, A., Pérez, J., and Garcia-Segura, L. M. (1993). Estradiol induces plasticity of GABAergic synapses in the hypothalamus. *Neuroscience* 53: 395–401.

Perlmutter, L. S., Hatton, G. I., and Tweedle, C. D. (1984). Plasticity in the neurohypophysis *in vitro*: Effects of osmotic changes on pituicytes. *Neuroscience* 12: 503–511.

Pérez, J., Luquín, S., Naftolin, F., and García-Segura, L. M. (1993). The role of estradiol and progesterone in phased synaptic remodelling of the rat arcuate nucleus. *Brain Res.* 608: 38–44.

Perlmutter, L. S., Tweedle, C. D., and Hatton, G. I. (1984). Neuronal glial plasticity in the supraoptic dendritic zone: Dendritic bundling and double synapse formation at parturition. *Neuroscience* 13: 769–779.

Perlmutter, L. S., Tweedle, C. D., and Hatton, G. I. (1985). Neuronal/glial plasticity in the supraoptic dendritic zone in response to acute and chronic dehydration. *Brain Res.* 361: 225–232.

Poulain, D. A., and Wakerley, J. B. (1982). Electrophysiology of hypothalamic magnocellular neurones secreting oxytocin and vasopressin. *Neuroscience* 7: 773–808.

Pow, D. V., and Morris, J. F. (1989). Dendrites of hypothalamic magnocellular neurons release neurohypophysial peptides by exocytosis. *Neuroscience* 32: 435–439.

Ramirez, V. D., Ramirez, A. D., Rodriguez, F., and Vincent, J. D. (1990). Positive feedback of vasopressin on its own release in the central nervous system: *In vitro* studies. *J. Neuroendocrinol.* 2: 461–465.

Renaud, L. P., and Bourque, C. W. (1991). Neurophysiology and neuropharmacology of hypothalamic magnocellular neurons secreting vasopressin and oxytocin. *Prog. Neurobiol.* 36: 131–169.

Rökaeus, A., Young, W. S. I., and Mezey, E. (1988). Galanin coexists with vasopressin in the normal rat hypothalamus and galanin's synthesis is increased in the Brattleboro (diabetes insipidus) rat. *Neurosci. Lett.* 90: 45–50.

Sáez, J. C., Berthoud, V. M., Kadle, R., Traub, O., Nicholson, B. J., Bennett, M. V. L., and Dermietzel, R. (1991). Pinealocytes in rats: Connexin identification and increase in coupling caused by norepinephrine. *Brain Res.* 568: 265–275.

Sakai, N., Blennerhassett, M. G., and Garfield, R. E. (1992). Intracellular cyclic AMP concentration modulates gap junction permeability in parturient rat myometrium. *Can. J. Physiol. Pharmacol.* 70: 358–364.

Salm, A. K., and Hatton, G. I. (1980). An immunocytochemical study of astrocytes associated with the rat supraoptic nucleus. *Soc. Neurosci. Abstr.* 6: 547.

Salm, A. K., Modney, B. K., and Hatton, G. I. (1988). Alterations in supraoptic nucleus ultrastructure of maternally behaving virgin rats. *Brain Res. Bull.* 21: 685–691.

Salm, A. K., Smithson, K. G., and Hatton, G. I. (1985). Lactation associated redistribution of the glial fibrillary acidic protein within the supraoptic nucleus: An immunocytochemical study. *Cell Tiss. Res.* 242: 9–15.

Sar, M., and Stumpf, W. E. (1979). Simultaneous localization of steroid and peptide hormones in rat pituitary by combined thaw-mount autoradiography and immunohistochemistry: Localization of dihydrotestosterone in gonadotropes, thyrotropes and pituicytes. *Cell Tiss. Res.* 203: 1–7.

Sar, M., and Stumpf, W. E. (1980). Simultaneous localization of [^{3}H]estradiol and neurophysin I or arginine vasopressin in hypothalamic neurons demonstrated by a combined technique of dry-mount autoradiography and immunohistochemistry. *Neurosci. Lett.* 17: 179–184.

Sawchenko, P. E., Swanson, L. W., and Vale, W. W. (1984). Corticotropin-releasing factor: Co-expression within distinct subsets of oxytocin-, vasopressin-, and neurotension-immunoreactive neurons in the hypothalamus of male rat. *J. Neurosci.* 4: 1118–1129.

Smithson, K. G., Suarez, I., and Hatton, G. I. (1990). β-Adrenergic stimulation decreases glial and increases neural contact with the basal lamina in rat neurointermediate lobes incubated in vitro. *J. Neuroendocrinol.* 2: 693–699.

Stumpf, W. G., and Sar, M. (1975). Hormone-architecture of the mouse brain with estradiol. In *Anatomical Neuroendocrinology*, ed. W. E. Stumpf and L. D. Grant, pp. 82–103. New York: Karger.

Taylor, C. P., and Dudek, F. E. (1984). Excitation of hippocampal pyramidal cells by an electrical field effect. *J. Neurophysiol.* 52: 126–142.

Theodosis, D. T. (1985). Oxytocin-immunoreactive terminals synapse on oxytocin neurones in the supraoptic nucleus. *Nature* 313: 682–684.

Theodosis, D. T., Montagnese, C., Rodriguez, F., Vincent, J.-D., and Poulain, D. A. (1986a). Oxytocin induces morphological plasticity in the adult hypothalamo-neurohypophysial system. *Nature* 322: 738–740.

Theodosis, D. T., Paut, L., and Tappaz, M. L. (1986b). Immunocytochemical analysis of the GABAergic innervation of oxytocin- and vasopressin-secreting neurons in the rat supraoptic nucleus. *Neuroscience* 19: 207–222.

Theodosis, D. T., and Poulain, D. A. (1984). Evidence for structural plasticity in the supraoptic nucleus of the rat hypothalamus in relation to gestation and lactation. *Neuroscience* 11: 183–193.

Theodosis, D. T., Poulain, D. A., and Vincent, J. D. (1981). Possible morphological bases for synchronisation of neuronal firing in the rat supraoptic nucleus during lactation. *Neuroscience* 6: 919–929.

Tweedle, C. D., and Hatton, G. I. (1977). Ultrastructural changes in rat hypothalamic neurosecretory cells and their associated glial during minimal dehydration and rehydration. *Cell Tiss. Res.* 181: 59–72.

Tweedle, C. D., and Hatton, G. I. (1980a). Glial cell enclosure of neurosecretory endings in the neurohypophysis of the rat. *Brain Res.* 192: 555–559.

Tweedle, C. D., and Hatton, G. I. (1980b). Evidence for dynamic interactions between pituicytes and neurosecretory axons in the rat. *Neuroscience* 5: 661–667.

Tweedle, C. D., and Hatton, G. I. (1982). Magnocellular neuropeptidergic terminals in neurohypophysis: Rapid glial release of enclosed axons during parturition. *Brain Res. Bull.* 8: 205–209.

Tweedle, C. D., and Hatton, G. I. (1984). Synapse formation and disappearance in adult rat supraoptic nucleus during different hydration states. *Brain Res.* 309: 373–376.

Tweedle, C. D., and Hatton, G. I. (1987). Morphological adaptability at neurosecretory axonal endings on the neurovascular contact zone of the rat neurohypophysis. *Neuroscience* 20: 241–246.

Tweedle, C. D., Modney, B. K., and Hatton, G. I. (1988a). Ultrastructural changes in the rat neurohypophysis following castration and testosterone replacement. *Brain Res. Bull.* 20: 33–38.

Tweedle, C. D., Smithson, K. G., and Hatton, G. I. (1988b). Evidence for dendritic and axonal hormone release in the rat hypothalamo-neurohypophysial system. *Soc. Neurosci. Abstr.* 14: 1176.

Tweedle, C. D., Smithson, K. G., and Hatton, G. I. (1989). Neurosecretory endings in the rat neurohypophysis are *en passant*. *Exp. Neurol.* 106: 20–26.

Tweedle, C. D., Smithson, K. G., and Hatton, G. I. (1993). Rapid synaptic changes in the supraoptic nucleus of the perfused rat brain. *Exp. Neurol.* 124: 200–207.

Vanderhaeghen, J. J., Lotstra, F., Vandesande, F., and Dierickx, K. (1981). Coexistence of cholecystokinin and oxytocin-neurophysin in some magnocellular hypothalamo-neurohypophyseal neurons. *Cell Tiss. Res.* 221: 227–231.

Wakerley, J. B. (1987). Electrophysiology of the central vasopressin system. In *Vasopressin: Principles and Properties*, ed. D. M. Gash and J. G. Boer, pp. 211–256. New York: Plenum Press.

Wakerley, J. B., and Ingram, C. D. (1993). Synchronisation of bursting in hypothalamic oxytocin neurones: Possible coordinating mechanisms. *News Physiol. Sci.* 8: 129–133.

Warembourg, M., and Poulain, P. (1991). Presence of estrogen receptor immunoreactivity in the oxytocin-containing magnocellular neurons projecting to the neurohypophysis in the guinea-pig. *Neuroscience* 40: 41–53.

Watson, S. J., Akil, H., Fischli, W., Goldstein, A., Zimmerman, E., Nilaver, G., and Van Wimersma Greidanus, T. B. (1982). Dynorphin and vasopressin: Common localization in magnocellular neurons. *Science* 216: 85–87.

Whitnall, M. H., Gainer, H., Cox, B. M., and Molineaux, C. J. (1983). Dynorphin A$_{(1-8)}$ is contained within vasopressin neurosecretory vesicles in rat pituitary. *Science* 222: 1137–1139.

Wittkowski, W. (1986). Pituicytes. In *Development, Morphology and Regional Specialization of Astrocytes*, ed. S. Fedoroff and A. Vernadakis, pp. 173–208. London: Academic Press.

Yamashita, H., Okuya, S., Inenaga, K., Kasai, M., Uesugi, S., Kannan, H., and Kaneko, T. (1987). Oxytocin predominantly excites putative oxytocin neurons in the rat supraoptic nucleus *in vitro*. *Brain Res.* 416: 364–368.

Yang, Q. Z., and Hatton, G. I. (1987). Dye coupling among supraoptic nucleus neurons without dendritic damage: Differential incidence in nursing mother and virgin rats. *Brain Res. Bull.* 19: 559–565.

Yang, Q. Z., and Hatton, G. I. (1988). Direct evidence for electrical coupling among rat supraoptic nucleus neurons. *Brain Res.* 463: 47–56.

Index

accessory olfactory bulb, vomeronasal afferents to, 90–91

acetylcholine
  in regulation of female sexual behavior, 184–206
  sex hormone–mediated regulation of, 120, 199

actin, 426

activational effects, of sex steroids, 302

activation during mating, of retrorubal field of midbrain tegmentum, 45

acyltransferase, role in brain neurosteroid metabolism, 283

adrenal glands, progesterone secretion by, 120

adrenal steroids, effects on brain function and behavior, 117

adrenocorticotropic hormone, 144

adrenodoxin, 282

adrenodoxin reductase, 282

aggression
  neurosteroid inhibition in male rat, 289, 290
  reproduction relation to, 303

aging, possible neurosteroid role in, 292

amitryptyline, effect on female sexual behavior, 186

amphibians
  sexually dimorphic brain areas in, 259
  vasopressin-immunoreactive projections in, 260

amygdala
  corticomedial area of, estrogen receptor mRNA in, 71, 72
  estrogen receptors in, 325
  extended
    anatomy, 10
    hormonal and chemosensory subcircuitry, 12, 20, 22–23, 31, 34
    neural circuitry through, 10
    sexual dimorphism, 29
  hormonal and chemosensory circuit projections to, 15
  olfactory and vomeronasal connections through, 10
  steroid receptors in, 14

amygdalohippocampal zone
  activation during mating, 45
  estrogen receptors in, 14

inputs to medial preoptic nucleus, 89
  subcircuitry of, 13

anatomical relationships, between chemosensory and hormonal circuits, 20

androgen(s)
  effect on brain androgen receptors, 25–26, 28
  effects on brain nuclei, 17–19
  effects on male copulatory behavior, estrogen comparison with, 6–8, 19
  stimulation of male sexual behavior by, 14

androgen receptors (AR)
  in chemosensory pathways, 13
  in medial preoptic area, 16
  neurons containing, 97
    mating stimulation of, 20, 21
  in tyrosine hydroxylase neurons in brain, 31, 33
  in ventral lateral septum, 14
  in ventral premammillary nucleus, 13

androstatrienedione (ATD), as testosterone-conversion blocker, 41, 284

anterior hypothalamus
  estrogen receptors in, 325
  magnocellular neuroendocrine cells of, 413–415

anteroventral periventricular nucleus of preoptic region (AVPv), 93–97
  estrogen receptors in, 93–94
  hormone secretion by, 89
  inputs to medial preoptic nucleus, 89
  sex differences in rats, 85
  sexual dimorphism of, 93
  steroid receptors in, 93, 98

antidiuretic receptor, in hippocampus, 266

antiestrogens, effects on circulating luteinizing hormone, 94

AP-1 binding sites, in genes, 356, 358, 398

AP-2 binding sites, in genes, 358, 385

apomorphine, as dopamine agonist, 238

arcuate nucleus (ARC), 16
  cholecystokinin receptors in, 170
  hormone secretion by, 89
  inputs to medial preoptic nucleus, 89
  *Jun*D expression in, 133

433